Stefanie Klein, Maximilian Kryschi, Carola Kryschi
Functional Nanomaterials

Also of Interest

Nanochemistry
From Theory to Application for In-Depth Understanding of Nanomaterials
Edited by Xuan Wang, Sajid Bashir, Jingbo Liu, 2023
ISBN 978-3-11-073985-5, e-ISBN 978-3-11-073987-9

Environmental Functional Nanomaterials
Edited by Qiang Wang, Ziyi Zhong, 2020
ISBN 978-3-11-054405-3, e-ISBN 978-3-11-054418-3

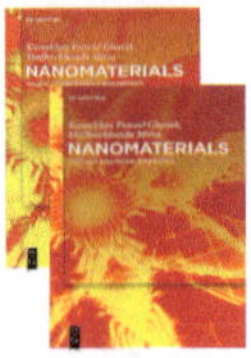

Nanomaterials
Engg Kamakhya Prasad Ghatak, Madhuchhanda Mitra
Volume I: Electronic Properties, 2019
ISBN 978-3-11-060922-6, e-ISBN 978-3-11-061081-9
Volume II: Quantization and Entropy, 2020
ISBN 978-3-11-065972-6, e-ISBN 978-3-11-066928-2

Handbook of Nanoethics
Edited by Gunjan Jeswani, Marcel Van de Voorde, 2021
ISBN 978-3-11-066923-7, e-ISBN 978-3-11-066928-2

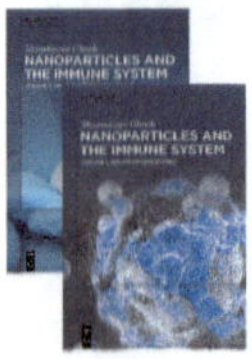

Nanoparticles and the Immune System
Ghosh Shyamasree, 2022
Volume I: Human Immune System
ISBN 978-3-11-065407-3, e-ISBN 978-3-11-065585-8
Volume II: Immune System of Animals
ISBN 978-3-11-065408-0, e-ISBN 978-3-11-065587-2

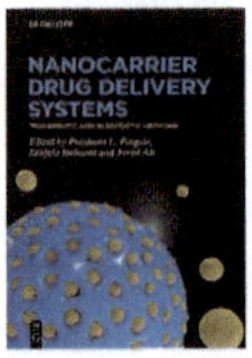

Nanocarrier Drug Delivery Systems
Therapeutic and Diagnostic Medicine
Edited by Prashant L. Pingale, Sanjula Baboota, Javed Ali, 2024
ISBN 978-3-11-132056-4, e-ISBN 978-3-11-132084-7

Stefanie Klein, Maximilian Kryschi, Carola Kryschi

Functional Nanomaterials

Applications in Medicine and Life Sciences

DE GRUYTER

Authors
Dr. rer. nat. Stefanie Klein
Friedrich-Alexander-Universität Erlangen-Nürnberg
Dept. Chemie u. Pharmazie
Egerlandstr. 3
91058 Erlangen
Germany
stefanie.klein@fau.de

Maximilian Kryschi
Universitätsklinikum Heidelberg
Klinik für Allgemein-, Viszeral- und
Transplantationschirurgie
Im Neuenheimer Feld 420
69120 Heidelberg
Germany
maximilian.kryschi@med.uni-heidelberg.de

Prof. Dr. Carola Kryschi
Friedrich-Alexander-Universität Erlangen-Nürnberg
Dept. Chemie u. Pharmazie
Egerlandstr. 3
91058 Erlangen
Germany
carola.kryschi@fau.de

ISBN 978-3-11-060545-7
e-ISBN (PDF) 978-3-11-060549-5
e-ISBN (EPUB) 978-3-11-060686-7

Library of Congress Control Number: 2024944667

Bibliographic information published by the Deutsche Nationalbibliothek
The Deutsche Nationalbibliothek lists this publication in the Deutsche Nationalbibliografie; detailed bibliographic data are available on the Internet at http://dnb.dnb.de.

Cover image: koto_feja / iStock / Getty Images Plus
Typesetting: VTeX UAB, Lithuania
Printing and binding: CPI books GmbH, Leck

MIX
Papier | Fördert gute Waldnutzung
FSC® C083411
FSC www.fsc.org

www.degruyter.com

Contents

1 Introduction

Due to rapidly expanding and deepening knowledge, cancer therapies based on targeting tumor-cell biology, for example, monoclonal antibodies, small-molecular inhibitors, and miRNAs, are already being used in clinical routine with promising success. Nevertheless, still more than 50 % of cancer patients receive tumor-removal surgery in combination with radiotherapy and chemotherapy. Unfortunately, cancer patients suffer from severe side effects during their treatments with radio- and chemotherapy [2]. Moreover, these therapy modalities do not completely eradicate tumor tissue and metastases because of radioresistance and multidrug resistance [3]. Over the past decade, a broad variety of nanoparticular therapeutics has been developed to enhance the efficacy of diverse cancer therapy modalities [4–8]. In addition to their therapeutic effects, the so-called theranostic nanoagents are also used as contrast agents in various biomedical imaging procedures [9]. Nowadays, nanoparticle synthesis routes allow to adjust their sizes and shapes and to control their surface structures. While size and shape are key determinants for the cellular uptake mechanism, surface structures can be tailored for reversible binding of anticancer drugs, miRNA, and tumor-targeting species like aptamers, small molecules, and proteins. Their composition (e. g., high-Z atoms or magnetic transition metal cations), long blood circulation time, and tumor delivery specificity make them highly efficient contrast agents for clinical imaging techniques such as MRI, MPI, CT, US, and PET [10]. Suitably (multi)functionalized nanocomposites can simultaneously operate as radio- or photosensitizer, magnetic hyperthermia or plasmonic agents, and targeted delivery vehicles for anticancer drugs and so exhibit a high clinical potential for multimodal image-guided cancer therapies.

The book begins with an introduction to the current understanding of the molecular and cellular processes involved in the initiation, progression, and dissemination of cancer cells (Chapter 2). Current diagnostic tools and cancer therapy modalities are outlined. In Chapter 3, key nanomaterials, including biocompatible gold, iron oxide, and silica nanoparticles, and their particular structural, optical, magnetic, and catalytic properties are described. Controllable synthesis routes and facile surface modification recipes are presented. Chapter 4 focuses on interactions between biological entities and nanoparticles. Nanoparticle size, shape, and surface chemistry determine the protein corona formation, tumor-targeting modes, cellular uptake mechanisms, and intracellular fates. Positively charged ultrasmall nanoparticles (i. e., quantum dots) can interact with mitochondrial and nuclear membranes, while Fe_3O_4 nanoparticles influence the formation of reactive oxygen species (ROS). Chapter 5 discusses the multiple routes of nanoparticle delivery into the human body, diffusivity, circulation half-life, drug release characteristics, pharmacokinetics and pharmacodynamics, and their accumulation in targeted tumor tissues. Chapter 6 provides a current overview of multifunctional nanomaterials tailored for cancer diagnosis and therapy. While the elemental composition, structure, and morphology of nanoparticles are essential for their application as various

https://doi.org/10.1515/9783110605495-001

contrast imaging and therapeutic agents, the surface chemistry and, in particular, specific surface functionalities determine the tumor-targeting mode, drug loading capacity, and intracellular drug release. Future perspectives of nano-oncology are based on multifunctional nanoparticles that incorporate tumor-targeting ability, contrast agents, and multiple therapeutic modalities enabling simultaneous detection and complete elimination of cancer cells which prevents their spread and increases the effectiveness of the treatment.

2 Biology of cancer

2.1 Tumorigenesis

Multicellular organisms maintain a state of equilibrium (tissue homeostasis) between cell proliferation, differentiation, and programmed cell death (apoptosis). Perturbations in this balance, caused for example by mutations in genes encoding enzymes that regulate cell division, can lead to neoplasia. The underlying process, known as tumorigenesis, involves a variety of genetic and epigenetic alterations and can lead to uncontrolled cell proliferation [11].

Carcinogenesis describes the development of malignant neoplasia or cancer as a multistep process whose major phases are initiation, promotion, malignant transformation, and progression. During the initiation phase, the DNA sequence of the normal cell genome undergoes changes that can be induced, for example, by environmental carcinogens, viruses, or ionizing radiation [12]. Genetic mutations can be small and affect only single genes, such as point mutations by insertion, deletion, or substitution of a single nucleotide, or large, such as amplification of entire genes, changes in the structure of chromosomes, or even duplication of entire chromosomes (polyploidy). Chromosomal mutations can change the structure of a chromosome by deletion, translocation, duplication, inversion, or insertion of a chromosomal segment. Mutations can occur spontaneously or can be induced by chemical, physical, or biological mutagens. Chemical mutagens are chemical compounds that can directly or indirectly alter DNA, such as base analogues, alkylating agents, and intercalating agents, while UV light, X-rays, and ionizing radiation act as physical mutagens. For example, UV light can induce the formation of cyclobutane pyrimidine dimers, while the interaction of DNA with ionizing radiation can cause DNA single- or double-strand breaks. Biological mutagens include viruses, such as some strains of human papillomavirus (particularly HPV 16 and HPV 18) or Epstein–Barr virus. These viruses can cause mutations in tumor suppressor genes, oncogenes, or DNA repair genes, leading to uncontrolled cell growth and division, disrupting the DNA repair mechanism, allowing further mutations, and preventing apoptosis.

Tumor suppressor genes, such as the famous TP53, play an important role in tumorigenesis because loss of function of these genes is critical for the proliferation of genetically altered cells. According to the two-hit hypothesis proposed by Alfred Knudson [13], both alleles of a tumor suppressor gene must be inactivated to have a phenotypic effect on the cell. Mutations in tumor suppressor genes, protooncogenes, or DNA repair genes lead to uncontrolled cell growth and division, disrupt DNA repair, allow further mutations, and inhibit apoptosis. For example, mutations in TP53, which encodes the tumor suppressor p53, the "guardian of the genome," are found in up to 50 % of samples from some cancers. Mutant p53 proteins lose their tumor suppressor functions, including the ability to induce cell cycle arrest, DNA repair, senescence, and apoptosis. Suppression of

https://doi.org/10.1515/9783110605495-002

the DNA repair mechanism results in a damaged genome. The p53 protein thus has critical regulatory functions at the cell cycle checkpoints, which represent control points for new DNA synthesis and thus the distribution of genetic material in mitosis. During the S phase, when new DNA synthesis takes place, nucleic bases can be incorrectly assembled. A state of drastically increased genetic instability is reached when the DNA mismatch repair system fails to recognize and repair an incorrect sequence of nucleotides.

Vogelstein and Kinzler showed for colorectal cancer that cell growth is fundamentally altered by mutations in 4 to 7 genes [14]. In a landmark paper in 1990, Fearon and Vogelstein postulated that colorectal cancer arises from the accumulation of mutations in protooncogenes and tumor suppressor genes [15]. Protooncogenes generally have a growth-promoting effect in normal cells and are subject to a complex control system [16]. In contrast to tumor suppressor genes, protooncogenes encode enzymes of growth-signaling pathways and generally have a growth-promoting effect in normal cells and are subject to a complex control system [17]. Gain-of-function mutations of such protooncogenes can result in oncogenes that bypass the control mechanisms, while the cell receives an increased growth stimulus that can lead to uncontrolled proliferation. An example of a protooncogene is the epidermal growth factor receptor (EGFR), which is an important target for cancer therapy.

The next stage of tumor development is promotion. During this phase, the initiated cells begin to proliferate, resulting in an increase in the size of the initiated cell population and the formation of a neoplastic lesion or tumor. In many cases, growth factors activate signaling pathways in the cancer cell that promote uncontrolled proliferation. The clinical correlate of this phase is a benign or even precancerous tumor, such as a neoplastic colorectal polyp (adenoma).

In contrast to malignant solid tumors, benign neoplasms are well-defined, often capsulated, and do not invade surrounding tissues. During malignant transformation, cells of benign neoplasms can acquire several characteristic abilities that contribute to their cancerous behavior by (re-)activating pathways that enable invasion of foreign tissues. By generating their own growth signals or altering growth signaling pathways, they can proliferate without the normal regulatory mechanisms. Loss of function, mutation, or inactivation of tumor suppressor genes leads to loss of genetic integrity, paving the way for further mutations. At the same time, malignant cells evade the programmed cell death. Somatic cells are physiologically unable to divide indefinitely because the noncoding ends of their DNA strands (telomeres) shrink with each duplication of DNA. Cancer cells often activate telomerase, an enzyme that maintains the ends of chromosomes (telomeres), leading to theoretically unlimited proliferation of cancer cells. Cancer cells also stimulate the formation of new blood vessels that supply nutrients and oxygen.

During tumor progression, cancer cells may acquire additional genetic mutations and alter their epigenetics and gene-expression patterns. Epigenetics refers to changes in genome expression that are not associated with changes in the underlying DNA sequence [17]. These changes include DNA methylation, histone modification, and regulation of noncoding RNA. Aberrant DNA methylation patterns, such as hypermethylation of tumor suppressor gene promoters and global hypomethylation, are found in many cancers. Histone modifications, including acetylation and methylation, can alter chromatin structure and gene expression, thereby influencing cancer initiation and progression. Genetic mutations and epigenetic alterations can change the behavior of cancer cells, making them more aggressive and invasive.

Hanahan and Weinberg [2, 18, 19] identified ten hallmarks of cancer that reflect specific capabilities: sustaining proliferative signaling, evading growth suppressors, immune escape, enabling replicative immortality, tumor-promoting inflammation, activating invasion and metastasis, inducing angiogenesis, genome instability and mutation, resisting programmed cell death (apoptosis), and deregulating cellular metabolism (Figure 1). These features provide a conceptual framework for understanding the complex phenotypic and genotypic variations in cancer. For example, cancer cells use multiple signaling pathways to maintain a continuous state of cell proliferation. In addition, silencing of tumor suppressor genes is required to disrupt cell cycle control and arrest senescence. Furthermore, during tumorigenesis, cells acquire the ability to escape programmed cell death (apoptosis). This ability is achieved, for example, by mutation of the caspase genes responsible for apoptotic pathways. The reproductive immortality of cancer cells requires unlimited cytokinesis, which can be achieved, for example, by expressing telomerase, which extends the ends of the DNA strand that would otherwise shorten after each mitosis.

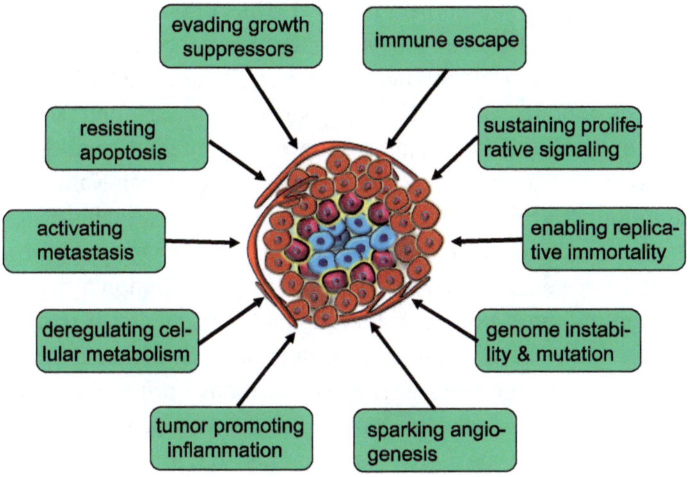

Figure 1: Schematic of Hanahan and Weinberg's 10 hallmarks of cancer.

Growing tumors require a blood supply to obtain nutrients and oxygen. Cancer cells accomplish this by secreting angiogenic factors that stimulate the formation of new blood vessels (angiogenesis). Activation of invasion and metastasis allows cancer cells to detach from the primary tumor, invade the surrounding tissue, travel, e. g., through the blood or lymphatic system, and form secondary tumors. This process involves several steps, including epithelial-mesenchymal transition, extracellular matrix degradation, intravascular migration, and extravascular infiltration into new tissue, followed by angiogenesis and tumor growth. Cancer cells significantly deregulate energy metabolism to survive and proliferate despite nutrient and oxygen deprivation. Cancer cells evade destruction by the immune system by expressing immune inhibitory molecules, such as PD-L1, which prevent immune cells from recognizing and eliminating them. They can alter the tumor microenvironment to promote immunosuppressive cells, such as regulatory T cells and myeloid suppressor cells that weaken the immune response. Tumor-promoting inflammation occurs when chronic inflammatory processes in tissues support tumor development and progression by releasing growth-promoting cytokines and chemokines. These inflammatory conditions promote cell proliferation, survival, and angiogenesis. Genomic instability is associated with an increased mutation rate in cancer cells, resulting in a variety of mutations that promote tumor growth and allow cancer cells to adapt to different environments. The accumulation of such mutations can activate oncogenes and inactivate tumor suppressor genes, further driving tumor progression.

Taking into account new findings in cancer research, Hanahan and Weinberg later decided to expand the previously proposed 10 hallmarks of cancer to fully capture the complexity of cancer pathogenesis [18]. In addition, the molecular and cellular mechanisms that contribute to the development of preneoplastic cells and their phenotypic capabilities during tumor progression should be addressed. Thus, cancer cells and tumors are endowed with the enabling features “development of phenotypic plasticity”, “nonmutative epigenetic reprogramming”, “polymorphic microbiome”, and “senescent cells”. Phenotypic plasticity of tumor cells is regulated by various transcription factors and signaling pathways to optimally adapt tumor cells to environmental conditions, enabling them to evade therapeutic intervention and ensure their survival. Nonmutative epigenetic reprogramming alters gene expression through processes such as DNA methylation, histone modification, and chromatin remodeling. Thus, epigenetic reprogramming can activate or silence genes, which can promote tumor initiation and progression. Senescent cells stop dividing, exit the cell cycle, and accumulate in tissues. These senescent cells are thought to be resistant to conventional systemic chemotherapy and play a critical role in cancer recurrence. They can also have pro-tumorigenic effects by secreting pro-inflammatory cytokines, growth factors, and proteases. This senescence-associated secretory phenotype (SASP) can also alter the tumor microenvironment.

Furthermore, the authors recognized the key role of the tumor microenvironment (TME) in tumor biology. The TME is a dynamic and complex network of noncancerous cells, extracellular matrix components, and signaling molecules that provides an interactive environment for tumor cells. The TME supports tumor growth, invasion, and metastasis. Key cellular players in the TME include cancer-associated fibroblasts (CAFs), immune cells, endothelial cells, and extracellular matrix proteins. CAFs promote tumor-cell proliferation, survival and metastasis by secreting growth factors, cytokines, and extracellular matrix proteins. Immune cells within the TME, such as macrophages, lymphocytes, and neutrophils, can play a dual role by either suppressing or promoting tumor growth, depending on their functional state. For example, tumor-associated macrophages (TAMs) often exhibit a pro-tumor M2 phenotype that supports angiogenesis, tissue remodeling, and immunosuppression. Endothelial cells contribute to tumor angiogenesis, the formation of new blood vessels that provide nutrients and oxygen to the growing tumor. This process is mediated by pro-angiogenic factors such as vascular endothelial growth factor (VEGF) and fibroblast growth factor (FGF) secreted by both tumor cells and stromal cells in the TME.

2.2 Metastasis

Metastasis, the spread of cancer cells from the primary tumor to other sites or organs, is the clinical correlate of the systemic stage of an initially localized malignancy and is often the final stage of cancer progression [20, 21]. It is the process of dissemination of cancer cells from the primary tumor to distant organs and tissues and ends with the formation of new metastatic tumors. These transformed cells often exhibit altered cell adhesion properties, increased motility, and resistance to apoptosis, and can proliferate uncontrollably and invade surrounding tissues. Autocrine and paracrine factors are released that further promote tumor growth and motility. Epithelial–mesenchymal transition (EMT) is the histopathologic correlate of cancer-cell dissemination and is widely regarded as the first stage of metastasis. During EMT, tumor cells lose the expression patterns of epithelial cells (e. g., loss of E-cadherin) and acquire the expression patterns of mesenchymal cells (e. g., expression of vimentin).

2.2.1 Epithelial–mesenchymal transition

Epithelial-to-mesenchymal transition (EMT) follows a (reversible) dedifferentiation program [22]. Epithelial cells gradually lose their characteristic apical–basal polarity and cell–cell contacts to become cells with mesenchymal characteristics, which are spindle-shaped, fibroblast-like with elongated cell bodies, tapered ends, and a prominent oval nucleus (Figure 2). These mesenchymal-like cells are characterized by increased motility, invasiveness, and the ability to produce extracellular matrix (ECM). Epithelial and mesenchymal cells differ dramatically in morphology and functionality. Epithelial cells are organized in cell layers with apical–basal polarity and adhere strongly to the basement membrane and to their neighboring cells. In contrast, the mesenchymal single cell state is characterized by much weaker adhesion properties and a significantly increased migration potential, as well as a highly invasive capacity. Epithelial cells are connected to their neighbors by tight junctions, adherens junctions, and desmosomes [23]. The gap junction channels between neighboring cells allow the exchange of small molecules. EMT leads to the disruption of these cell–cell junctions and the degradation or delocalization of the connecting proteins involved. Tight junctions are localized near the apical surface of epithelial cells. Claudins and occludins are part of the intercellular tight junction strands and are associated with the cytoplasmic proteins zonula occludens 1, zonula occludens 2 and zonula occludens 3 (ZO-1, ZO-2 and ZO-3), which bind to the actin cytoskeleton to stabilize the junctions [24]. During EMT, the expression of claudins and occludins is repressed and ZO-1 translocates to the cytoplasm. Adherens junctions are mediated by cadherin proteins that form calcium-dependent homophilic interactions.

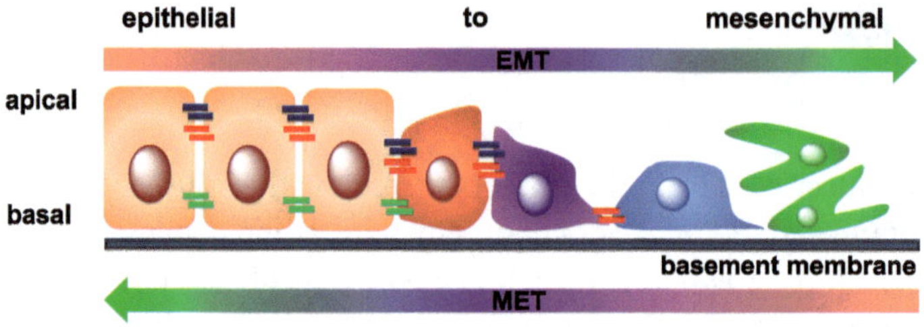

Figure 2: Illustration of endothelial–mesenchymal transition (EMT) and its reverse process mesenchymal–endothelial transition (MET); tight junction (blue double strip), adherens junction (red double strip), desmosome (green double strip).

Epithelial cadherin (E-cadherin) is associated with the actin cytoskeleton in a complex with β-catenin, α-catenin, and p120-catenin and is a major contributor to epithelial cell adhesion [25]. During EMT, cells lose their epithelial markers and acquire mesenchymal markers. Other signaling pathways involved in EMT include the transforming growth factor-beta (TGF-β) signaling pathway, the Wnt/β-catenin signaling pathway, the Notch signaling pathway, the Hedgehog signaling pathway, and signaling pathways activated by growth factors such as epidermal growth factor (EGF) and fibroblast growth factor (FGF) [26]. TGF-β is one of the most prominent inducers of EMT. By binding to its receptors, TGF-β activates SMAD-dependent and -independent signaling pathways, leading to the transcription of mesenchymal markers such as vimentin, N-cadherin, and fibronectin, while epithelial markers such as E-cadherin are downregulated. N-cadherin increases mesenchymal cell motility, proliferation, and invasion. This "cadherin switch" leads to reorganization of the cytoskeleton and remodeling of cell-matrix interactions, making the cells more mobile and invasive. During EMT, the actin cytoskeleton is remodeled into a network of highly dynamic and contractile actin stress fibers that are critical for the directional migration of mesenchymal cells [27]. Mesenchymal cell migration occurs through the formation of cell protrusions and oscillatory cell contraction that pushes the cell forward. The intermediate filament system, which increases the mechanical strength of cells and mediates the transport of organelles and membrane proteins, is also remodeled during EMT. Epithelial cytokeratin is replaced by vimentin filaments, which are critical for the shape, motility, and adhesion of mesenchymal cells. The extracellular matrix (ECM) consists mainly of proteoglycans and fibrous proteins such as collagens, fibronectins, and laminins, which form the interstitial matrix and basement membrane. Signals from the ECM are important regulators of many cellular processes, including proliferation and migration.

Adhesion of epithelial cells to the extracellular matrix (ECM) and signaling from the ECM to epithelial cells is mediated by integrins. During EMT, epithelial integrins are downregulated and mesenchymal integrins are reexpressed. These mesenchymal integrins also contribute to EMT and cell migration [27]. EMT also alters the composition of the ECM by increasing the secretion of ECM components, including fibronectin, and the expression of proteases, particularly matrix metalloproteinases (MMPs). MMPs facilitate invasion by degrading ECM components. In addition, MMPs cleave cell receptors on the cell surface and release growth factors that remain in the ECM. EMT confers increased migratory and invasive capabilities and stem cell-like properties to epithelial cancer cells, thereby contributing significantly to metastasis and therapy resistance [25].

Understanding the underlying mechanisms and signaling pathways of EMT offers potential for the development of new therapeutic approaches for the treatment of fibrotic diseases and cancer. For example, targeted intervention in EMT pathways could reduce tumor-cell invasion and metastasis.

2.2.2 Invasion and intravasation

The invasion of cells from a primary tumor into the stroma is facilitated by a variety of cellular mechanism [28]. Cells use proteolytic enzymes, such as matrix metalloproteinases (MMPs) and serine proteinases, to degrade components of the extracellular matrix (ECM), thereby increasing cell motility. Adhesion of tumor cells to the ECM is mediated by integrins and other cell receptors for ECM proteins. Tumor-cell invasion into the stroma occurs by either single cell or collective cell migration. In collective invasion, few tumor cells retain epithelial characteristics such as E-cadherin-mediated cell–cell adhesion. Collectively migrating tumor cells adopt a partial (hybrid) EMT phenotype (Figure 3). This means that the cells at the top of the migrating cell cluster (leader cells) have a more pronounced mesenchymal phenotype than the so-called follower cells. Accordingly, the ability of tumor cells to migrate appears to depend on the activity of transforming growth factor (TGF)-beta-Smad signaling. Other important determinants of collective migration are intrinsic factors of tumor cells, such as the production of proteases, and extrinsic factors, which include the activities of pro-tumor neutrophils (N2-type), fibroblasts (CAFs), and macrophages (M2-type). Invasion is often followed by intravasation: cancer cells appear to enter the peripheral blood circulation or lymphatic system either individually or in aggregates. The aggregated tumor cells are referred to as multicellular circulating tumor-cell clusters (CTC) after entering the bloodstream in the lymphatic system. Once in the bloodstream, individual tumor cells and CTCs are exposed not only

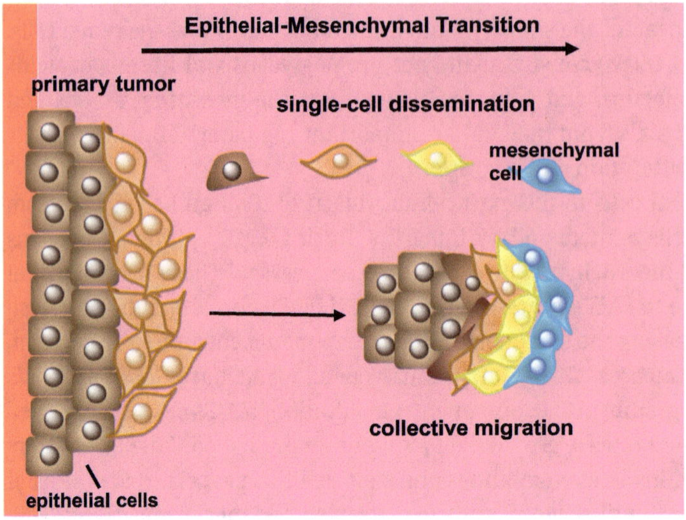

Figure 3: Illustration of single-cell and collective-cell migration mediated by epithelial–mesenchymal transition.

to shear stress but also to immune attack. Surviving tumor cells can then adhere to the endothelial lining of distant capillaries, a process facilitated by interactions between cell-surface molecules on the cancer cells and endothelial cells.

2.2.3 Stasis, adhesion, and extravasation

The ability of CTCs to leave blood vessels may be critical for disease progression [29, 30]. This process is referred to as extravasation. The release of cytokines into the systemic circulation facilitates the extravasation of cancer cells to distant organs. Platelet activity also appears to be essential for tumor-cell extravasation. Platelets provide coagulation and weaken the endothelial barrier, possibly by releasing tumor growth factor β (TGF-β). It has also been reported that the tumor microenvironment in distant organs also sends signals, such as the vascular endothelial growth factor (VEGF), that promote extravasation of individual cancer cells or CTCs. Extravasation is also a multistep process. CTCs migrate through the circulation and eventually adhere to capillary beds in distant organs. The tumor cells then bind to the endothelium via receptor proteins such as selectins, cadherins, and integrins anchored to the membrane surface of the CTCs. After adhesion, the cancer cells can extravasate through the endothelial lining into the parenchyma of the target tissue. Transendothelial migration (TEM) could be prepared by proteolytic degradation of the vascular basement membrane.

2.2.4 Proliferation and growth at secondary sites

After extravasation, cancer cells must adapt to the new tissue environment [29], and they must proliferate to form secondary tumors. This process, known as colonization, involves the re-establishment of cell–cell and cell–matrix interactions. Often, cancer cells must undergo a mesenchymal–epithelial transition to regain their proliferative capacity and evade immune recognition. Successful metastasis requires the establishment of a supportive microenvironment involving the recruitment of stromal cells and immune cells and the formation of new blood vessels (angiogenesis) to supply the growing tumor with nutrients and oxygen. Angiogenesis can be stimulated by factors such as vascular endothelial growth factor (VEGF) and basic fibroblast growth factor (bFGF).

2.3 Tumor metabolism

Oxygen and nutrient consumption is often increased in rapidly growing tumors and metastases [11]. Reprogramming of cellular metabolic pathways enables the required abnormal nutrient cycling and adaptation of cancer cells to hypoxic environmental conditions. In differentiated cells, oxidative phosphorylation can be replaced by glycolysis under hypoxic conditions [31]. Other pathways for energy production include mitochondrial glutaminolysis and fatty acid oxidation, as in all eukaryotic cells [32, 33]. Metabolic reprogramming affects not only gene expression in cancer cells, but also cellular differentiation and the microenvironment of the tumor tissue [34]. Tumor metabolism is characterized by one or more of the following: uncontrolled uptake of glucose and amino acids, especially glutamine; opportunistic nutrient acquisition; utilization of glycolysis and tricarboxylic acid cycle (TCA) intermediates for biosynthesis and NADPH production; increased nitrogen demand; alterations in metabolic gene regulation; and metabolic interactions with the microenvironment. NADPH is the phosphorylated form of the coenzyme nicotinamide adenine dinucleotide (NADH). Rapid tumor growth leads to oxygen depletion in the tumor environment, which renders the metabolism of healthy cells, oxidative phosphorylation (OXPHOS), inefficient. OXPHOS is essential for energy production in aerobic eukaryotic cells and takes place in mitochondria [35, 36]. Mitochondria enable the conversion of O_2 into H_2O and adenosine triphosphate (ATP), the primary energy store in cells. OXPHOS uses the mitochondrial electron-transport chain (ETC) in combination with chemiosmotic coupling to drive ATP synthesis catalyzed by ATP synthase. The protein complexes of the ETC are NADH dihydrogenase (I), succinate dehydrogenase (II), cytochrome bc_1 complex (III), cytochrome c oxidase (IV), and ATP synthase (V). Electron carriers such as NADH and dihydroflavine adenine dinucleotide (FADH2) introduce electrons into the ETC (Figure 4).

The ETC begins with the transfer of electrons from NADH to coenzyme Q (CoQ) via complex I, while electrons from $FADH_2$ reach CoQ via complex II. CoQ transfers the electrons to complex III, where they are transferred by cytochrome c (Cyt c) to complex IV.

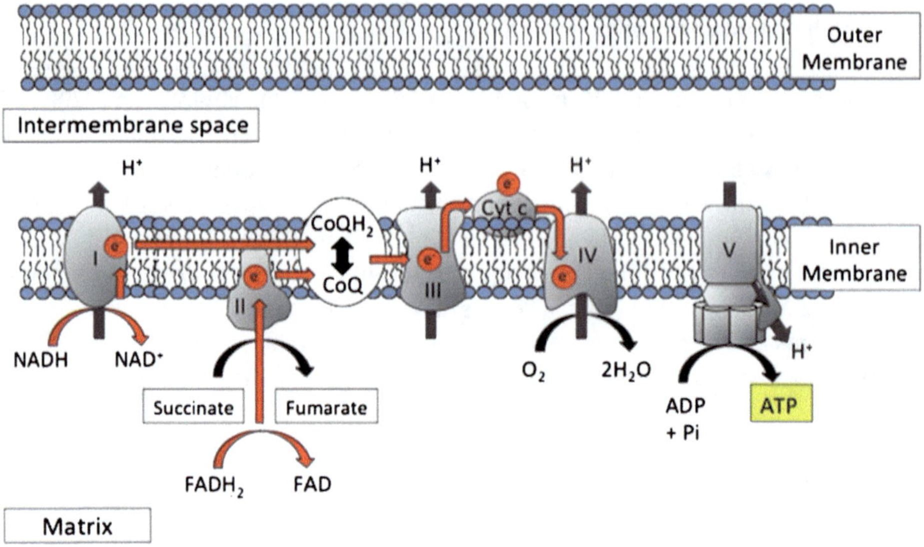

Figure 4: The mitochondrial electron-transport chain (ETC): electrons (e-) are transferred through complex I to IV, resulting in reduction of O_2 to H_2O; the proton (H^+) gradient drives the synthesis of ATP (adenosine triphosphate) from ADP (adenosine diphosphate) and inorganic phosphate (Pi) by complex V (ATP synthase) [35].

Complex IV reduces O_2 to 2 O^{2-}, which reacts with 4 protons (H^+) to form 2 moles of H_2O. The electron transport chain pumps H^+ from the mitochondrial matrix into the intermembrane space using the energy released during electron transfer by complexes I, III, and IV. This process creates an electrochemical proton gradient, also called the proton motor force, with a high proton concentration in the intermembrane space and a lower concentration in the matrix. This proton gradient activates ATP synthase, which drives protons back into the matrix, allowing phosphorylation (Pi) of adenosine diphosphate (ADP) to ATP. The overall reaction equation for oxidative phosphorylation is

$$10\,\mathrm{NADH} + 2\,\mathrm{FADH_2} + 28\,\mathrm{Pi} + 10\,\mathrm{H^+} + 6\,\mathrm{O_2} + 28\,\mathrm{ADP} \longrightarrow 10\,\mathrm{NAD^+} + 2\,\mathrm{FAD} + 12\,\mathrm{H_2O} + 28\,\mathrm{ATP}.$$

Due to the often increased energy consumption and lack of oxygen, cancer cells may switch their metabolism to glycolysis, resulting in an increased demand for glucose [34]. In glycolysis, one molecule of glucose is broken down into two molecules of pyruvate and reduced to lactate in the absence of oxygen (anaerobic) or oxidized to acetyl-coenzyme A via the citric acid cycle in the presence of oxygen (aerobic) and then completely oxidized to CO_2 and H_2O (Figure 5). This implies that glycolysis occurs not only in hypoxic cancer cells but also under normoxic conditions. The conversion of glucose to lactic acid in the presence of oxygen is known as aerobic glycolysis and was first described by Warburg (Warburg effect) [37]. Most tumor cells also have normal mitochondrial function and use it for ATP production. Glycolysis promotes cancer-cell survival and growth, although this pathway is energy inefficient compared to OXPHOS. However, the ATP production rate of glycolysis is much faster than that of OXPHOS. To meet the high demand for glucose, tumor cells upregulate the expression of cell membrane-embedded glucose transporters (GLUTn, n = 1–14). The uptake of glucose across the cell membrane is carried out passively by GLUT transporters, which allow concentration-dependent glucose transport. Once inside the cell, glucose is phosphorylated by hexokinase or glucokinase (in the liver) to form glucose-6-phosphate, which is converted to pentoses via the pentose phosphate pathway. Ribose 5-phosphate and NADPH are formed. The pentose phosphate pathway thus controls anabolic biosynthesis, redox homeostasis, and resistance to chemotherapeutic agents.

Proliferating cancer cells benefit from a high glycolytic rate because they can use the major extracellular nutrient, glucose, to produce ATP [31]. The yield of ATP per glucose consumed, at 2 mol ATP/1 mol glucose (anaerobic glycolysis) and 4 mol ATP/1 mol glucose (aerobic glycolysis), is low compared to the ATP yield of OXPHOS (38 mol ATP/1 mol glucose). However, the rate of glycolysis is much higher than that of OXPHOS, resulting in a much higher concentration of cellular ATP. In addition, the rapid breakdown of glucose provides cells with intermediates needed for biosynthetic pathways. These intermediates include ribose, for nucleotide synthesis, glycerol, and citrate for lipid production, and amino acids.

The molecular mechanisms leading to the constitutive upregulation of aerobic glycolysis are not fully understood. An important regulator of glycolysis is the transcription factor hypoxia-inducible factor-1 (HIF-1), which consists of two subunits, HIF-1α and HIF-1β [38]. This factor mediates a pleiotropic response to hypoxic stress by activating survival genes, including glucose transporters, angiogenic growth factors (e. g., VEGF), hexokinase II59, and hematopoietic factors (e. g., transferrin and erythropoietin). Constitutively elevated levels of HIF-1α often correlate with high glucose consumption in certain cell lines. Cancer cells harboring a mutation in the von Hippel–Lindau (VHL) ubiquitin ligase have constitutively high levels of HIF-1α. At physiological oxygen levels, the HIF-1α subunit is hydroxylated by prolyl hydroxylases (PHDs) and subjected to proteasomal degradation. Increased levels of reactive oxygen species (ROS) in hypoxic cancer cells can inhibit PHD and stabilize the HIF-1α subunit [39]. In addition to its function as an activator of aerobic glycolysis and lactate production, HIF-1α plays an important role in the transcription of glucose transporters and glycolytic enzymes, including hexokinase II (HKII), pyruvate dehydrogenase kinase (PDK), and pyruvate kinase M2 (PKM2). Anaerobic glycolysis may ensure the maintenance of HIF-1α activity. This would represent a novel feed-forward mechanism of glycolysis-HIF-1α signaling.

Another important metabolic process described in cancer cells is mitochondrial glutaminolysis, in which glutamine is converted to glutamate and then to α-ketoglutarate in the mitochondrial matrix to support the citric acid (TCA) cycle (Figure 5) [33]. Glutamine acts as an essential source of carbon and nitrogen, which can promote cancer-cell proliferation. Glutamine nitrogen fuels the de novo biosynthesis of purine and pyrimidine

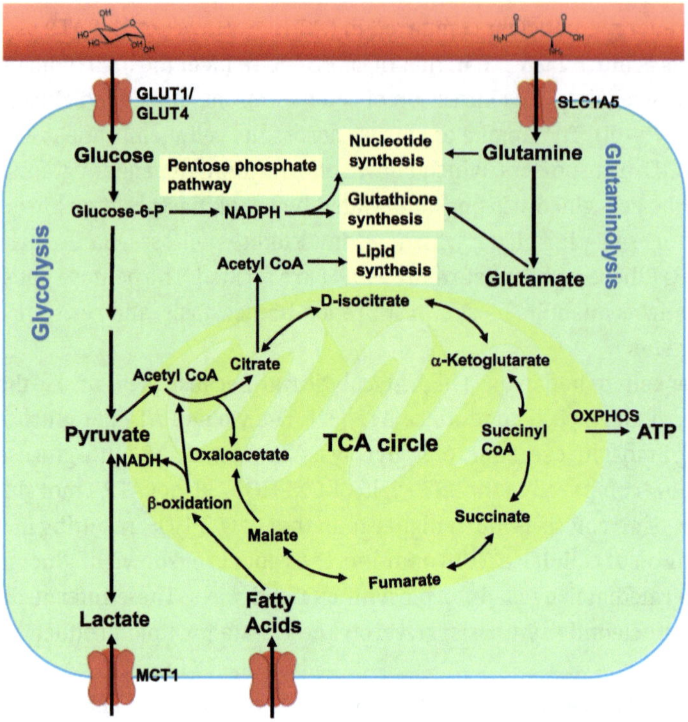

Figure 5: Metabolic pathways of cancer cells: cellular uptake of glucose and glutamine to fuel the tricarboxylic acid (TCA) cycle, oxidative phosphorylation (OXPHOS), and the pentose phosphate pathway (nucleotide synthesis); with ATP: adenosine triphosphate; CoA: coenzyme A, GLUT1/GLUT4, glucose transporter 1 or 4, MCT1: monocarboxylate transporter 1; NADH, nicotinamide adenine dinucleotide (reduced); NADPH, nicotinamide adenine dinucleotide phosphate (reduced); P: phosphate, SLC1A5, solute carrier family 1 member 5.

nucleotides, glucosamine 6-phosphate, and nonessential amino acids. Cancer cells often use glutaminolysis as an alternative energy source, especially under conditions where glucose is scarce. Glutamine, like glucose, is taken up by cancer cells from the extracellular environment. Glutamine also mediates the uptake of essential amino acids such as leucine, isoleucine, valine, tyrosine, tryptophan, and phenylalanine. The import of such amino acids is mediated by the plasma membrane-localized neutral amino acid antiporter (large neutral amino-acid transporter 1) LAT1, which is promoted by the simultaneous efflux of glutamine. The most important regulator of glutamine utilization is the transcription factor c-myc, which is often upregulated in proliferating cells. Indeed, c-myc might induce the transcription of glutamine transporters, such as SLC1A5/ASCT2, and also drives the expression of glutamine-degrading enzymes, such as glutaminase 1 (GLS1) and carbamoyl phosphate synthetase 2—aspartate transcarbamylase—dihydroorotase (CAD), to promote glutamine uptake by converting glutamine to glutamate. In addition, glutamine uptake can be negatively regulated by the retinoblastoma (Rb) family of tumor suppressors. Deletion of this tumor suppressor family increases glutamine uptake via E2F-dependent upregulation of SLC1A5/ASCT2 and GLS1. Thus, glutamine consumption is supported by the activity of the transcription factors c-myc and E2F, which also regulate the cell cycle.

2.4 Multimodal tumor therapy

The only curative treatment for solid malignancies is early surgical resection, often combined with adjuvant (after surgical resection) and increasingly neoadjuvant (before surgical resection) systemic therapy. However, when cancer is diagnosed at an advanced stage, multimodal treatment approaches are increasingly used. Although complete cure is often not achievable, multimodality therapy can "chronicize" malignant disease and significantly prolong patient survival. Multimodal cancer therapy is an integrative approach to cancer treatment that combines the benefits of different therapeutic modalities to achieve the best possible outcomes. The therapeutic effect of the combined approaches should be greater than the sum of their individual effects. The choice and sequence of therapies depend on the type of cancer, its stage, the genetic and molecular characteristics of the tumor, and the patient's overall condition and preferences. Individualized treatment planning based on detailed diagnostic information is critical to the success of multimodal cancer therapy. Resection of cancerous tissue (surgery) is combined with chemotherapy, radiation therapy, and immunotherapy when possible. Each of these modalities targets different aspects of the tumor and its environment to synergistically maximize tumor control and improve patient survival.

Tissue samples obtained from surgery or interventional biopsies are analyzed by pathologists, paving the way for appropriate systemic therapy. In multimodal therapy, surgery is performed whenever indicated to remove tumor tissue and metastases and to improve the efficacy of subsequent therapies. Cytotoxic (cell-damaging) chemotherapy drugs are usually given intravenously to the patient to eliminate cancer cells or inhibit their growth [40, 41].

Because chemotherapeutic agents have an untargeted effect on the entire organism, they can also be used to treat metastatic disease. The combination of different chemotherapeutic agents can produce synergistic effects and prevent the development of resistance. Radiation therapy uses ionizing radiation to destroy cancer cells or interfere with their ability to divide.

The term ionizing radiation can be categorized as either (X-rays and γ rays) or particle radiation (e. g., alpha, beta, and neutron radiation), while modern techniques such as intensity-modulated radiation therapy (IMRT) and stereotactic radiation therapy (SRT) allow precise targeting of the tumor while sparing surrounding healthy tissue. Radiation therapy damages DNA in the cell nucleus, and cancer cells are particularly susceptible due to their rapid division and inefficient DNA repair mechanisms.

Radiation therapy is used in curative, adjuvant, neoadjuvant, and palliative settings. In curative radiotherapy, tumor tissue and metastases are completely destroyed. Adjuvant radiotherapy is used after surgery to reduce the risk of recurrence. On the other hand, neoadjuvant modality is given to the patient before surgery to reduce the size of the tumor to make complete resection more feasible or even possible. Palliative radiotherapy is given to patients with advanced cancer to relieve symptoms and improve the quality of life. Antibody therapy uses antibodies to selectively target cancer cells that express the appropriate antigens [42]. (Monoclonal) antibodies recognize and bind to specific antigens on the surface of cancer cells. This type of therapy takes advantage of the highly specific binding property of antibodies, which not only allows for a targeted attack on cancer cells, but can also activate the immune system. Binding of antibodies to antigens on the surface of cancer cells can inhibit the signaling cascades required for cancer-cell proliferation and survival. A famous example is trastuzumab (Herceptin), which binds to HER2 receptors on HER2+ breast-cancer cells. Other antibodies work by blocking growth factors or their receptors that are necessary for tumor growth and angiogenesis. An example is bevacizumab (Avastin), which inhibits vascular endothelial growth factor (VEGF). There are also conjugated antibodies that are linked to cytotoxins, radioisotopes or other agents. These agents are released after binding to cancer cells to specifically kill the tumor cells. One example is brentuximab (Vedotin), a monoclonal antibody against CD30 that is linked to a microtubule inhibitor. Antibody therapy is used in a variety of cancers, alone or in combination with chemotherapy, radiation therapy, or other targeted therapies to achieve synergistic effects and improve treatment outcomes. The therapeutic goal of immunotherapy is to activate or modify the patient's immune system to better target cancer cells. Checkpoint inhibitors such as pembrolizumab (Keytruda) and nivolumab (Opdivo) block inhibitory pathways. T lymphocytes (T cells) are reactivated and can attack cancer cells. Another checkpoint inhibitor, ipilimumab (Yervoy), blocks CTLA-4, another protein that inhibits T-cell activation. One example of emerging personalized therapies is chimeric antigen receptor T-cell therapy (CAR T-cell therapy). The patient's T cells are genetically engineered to express a chimeric antigen receptor (CAR) that specifically binds to an antigen on cancer cells. Once modified, the CAR T cells are reinfused into the patient, where they specifically recognize and attack the cancer cells. The growth and activity of immune cells can also be stimulated by cytokines such as interleukin-2 and interferons.

2.5 Diagnostic imaging modalities

The diagnosis of cancer involves not only the visualization of tumor tissue using imaging modalities, but also the identification of the cancer cells involved. Imaging techniques, such as X-rays, computed tomography (CT) scans, and positron emission tomography (PET) scans, are used to visualize the spatial extent, location, and shape of tumor tissue in the body and to identify cancer cells, for example, by receptor-specific radioactive labeling [43]. Other diagnostic methods that can be used to detect cancer include biopsies, blood and urine tests, and molecular pathology laboratory tests [44]. In a biopsy, tissue is removed from a potentially cancerous tumor and examined under a microscope for the presence of cancer cells. For example, a biopsy may be performed by fine-needle aspiration, core-needle biopsy, or surgical biopsy [45]. Laboratory tests, such as blood and urine tests, are used to determine the levels of certain hormones or tumor markers in the body, while molecular pathology studies of tumor-tissue samples can identify cancer cells [46]. The choice of diagnostic test depends on the type and stage of the cancer, as well as the patient symptoms and medical history.

Imaging modalities make a valuable contribution to cancer diagnosis by providing detailed anatomical and functional information about the location, size, and extent of malignant tumors. The most important imaging modalities are X-ray, computed tomography (CT), magnetic resonance imaging (MRI), ultrasound, positron emission tomography (PET), scintigraphy, mammography, and fluorescence imaging [40, 47]. X-rays provide fast and cost-effective views of the anatomical structures of the body [48, 49]. They are often instrumental in making an initial diagnosis and are used to monitor disease progression and assess the effectiveness of treatment for various types of cancer. X-rays can visualize abnormal bone structures such as primary bone cancer and metastases, fractures, and osteolytic (bone-destroying) or osteoblastic (bone-building) lesions. However, neither the contrast nor the lateral resolution of X-rays is sufficient to detect tumor tissue in soft tissues such as muscles, tendons, joint capsules, and ligaments. Since its introduction in 1973, X-ray CT has revolutionized radiological imaging by enabling 3D imaging of a wide range of tissue types and organ systems, including the gastrointestinal tract, cardiovascular system, renal tract, liver, lung, bone, cartilage, and tumor tissue [50, 51]. In its simplest form, a high-resolution CT image is obtained by rotating an X-ray source around an object with a detector positioned directly opposite the radiation source. In most cases, X-ray scans are generated in small angular steps over 360° during the rotation around the object. The fourth generation of CT scanners used a stationary 360° detector ring and a rotating X-ray tube. Spiral CT, introduced in 1989, continuously scans larger volumes of the patient during a single breath hold. Today, multidetector CT (MDCT) captures 3D volume images of the human body, which also significantly improves scanning speed compared to earlier systems (Figure 6). The X-ray beam is wider and can scan multiple layers of the body simultaneously. Several rows of detectors are used to acquire more than one CT slice in a single tube rotation. Modern MDCT systems have wide detector arrays consisting of 320 detectors that cover the entire region of interest per rotation, enabling submillimeter scanning [53, 54]. The X-rays emitted during the CT scan are attenuated as they pass through the human body and are then received by the detectors. The data obtained from the scanning process contains multiple projections that are processed using appropriate algorithms to define each CT slice. The resulting 3D CT image is divided into a matrix of voxels. The X and Y dimensions of each voxel, defined in the image plane, correspond to the pixel dimensions. The voxel

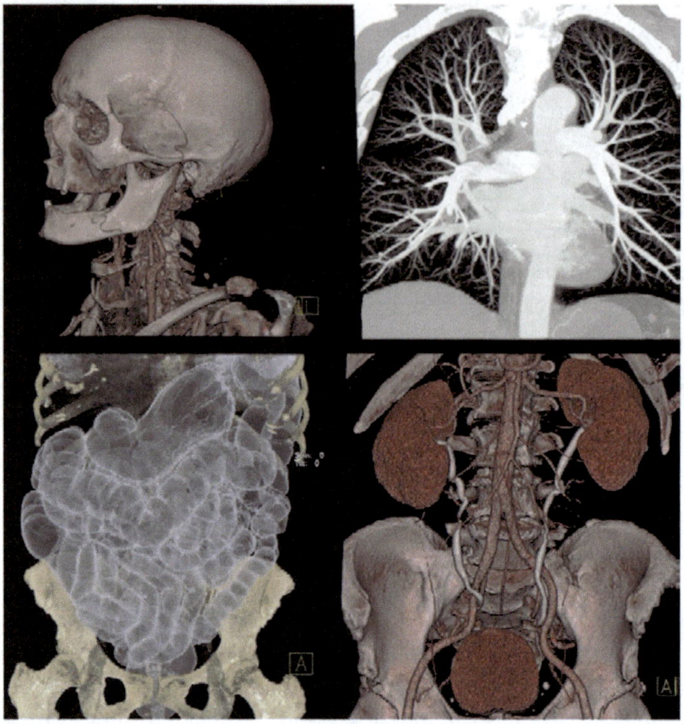

Figure 6: Multislice spiral CT scans virtually all regions of the body with submillimeter isotropic spatial resolution and scan times of 5 to 15 s. [52].

dimension in the Z direction corresponds to the slice thickness. For each image pixel, a CT number is calculated and displayed. This calculation is always based on the X-ray attenuation properties of the tissue within the voxel. The CT number assigned to each pixel determines the gray value of that pixel.

Shortly after its introduction in the early 1980s, MRI became one of the most important diagnostic imaging techniques [55]. Unlike CT, MRI is excellent at imaging soft tissue and can also image organic structures in 2D and 3D in oblique orientations. In addition, MRI is considered a noninvasive diagnostic imaging modality that produces cross-sectional images of the human body without the use of ionizing radiation, providing morphological and structural information about different types of tissue. The MR signal depends on tissue properties such as diffusion, perfusion, blood flow, temperature, and magnetic susceptibility, which can be detected by MRI. For example, metabolic changes surrounding tumors and degenerative brain diseases caused by metabolic syndrome can be visualized using MR spectroscopy (^{1}H, ^{13}C, ^{31}P) [56]. On the other hand, whole-body MRI allows for whole-body tumor staging and metastases screening.

Magnetic resonance imaging (MRI) is based on nuclear magnetic resonance. The nuclei of ^{1}H, ^{19}F and ^{31}P possess an odd number of protons and therefore a magnetic nuclear spin moment. Organic tissues such as muscles, organs, and adipose tissue are composed of water, hydrocarbons, lipids, and phosphates, and therefore contain MRI-active ^{1}H and ^{31}P [57]. In a static homogeneous magnetic field, unpaired nuclear spins align parallel to the magnetic flux density B_0 due to their magnetic moment. The precession motion of the nuclear spins takes place at the Larmor frequency, which is element-specific and, in addition, depends on the strength of the external magnetic field ($\vec{H}_0$). The Larmor frequency of ^{1}H, ^{19}F, and ^{31}P is 42.6 MHz/T, 40.0 MHz/T, and 10.8 MHz/T, respectively, which lies in the FM radio-wave range. The MRI measurement principle is based on the magnetization of tissue, the perturbation of the magnetization by the application of two and more radio frequency (RF) pulses and the detection of the MR signal. The magnetization $\vec{M}$ results from the alignment of the unpaired nuclear spins of the examined tissue along the magnetic field flux density (i. e. $\vec{M} \parallel \vec{B}_0$) and thus also represents a vector field. The aligned nuclear spins can be excited by RF pulses when they resonate at the average Larmor frequency of the nuclear spins of the respective nuclei. The RF pulses are applied orthogonal to the external magnetic field $\vec{B}_0$. Acting as a time-dependent pulsed magnetic field $\vec{B}_1$, the RF pulses tilt the magnetization vector into the x, y plane (90° RF pulse) and subsequently flip the magnetization vector field from the x, y plane by 180° into the –x, –y plane (180° RF pulse) (Figure 7). In the x, y plane, the nuclear spins dephase by spin–spin relaxation (transverse relaxation) due to interactions with proton spins and other magnetic fields in the environment. The 180° RF pulse inverts the nuclear spins and thus their direction of motion (rephasing), resulting in the generation of the spin echo (MR signal). In a conventional spin echo sequence, the 180° pulse is emitted half the echo time (1/2 TE) after the 90° pulse is applied, and the spin echo is generated after another 1/2 TE.

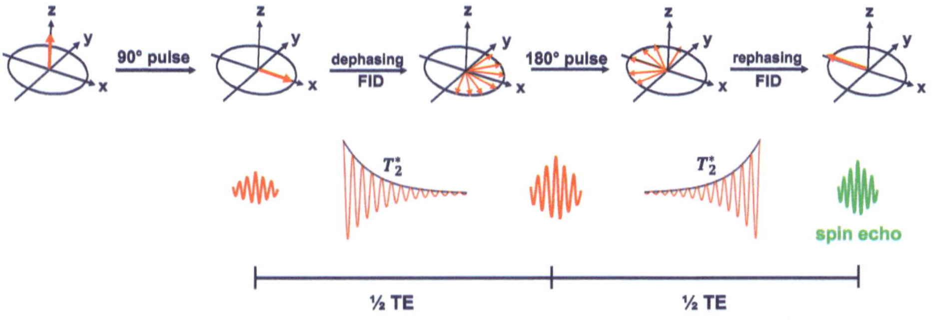

Figure 7: Spin echo sequence scheme sketching the effect of 90° and 180° RF pulses, dephasing and rephasing by free induction decay (FID) with the effective dephasing time T_2^*; the time interval between the 90° and 180° pulses and between the 180° pulse and spin echo is half the echo time (1/2 TE).

Dephasing and rephasing undergo free induction decays (FIDs) with the effective dephasing time T_2^* (Figure 8). The recovery of the longitudinal magnetization (M_z) by realignment of the nuclear spins occurs by spin-lattice relaxation with the time constant

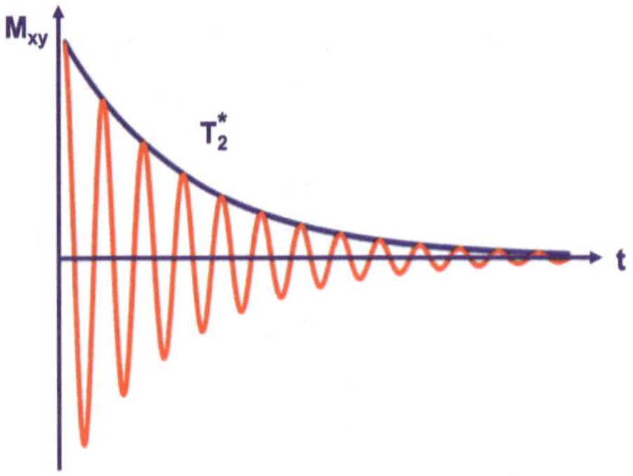

Figure 8: Dephasing of the transversal magnetization through free induction decay with the time constant T_2^*.

T_1. During longitudinal relaxation, energy is dissipated into the lattice. The decrease of the transverse magnetization (M_{xy}) occurs due to phase loss (dephasing) of the nuclear spins (Figure 8). The spin–spin relaxation time T_2 is a measure of how fast the spins exchange energy in the "xy" plane. The effective dephasing time T_2^* is always shorter than T_2 as can be seen from the following relationship:

$$\frac{1}{T_2^*} = \frac{1}{T_2} + \frac{1}{T_2'}, \tag{1}$$

where T_2' accounts for the loss of phase coherence occurring through inhomogeneities in the external static magnetic field $\vec{B}_0$. The temporal evolution of longitudinal and transverse magnetization is described by the Bloch equations

$$\frac{dM_z(t)}{dt} = \gamma(\vec{M}(t) \times \vec{B})_z - \frac{M_z(t) - M_0}{T_1}, \tag{2}$$

$$\frac{dM_y(t)}{dt} = \gamma(\vec{M}(t) \times \vec{B})_y - \frac{M_y(t)}{T_2}, \tag{3}$$

$$\frac{dM_x(t)}{dt} = \gamma(\vec{M}(t) \times \vec{B})_x - \frac{M_x(t)}{T_2}, \tag{4}$$

where M_0 is the longitudinal equilibrium magnetization. According to the Bloch equations, the longitudinal magnetization M_z increases as an exponential saturation function, while the transverse magnetization M_{xy} decays exponentially (Figure 9).

A conventional MRI device consists of three hardware groups: (1) the main magnet with the static homogeneous magnetic field $\vec{B}_0$ over the patient body; (2) the linear gradient system over the imaging volume; and (3) the RF pulse system providing the rotating magnetic field ($\vec{B}_1$) (Figure 10). A superconducting magnet with a high magnetic-field strength is used to magnetize patient tissue. Three orthogonal gradient fields (G_x, G_y, and G_z) are used for signal localization. The RF system emits RF pulses and provides low-loss detection of MR signals. The phased-array surface RF coil technology currently

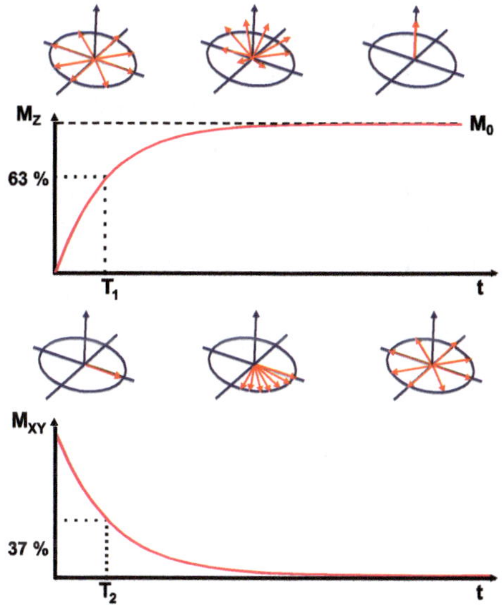

Figure 9: Scheme illustrating the recovery of the longitudinal magnetization M_z with the spin-lattice relaxation time T_1 (top) and the decay of the transversal magnetization M_{xy} through dephasing by spin-spin relaxation (T_2).

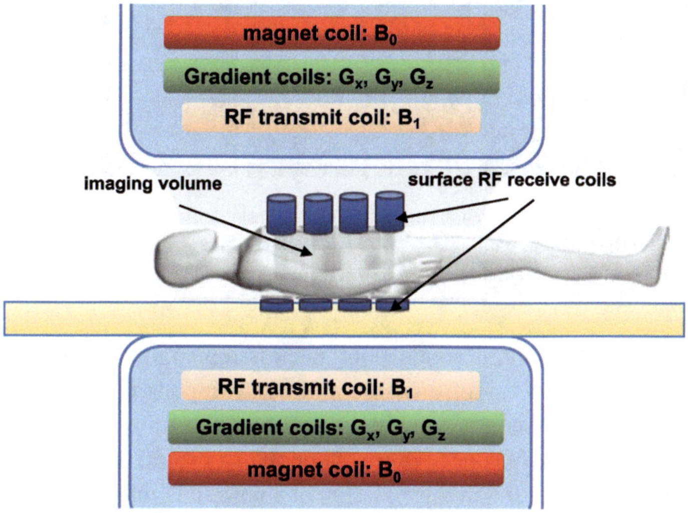

Figure 10: Scheme of a MRI system: magnetic coil for the generation of the static homogeneous magnetic field $\vec{B}_0$ for nuclear spin alignment; gradient coils to produce the time-varying gradient fields G_x, G_y and G_z for spatial encoding of the MR signals; and transmitter radiofrequency (RF) coil to generate an electromagnetic RF field $\vec{B}_1$ for nuclear spin excitation.

in use makes possible short acquisition times and great flexibility in data acquisition by using parallel imaging strategies in combination with whole-body RF coil coverage of the patient. The RF system typically consists of a RF transmitter coil installed in the MR housing and an RF receiver for detecting, for instance, the weak spin echo.

MRI systems with field strengths from 1 T to 3 T are currently accepted as the clinical standard. A cylindrical main magnet with a homogeneous volume at the center of the central axis meets all these requirements and is therefore the most common magnet shape today. The two most important MR parameters that determine almost every MR sequence are TE and the repetition time (TR). TE defines the time interval between RF pulse excitation and signal acquisition, and TR is the time between successive RF pulses. In combination, TE and TR determine the relative contrast of different tissues. T_1-weighted MR images are produced with short TR and TE, and T_2-weighted images are produced with long TR and TE. The combination of long TR and short TE results in proton-density-weighted images in which the image signal is determined solely by the hydrogen content of the voxels (Figure 11).

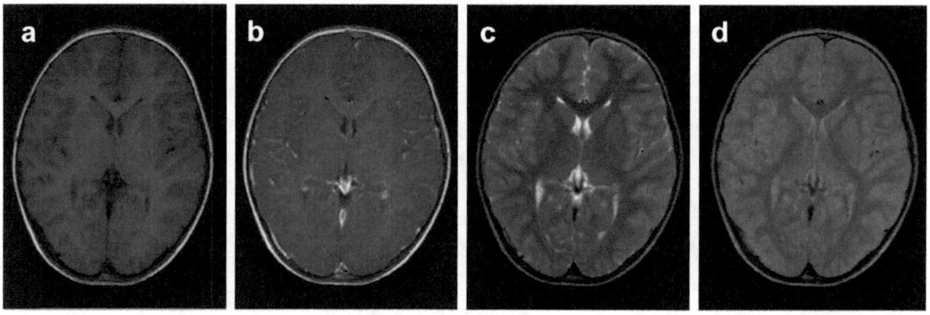

Figure 11: Repeated MR imaging of an axial cross section through a human brain demonstrates the dependence of the soft tissue contrast on the imaging parameters: (a) T_1 weighted MR image, (b) T_1 weighted MR image with a contrast agent, (c) T_2 weighted MR image, (d) proton-density-weighted MR image [55].

Positron emission tomography (PET) and single photon emission-computed tomography (SPECT) are based on the β^+ decay of unstable and neutron-poor nuclides (radionuclides) [58]. An isotope of the element (^A_ZX) with the number of protons (Z) and the total number of protons and neutrons (A) decays by conversion of a proton to a neutron into the isotope or element $^A_{Z-1}Y$, resulting in the emission of a positron and an electron neutrino, which can be written as Eqs. (5)–(6),

$$^A_ZX \rightarrow {}^A_{Z-1}Y + e^+ + \nu_e, \tag{5}$$

$$e^+ + e^- \rightarrow 2\gamma. \tag{6}$$

The radionuclide emits a positron, which collides with an electron in the surrounding tissue. The annihilation reaction produces two gamma (γ) photons that travel in oppo-

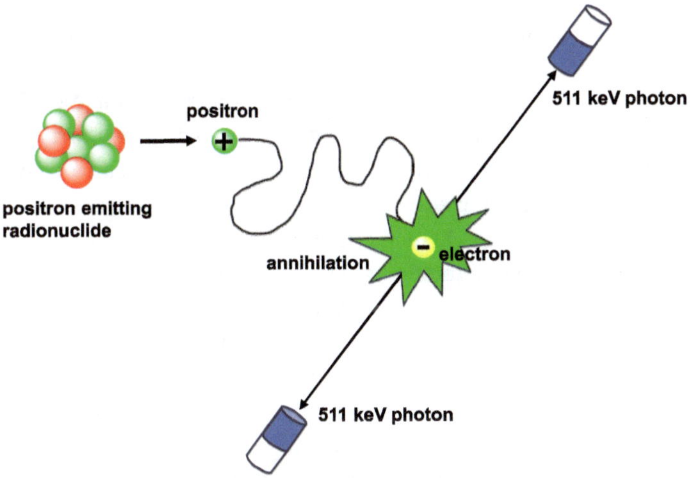

Figure 12: Schematic of the positron–electron annihilation event resulting in the generation of two 511 keV γ photons traveling in opposite directions and being registered by coincidence detectors.

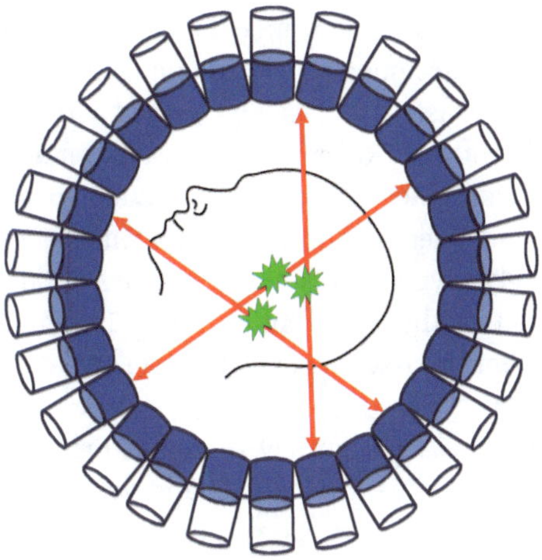

Figure 13: Circular arrangement of coincidence detectors for simultaneous registration of γ photons per annihilation event in target tissue.

site directions (180°) and have an energy of 511 keV (Figure 12). The γ photons from an annihilation event are measured by opposing coincidence detectors arranged in a circle around the target (Figure 13).

The measurement principle of PET is to register coincidences between two opposing detectors to localize the annihilation event [59]. The typical time windows of the detection electronics are 4.5 to 15 nanoseconds. From the temporal and spatial distribution of these registered decay events, the spatial distribution of the radionuclide inside the body is deduced, and a series of sectional images are calculated. The radionuclides used in diagnostic PET are ^{11}C ($t_{1/2}$ ~ 20.38 min), ^{18}F ($t_{1/2}$ ~ 109.8 min), ^{64}Cu ($t_{1/2}$ ~ 12.7 h), and ^{86}Y ($t_{1/2}$ ~ 14.7 h).

PET scans are used for the (early) detection of cancer, to assess the stage of the disease and to monitor the response to treatment. For example, cancer cell-targeted bio(molecules) or metabolic substrates are administered to the patient. Cancer cell-targeted compounds and suitable metabolic substrates are taken up by cancer cells in higher concentrations than by healthy cells, which enables the visualization of tumor tissue [60, 61]. SPECT, on the other hand, detects single γ photons emitted directly by a radionuclide or radionuclide-labeled substance. This radiopharmaceutical is administered intravenously to the patient at the beginning of the examination. The emitted γ photons are monitored by scintillator detectors equipped with collimators. ^{99}mTc ($t_{1/2}$ ~ 6.02 h), ^{123}I ($t_{1/2}$ ~ 13.2 h), and ^{111}In ($t_{1/2}$ ~ 2.8 d) are among the most commonly used radionuclides for SPECT. However, PET offers several advantages over SPECT, including high sensitivity, better spatial resolution, and accurate activity quantification. One of the most successful radiotracers in PET is ^{18}F-fluorodeoxyglucose (FDG), as it can be used to detect the metabolic activity of cancerous tissue [62]. FDG is obtained by radiolabeling glucose by replacing an OH^- group with a β^+ decaying ^{18}F. FDG is injected into the patient's body and enters the glucose metabolism. Cancer cells consume much more glucose than healthy cells, which leads to an accumulation of radionuclides and thus more activity in the tumor tissue. This activity distribution can be imaged with the PET system. PET, in combination with the radiotracer FDG, can also be used to monitor activity in the brain, as activity correlates with glucose consumption. Whole-body PET in combination with MRI is particularly indicated when both the advantages of PET in detecting cancer tissue with increased glycolytic metabolism and those of MRI in recognizing bone metastases can be exploited for the treatment of cancer. This new technology is therefore an attractive alternative for the assessment of multiple myeloma and bone metastases (Figure 14).

Sonography or ultrasound imaging, also known as echography, is an imaging technique that uses ultrasound waves to examine soft tissue in medicine and veterinary medicine. In a homogeneous medium, ultrasound waves propagate as longitudinal material waves in a straight line. In an inhomogeneous medium, such as the human body, the propagation speed of ultrasound waves depends on the density and compressibility of the tissue being examined (Figure 15). The generation of an ultrasound image in medicine is the response of ultrasound waves to differences in density and compressibility of adjacent tissues in the human body. Ultrasound imaging is performed using the pulse-echo technique [63]. Ultrasound pulses are generated in the transducer, which contains an array of piezoelectric crystals. Electrical pulses from a high-frequency generator are converted into ultrasound pulses in the transducer by the reciprocal piezoelectric effect and emitted. On the other hand, ultrasound waves received by the transducer periodically deform the piezoelectric crystals, causing an alternating voltage change and thus generating an electrical signal. The piezo crystals in the transducer thus act as both transmitter and receiver. The ultrasound beam can be shaped by the geometry of the

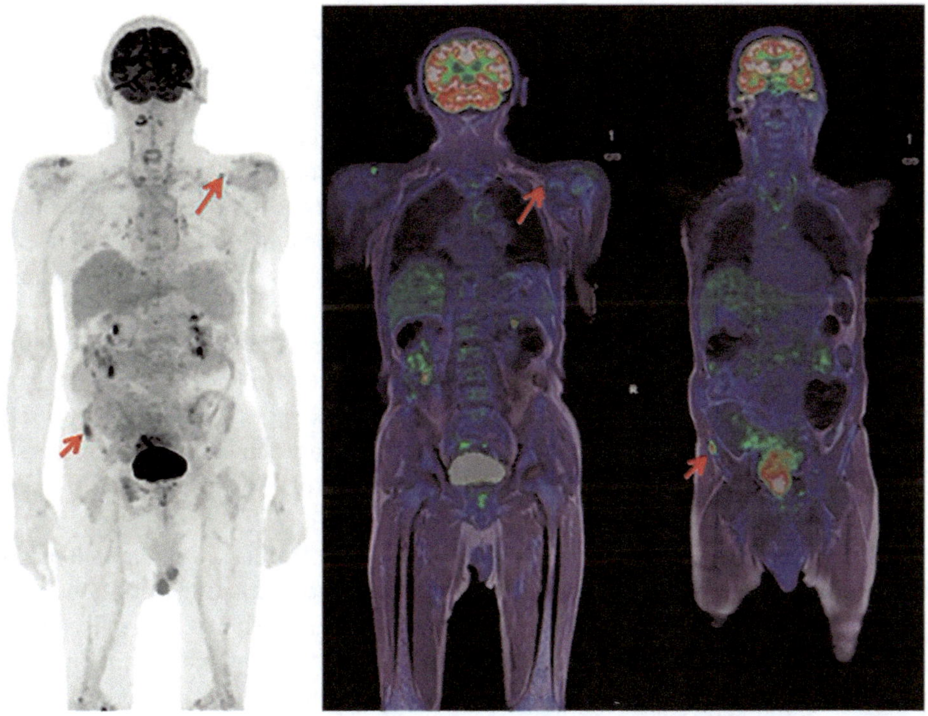

Figure 14: Whole-body ^{18}F-FDG PET/MRI images of a male patient with multiple myeloma after chemotherapy showing residual metabolic activity in the right iliac crest and a hypermetabolic left supraclavicular lymph node (red arrows): FDG-PET image (A) fused PET/T_1-weighted MRI images (B and C) [1].

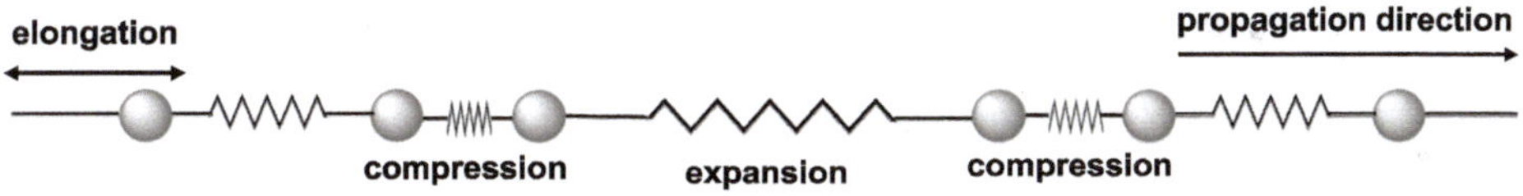

Figure 15: Longitudinal wave propagating through solid matter.

piezo crystals. Ultrasound pulses coupled into the body pass through it as ultrasound waves with frequencies of 1–15 MHz. They are partially or completely scattered and attenuated by density inhomogeneities in the tissue. Since these scattering centers are smaller than the wavelength of the ultrasound wave, they are scattered in all spatial directions, which appears as a bright veil in the ultrasound image. Ultrasound waves are partially reflected at interfaces between tissues of different densities. 100 % reflection occurs at interfaces between air, bone, or calcareous structures and other tissues. Refracted ultrasound waves change their direction of propagation. By absorbing the ultrasound wave, its energy is completely converted to heat. The reflected ultrasound pulses, called echoes, mainly contribute to structure formation in ultrasound imaging. When some of the ultrasound waves pass through a boundary surface, further echoes can be generated by reflections from deeper boundary surfaces. The echoes travel back to the transducer, where they are converted into an electrical signal. In ultrasound imaging, an ultrasound pulse is emitted that is partially reflected and partially transmitted by the interface between two tissue structures. The amount of reflection depends on the impedance differences between the two tissues and is expressed as the intensity of the echoes received by the transducer. The depth of the various reflector interfaces along the beam is calculated by determining the time interval between the emission of the ultrasound pulse and the reception of the echo. From this transit-time value, the distance between the transducer on the body and the reflective layer inside the body is calculated using the ultrasound velocity in tissue. This distance information is then processed into an ultrasound image. The resolution of an ultrasound image depends on the ultrasound wavelength and the geometry of the transducer. The smallest distance between two separately imaged interfaces in the direction of sound propagation defines the axial resolution, which is 200–300 μm for a sound wave frequency of 7.5 MHz. Lateral resolution is perpendicular to the direction of propagation.

Four standard ultrasound techniques are used in medical imaging: A-mode, B-mode, M-mode, and Doppler ultrasound [64, 65]. The amplitude scan or A-mode technique measures the temporal evolution of the amplitude of the echoes relative to the distance of the transducer. A single transducer scans a line through the body, and the amplitude of the echoes is displayed on the screen as a function of depth. The brightness (B) mode method is an advancement of the A mode method. A linear array of transducers is used to scan a plane through the body so that a two-dimensional image can be displayed on the screen. In motion (M) mode, the echoes are measured with two-dimensional temporal resolution, equivalent to a rapid sequence of B-mode scans. This makes it possible to visualize dynamic processes such as the movement of heart valves. The M-mode technique is therefore mainly used in cardiology. The Doppler-mode technique is applied in vascular and cardiac diagnostics because it enables visualization of blood flow. Doppler measurements of the frequency shifts of blood cells moving as reflectors along the blood flow are used to determine the velocity and direction of blood flow. These frequency shifts can be displayed in color according to the direction of blood flow. In principle, many blood cells with various speeds are registered in the blood, providing

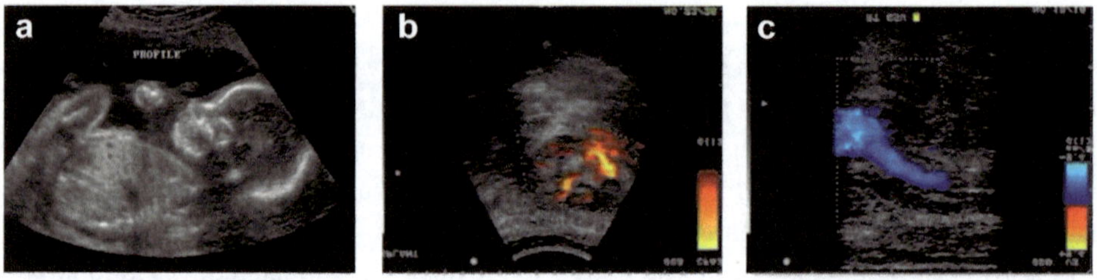

Figure 16: Ultrasound images of a fetus in the uterus (B mode), an artery of a transplanted kidney (B mode combined with color-coded duplex) and a vein (M mode combined with color-coded duplex) [66].

information about constrictions and occlusions of the vessels. There are two different Doppler ultrasound techniques. The spectral Doppler method determines the velocity distribution of the blood. The blood vessel is scanned by an ultrasound pulse beam at a fixed point in time. In contrast, the color Doppler method measures the blood velocity at different locations simultaneously. Echoes from moving scatterers are analyzed by an autocorrelation detector to assess average changes in flow velocity and displayed as color pixels, with blue and red hues used to indicate positive and negative Doppler shifts, respectively. Color-coded duplex sonography combines the Doppler method with the B-mode method (Figure 16). In color-coded duplex sonography, each color window is assigned either a gray value or a color value. The gray values (B-scan) represent the morphology of the tissue, and the color values represent the blood flow. The colors in the frequency or velocity display visualize the velocity-weighted Doppler shift, the lighter the faster.

3 Biocompatible nanomaterials

For in vivo applications, inorganic nanoparticles should possess fundamental properties such as water solubility, biocompatibility, and colloidal stability under physiological conditions. Biocompatible nanoparticles should not cause undesirable side effects in the human body. Among other things, this means neither the nanoparticle core nor the surface coating should consist of toxic material. For example, gold nanoparticles are inert and, if sufficiently small, are excreted through the kidneys. Iron oxide nanoparticles are metabolized to Fe^{3+} cations, which are added to the iron store in humans. In acidic medium (e. g., the stomach), silica and silicon nanoparticles are converted to water-soluble bioavailable silicic acid. Polymeric surface coatings consisting of, for instance, polyethylene glycol (PEG) or polyethyleneimine (PEI), are biocompatible, and double layers of cetyltrimethylammonium bromide (CTAB) are unstable and release cytotoxic CTAB into biofluids. For applications such as nanocarriers in chemotherapy, gene therapy, or immunotherapy, nanoparticles must have a long half-life in the bloodstream and be equipped with surface functionalities that reversibly bind drugs, DNA, siRNA or mRNA, proteins, or peptides. The plasma half-life of nanoparticles can be tuned by size, shape, and surface stabilization. Inert polymer coatings, such as PEG, provide nanoparticles with long circulation times by hindering adsorption of proteins in biofluids and thus recognition by macrophages by evading macrophage-mediated uptake. However, the thickness and density of polymer coatings are almost impossible to control during synthesis. In addition, polymer coatings are difficult to functionalize, making them rather unsuitable for coating drug nanocarriers.

Essential for controlled functionalization of nanoparticles by covalent binding of small functional units, bioconjugation, or even by self-assembly of ordered lipid molecule-based shells is the structural elucidation of nanoparticles, surface chemistry, and surface coating. For example, the size and shape of nanoparticles determine their surface curvature and thus the topology, which affects the surface reactivity, charge arrangement, physisorption, and chemosorption.

The morphology, sizes, and size distribution of nanoparticles are usually characterized by transmission electron microscopy (TEM). The dynamic light scattering method makes it possible to determine the hydrodynamic diameter. However, the reliability of the results depends sensitively on the solubility of the nanoparticles in the respective solvent and cannot be assumed in the case of agglomeration. X-ray photoelectron spectroscopy, X-ray Powder Diffraction (XRD), high-resolution TEM (HRTEM), UV-Vis absorption, and fluorescence spectroscopy are used to study the composition and structural and optical properties of the nanoparticles. Information on surface structures and charges is accessible by FTIR transmission spectroscopy and zeta potential measurements. The nanoparticle surface composition can be elucidated by small-angle X-ray scattering (SAXS) and small-angle neutron scattering (SANS).

For biomedical applications the as-synthesized nanoparticles must be diligently purified and then endowed with biocompatible surface functions. Purification usually in-

https://doi.org/10.1515/9783110605495-003

volves multiple ultracentrifugation–washing cycles and, subsequently, dialysis, membrane filtration, or chromatography.

3.1 Plasmonic noble metal nanoparticles

Noble metal nanoparticles, especially gold and platinum nanoparticles, are gaining importance due to their exceptional application potential in heterogeneous catalysis, biotechnology, and medicine [67–71]. Silver and palladium nanoparticles play a minor role in medical applications as releasing cytotoxic Ag^+ and Pd^{2+} cations in the cytoplasm [72, 73]. The uniqueness of noble metal nanoparticles is characterized by their size- and shape-dependent optical absorption, light scattering, and local electric field enhancement due to surface plasmons as being confined in nanoparticles [74]. The associated localized surface plasmon resonance (LSPR) arises from coherent, collective oscillation of conduction band electrons in nanoparticles with sizes much smaller than the wavelength of the incident light [75–77]. The spectrum and electromagnetic field strengths of the LSPR, and thus the sensitivity to surface adsorbates, can be fine-tuned and enhanced by the size and shape of their noble metal nanostructure.

Noble metal nanoparticles can be obtained as nanospheres, nanocubes, nanoprisms, nanorods, nanostars, and nanoshells by suitable chemical reactions [78–82]. The tunability of the morphology of noble metal nanostructures via chemical synthesis routes paves the way for the development of effective biosensors and nanotheranostics. Moreover, the surfaces of noble metal nanoparticles can be easily functionalized with a wide variety of organic and biomolecular ligands, preferably with those containing thiol groups.

Platinum nanoparticles (PtNPs) exhibit LSPR in the UV range and therefore have a limited applicability as optical biosensors [83]. These high-Z metal nanoparticles exhibit catechol oxidase enzyme-like activity [84, 85]. Furthermore, PtNPs have been reported to act as radiosensitizers in X-ray radiation therapy and as photosensitizers in photothermal therapy of cancer and pathogenic diseases. However, the applicability of Pt nanostructures as both optical biosensors and photosensitizers is severely limited by their absorption in the UV range. In addition, the adjustability of Pt nanoparticle shapes is restricted to the synthesis of nanospheres and nanorods [86, 87].

Gold nanostructures play a key role in plasmonic biomedical applications as they are biocompatible and have an LSPR whose spectral position can be varied across the visible and NIR spectrum through size and shape controlling synthesis routes [88, 89]. The easy accessibility of versatile shapes (e. g., nanoprisms, nanorods, or nanoshells) enables the design of biosensing agents and plasmonic heaters. Gold nanospheres with sizes between 4 and 50 nm exhibit a LSPR band around 530 nm in the absorption spectrum. In comparison, the spectral position of the longitudinal LSPR of nanorods can be tuned via their aspect ratio and the LSPR of spherical nanoshells via their shell thickness and diameter from the visible to the near-infrared spectral range. The spectral variability of the LSPRs of gold nanorods and gold nanoshells enables applications in in vivo imaging and photothermal cancer therapy.

3.1.1 Plasmon resonance in noble metal nanostructures

When irradiated with electromagnetic waves, the conduction band electrons in noble metal nanoparticles will be periodically displaced relative to the positively charged metal cations in phase with the incident light. This coherent collective oscillation of the electrons is confined to the nanoparticle and is therefore referred to as localized surface plasmons. Localized surface plasmon resonance occurs when the frequency of the incident light matches the natural frequency of the electron oscillation. At this specific resonance frequency, the oscillation amplitude reaches a maximum, which results in a strong enhancement of the electric field at the surface of the nanoparticle. The optical absorption of nanoparticles peaks at the plasmon resonance frequency. The localized surface plasmon resonance sensitively depends on the elemental composition, shape, size, and environment of the nanoparticles [90].

In nanoparticles much smaller than the wavelength of visible light, defining the quasistatic approximation, the phase of the harmonically oscillating electromagnetic field remains approximately constant over the whole nanoparticle. The electric field-driven oscillation of the negative electron cloud against the positive metal cations acts as a Hertz dipole that resonantly absorbs and emits or scatters electromagnetic waves (Figure 17) [88, 90].

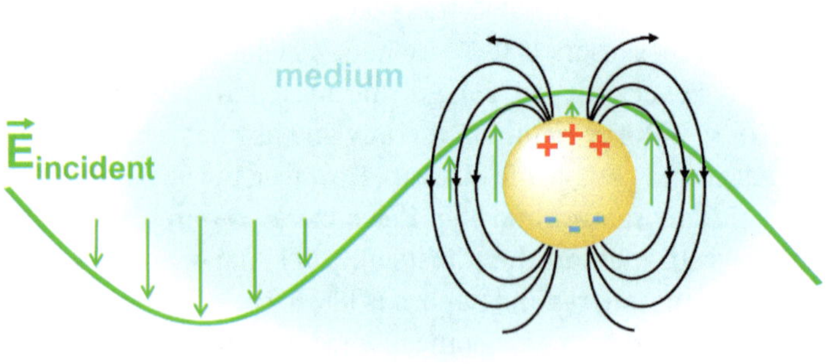

Figure 17: Schematic illustration of surface plasmon resonance of a spherical noble metal nanoparticle due to interaction with an oscillating electrical field.

Localized surface plasmons decay by emitting radiation, or they decay nonradiatively by converting absorbed light to heat (photothermal conversion). The electromagnetic field intensity and the scattering and absorption cross sections of noble metal nanoparticles are maximum at the localized surface plasmon resonance (LSPR), which peaks around 520 nm and 400 nm for gold nanospheres and silver nanospheres, respectively.

The absorption and scattering spectra of gold and other noble-metal nanoparticles can be calculated in terms of Mie's theory [91]. Maxwell's equations are solved with appropriate boundary conditions using multipole expansions of the incoming and outcoming electromagnetic fields in spherical coordinates. However, Mie's theory is limited to the LSPR calculation of only a few nanoparticle shapes (spheres, concentric spherical shells, and spheroids). Nowadays, for nanoparticles with arbitrary geometric shapes, numerical methods are available for simulating the interaction between light and nanoparticles. In the discrete dipole approximation (DDA), the polarization induced in nanoparticles is approximated by the response of a cubic lattice of polarizable dipoles [81, 92–94]. This involves the replacement of a nanoparticle by a cubic array of polarizable elements

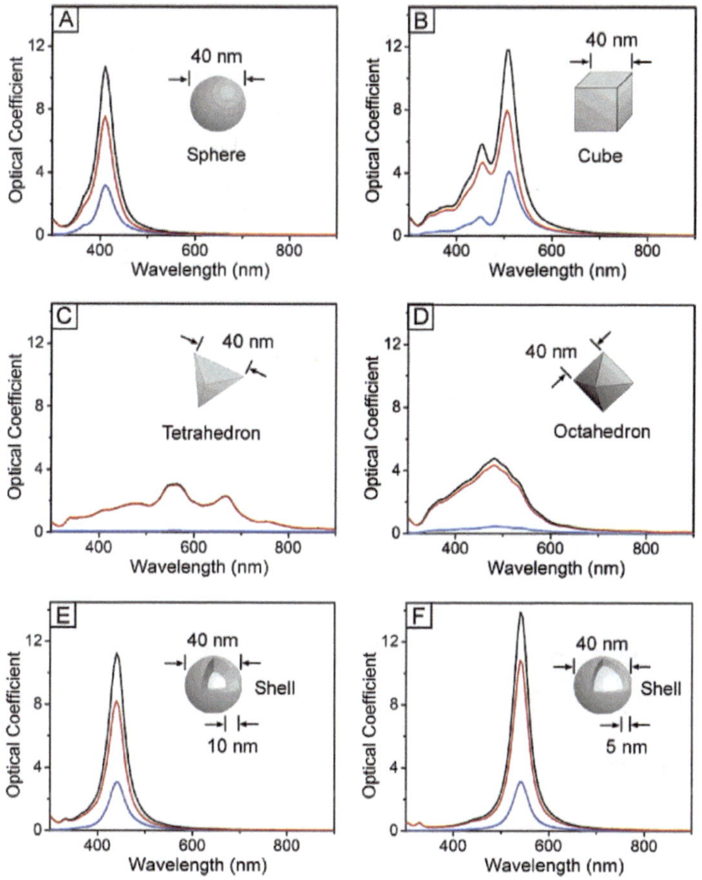

Figure 18: Calculated UV-Vis extinction (black), absorption (red), and scattering (blue) spectra of Ag nanosperes (A), Ag nanocube (B), Ag tetrahedron (C), Ag octahedron (D), 10 nm-thick Ag nanoshell (E), and 5 nm-thick Ag nanoshell (F) [95].

whose polarizabilities are determined by the nanoparticle dielectric function. The polarizable elements are so small that the mathematical problem is reduced to the consideration of dipole interactions between the applied electromagnetic field and the induced electromagnetic fields in other elements. Figure 18(A) shows the UV-Vis extinction, absorption, and scattering spectra of a 40 nm-sized Ag sphere in water calculated using the Mie theory. The extinction, absorption, and scattering spectra of a 40 nm-sized Ag cube (B), Ag tetrahedron (C), and Ag octahedron (D) exhibit several pronounced symmetries for dipole resonance, leading to more LSPR peaks (Figure 18). The significantly red-shifted LSPR peaks are ascribed to the emergence of higher surface-charge densities at the corners of the cube, tetrahedron, and octahedron. Surface charges concentrated at corners or sharp edges of nonspherical noble metal nanoparticles increase charge separation and thereby weaken the restoring forces of the light-field-induced electron displacements. On the other hand, the calculated single resonance peak of Ag shells appears red shifted in the extinction, absorption, and scattering spectra compared to those of the equally sized Ag sphere (Figures 18(E) and (F)). The incident electromagnetic field induces the formation of surface charges outside and inside the shell. These surface charges have the same sign to cancel out the interacting light field. The enlarged charge separation is associated with the diminishment of the surface plasmon frequency. As the shell thickness decreases, the charge separation decreases, resulting in a redshift of the LSPR peak. The calculated UV-Vis absorption spectra of noble metal nanoparticles resemble the characteristic features of the experimental spectra. Under light irradiation, spherical noble metal nanoparticles exhibit a single LSPR band the width, shape, and spectral position of which crucially depend on the nanoparticle diameter [96]. Figure 19 shows TEM images (left panel) and the associated UV-Vis absorption spectra (right panel) of gold nanoparticles of different sizes.

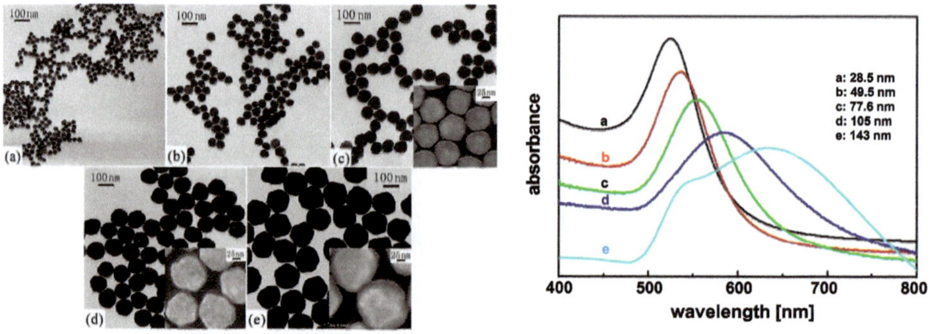

Figure 19: TEM images (left panel) and associated UV-Vis absorption spectra (right panel) of variously sized gold nanoparticles [96].

Spherical gold nanoparticles larger than 28.5 nm do not only show increasingly red-shifted LSPR bands, but also significantly altered spectral properties with diameters larger than 80 nm. Both, the size-dependent redshift and spectral broadening of the LSPR band for nanoparticle sizes larger than 10 % of the incident light wavelength are explained with electromagnetic retardation and the emergence of higher multipolar surface-plasmon resonances besides the dipolar LSPR. According to the so-called skin-depth effect, electromagnetic fields cannot penetrate deeper than 15 nm inside noble metallic materials [75]. On the other hand, the medium surrounding the nanoparticles causes a redshift in the LSPR band with increasing refractive index. This medium effect is attributed to the screening of Coulomb attraction between the oscillating electrons and the positively charged gold nuclei inside the nanoparticle. On the other hand, hollow spherical gold nanoshells and gold nanoshells grown around a spherical silica, metal oxide, or polymer cores possess LSPRs tunable from the visible to the NIR (600–1,000 nm) [97]. The spectral position and width of the LSPR band depend on the ratio of core diameter to shell thickness, which can be tuned via facile wet-chemical synthesis routes [80, 98]. Suitably sized spherical gold nanoshells absorb NIR light, which facilitates their application in photothermal therapy of cancer [99, 100]. UV-Vis spectra and associated TEM images of hollow gold nanoshells with diameters between 37 and 96 nm and a shell thickness of 10 nm demonstrated the tunability of the LSPR of gold nanoshells being shifted from 657 nm to 957 nm for diameter-to-nanoshell thickness ratios varied from 3.7 to 9.6. These gold nanoshells were synthesized by reducing $HAuCl_4$ with 3-aminopropyltriethoxysilane (APTES) in water, and the $HAuCl_4$-to-APTES ratio were shown to determine the diameter of the gold nanoshells, while the shell thickness with 10 nm remained constant.

Nonspherical gold nanoparticles, such as nanobranches, nanopyramids, or nanorods (Figure 20), exhibit two or more nondegenerate surface plasmon modes with different oscillation directions (Figure 21) [76]. The UV-Vis absorption of gold nanocubes (Fig-

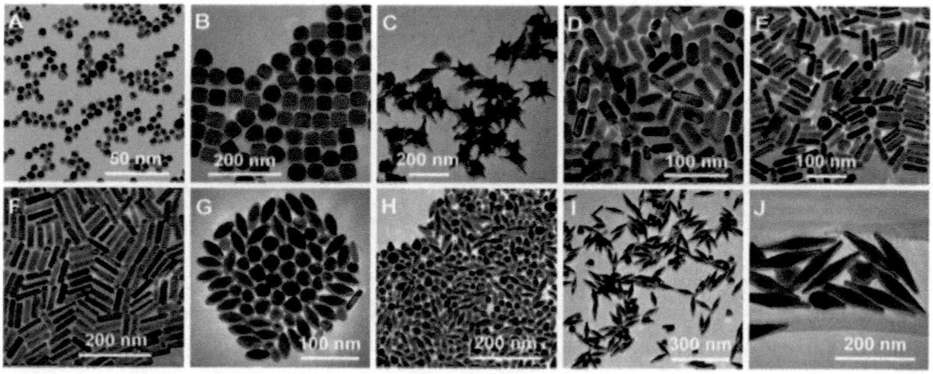

Figure 20: TEM images of gold nanoparticles with different shapes and sizes: (A) nanospheres, (B) nanocubes, (C) nanobranches, (D) nanorods (AR = 2.4 ± 0.3), (E) nanorods (AR = 3.4 ± 0.5), (F) nanorods (AR = 4.6 ± 0.8), (G) nanobipyramids (AR = 1.5 ± 0.3), (H) nanobipyramids (AR = 2.7 ± 0.2), (I) nanobipyramids (AR = 3.9 ± 0.2), (J) nanobipyramids (AR = 4.7 ± 0.2) [76].

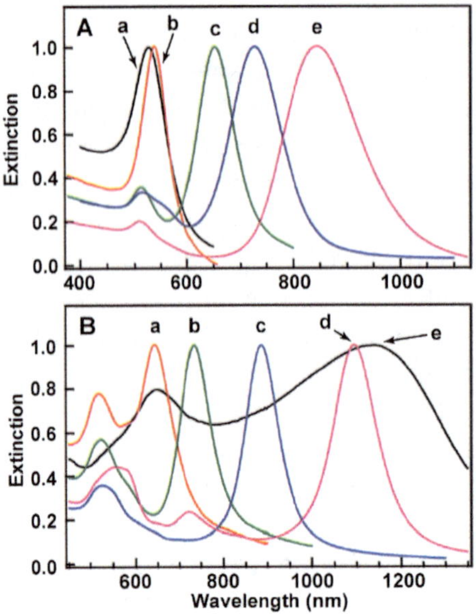

Figure 21: Normalized UV-Vis absorption spectra of gold nanoparticles with various shapes and sizes: (A) gold nanospheres (a), gold nanocubes (b) and gold nanorods with aspect ratios of 2.4 ± 0.3 (c), 3.4 ± 0.5 (d), and 4.6 ± 0.8 (e), respectively; (B) gold nanobipyramids with aspect ratios of 1.5 ± 0.3 (a), 2.7 ± 0.2 (b), 3.9 ± 0.2 (c), and 4.7 ± 0.2 (d), and gold nanobranches (e), respectively [76].

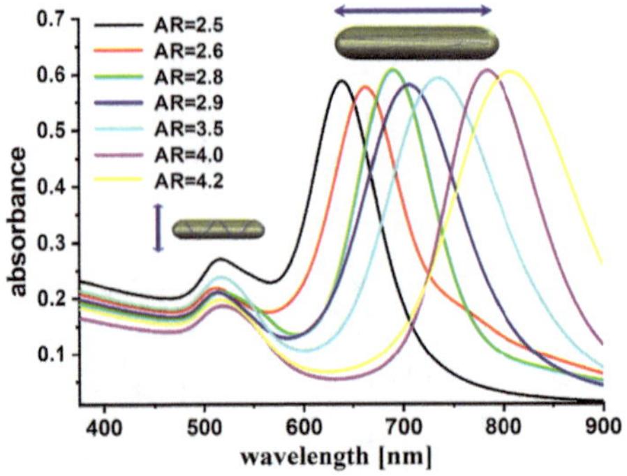

Figure 22: UV-Vis absorption spectra of AuNRs with aspect ratios (ARs) between 2 and 3.7.

ure 20: TEM image B, Figure 21: UV-Vis spectrum b) reflecting that of gold nanospheres (Figure 20: TEM images A and B, Figure 21: UV-Vis spectra a and b) which reflects their common structural isotropy. On the other hand, anisotropic gold nanostructures such as nanobranches, nanorods, and nanobipyramids (Figure 20: TEM images C, D, E, F, G, H, I, J) exhibit two LSPR bands (Figure 21: UV-Vis spectra A and B). Two surface plasmon resonances of gold nanorods (AuNRs) and gold nanobipyramids (AuNBPs) were rationalized to arise from coherent electron oscillation along the short axis (transversal) and the other one from electron oscillation along the long axis (longitudinal). The transversal LSPR band peaks between 520 and 530 nm. On the other hand, the spectral position of the longitudinal LSPR band can be tuned from the visible region to the NIR by varying the nanorod and bipyramid aspect ratio (AR), where the AR is defined as the length of the anisotropic shape (nanorod or bipyramid) divided by its width (Figures 21 A and B, Figure 22) [77, 78]. With decreasing volume of gold nanoparticles the ratio between the absorption and scattering cross section increases [90, 101]. The increase in absorption at the expense of light scattering increases the photothermal heat conversion of AuNRs and AuNBPs, in particular, thus potentially enhancing their plasmonic photothermal therapy efficacy [102–106]. Photoexcitation of gold nanoparticles generates via electron–electron scattering a hot electron distribution [107]. Thermalization occurs within 1–10 ps by electron–phonon scattering and leads to heating of the gold lattice. Subsequent phonon–phonon scattering (~100 ps) provides heat dissipation to the environment.

3.1.2 Synthesis routes for gold nanostructures

The preparation of biocompatible monodisperse gold nanoparticles (AuNPs) with defined morphologies and surface functionalities remains a challenge but is essential for all biomedical applications. There are two strategies for AuNP synthesis, which consist of top–down and bottom–up methods [79, 108–110]. Top–down strategies employ lithographic and laser-ablation techniques providing successive degradation of material blocks to gold nanoparticles [111–113]. The most promising technique is the pulsed laser ablation in liquid (PLAL): irradiation of gold plates, wires, or gold in diluted electrolytes with intense laser pulses yields spherical surfactant-free AuNPs with defined size distributions [114–122]. The laser-synthesized AuNPs are negatively charged as containing $Au(I)O^{1-}$ surface states and, therewith, can be easily functionalized with a broad variety of biomolecules (e. g., oligosaccharides, proteins, and oligonucleotides) [123].

Bottom–up strategies for the synthesis of AuNPs are based on wet-chemistry procedures requiring gold acid ($HAuCl_4$) or gold chloride ($AuCl_3$) as precursor, surfactants, and mild reducing agents [79, 82, 109]. Suitable reducing agents are for instance borohydrides, aminoboranes, hydrazine, formaldehyde, hydroxylamine, citric, ascorbic, and oxalic acids. The redox reaction in aqueous solution (e. g., $2\,Au^{3+} + 6\,BH_4^- \rightarrow 2\,Au^0 + 3\,B_2H_6 + 3\,H_2\uparrow$) must be carried out in the presence of a surfactant in excess. Suitable surfactants are polyvalent carboxylic acids, phosphonic acids, phosphines alcohols, thiolates, polymers, or amphiphilic compounds (e. g., cetyltrimethylammonium bromide (CTAB)). Shapes and size distributions of AuNPs can be tuned by the concentration ratio of surfactant and precursor, the surfactant structure and binding mode, the pH value, the reaction time, and temperature. Supersaturation of the precursor solution that initiates the nucleation process, followed by precursor-driven and surfactant-limited growth of the seeds into nanoparticles. The preparation of nonspherical AuNPs requires suitable surfactants that bind preferentially to specific crystallographic facets. One such surfactant is CTAB, which forms a bilayer template to promote the anisotropic growth of gold nanostructures [77, 124–127]. The so-called soft template technique has been successfully used for the preparation of gold nanorods, nanowires, and nanobipyramids. The disadvantage is the cytotoxicity of CTAB and the instability of the bilayer in cell media [128]. Therefore, CTAB-coated gold nanostructures need to be enclosed in a second SiO_x or polymer shell, which has the disadvantages of significantly increasing the nanoparticle volume and shielding their plasmonic function.

3.1.2.1 Top–down strategy: pulsed laser-ablation in liquid (PLAL)

The pulsed laser ablation in liquid technique makes possible producing surfactant-free or in situ stabilized spherical AuNPs with wide size distributions which requires a subsequent size-dependent separation by means of differential centrifugation [107, 115–118, 120, 122, 129–131]. For biomedical applications, a gold target being a wire, plate, or block is immersed in aqueous solution and irradiated with a pulsed laser of suitable wavelength, repetition rate (100–1,000 Hz) and energy fluence (60 mJ/cm^2–1,000 mJ/cm^2) for a certain time span. Lasers used for PLAL are nanosecond pulsed Nd:YAG laser, operating at 1064 nm, 532 nm, or 355 nm, femtosecond pulsed Ti:sapphire laser (800 nm) or Yb:KGW laser (1025 nm). The wavelength, repetition rate, pulse duration, and energy fluence of the laser pulses, as well as the irradiation time and composition of the aqueous solution, are key determinants for adjusting the yield, sizes, and size distribution of the gold nanoparticles.

The pulse wavelength determines the penetration depth into the gold target, the laser fluence and pulse repetition rate influence the ablation yield, while the pulse duration controls the heat dissipation and thereupon the propagation of the plasma plume emerging from laser pulse induced photoionization and atomization. For example, nanosecond Nd:YAG laser pulses at 1,064 nm produce larger AuNPs with broader size distributions compared to the frequency-doubled pulses (532 nm). The concentration of laser-pulse synthesized AuNPs generated by laser pulses increases with the repetition rate, whereas increasing laser fluence enlarges the nanoparticle size. Femtosecond laser pulses heat up the electron gas in the gold target on a time scale much shorter than the subsequent thermalization by electron–phonon and phonon-phonon scattering (10^{-11}–10^{-8} s), resulting in cold material ablation and controlled nanoparticle growth. In comparison, with nanosecond laser-pulse irradiation, the electronic excitation, heat dissipation, and material ejection occur within the pulse duration. AuNPs formed in aqueous solution have partially oxidized surfaces containing AuO^- and Au–OH moieties and are thus negatively charged. Hence, the growth and thereby, the sizes and size distributions of bare AuNPs can be controlled by the composition and concentration of diluted electrolytes. The preparation of functionalized AuNPs is achievable by suitable surfactants that are either already present ("in situ") in the solution during the formation of the AuNPs or are added in a subsequent procedure ("ex-situ").

Suitable surfactants for the preparation of biocompatible AuNPs are polymers such as polyvinylpyrrolidone (PVP), polyethylene glycol (PEG), biopolymers (dextrane, starch or chitosane), oligonucleotides, and oligosaccharides. The formation and growth of AuNPs can be controlled by surfactants since they modify the surface chemistry of the AuNP in dependence on their binding mode and structure.

Laser ablation of gold atoms or cations in liquids is performed by irradiating a gold target with intense laser pulses. The laser-pulse deposited energy provides local melting, heating, sputtering, photoionization, and vaporization [107, 115]. A dense, high-energy plasma plume is generated, which cools down by rapid expansion against the surrounding aqueous liquid and target, resulting in the formation of a cavitation bubble. Supersaturation of the aqueous solution of gold atoms and cations inside arises which launches nucleation. The gold nuclei coalesce to form gold nanoparticles. Continued growth of the gold nanoparticles occurs during their diffusion throughout the expanding cavitation bubble that finally release the nanoparticles into the surrounding liquid. Various setups exist for the implementation of the PLAL technique. Figure 23(a) illustrates nanosecond-laser pulse generation of bare gold nanoparticles in a diluted

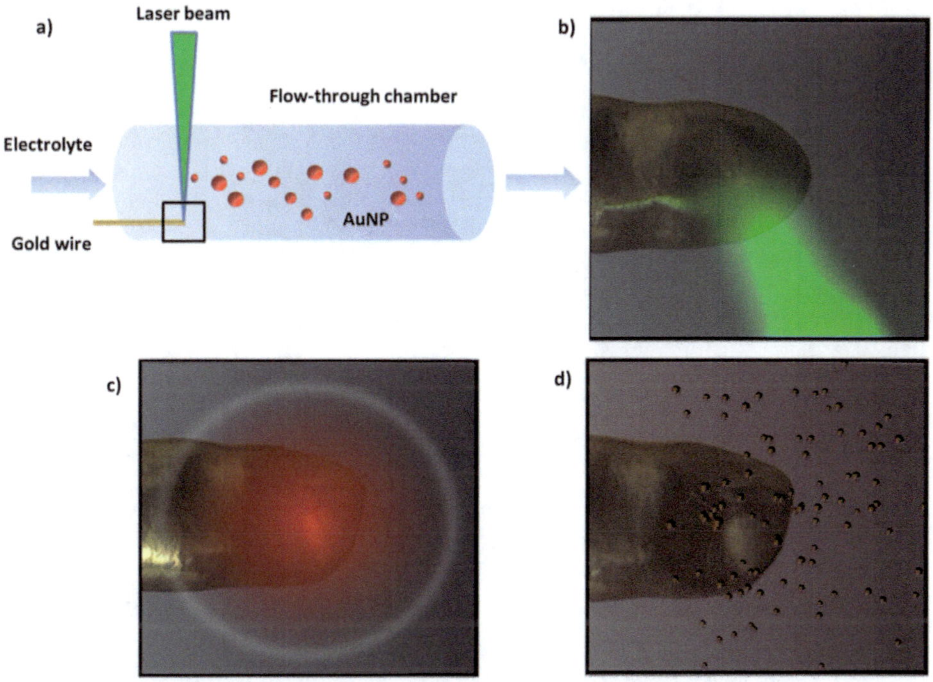

Figure 23: (a) Flow-through setup for preparation of AuNPs in electrolytes via the laser-ablation process; three stages of AuNP formation: (b) absorption of 10 laser pulse, (c) formation of plasma plume, and (d) nanoparticle formation by coalescence and growth during and after cavitation bubble formation and collapse in the micromolar electrolyte [131].

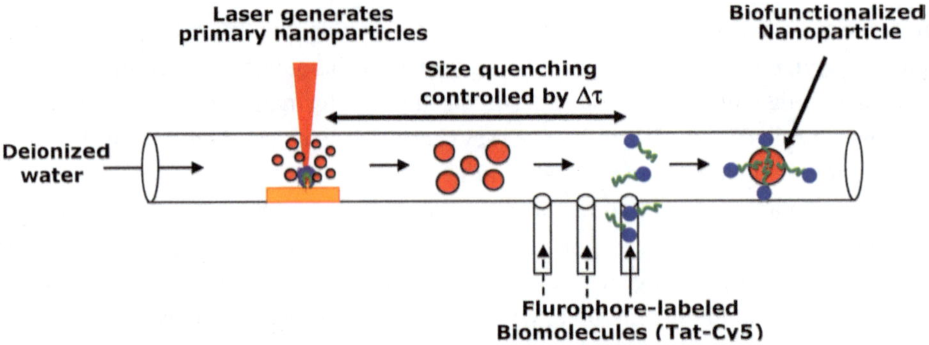

Figure 24: Experimental setup for nanosecond laser-pulse ablation-induced production of AuNPs followed by ex situ bioconjugation in flow chamber; the growth of the initially formed gold nuclei into gold nanoparticles occurs along the flow direction of a TRIS-buffered aqueous solution and may be adjusted by injection of the bioconjugation agent through one of the three side-mounted tubes [132].

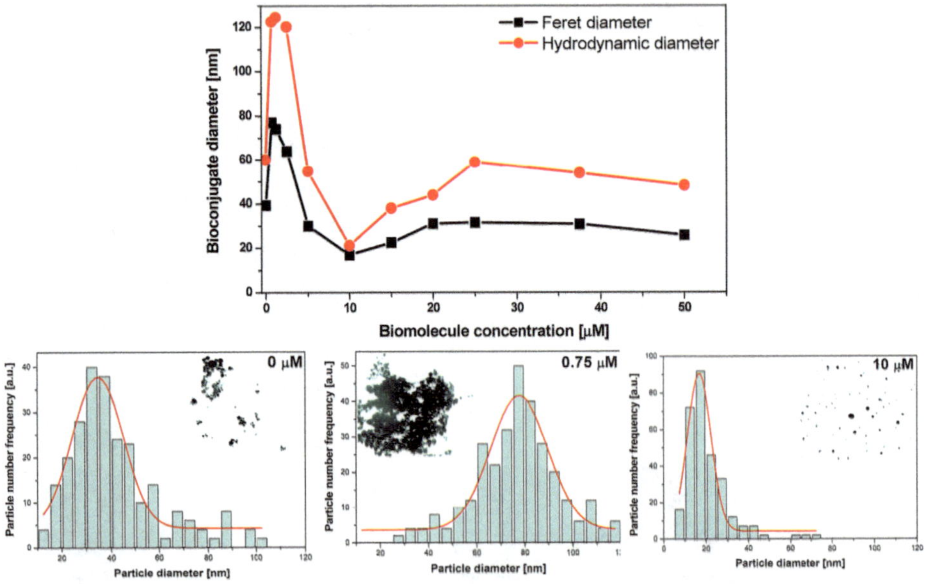

Figure 25: Influence of the Tat-peptide concentration on Feret and hydrodynamic diameter of the bioconjugated gold nanoparticles (top): the TEM images and associated histograms of ligand- For "ex-situ" functionalization or bioconjugation, one of the three side-mounted injection tubes free AuNPs (left), AuNPs conjugated with a Tat-peptide concentration of 0.75 μM (center) and 10 μM (right) (bottom) [132].

electrolyte in a flow-through chamber [131]. Three stages of the AuNP formation are illustrated by snapshots of the irradiation of a gold wire with 1,064 nm Nd:YAG laser pulses (b), formation of the plasma plume (c), and the coalescence of Au seeds to AuNPs and their release from cavity bubble (d).

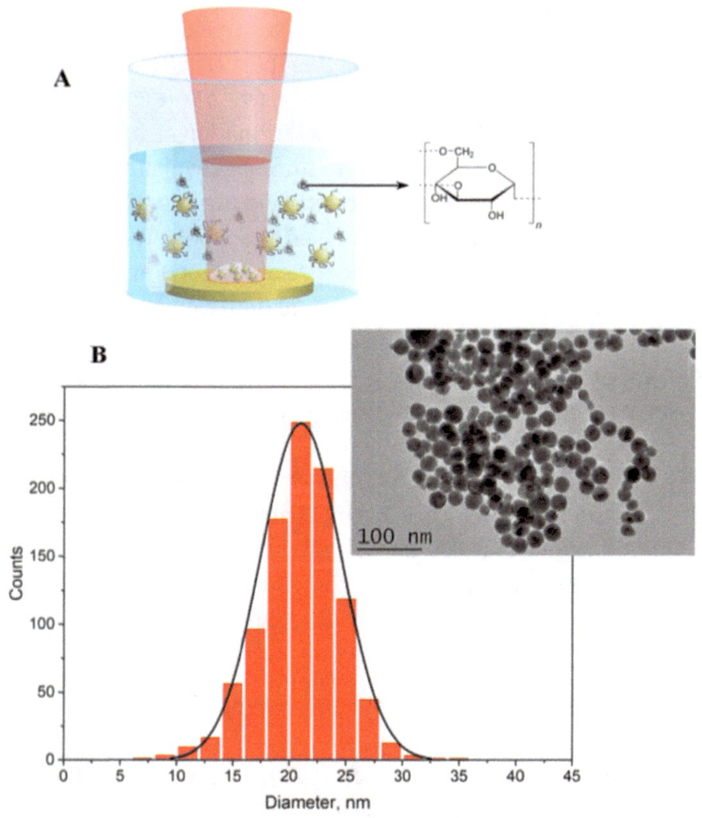

Figure 26: 480-fs laser pulses are focused onto a gold disc in glass vessel filled with an aqueous solution (A); transmission electron spectroscopy (TEM) image of dextran-coated AuNPs and histogram of the nanoparticle-size distribution from TEM analysis [107].

In an alternative theoretical approach, the nanoparticles are found to be formed instantaneously (within nanoseconds) by the accumulation of a hot metal cation and atom at the interface between the plasma plume and water [133]. However, the experimental evidence of the instantaneous formation of AuNPs at the plume-water interface only confirms that the nucleation process takes place at the plume-water interface. Unfortunately, this experimental result is not sufficient to explain the continuous growth of the nanoparticles during the expansion of the plasma plume. The PLAL technique in dilute electrolytes yields ligand-free or bare AuNPs with partially oxidized surfaces that enable subsequent functionalization with a large variety of biomolecules containing amino or carboxyl groups. To avoid time-consuming two-step or multistep procedures for bioconjugated AuNPs, Sajti et al. developed the so-called "ex-situ" flow-through and injection process (Figure 24) [132]. The gold target, a gold plate is placed in a flow cell through which a TRIS-buffered aqueous solution at pH 8.5 flows. Forty-nanosecond laser pulses (1,064 nm, 3 kHz, 6 mJ) are focused onto a gold plate and, therewith, initiate the formation of gold nuclei by ablation during the expansion and propagation of the cavity bubble along the flow direction of the gold nuclei into gold nanoparticles. Three injection slots make it possible to add the respective bioconjugation agent at different growth times being 200 ms, 1.4 s, or 7 s after the laser-pulse initiated formation of gold nuclei.

Effective bioconjugation or functionalization occurs through covalent or electrostatic binding of ligands at the nanoparticle surfaces, restricting further growth of the AuNPs. For ex situ bioconjugation of the AuNPs, a Cy5-labeled Tat-peptide was used that binds to the AuNP surface via a thiol group. The dependence of the Feret and hydrodynamic diameter on the Tat-peptide concentration in Figure 25 (top) reflects the binding efficiency of the Tat-peptide. The highest Tat peptide concentration (10 μM) correlates with the smallest AuNP@Tat-peptide conjugate and the narrowest size distribution, indicating complete coverage of the AuNP surface and thus limiting the growth (Figure 25, bottom).

In situ functionalization of laser-pulse-ablated AuNPs requires the presence of the surfactant in the solution around the gold target. Exemplarily, for the preparation of dextran-coated AuNPs, 480-fs laser pulses (1,025 nm) were focused onto the surface of a gold target in a glass vessel containing an aqueous dextran solution, and biocompatible AuNPs with a mean size of 21 nm were obtained (Figure 26) [107].

3.1.2.2 Bottom–up strategy: redox-based wet-chemistry procedures

Turkevich synthesis of gold nanospheres

The synthesis technique developed in 1951 by Turkevich et al. [134, 135] is carried out in aqueous solution containing $HAuCl_4$ and sodium citrate ($C_6H_5Na_3O_7$) (Figure 27). Sodium citrate acts as reducing agent and surfactant. Gold seeds of sizes around 1.5 nm (*ca.* 800 Au atoms) are formed by reduction of Au^{3+} cations via Au^+ cations to Au^0 atoms, initially coalescing to Au clusters. The surfactant citrate anions electrostatically bind through two carboxyl groups to the gold seeds, whereas citrate anions, when acting as the reducing agent, are oxidized to acetonedicarboxylic acid. Undergoing a seed-mediated mechanism, the gold seeds grow into spherical citrate-stabilized AuNPs with mean sizes be-

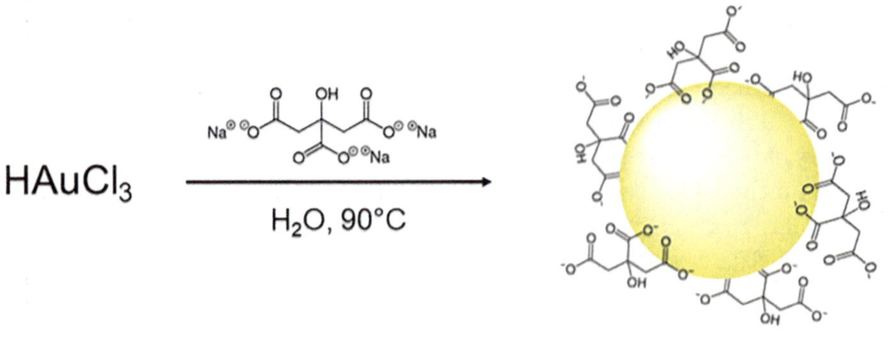

Figure 27: Schematic representation of the Turkevich reaction.

tween 10 and 50 nm. Size distributions and mean sizes of the AuNPs can be adjusted by varying the concentration ratio of citrate to $HAuCl_4$, pH, reaction time, and water temperature. The disadvantages of Turkevich' gold nanoparticle synthesis are the relatively broad size distributions of the AuNPs, the small reaction yield, and the too low stability of the as-prepared gold colloids [136, 137]. The latter is due to the weak electrostatic interactions between the citrate anions and the only weakly positively charged AuNP surfaces. Thus, citrate-stabilized AuNPs in aqueous solution undergo the process of Ostwald ripening, which leads to the growth of large AuNPs at the expense of small AuNPs and ultimately to the formation of large agglomerates that precipitate. Because of their low stability, citrate-stabilized AuNPs require suitable surface modification. This can be achieved, for example, through ligand exchange with thiol-containing compounds or also by coating with SiO_x or biocompatible polymers.

Brust–Schiffrin synthesis of gold nanospheres

For the preparation of thiolated AuNPs, Brust and Schriffrin [138–140] developed a multistep synthesis technique using a two-phase reaction in combination with the phase transfer reagent tetraoctylammonium bromide (TOAB) (Figure 28). The precursor gold acid ($HAuCl_4$) is dissolved in the aqueous (aq) phase, while the TOAB (TOA^+Br^-) in toluene (organic phase) provides the exchange of Cl^- for Br^- in the gold salt, enabling the transfer of the gold salt to the toluene phase (Figure 29). Addition of a thiol derivative (RSH) reduces Au^{3+} to Au^+, while complete reduction to Au^0 is achieved by dissolving $NaBH_4$ in the aqueous phase. As the formation of covalent Au-S bonds limits the size growth, thiolated AuNPs with sizes between 0.8 and 3 nm emerge. The hydrophobic spherical AuNPs can be made water-soluble and thus biocompatible through suitable subsequent surface modification. The mean AuNP diameter can be adjusted by varying the RSH:$HAuCl_4$ concentration ratio. What is disadvantageous for applications in medicine are the very (too) small sizes of the AuNPs that do not exhibit surface plasmon resonance. These gold nanoclusters are too small to possess a crystalline core, but they have discrete electronic energy states and emit size-dependent fluorescence [141, 142].

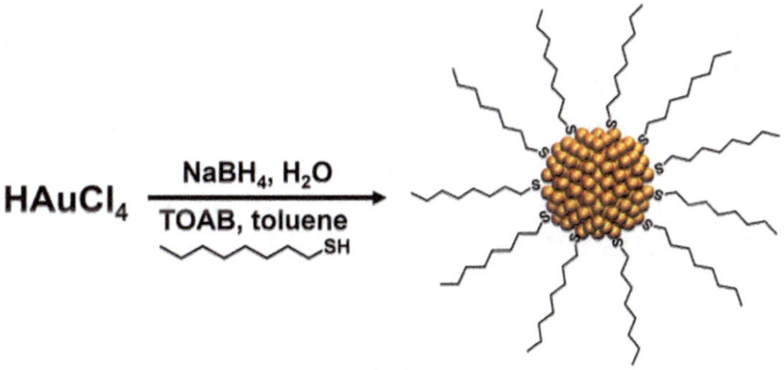

Figure 28: Schematic illustration of the Brust–Schiffrin reaction.

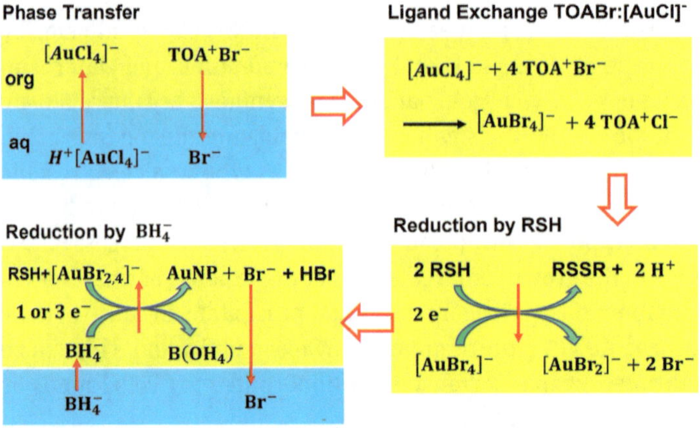

Figure 29: Schematic illustration of the reaction steps involved in Brust–Schiffrin's two-phase synthesis method: the phase transfer agent TOAB in the toluene phase (yellow) provides the transfer of $AuCl_4^-$ from the aqueous (blue) to the toluene phase, reduction of Au^{3+} takes place via Au^+ to Au^0 by RSH and $NaBH_4$ and results in the formation of AuNP.

Seed-mediated and seed-less synthesis of gold nanorods

In 2001, Murphy et al. [143] developed the so-called seed-mediated or seeded growth method for the synthesis of gold nanorods (AuNRs), which is carried out in two steps in the presence of the amphiphilic compound CTAB. In an aqueous solution, CTA^+ binds through its positively charged trimethyl ammonium group (A^+) via the Br^- anion to the gold surface and forms a template-like bilayer around the AuNP [125, 126, 144]. The preferential binding of CTAB to the crystallographic {110} and {100} facets of AuNPs steers the anisotropic growth to AuNR. The seed-mediated synthesis procedure starts with the synthesis of 1.5–4 nm Au nanoclusters (seeds) obtained by reduction of $HAuCl_4$ with $NaBH_4$ in the presence of an excess of CTAB or sodium citrate as surfactant (Figure 30). In the second reaction step, the Au seeds are added to the growth solution containing $HAuCl_4$, ascorbic acid as a mild reducing agent and CTAB in excess. The growth process takes 4 to 24 h.

As M. A. El-Sayed [145] and C. J. Murphy [146] independently ascertained, the anisotropic growth to AuNR growth can be controlled by adding Ag^+ cations. The Ag^+ free synthesis procedure yields AuNRs grown as pentahedral twins, whereas the growth solution containing $AgNO_3$ fosters the formation of nearly monodisperse single crystalline AuNRs with {110} facets at the sides and {111} facets at the end of the rod, respectively, and ARs between 1.5 and 6 [146, 147]. The specific role of Ag^+ for the anisotropic growth is understood to arise from the replacement of Au/Au^+ atoms/cations by Ag^+ cations in the side Au {110} facets of the AuNRs. Surface Ag atoms bind more efficiently the Br^- anions than the Au atoms, which strengthens the electrostatic binding to CTA^+ on the side faces (Figure 31). The final AR of the AuNRs can be adjusted by varying the

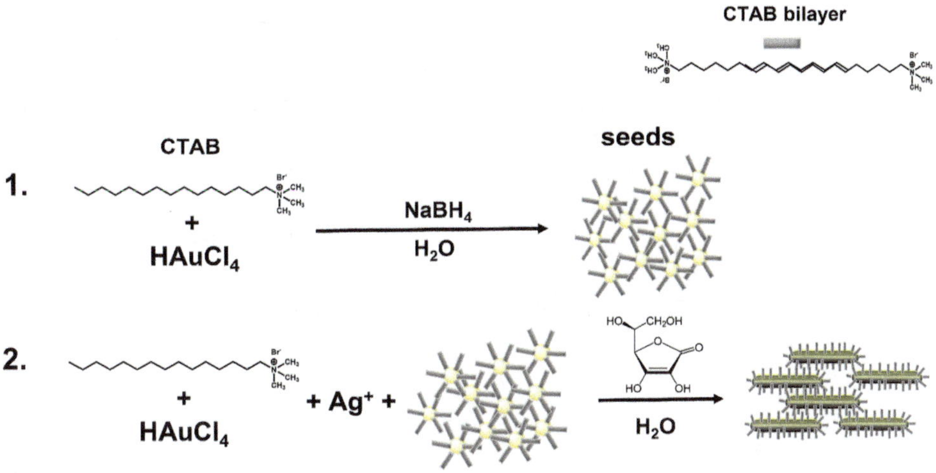

Figure 30: Seed-mediated procedure for the synthesis of AuNRs: in the first reaction step (1) Au nanoclusters ("seeds") are synthesized by reducing $HAuCl_4$ with $NaBH_4$ in the presence of excess CTAB; the second reaction step (2) consists of the addition to the growth solution containing $HAuCl_4$, ascorbic acid, $AgNO_3$, and CTAB; the growth to AuNRs occurs under ambient conditions within 4–24 h.

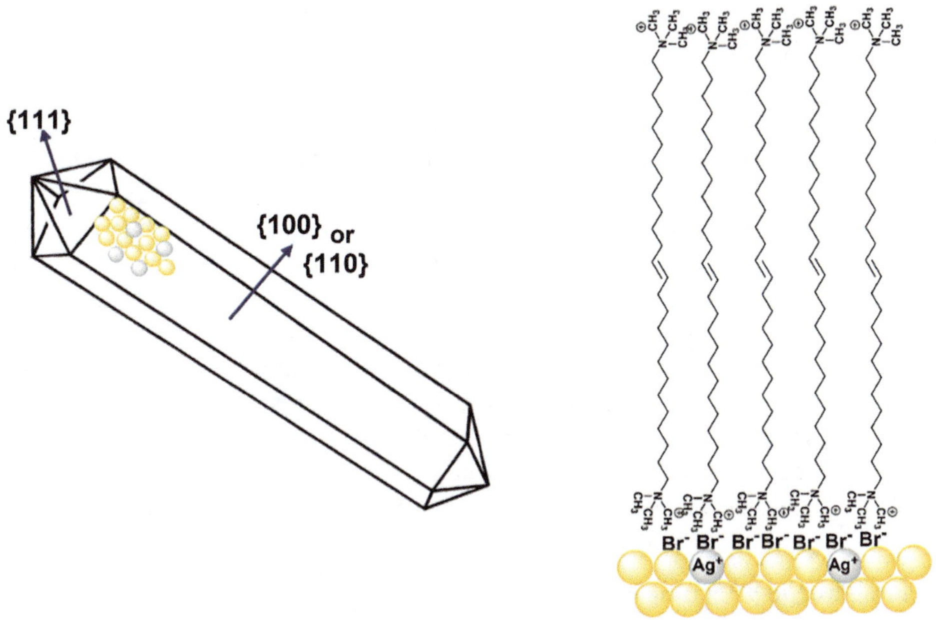

Figure 31: Crystallographic {110} or {100} facets and {111} facets at the sides and ends of a single crystalline AuNR (left); electrostatic binding of CTA^+ via Br^- to the Ag doped Au {110} facets (right).

Ag^+ concentration. Moreover, the presence of Ag^+ in the growth solution increases the yield of AuNRs to 100 %.

El-Sayed's one-step seedless method for the synthesis of small AuNRs represents an elegant shortcut of the seed-mediated technique and exploits the pH dependence of Au nucleation [148]. The pH value of the growth solution containing $HAuCl_4$, $AgNO_3$, and CTAB is adjusted to 1–1.15 by adding HCl. Subsequently added ascorbic acid reduces Au^{3+} to Au^{1+}. Addition of $NaBH_4$ causes the reduction to Au^0 which initiates the nucleation process. The presence of Ag^+ cations and CTAB in the growth solution guide the anisotropic growth to AuNRs with ARs between 1.5 and 6 (Figure 32).

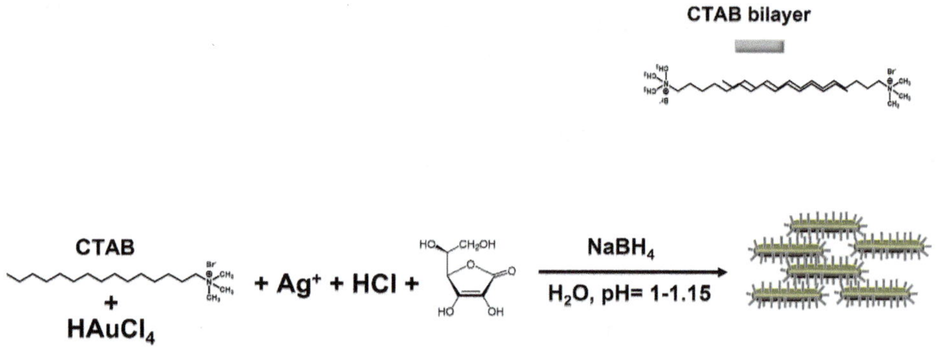

Figure 32: Schematic illustration of El-Sayed's one-step seedless procedure for AuNR synthesis.

Unfortunately, CTAB-stabilized AuNRs are not biocompatible because the weakly bound CTAB may easily cross the cell membrane and cause mitochondrial damage that induces apoptosis [128, 149–151]. In addition, aqueous solutions of CTAB-stabilized AuNRs are only stable if there is a sufficiently high concentration of free CTAB molecules in the solution. The CTAB molecules in the bilayer will be continuously exchanged for free CTAB molecules in the solution. This does not only complicate the purification of CTAB-stabilized AuNRs but also makes a 1:1 ligand exchange practically impossible. Thus, biocompatibility of AuNRs can only be achieved either by overcoating the CTAB bilayer with biocompatible compounds such as sodium dodecyl sulfate or by completely encapsulating them with polymers (e. g., PEG, PAA, PSS) or silicates [78, 128, 149]. The disadvantages of these coatings are the formation of AuNR agglomerates, too-large sizes of single AuNRs, and the partial or complete loss of plasmonic properties.

Synthesis of gold nanoshells

The most reliable methods for the preparation of gold nanoshells with adjustable diameters and shell thicknesses consist of gold coating of spherical silica or polymer cores via one-step or multi-step reactions [97, 152, 153]. The surfactant-assisted seeding method starts with the synthesis of monodisperse silica nanospheres ($NH_4OH + TEOS \rightarrow$

SiO_2 NP) which are grafted with amine-group terminated silanes (3-aminopropyltriethoxysilane (APTES) or 3-aminopropyltrimethoxysilane (APTMS)) (Figure 33) [154]. Negatively charged Au nanoclusters (Au_{55}) with sizes of around 2 nm are prepared by reducing $HAuCl_4$ with tetrakis (hydroxymethyl) phosphonium chloride (THPC). Mixing the ethanolic SiO_2 NP solution with the aqueous gold colloid leads to electrostatic binding of the AuNC via the terminal amine groups to the SiO_2 NP surface. The surface bound AuNCs serve as nucleation sites for the growth of the Au shell.

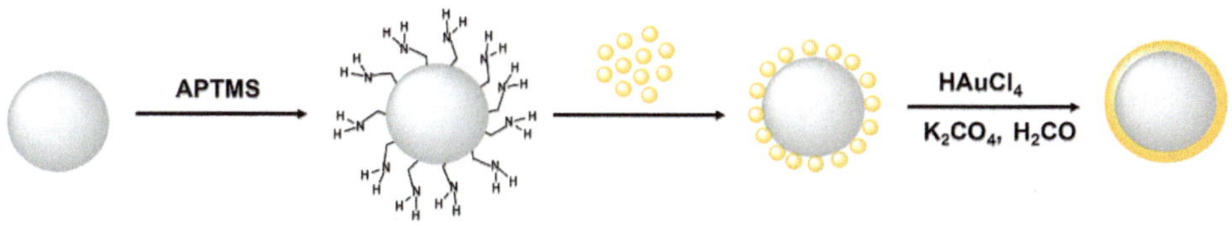

Figure 33: Schematic illustration of surfactant-assisted seeding method for the preparation of gold nanoshells encapsulating a spherical SiO_2 core.

The formation of the Au nanoshell is achieved by exposing the gold-seeded SiO_2 surface with an aged growth solution containing $HAuCl_4$ and K_2CO_3, and formaldehyde is added as reducing agent. The optical spectra of the gold nanoshells can be fine-tuned by adjusting the ratio of gold-shell thickness to the diameter. Another reliable synthesis technique for gold nanoshells is the deposition–precipitation technique wherein the gold seeds are directly generated on APTES stabilized silica nanoparticles [155]. $Au(OH)_3$ nanoclusters are formed at the terminal NH_2 goups of the SiO_2 NP surfaces when treated with an alkaline solution of $HAuCl_4$. The nanoshell growth is initiated by adding a solution of $HAuCl_4$ and K_2CO_3 and subsequent reduction with $NaBH_4$.

Hollow gold nanoshells with sizes between 37 and 96 nm can be prepared by reducing $HAuCl_4$ in an aqueous APTES suspension with $NaBH_4$, where the gold nanoshells are formed by growing onto the hydrophobic APTES droplets [97].

3.2 Superparamagnetic iron oxide nanoparticles

Iron oxides occur naturally in many crystalline phases with magnetite (Fe_3O_4), maghemite (γ-Fe_2O_3), and hematite (α-Fe_2O_3) being the most common crystallographic types and exhibiting ferrimagnetism or ferromagnetism as bulk material [156–158]. Ferro- and ferrimagnetic bulks consist of magnetic domains that differ in the orientation of their magnetic dipoles and are thus separated by Bloch or Néel walls. Iron oxide nanoparticles, with sizes much smaller than the magnetic domains (<100 nm), are superparamagnetic. Due to their multiple functionality and good biocompatibility, **s**uper**p**aramagnetic **i**ron **o**xide **n**anoparticles (SPIONs) have stimulated intense research activities in medicine and biomedical research because of their potential applicability as contrast agents for magnetic resonance imaging, magnetic particle imaging (MPI), and photoacoustic imaging [159–180]. Moreover, their versatile biomedical potential has been demonstrated as mediators for magnetic drug delivery, magnetic hyperthermia, radiotherapy, and cell-separation techniques [181–194]. In addition, Fe_3O_4 nanoparticles act through their surface Fe^{2+} cations as "nanozymes" by mimicking oxidoreductases, including peroxidase, catalase, superoxide dismutase, and oxidase [188, 195–197]. The unique property of SPIONs, i. e., their superparamagnetism, arises from the single-crystalline Fe_3O_4 or Fe_2O_3 core with sizes much smaller than the magnetic (Néel) domain (<100 nm). Magnetite crystallizes in the cubic (fcc) inverse spinel structure, with O^{2-} anions forming the fcc lattice and the Fe cations occupying interstitial octahedral and tetrahedral sites.

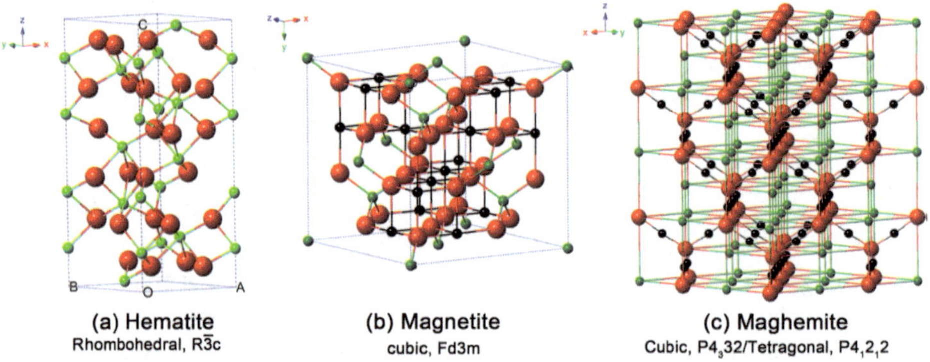

Figure 34: Crystal structure and crystallographic data of the iron oxide phases with α-Fe_2O_3 (hematite), γ-Fe_2O_3 (maghemite). and Fe_3O_4 (magnetite): Fe^{3+} (green spheres), Fe^{2+} (black spheres), and O^{2-} (red spheres) [157].

Maghemite is formed by oxidization of magnetite under ambient conditions. In magnetite, one-third of the Fe^{3+}cations are tetrahedrally coordinated by O^{2-}, whereas two-thirds of all iron ions (Fe^{3+} and Fe^{2+}) occupy the octahedral lattice sites of the inverse spinel lattice (Figure 34) [157, 188]. Maghemite is a Fe^{2+} deficient magnetite that also crystallizes in the inverse spinel structure, but contains unoccupied octahedral sites (vacancies). Hematite crystals possess a rhombohedral structure. Suitably surface-modified SPIONs are biocompatible, whereas uncoated SPIONs exhibit size-dependent cytotoxicity.

The magnetic properties of materials are examined by subjecting powder samples to an external variable magnetic field and by measuring the magnetization M (emu/g) of the sample as function of magnetic field strength H (T). Iron cations (Fe^{2+} and Fe^{3+}) possess four or five unpaired d electrons whose spins, when performing as magnetic dipoles, may be aligned along the direction of an external magnetic field. Magnetite and maghemite bulk material consist of uniformly magnetized domains which differ in the magnetic dipole orientation [198]. Superexchange interactions between electron spins provide their antiparallel alignment in antiferromagnetic and ferrimagnetic materials. While the magnetization of magnetic bulk material and nanoparticles increases with the magnetic field strength H, temperature-assisted motions counteract the magnetic ordering and thus decrease the magnetic susceptibility. The magnetization curves of magnetic (i. e., ferro-, ferri-, and antiferromagnetic) bulk materials take a sigmoidal course to a saturation magnetization M_S and display a more or less pronounced hysteresis loop (Figure 35, left panel). The development of hysteresis, the associated coercivity, and remanence values reflect the irreversibility of the magnetization process that depends on the number and sizes of magnetic domains. M_S indicates the complete alignment of all magnetic dipoles in an external magnetic field, whereas the remanence is the residual magnetization left over after the removal of the applied magnetic field. The coercivity is the magnetic field strength required to force the magnetization to zero. SPIONs are characterized by hysteresis-less sigmoidal magnetization curves, indicating the reversibility of magnetization of single-crystalline nanoparticles (Figure 35, right panel).

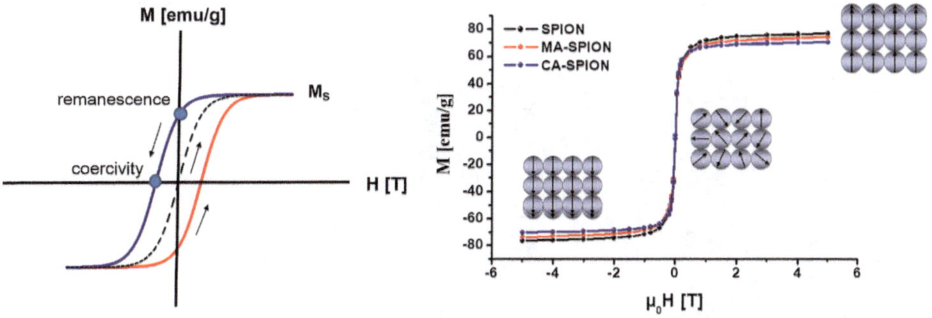

Figure 35: Schematic illustration of the magnetization curve of ferrimagnetic bulk material (left); magnetization curves of uncoated SPIONs (black solid line), citrate-coated SPIONs (blue solid line), and malate-coated SPIONs (red solid line) at 300 K.

SPIONs can be accumulated in biological tissue through an external inhomogeneous magnetic field which enables, for example, magnetic tumor targeting useful for drug delivery, magnetic hyperthermia, and cancer radiotherapy. Adequately surface-modified SPIONs are biocompatible, whereas uncoated SPIONs exhibit size-dependent cytotoxicity [187, 188].

3.2.1 Synthesis routes for SPIONs

In recent decades, intensive research has been devoted to the development of efficient synthetic routes for monodisperse SPIONs with defined shapes and narrow size distributions [156, 199–204]. The major goals of these synthetic routes include high scalability, single crystallinity, monodispersity, catalytically activatable, or at least biocompatible, surfaces. The preferred synthesis techniques for biomedical applications are the alkaline co-precipitation, thermal decomposition and/or reduction, polyol synthesis, and sol-gel technique. Alkaline co-precipitation in aqueous phase, first proposed by Massart in 1981 [205], has now been developed to a level that enables the preparation of biocompatible SPIONs with narrow size distributions and mean sizes in the range from 5 to 25 nm [206]. Nevertheless, mostly SPIONs, independently of possessing magnetite or maghemite composition, are obtained with more potato-like than spherical shapes and size distributions between 4 and 17 nm (Figure 36(A)) [187, 188]. In contrast, the thermal decomposition of organic Fe^{3+} precursors (iron acetylacetonate ($Fe(acac)_3$), iron oleate ($Fe(oleate)_3$) and $Fe(CO)_5$) at temperatures between 280 and 320 °C, in the presence of oleic acid and oleylamine, yields monodisperse hydrophobic oleylamine/oleic acid stabilized Fe_3O_4 nanoparticles with cubic, octahedral, or spherical shapes and sizes, adjustable in the range of 10 to 50 nm (Figure 36(B)) [199, 204, 207, 208]. Biocompatible ferrite NPs, either $ZnFe_2O_4$ or $MnFe_2O_4$ NPs, can be prepared analogously by thermal decomposition of a mixture of $Zn(acac)_2$ or $Mn(acac)_2$ and $Fe(acac)_3$ in a stoichiometric

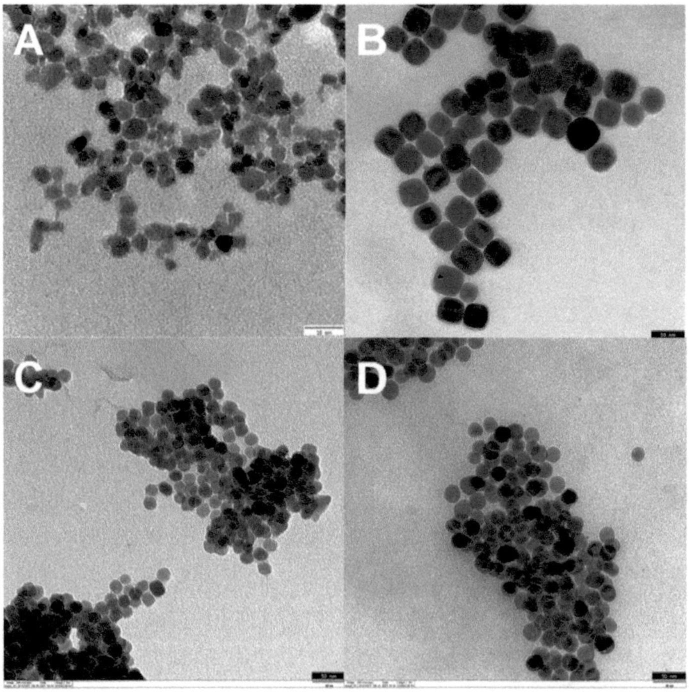

Figure 36: TEM images of malate-stabilized SPIONs (A), synthesized through alkaline co-precipitation; oleic-acid stabilized SPIONs (B), $MnFe_2O_4$ NPs (C), and $ZnFe_2O_4$ NPs (D), synthesized via the thermal decomposition route.

ratio of 1:2 (Figures 36(C) and (D)) [209–211]. Another advantage of the thermal decomposition procedure is the production of single-crystalline nanoparticles. The drawback of this synthesis technique is the complicacy of subsequent surface modification. The exchange of the oleic acid ligands for hydrophilic surfactants or functional agents is either incomplete or does not occur at all. Overcoating with hydrophilic polymers such as polyethylene glycol (PEG), polyethyleneimine (PEI), or polyacrylic acid (PAA) involves the risk of agglomeration and, simultaneously, the complete loss of catalytic activity. In the polyol synthesis, an iron (II) of iron (III) precursor, either a water-soluble Fe(II) salt or an organic Fe(III) complex, is chelated and oxidized and partially reduced in case of the iron (III) precursor at temperatures between 200 and 260 °C [212–216]. The polyols used in this synthesis are diethylene glycol (DEG), triethylene glycol (TREG), or tetraethylene glycol (TEG) which act as solvent, chelating surfactant, and reducing agent. Regrettably, the SPIONs obtained have relatively small sizes and polyol surface structures inducing self-organization to flower-like or raspberry-like agglomerates with sizes larger than 60 nm. Polyol-stabilized surfaces hamper subsequent ligand exchange, and, thus, any kind of functionalization with, for instance, tumor-targeting surfactants.

Alkaline co-precipitation

Preferably, for biomedical applications, SPIONs and ferrite nanoparticles ($MnFe_2O_4$, $ZnFe_2O_4$, or $CoFe_2O_4$ NPs) are prepared by alkaline co-precipitation in aqueous solution at a pH value in the range of 8–12 [217–219]. For the synthesis of Fe_3O_4 NPs, Fe(III), and Fe(II) salts (e. g. $FeCl_3/FeCl_2$, $FeSO_4/Fe_2(SO_4)_3$, $Fe(NO_3)_3/Fe(NO_3)_2$) are dissolved in water at a stoichiometric ratio of 2:1 under a N_2 flow and adjusted to the desired pH values using an alkaline solution (NaOH or NH_4OH in water) (Figure 37). The as-synthesized SPIONs are magnetically separated and washed with deionized water. For subsequent surface stabilization, the SPIONs are dispersed in an aqueous solution of, for example, sodium citrate and stirred at 60 °C for 1 h or longer.

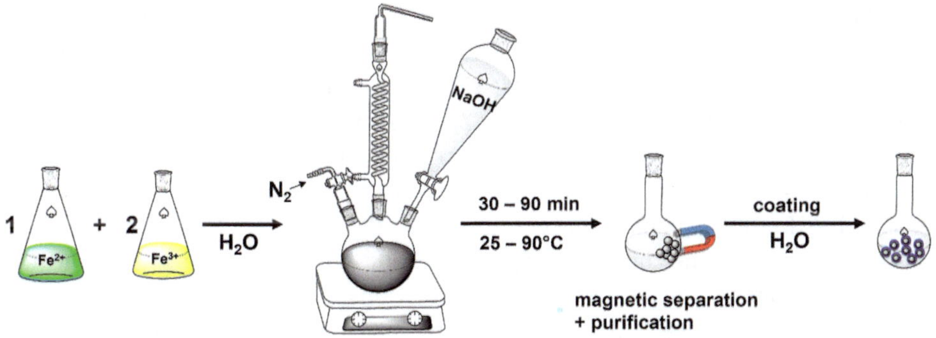

Figure 37: Schematic illustration of alkaline co-precipitation of Fe_3O_4 NPs and subsequent surface stabilization.

The Fe_3O_4 NPs oxidize to γ-Fe_2O_3 NPs under ambient conditions. $MnFe_2O_4$, $ZnFe_2O_4$, or $CoFe_2O_4$ NPs and other M(II)Fe_2O_4 NPs (M^{2+}: Mg^{2+}, Cu^{2+}, Ni^{2+}) are synthesized analogously, replacing the Fe(II) salt by the equal concentration of the Mn(II), Zn(II) or Co(II) salt in the aqueous mixture with the Fe(III) salt. The two-step synthesis route to surface-stabilized SPIONs or ferrite NPs can be shortened to a one-pot reaction by adding, in a suitable concentration ratio, the desired surfactant to the aqueous Fe^{3+}/Fe^{2+} solution before alkalizing the solution.

The chemical reaction for the formation of Fe_3O_4 NPs is described by the following reaction Eqs. (7)–(9):

$$Fe^{3+} + 3\,OH^- \rightarrow Fe(OH)_3 \quad \text{and} \quad Fe^{2+} + 2\,OH^- \rightarrow Fe(OH)_2, \tag{7}$$

$$Fe(OH)_3 \rightarrow FeOOH + H_2O, \tag{8}$$

$$2\,FeOOH + Fe(OH)_2 \rightarrow Fe_3O_4 + 2\,H_2O. \tag{9}$$

The process of alkaline co-precipitation starts with the formation of the iron hydroxy intermediates. The generation of the Fe_3O_4 nuclei and growth to SPIONs are driven by the pH-dependent dehydration reaction. The final size distribution of the Fe_3O_4 NPs also depends on the reaction temperature and duration, the base (NaOH or NH_4OH), the concentrations of Fe(II) and Fe(III) salts, and their anions. Mostly, SPIONs with size distributions between 5 and 17 nm are produced. The higher the pH value and ionic strength are, the smaller SPION sizes and the narrower the size distribution become. Biocompatible, surface stabilized SPIONs may be directly prepared by adding chelating organic anions such as citric acid, malate acid, or gluconic acid or polymeric surfactants (dextran, starch, or polyvinyl alcohol) during the SPION formation process [183, 186, 188, 220]. In situ surface modification affects the growth process of the SPIONs and thus shapes the size distribution.

Thermal decomposition

Monodisperse single-crystalline SPIONs (Fe_3O_4/Fe_2O_3 NPs) and ferrite NPs (MFe_2O_4 NPs, with M = Zn, Mg, Mn, Cu or, Co) with adjustable shapes and sizes can be obtained by thermal decomposition of organometallic precursors in a high boiling solvent at 220–320 °C and in the presence of hydrophobic surfactants such as oleic acid and oleylamine (Figure 38) [199, 208, 209, 211, 221, 222]. The most commonly used organometallic precursors are iron acetylacetonate ($Fe(acac)_3$), iron oleate ($Fe(oleate)_3$), and $Fe(CO)_5$. For the synthesis of ferrite NPs, $Fe(acac)_3$ or $Fe(oleate)_3$ and $M(acac)_2$ or $M(oleate)_3$ are used in the stoichiometric concentration ratio (Fe:M) of 2:1. Metal oleates are synthesized by reacting the corresponding metal cation salt with oleic acid and NaOH in methanol or ethanol. The purification is very laborious, but essential for reproducibility of nanoparticle synthesis. Suitable high-boiling solvents used are 1-octadecene (ODE) and dibenzyl ether (DBE). The most used surfactants are oleic acid, oleylamine, and decanoic acid. In the reductive thermal decomposition synthesis 1,2-hexadecanediol or

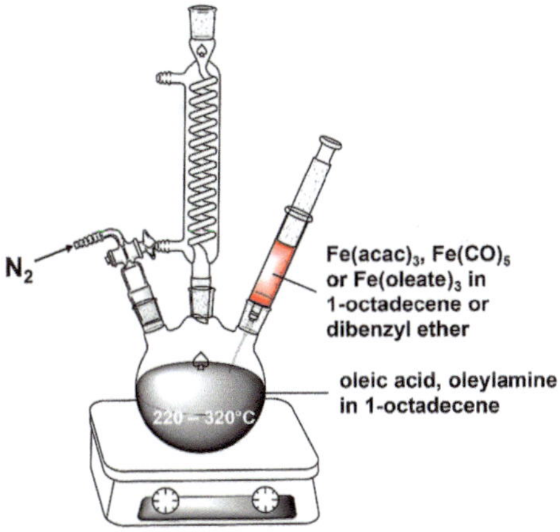

Figure 38: Schematic illustration of the thermal decomposition route for the preparation of oleic acid/oleylamine stabilized SPIONs.

hydrazine are added as reducing agents. Oleylamine, also, acts as a reducing agent at high temperatures by partially reducing $Fe(acac)_3$ to Fe_3O_4. The sizes of SPIONs and MFe_2O_4 NPs can be tuned in the range of 14–100 nm by the reaction temperature, heating rate (220 °C–300 °C) and reaction time, while the concentration ratio of oleic acid to oleylamine controls the shape of the nanoparticles. For instance, 14-nm spherical SPIONs can be synthesized by thermal decomposition of $Fe(acac)_3$ in 1-octadecene at 220 °C (30 min) and 300 °C (30 min) in the presence of oleic acid and oleylamine used in a concentration ratio of 2:3. Larger spherical SPIONs (with sizes up to 100 nm) and octahedral SPIONs are obtained by longer reaction times at 220 °C or/and 300 °C. Cubic SPIONs are formed by adding sodium oleate as a co-surfactant for oleic acid and oleylamine in 1-octadecene. Thermal decomposition of $Fe(acac)_3$ in dibenzyl ether (DBE) with oleic acid as the only surfactant yields cubic SPIONs, the sizes of which can be varied between 20 and 160 nm by varying the reaction temperature and time [223]. In this synthesis process $Fe(acac)_3$ is reduced to Fe^{2+} through benzaldehyde formed by partial thermal decomposition of dibenzyl ether. The as-synthesized SPIONs or ferrite NPs can be purified either by repeated ultracentrifugation–washing cycles or magnetic separation-washing cycles by using *n*-hexane and ethanol as solvents, where n-hexane serves to redisperse the nanoparticles and ethanol dissolves excess oleylamine, oleic acid, and unreacted precursors.

Polyol synthesis

In the reductive polyol synthesis of SPIONs, ethylene glycol (EG), diethylene glycol (DEG), triethylene glycol (TREG), and tetraethylene glycol (TEG) are used [212, 213, 215, 216]. These polyols do not only act as high-boiling solvents (200–320 °C), but also chelate the Fe^{3+} cations of ferric chloride or $Fe(acac)_3$ precursors. At high temperatures (>160 °C), the polyol partially dehydrates to the respective aldehyde derivative that reduces Fe^{3+} cations to Fe^{2+} cations and is oxidized to diacetyl (Figure 39).

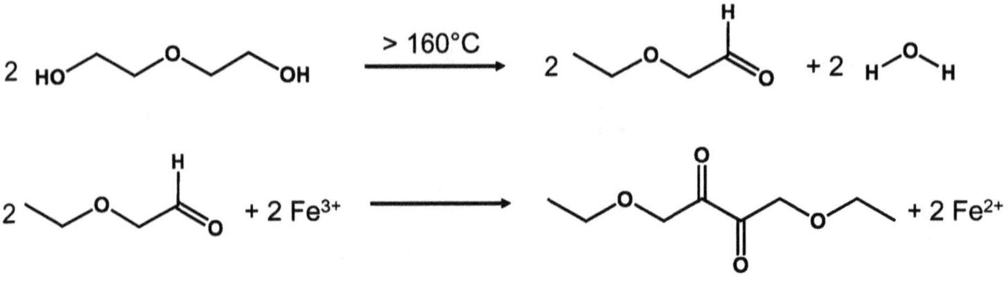

Figure 39: Reaction scheme of the thermal decomposition of ethylene glycol to the aldehyde derivative and water; the aldehyde derivative reduces Fe^{3+} to Fe^{2+} and is oxidized to diacetyl.

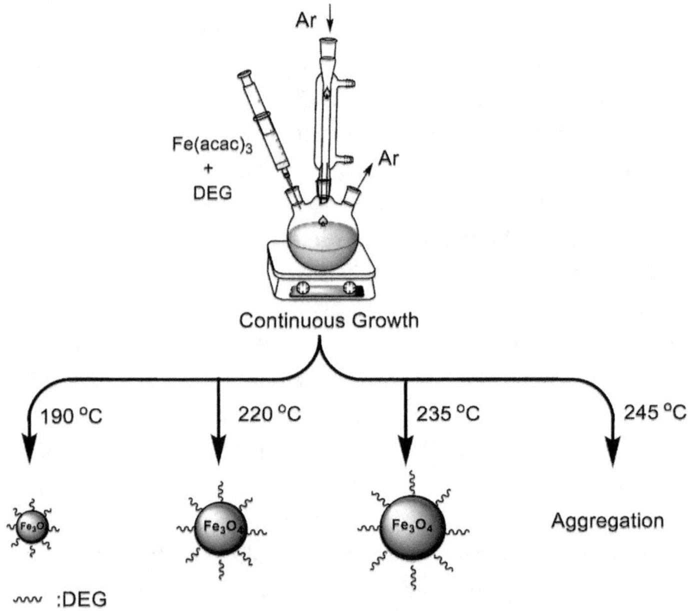

Figure 40: Schematic illustration of the SPION-size dependence on the reaction temperature of the reductive polyol synthesis using $Fe(acac)_3$ as precursor and DEG as solvent while being conducted at 190 °C, 220 °C, 235 °C, and 245 °C [216].

High reaction temperatures promote the hydrolysis of the metal oxide–chelate complexes and induce the formation of nanocrystals. The chelate effect of polyols ochestrates the formation of nuclei, the growth phase, and agglomeration of the SPIONs. In a typical polyol synthesis of nonagglomerated spherical SPIONs, $Fe(acac)_3$ is used as a precursor, and DEG serves as a high-boiling solvent (b. p. = 245 °C), surfactant, and reducing agent. The decomposition temperature of $Fe(acac)_3$ is 186 °C. The SPION sizes can be adjusted between 3 and 11 nm by varying the concentration ratio of $Fe(acac)_3$ to DEG, reaction temperature, and time (Figure 40) [216]. On the other hand, heating the reaction mixture to 245 °C under reflux leads to agglomeration.

3.3 Nanoheterostructures

The production of nanoscale heterostructures with defined morphology, elemental composition, and surface chemistry is one of the greatest challenges of recent research in nanochemistry. Nanoheterostructures (NHSs) are composed of various nanocrystalline materials such as noble metal, metal oxide, and semiconductor materials [224–226]. These nanomaterials are assembled into a three-dimensional structure by epitaxial growth. The optical, electronic, and magnetic properties of such nanocomposites not only complement each other, but also influence and reinforce each other, which may generate novel physical phenomena. Since biomedical applications require at least biocompatibility of NHSs and, at best, complete adverse effect-free degradation of NHSs in the human organism, metal oxide nanostructures containing copper, nickel, cobalt, cadmium, or lead-based semiconductor quantum dots and the noble metals palladium and silver, for example, are ruled out. The physical and catalytic properties of NHSs can be tuned by the elemental composition, morphology, and surface chemistry of the nanocrystal components. Their epitaxial interfaces, as compensating lattice mismatches and energy band-structure differences between adjacent nanocrystals, mediate, for example, heterogeneous electron and spin-polarization transfer [227, 228]. Large surface curvatures of complex nanoarchitectures, incomplete surface coverages, surface plasmons and plexcitonic interactions enhance the catalytic activity and other surface-driven processes. Exchange bias fields and magneto–plasmonic couplings may be employed for modulation of optical, electrical, and thermal properties [229, 230]. The most promising NHSs for medical applications consist of biocompatible iron oxide and gold nanoparticles [231–237]. The structural, chemical, magnetic and plasmonic properties of Au–Fe_3O_4 nanoheterodimers (Au–Fe_3O_4 NHDs) have been intensively studied and are considered to be largely elucidated since the seed-mediated growth method was successfully established by Sun et al. in 2005 [238].

3.3.1 Multifunctional magneto–plasmonic nanoheterostructures

Among the multitude of possible multifunctional nanocomposites, Au–Fe_3O_4 NHDs have so far proven to be excellent candidates for biomedical applications and especially for multimodal imaging [228]. Superparamagnetism of the magnetite component enables the accumulation of the Au–Fe_3O_4 NHDs in diseased tissue by means of an external inhomogeneous magnetic field. In addition, the magnetic properties of the NHDs make them suitable for applications as MRI and MPI contrast agents, and as mediators in magnetic hyperthermia therapy. Moreover, X-ray activated Fe_3O_4 surfaces perform as X-ray dosage enhancing agents [187, 188]. The gold component may act as an excellent CT contrast agent [50, 239], as well as photosensitizer and plasmonic heater in photodynamic and photothermal therapy [99, 104, 105, 240–243], respectively. Magneto–plasmonic Au–Fe_3O_4 NHDs are composed of epitaxially interconnected gold (Au) and magnetite (Fe_3O_4) nanospheres with equal or different diameters. The coalescence of the two nanomaterials creates new heterogeneous electron transfer pathways and interactions between surface plasmons and magnetic moments. The most reliable and reproducible synthesis technique for preparation of monodisperse Au–Fe_3O_4 NHDs with controllable morphologies is the seed-mediated growth method which involves the epitaxial growth of Fe_3O_4 nanodomains on presynthesized AuNPs nanoparticles (Figure 41) [225, 231, 236, 238, 244, 245]. For this purpose, organic ferric precursors are thermally decomposed under a N_2 atmosphere at temperatures between 300 and 320 °C in high-boiling solvents containing, in addition, oleic acid, oleylamine, and 1,2-hexanedecanediol as surfactants and reductants. The size distribution and morphology of the as-synthesized Au–Fe_3O_4-NHDs are tuned by varying the reaction parameters. The essential parameters for the synthesis are the sizes of Au seeds, the organic iron precursor, the ratio of the iron precursor concentration to the number of gold nanoparticles, the respective concentrations of surfactants and reducing agents, the reaction temperature profile, and the reaction time. The Au–Fe_3O_4 NHDs are precipitated by adding isopropanol to the reaction solution and are isolated by centrifugation or magnetic separation. For purification, the nanoparticles are redispersed in *n*-hexane and precipitated by addition of ethanol and

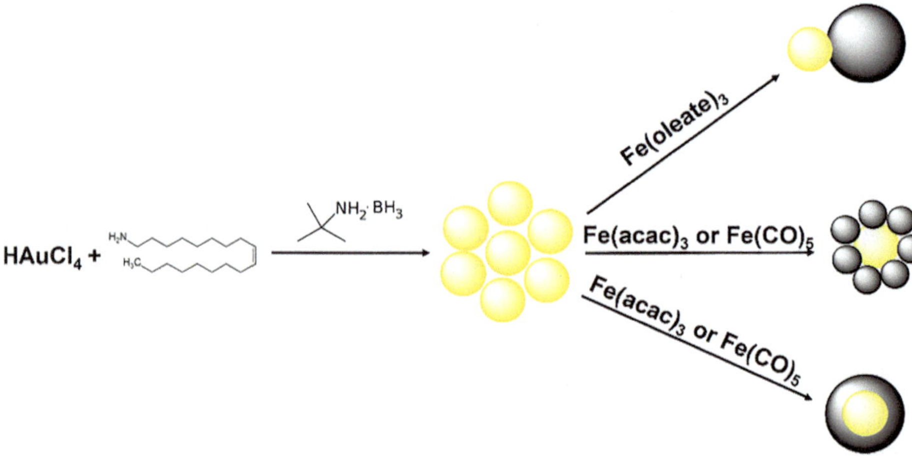

Figure 41: Schematic illustration of the preparation of dumbbell-shaped, flower-shaped, and core-shell Au–Fe_3O_4 NHDs via the seed-mediated growth route.

centrifuged again. For biomedical applications, the hydrophobic oleic acid-terminated nanoparticles are surface-modified via ligand exchange with water-soluble compounds or functional groups, while maintaining sizes and morphologies [233, 244, 246, 247]. Another surface modification technique involves coating the Au–Fe_3O_4 NHDs with water-soluble polymers or silicates.

The lattice constant of cubic Fe_3O_4, with 8.39 Å, is slightly more than twice as large as that of gold (4.08 Å), crystallizing in the fcc structure. This allows nucleation and epitaxial growth of Fe_3O_4 on the gold surface, which results in the entry of positive charges on the Fe_3O_4-Au interface. Charge-carrier balance occurs by electron transfer from the gold component. In small AuNPs, electron deficiency hampers multiple nucleation. Thus, AuNPs with sizes smaller than 5 nm can provide only one nucleation site. With increasing size and decreasing curvature, the formation probability of the crystallographic {111} and {200} facets of the AuNP surface becomes larger, fostering heterogeneous nucleation and thus the epitaxial growth of two or more crystalline Fe_3O_4 domains with Fe_3O_4{111}/Au{111} or Fe_3O_4{400}/Au{200} interfaces [229, 231, 238]. Oleylamine-stabilized AuNPs with sizes tunable between 3 and 12 nm are usually prepared by reduction of gold acid ($HAuCl_4$) with the *tert*-butylamine-borane complex in organic solvents (tetraline, cyclohexane or $CHCl_3$), containing oleylamine and eventually 1,2-hexanedecanediol. Frequently used high-boiling solvents are 1-octadecene, diphenyl ether, benzyl ether, and dibenzyl ether. The main ferric precursors are iron pentacarbonyl ($Fe(CO)_5$), iron acetylacetonate ($Fe(acac)_3$), and the iron oleate complex ($Fe(oleate)_3$) [231]. Snowman- and dumbbell-shaped Au–Fe_3O_4 NHDs are formed by thermal decomposition of $Fe(oleate)_3$ on 5–8 nm-sized AuNPs at 320 °C in 1-octadecene, oleic acid, and oleylamine after a 4-h reaction time [244]. An alternative synthesis route is followed by thermally decomposing $Fe(CO)_5$ on 3–8-nm sized AuNPs in the same solvent-surfactant mixture at 300 °C for 1 h [231]. Figures 42(A) and (B) present the TEM images of snowman- and dumbbell-shaped Au–Fe_3O_4 NHDs [238]. The TEM image contrast arises from differences in the electron density of the Fe_3O_4 and Au component, which is reflected by the appearance of the relatively dark Au and lighter Fe_3O_4 nanodomain. Snowman-shaped NHDs were prepared by epitaxially growing 14-nm Fe_3O_4 nanocrystals on 3-nm-sized AuNPs, while the dumbbell-shaped NHDs contain significantly larger AuNPs (8 nm). Figure 42(C) depicts an angle annular dark field scanning transmission electron microscopy (HAADF–STEM) image of three dumbbell-shaped Au–Fe_3O_4 NHDs. The brightness of the image scales with the intensity of the scattered electrons and is thus directly proportional to the atomic number (Z). The HRTEM image of a dumbbell-shaped NHD consisting of a 12 nm-Fe_3O_4 and a 8 nm Au nanosphere, maps the (111) plane of each component, demonstrating the single crystallinity of the NHDs (Figure 42(D)).

For the preparation of flower-shaped Au–Fe_3O_4 NHDs, 8–12-nm-sized AuNPs, $Fe(acac)_3$ are used as seed and precursor, respectively. 1-octadecene or dibenzyl ether was mixed with oleic acid as a surfactant and oleylamine as a reductant [238]. $Fe(acac)_3$ was thermally decomposed, and Fe_3O_4 nuclei and nanocrystals were grown on the AuNPs at 310–320 °C for 60–90 min. The number of nucleation sites was shown to

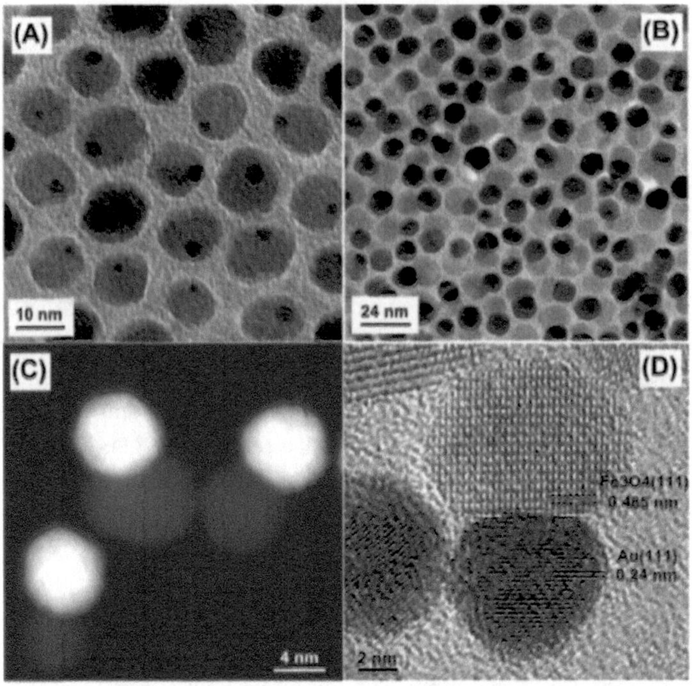

Figure 42: TEM images of snowman-shaped (A) and dumbbell-shaped Au–Fe_3O_4 NHDs (B), the HAADF-STEM image (C) and HRTEM image (D) of dumbbell-shaped Au–Fe_3O_4 NHDs [238].

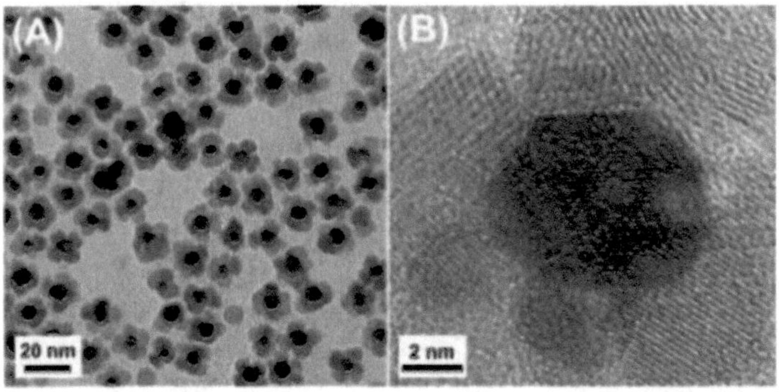

Figure 43: TEM (A) and (B) HRTEM images of flower-like Au–Fe_3O_4 NHDs [238].

increase by preheating the reaction mixture at 200–220 °C for 30 min. Alternatively, flower-shaped Au–Fe_3O_4 NHDs were synthesized by dissolving $Fe(CO)_5$ in diphenyl ether and oleylamine and adding 8-nm-sized AuNPs (Figure 43) [238]. The reaction mixture was heated to reflux for 90 min. Slightly polar solvents such as dibenzyl ether and diphenyl ether and the reductants hexadecanediol or dodecanediol can compensate for the positive charges generated at the growth interfaces which facilitates the formation of two and more nucleation sites on small AuNPs.

AuNPs larger than 12 nm provide enough suitable nucleation sites for Fe_3O_4 nuclei to coalesce into a shell around the gold core [231]. Both large-curvature and associated surface-defect structures of the gold surfaces impede homogeneous nucleation leading most often to the formation of polycrystalline, irregularly shaped Au–Fe_3O_4 core/shell structures (Au@Fe_3O_4 NHDs). Nearly monocrystalline Au@Fe_3O_4 NHDs were prepared by thermal decomposition of Fe(acac)$_3$ on 5-nm-sized AuNPs dissolved in 1-octadecene, oleic acid, oleylamine, and 1,2- hexadecanediol (Figure 44). A two-stage heating protocol

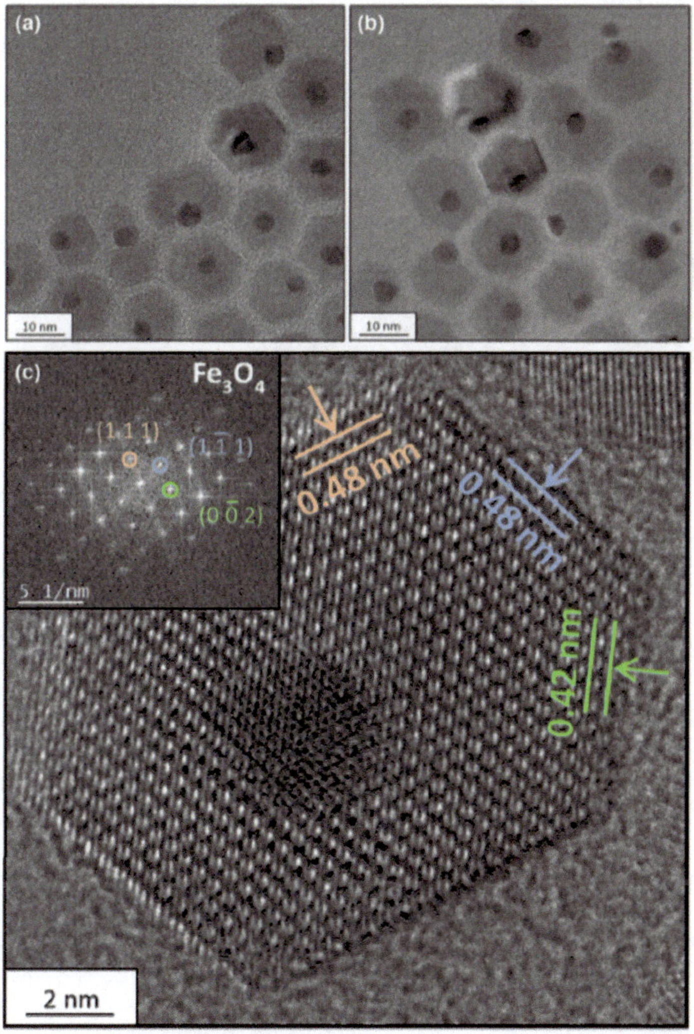

Figure 44: (a) and (b) TEM images and (c) HRTEM image of the Au@Fe_3O_4 NHDs [231].

was used. In the so-called soaking phase, which promotes the multiple nucleation process, the reaction mixture was heated at 200 °C for 120 min. Subsequently, the reaction mixture was refluxed at 315–320 °C for 90 min.

Addition of 1,2-hexadecanediol supports the multiple nucleation process by reducing Fe^{3+} to Fe^{2+} and undergoing ligand exchange process. Fe(II)alkoxy groups, which react with $Fe(acac)_3$, are generated, and Fe-O-Fe moieties are formed.

Star-shaped, 60-nm-sized Au–Fe_3O_4 NHDs were synthesized by thermal decomposition of $Fe(acac)_3$ on AuNPs (∅ = 20 ± 5 nm) in dibenzyl ether, oleylamine, and oleic acid [237]. For the thermal decomposition and nucleation process, the reaction temperature was set at 220 °C for 2 h, while the subsequent growth phase of Fe_3O_4 nanocrystals at 300 °C was allowed to last 2 h (Figure 45).

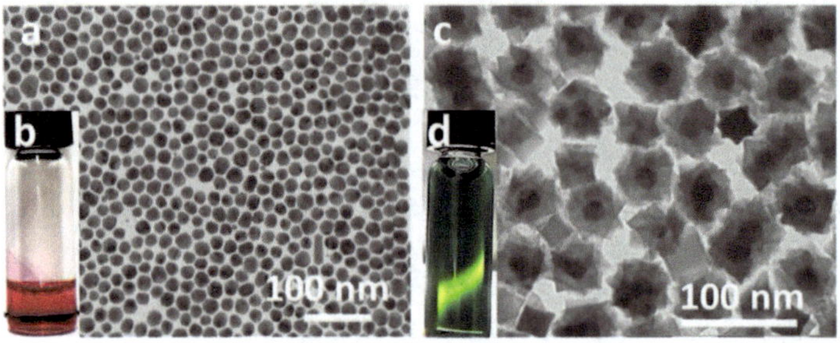

Figure 45: TEM images of the Au seeds (20 nm) (a) and star-shaped Au–Fe_3O_4 NHDs (c), photographs of the toluene solutions of the AuNPs (b) and Au–Fe_3O_4 NHDs (d) [237].

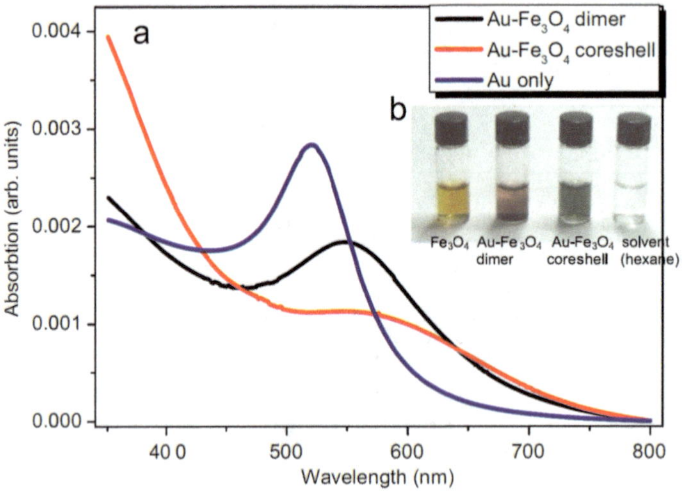

Figure 46: (a) UV-Vis absorption spectra of 8 nm-sized AuNPs, dumbbell-shaped Au–Fe_3O_4 NHDs und Au@Fe_3O_4 NHDs [236].

Epitaxial growth of Fe_3O_4 nanocrystals on AuNP generates positive charges on the growth interface which induces electron spill out of the AuNP [236, 238]. The electron transfer across the Au/Fe_3O_4 interface does not only maintain the Fe_3O_4 composition but also gives rise to the formation of an antiferromagnetic FeO phase alongside the ferrimagnetic Fe_3O_4. Figure 46 shows the UV-Vis absorption spectra of 8-nm AuNPs, dumbbell-shaped Au–Fe_3O_4 NHDs, and Au@Fe_3O_4 NHDs and the respective colloidal solutions. The LSPR of the AuNPs peaks around 520 nm, while the Au–Fe_3O_4 NHDs and Au@Fe_3O_4 NHDs exhibit broadened LSPR bands red-shifted by 30 and 70 nm, respectively. The interface between antiferromagnetic FeO and ferrimagnetic Fe_3O_4 provides the exchange bias effect. The strength of exchange bias fields depends on the interfacial area and lattice mismatch between both phases. Moreover, the Fe_3O_4/FeO nanodomain induces their magnetic moments into the AuNP substrate. The spin-polarization transfer from the Fe_3O_4 phase to the AuNP is driven by the interfacial electron transfer from the reverse direction.

3.4 Mesoporous silica nanostructures

After oxygen, elemental silicon is the second most abundant element in the earth crust and occurs there in silica and various silicate minerals. Silica is biocompatible and is absorbed by humans as orthosilicic acid [248]. Orthosilicic acid is involved in several biochemical processes, such as bone and connective-tissue metabolism, biosynthesis of collagen and glycosamino-glycans, which are required for the formation of an organic carbon matrix. Silica nanoparticles are degraded in the human organism by hydrolysis to monosilicic acid $Si(OH)_4$ [249].

Due to their unique surface texture and chemistry, mesoporous silica nanoparticles (MSNPs) enable selective adsorption and triggerable (controllable) release of drugs in tumor tissue or other target sites in patients [250–253]. Complementary applications of MSNPs in diagnostics use the encapsulation of fluorescent probes, CT (AuNPs), or MRI contrast agents (Gd complexes or SPIONs) to visualize the distribution of the nanocarriers in diseased tissue [254–261]. Mesoporous silica nanoparticles (MSNPs) ideally possess a periodically ordered pore-surface texture with pore sizes that can be tuned with diameters ranging from 2 to 30 nm [262, 263]. MNSP with narrow size distributions and average sizes between 20 and 200 nm are producible via a modified Stöber sol-gel synthesis technique [265, 266]. In the Stöber process, controlled growth of monodisperse spherical silica NPs is achieved by hydrolysis of an ester of orthosilicic acid and subsequent condensation of silica in alcoholic solutions, using ammonia for size control. The reaction mechanism of the Stöber synthesis, shown in Figure 47, involves hydrolysis of alkoxysilane ($Si(OR)_4$, R = alkyl) into silanol monomers ($Si(OR)_{4-n}(OH)_n$) and subsequent condensation of the silanol monomers into silica NPs [264]. Typically, tetraethyl orthosilicate (TEOS), tetramethyl orthosilicate (TMOS), and trimethoxymethylsilane (TMOMS) are used as alkoxysilane precursors. Hydrolysis of alkoxysilane to yield silanol

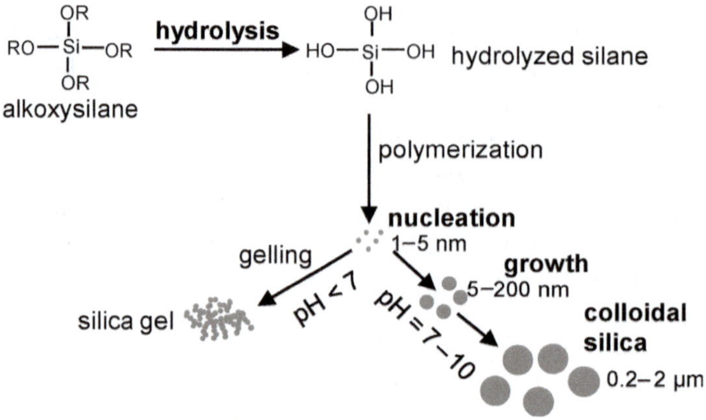

Figure 47: Schematic illustration of the silica nanoparticle synthesis using the Stöber method [264].

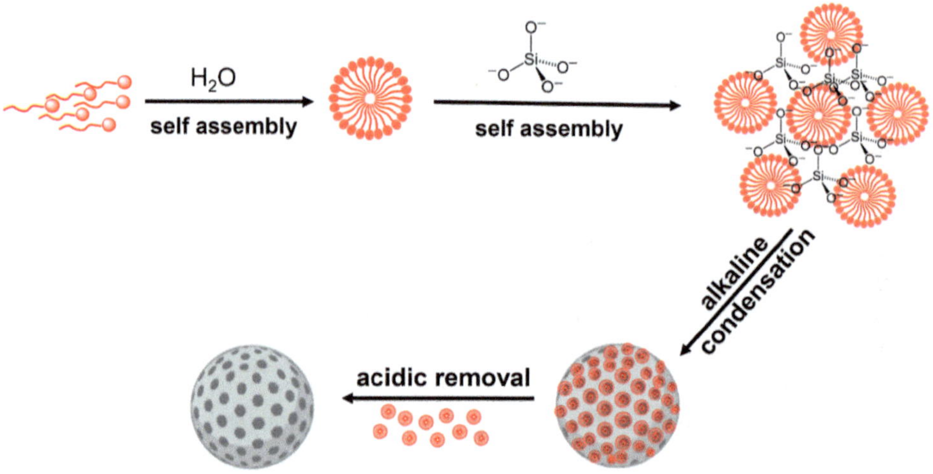

Figure 48: Schematic illustration of the synthesis of MSNPs.

monomers induces the nucleation process, while the ammonia-mediated condensation of the silanol monomers to the siloxane network controls the growth process. The silica NP sizes increase with rising ammonia concentration, while the corresponding size distribution decreases.

The synthesis of MSNPs requires the involvement of a template that controls the polymerization of silicates (SiO_4^{4-}) from alkoxysilane (Figure 48) [263, 267]. The template consisting of micelles is formed by dissolving quaternary cationic surfactants ($C_nH_{2n+1}N(CH_3)_3Br$ with n = 8–22) in an acidic or alkaline aqueous solution. The micelles function as structure-directing agents ('porogens'). The respective silicate precursor (alkoxysilanes) is then added to the micelle solution, where hydrolyzation by an acid or base catalyst yields the formation of a sol of silicate oligomers. Frequently

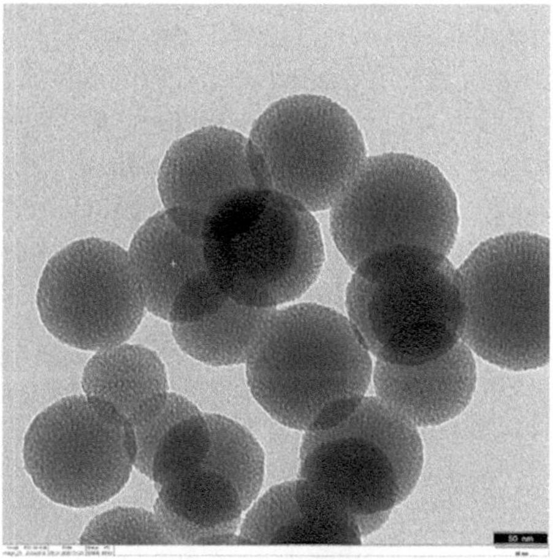

Figure 49: TEM image of MSNPs synthesized via a modified Stöber synthesis route using CTAB as template, TEOS as silicate precursor, and NaOH as base catalyst.

used base catalysts are NH_4OH, NaOH, tertiary amines, and amino acids. Hydrophobic interactions between the silicate oligomers and surfactant micelles mediate aggregation leading to precipitation of the gel. The silicate oligomers condense into nanoparticles. On the other hand, the surface-adsorbed micelles limit the nanoparticle growth and, simultaneously, provide the periodically ordered pore surface texture. The resulting MSNPs are composed of disordered polycondensed SiO_4 tetrahedra interlinked by siloxane bonds (–Si–O–Si). After removal of the surfactant micelles, the MSNP surfaces are made of a network of pores with tunable diameters (from 2 to 50 nm). The resulting large surface areas (1,000 m^2/g) are suitable for adsorption of drugs, quantum dots, siRNA, DNA, proteins, and fluorescent probes. Figure 49 depicts a TEM image of MSNPs synthesized by a modified Stöber method using CTAB as porogen, TEOS as silicate precursor, and NaOH as base. The arrangement of the pores depends on the diameters of the MSNPs: The larger the MSNPs are, the higher the degree of order of the pore network becomes. The surface chemistry consists of silanol groups that may be easily chemically modified. Amorphous MSNPs degrade faster in the human organism than crystalline and nonporous SiO_2 NPs. Silica (SiO_x) is degraded by hydrolysis to water-soluble silicic acid, which is excreted via the kidneys. Exhibiting excellent biocompatibility, large surface areas, and adjustable pore structures and morphologies, MSNPs show encouraging potential as imaging agents and nanocarriers for drug delivery in medicine [251, 268]. Pharmaceutics or imaging agents are embedded in the pores of the MSNP surface. The adsorption and release kinetics are determined by interactions between surface silanol groups and functional groups of the drugs. Chemisorption of drugs and biomolecules can be tailored for specific functional groups by coating the MSNPs with amine or carboxy donating surfactants (APTES and 5-(triethoxysilyl)pentanoic acid) [250, 267]. On the other hand, premature desorption of the drug can be prevented by grafting with stimulation-responsive pore blockers. For this purpose, the embedded drugs are protected with stimuli-responsive pore caps consisting of polyelectrolytes, copolymers, or drug chelating agents [269, 270]. As soon as drug loaded MSNPs are cellularly uptaken in tumors or otherwise diseased tissue, the drug is released in a controlled fashion by the application of internal or external stimuli (e. g., pH, enzymes, NIR light, ultrasound). A specific and very promising kind of external stimuli is an alternating magnetic field that acts on MSNPs when they encapsulate SPIONs or ferrite NPs [250, 255, 256]. An applied static inhomogeneous magnetic field makes possible guiding magnetic MSNPs to diseased (tumor) tissue, where the application of an alternating magnetic field act as a thermal stimulus for the drug release from the MSNP pores. Under the action of an alternating magnetic field, MSNPs undergo Brownian and Néel relaxation, which generates frictional heat. Simultaneously, magnetic MSNPs serve as MRI contrast agents and, therewith, enable MR imaging of cancerous tissue during the treatments.

Magnetic MSNPs are synthesized via a two-step procedure (Figure 50) [256]. First, hydrophobic (oleic acid terminated) SPIONs or ferrite NPs with 7–15-nm sizes are synthesized by thermal decomposition of organic Fe^{3+} complexes and eventually additional metal cation precursors. Alternatively, ammonium-cation stabilized SPIONs are prepared by alkaline co-precipitation. Subsequently, the hydrophobic or hydrophilic SPIONs are sonicated in an alkaline aqueous CTAB solution. In the second step, the

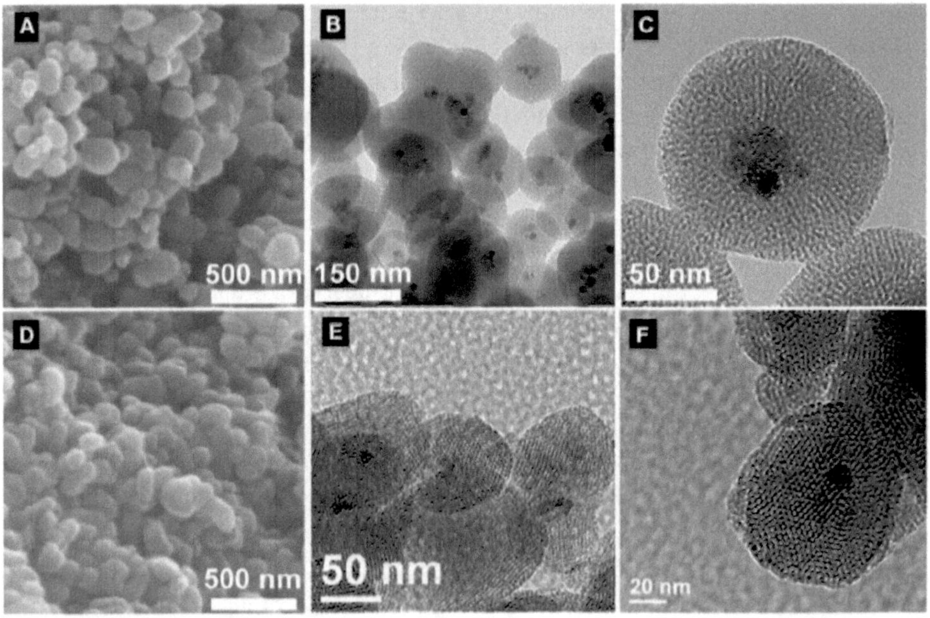

Figure 50: SEM image (A), TEM image (B) and HRTEM image (C) of SPION encapsulating MSNPs synthesized using TEOS as silicate precursor; SEM image (D); and HRTEM images (E and F) of SPION encapsulating MSNPs synthesized using TEOS and Ph-TMS as silicate precursors [256].

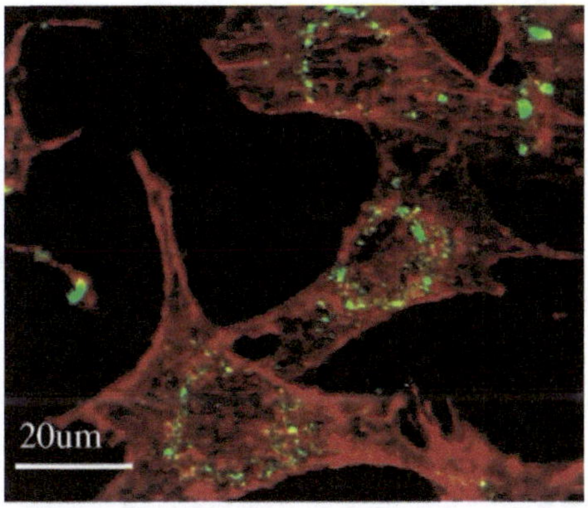

Figure 51: Confocal fluorescence microscope image of 3T3-L1 fibroblast cells containing green emitting FITC-MSNPs; the cell skeleton was stained with rhodamine phalloidin (red) [262].

CTAB coated SPIONs are used as seeds for the formation of MSNPs. This is achieved by dispersing the CTAB coated SPIONs in an alkaline aqueous (or partially ethanolic) CTAB solution and then adding drop wisely or by injection the silicate precursor(s): TEOS or phenylethyltrimethoxysilane (Ph-TMS) and TEOS. The CTAB template is conveniently removed either by calcination at 550 °C or by acidic etching using an ethanolic HCl solution.

Fluorescent MSNPs serving as imaging agents for confocal laser scanning microscopy (CLSM) are prepared by incorporating molecular fluorescent probes as, for instance, fluorescein isothiocyanate (FITS), during condensation of the silicate precursor [262]. In a typical synthesis, green fluorescent FITS incorporating MSNPs (FITS-MSNPs) are obtained by first preparing N-1-(3-trimethoxysilylpropyl)-N′-fluoresceyl thiourea (FITC-APTMS) by stirring FITC in ethanolic 3-aminopropyltrimethoxysilane (APTMS) solution. Then, FITC-APTMS is dissolved in alkaline CTAB solution, and ethanolic TEOS solution is added while stirring. The as-synthesized FITC-MSNPs are freed from the CTAB template by extraction in acidic ethanol. The cellular uptake of FITC-MSNPs was confirmed by CLSM imaging of 3T3-L1 fibroblast cells incubated with these MSNPs (Figure 51).

3.5 Luminescent semiconductor quantum dots

Due to their unique photoluminescence (PL) properties, semiconductor quantum dots (QDs) and nanoclusters have made inroads into the biomedical sciences since their discovery 40 years ago [271–273]. The uniqueness of semiconductor QDs lies in the fact that both the PL excitation and emission spectra can be tuned in the biologically relevant spectral range, from the visible to the NIR, by varying the size, composition, and doping of the single-crystal core. Biocompatibility, high PL quantum yields, and size-dependent PL spectra qualify semiconductor QDs for applications as biosensors, bioassays, and in

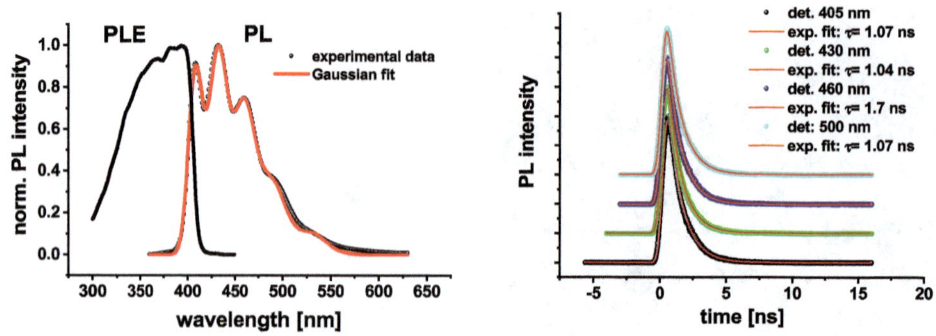

Figure 52: Photoluminescence excitation (PLE) and emission (PL) spectra (left) and photoluminescence decay curves of APTES stabilized SiNCs.

optical imaging techniques. Biocompatible semiconductor QDs consist of ZnS, Ge, or Si, for example, while most of the intensely luminescent QDs (e. g., CdSeQDs, CdTeQDs, PbSQDs, PbTeQDs) contain cytotoxic heavy elements such as cadmium, lead, mercury, or arsenic [274, 275].

Semiconductor nanoclusters (NCs) as being smaller than QDs do not possess a crystalline core and thus lack solid-state properties. With nearly all the atoms or ions in the surface, they exhibit molecular properties, including discrete electronic energy levels and molecular vibrations [276]. Silicon nanoclusters (SiNCs) are smaller than 2 nm and emit blue luminescence [277–279]. The PL spectrum of APTES-stabilized SiNCs displays a vibronic structure beginning with the zero-phonon peak at 405 nm (Figure 52). The mean peak spacing of 1,380 cm^{-1} is ascribed to a combination of the O–Si–O stretching and O–Si–O deformation vibration [277]. Moreover, the intensity distribution of the PL spectrum shows no dependence on the excitation energy. Consistently, the PL lifetime of 1.07 ns was found to be independent of the PL emission wavelength. These particularly spectral features indicate the monodispersity of the sample. The APTES-stabilized SiNCs were prepared wet-chemically using the inverse micelle reduction technique developed by Wilcoxon et al. [280] and Tilley and coworkers [281, 282]. Inverse micelles were formed by ultrasonication of a solution of tetraoctylammonium bromide (TOAB) in toluene. The reduction of $SiCl_4$ with $LiAlH_4$ inside the micelles to SiNCs was also carried out in the ultrasonic bath. Silanization of the purified SiNCs with APTES in toluene was conducted by boiling under reflux for 24 hours. Storage of the APTES-stabilized SiNCs for several weeks under ambient conditions caused partial surface oxidation. The aging process reduced the size of the SiNCs to a magic number of Si atoms. The zero-phonon peak at 405 nm corresponds to the energy band gap of 3.06 eV of Si_{123}NCs with a diameter of 1.7 nm [277]. In comparison, Wilcoxon et al. employed the inverse micelle technique to synthesize H-terminated Si nanocrystals in a size range from 1.8 to 10 nm [283]. The Si nanocrystals were size separated using HPLC combined with online optical diagnostics. The 1.8 nm-sized SiNCs were found to emit PL in the blue with maximum intensity around 365 nm and a PL lifetime of 1.5 ns. The vibronically structured absorption spectrum indicates to molecular-like discrete energy states of the SiNCs. The larger SiNP (1.8 < ⌀ < 5 nm) exhibited PL emission in the red (600 nm). Tilley and Warner advanced Wilcoxon's synthesis protocol to synthesize monodisperse blue luminescent *n*-heptyl- and aminopropyl-terminated SiNCs with average sizes of 1.8 ± 0.2 nm and 1.4 ± 0.3 nm, respectively [281, 282]. The aminopropyl-terminated SiNCs, which were water-soluble and thus suitable for biomedical applications, were photostable and emit PL at 480 nm with a quantum yield of 10 %. Surface functionalization of the H-terminated SiNCs with aminopropane ligands was performed by catalyzed hydrosilylation in isopropanol using H_2PtCl_6 as catalyst (Figure 53).

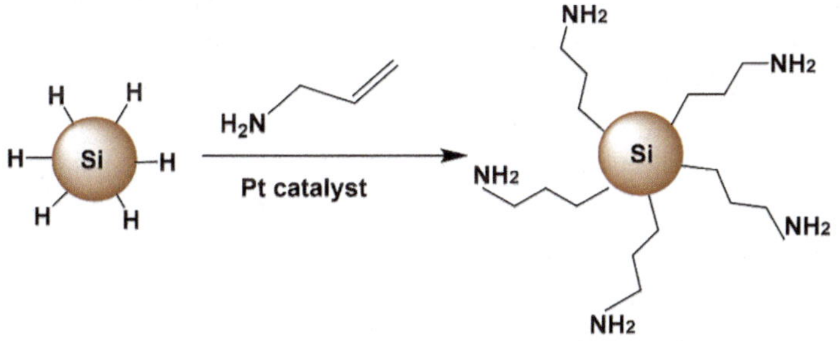

Figure 53: Schematic of catalyzed hydrosilylation of H-terminated SiNC with allylamine.

In comparison, semiconductor QDs consist of a single-crystal core smaller than Bohr's exciton radius r_{Bohr} [284]. The luminescence of semiconductor QDs arises from radiative recombination of charge carriers (i. e., electrons and holes) confined in the core. Hole and electron wavefunctions have nonzero amplitudes at the QD surfaces. Therefore, semiconductor QDs facilitate the tuning of the PL wavelength in the desired spectral range by appropriately adjusting the QD size between the NC size and Bohr's exciton radius. Direct-gap semiconductor QDs, such as CdSQDs, CdSeQDs, and CdTe-QDs, feature large PL quantum yields in the visible spectral region [273, 285–287]. As previously mentioned, unfortunately these cytotoxic semiconductor QDs are not suited for biomedical applications. Therefore, the focus here is laid on biocompatible silicon quantum dots (SiQDs). Silicon and silica are nontoxic and environmentally safe materials. Bulk silicon crystallizes in the diamond cubic crystal structure and emits as an indirect band-gap semiconductor weak luminescence in the NIR [288, 289]. The energy dispersion relationship of indirect semiconductors is characterized by the difference in the wave vectors of the bottom of the conduction band and the maximum of the valence band in the reciprocal space. Therefore, radiative electron–hole recombination across the band gap of the infinite crystal is dipole forbidden and requires phonon assistance to conserve the crystal momentum. Therefore, the PL quantum yield is rather small. By size reduction to the quantum confinement regime, the dispersion relation of Si nanocrystals changes such that the energy bands become pseudo-continuous and the band-gap energy increases with decreasing size [290]. Si nanoparticles (SiNPs) with sizes smaller than Bohr's exciton radius (4.3 nm), and larger than 1.7 nm are SiQDs that emit size-dependent PL in the visible spectral range [272, 288]. The essential prerequisite for this is the prevention of surface oxidation by suitable surface stabilization. Surface oxidation instantaneously occurs in the presence of even very small amounts of oxygen. Hence, as synthesized, SiNPs are surrounded by an amorphous SiO_x layer, the so-called native passivation shell, which offers a broad variety of defect structures including dangling bonds of surface Si atoms and oxygen vacancies [279, 288, 291, 292]. These defect structures reduce the PL quantum yield by capturing charge carriers and acting as nonradiative recombination centers for electron–hole pairs. Therefore, oxidized SiQD surfaces need to be modified, at least by saturating all dangling bonds and oxidizing the oxygen vacancies. This can be partially achieved by long-term aging, eventually followed by silanization [277, 278, 293]. An alternative and even more tedious method consists in the complete removal of the SiO_X shell using 5–10 % ethanolic HF solutions [291, 292, 294, 295]. Since H-terminated SiNPs are instable, subsequent hydrosilylation with unsaturated organic compounds requires forming gas (90 % N_2 + 10 % H_2) atmosphere. The preparation of biocompatible functionalized SiQDs is mostly carried out as a multistep procedure. Hydrosilylation of H-terminated Si surfaces consists in the formation of covalent Si-C bonds, which facilitates stable binding of functional ligands. The large surface curvature of SiQDs enables the complete saturation of surface Si atoms with Si-C bound ligands [290]. This does not only allow the generation of defined surface structures, but also prevents surface oxidation and, therewith, the formation of defect structures. Such functionalized and stabilized SiQDs should exhibit high PL quantum yields in the visible and with long-term stability [296, 297].

The main methods for fabricating H-terminated SiQDs include electrochemical HF etching of Si wafers [298], CO_2 laser induced pyrolysis of silane (SiH_4) [299], radio wave or microwave plasma-induced dissociation of SiH_4 [296, 297, 300–302], and wet-chemical

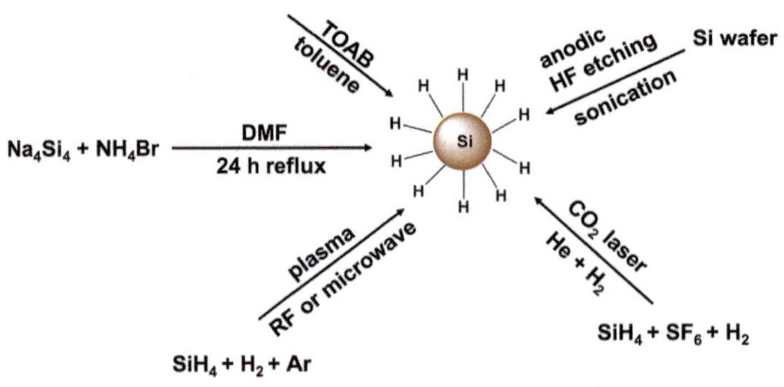

Figure 54: Schematic illustration of the main synthesis routes for the preparation of H-terminated SiQDs.

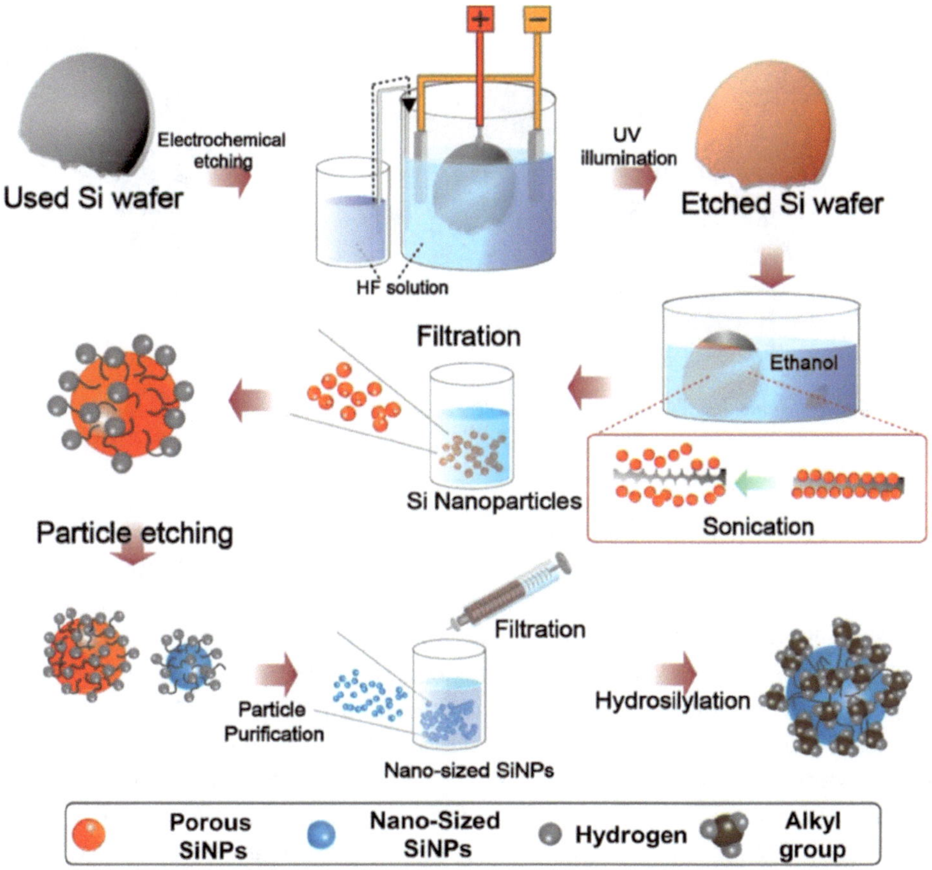

Figure 55: Schematics of the electrochemically mediated HF etching of Si wafers starting with anodic HF etching of used Si wafers, followed by UV irradiation and sonication; the resulting SiNPs are etched with a HF/HNO_3 mixture; blue-luminescent SiNCs are separated from red-luminescent SiNPs by filtration and then hydrosilylizated [298].

reduction and oxidation reactions [279, 281, 303–307] (Figure 54). Electrochemically mediated HF etching of Si wafers to prepare H-terminated SiNPs was successfully carried out by anodically etching n- or p-type Si wafers in a HF/H_2O/ethanol (1:1:4) solution, subsequent UV irradiation treatment, and then decomposition of the etched Si wafer to SiNPs in an ultrasonic bath [298] (Figure 55). Nevertheless, the SiNPs were obtained with partially oxidized surfaces and 4–8-nm diameters, which are larger than Bohr's exciton radius. Hence, the SiNPs were subsequently etched with a HF/HNO_3 acid mixture, which should allow the final size distribution of the H-terminated SiQDs to be adjusted via the etching time. Unfortunately, blue-luminescent SiNPs and red-luminescent porous SiNPs were simultaneously obtained. The smaller, blue-luminescent SiNPs were separated from the porous SiNPs by filtration.

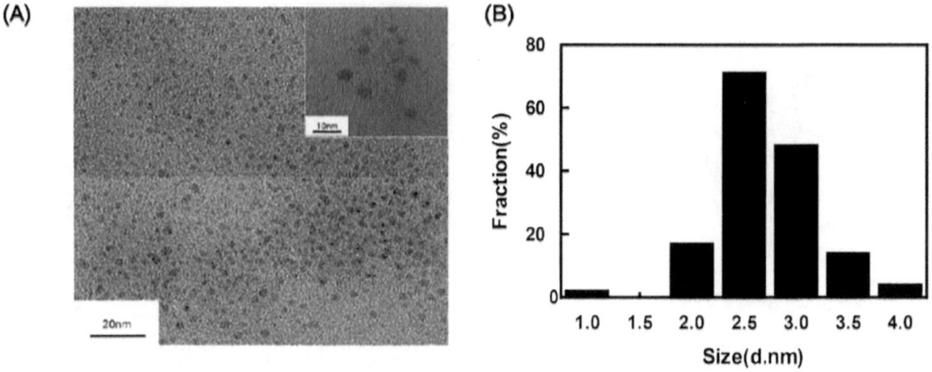

Figure 56: (A) TEM images of SiNPs after the HF/HNO_3 acid-etching process, (B) histogram of diameters for the etched SiNPs [298].

The size distribution of the blue luminescent SiNPs indicates that a small amount of 1-nm-sized SiNCs and a significantly larger fraction of 2–4-nm-sized SiQDs (Figure 56) were obtained, which is not consistent with the observation of blue PL.

More promising techniques for the synthesis of SiNPs with narrow band-size distributions are conducted in the gas phase and include CO_2-laser induced pyrolysis of SiH_4, and microwave or radio-wave induced generation of $Ar/H_2/SiH_4$ or $Ar/H_2/SiH_4$ plasmas. Li et al. developed a CO_2 laser-driven aerosol reactor for the gas-phase synthesis of SiNPs [308] (Figure 57). The reactor was designed in a cross-arm configuration, with the horizontal arm for passing the CO_2 laser beam, the vertical arm with inlets for the reaction gases (SiH_4, SF_6, H_2), and the carrier gas He. The total pressure in reactor was 540 mbar, and the flow rates of He, H_2, SiH_4 and SF_6 were adjusted at 1,500 cm^3/min, 700 cm^3/min, 24 cm^3/min, and 2 cm^3/min, respectively. IR radiation from a continuous 60-W CO_2 laser provides thermal decomposition of SiH_4 to Si and H_2 in the reaction zone. The gas-heating process is assisted by SF_6, because it has a large absorption cross section at the CO_2 laser wavelength of 10.6 µm and thus acts as photosensitizer. The nucleation and SiNP growth process were tuned by the flow rates of the gas components as given above to obtain SiNPs with sizes between 5 and 20 nm. These SiNPs were separated from the gas mixture using cellulose nitrate membrane filters. Since these SiNPs were larger than Bohr's exciton radius and most likely have partially oxidized surfaces, they do not show photoluminescence. An aqueous HF/HNO_3 etching solution was used to reduce the SiNP sizes to the quantum-confinement region. The reduction of SiNP size occurs in a two-step process consisting of oxidation of SiNP surface by HNO_3 and dissolution of SiO_x by HF. As reported by the authors, a controlled etching method using an aqueous solution of 0.5 %–20 % HF and 10 %–40 % HNO_3 was developed to adjust specific SiQD sizes to tune PL spectra in the visible range from red to green. Figure 57 (right, upper panel) shows the photography of SiQD powders exhibiting size dependent PL under UV illumination. With increasing etching time, the SiQDs became smaller which correlates with the blue shift

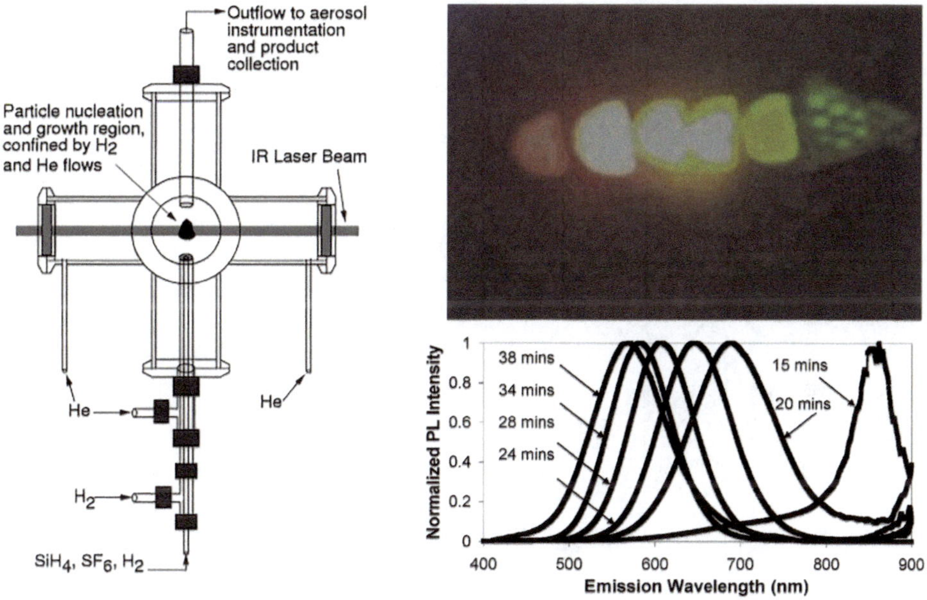

Figure 57: Schematic of CO_2 laser-driven aerosol reactor (left), photography of SiQD powder samples under UV illumination; the etching time of SiNPs increases from the red luminescent sample to the green luminescent one (right, upper panel); PL spectra of methanolic SiQD dispersions etched at different times (right, lower panel) [308].

of the PL spectra (Figure 57, lower panel). The dependence of the PL spectra on the etching time demonstrates the quantum-confinement effect. Unfortunately, the luminescent SiQDs oxidized as soon as they were exposed to air, which resulted in the loss of luminescence. Hua et al. optimized the HF/HNO_3 etching procedures by using a mixture of concentrated HF (48 wt %) and HNO_3 (69 wt %) in a volume ratio of 10:1 [309]. Subsequently, the freshly etched red, yellow, and green luminescent SiQDs were stabilized by photoinitiated hydrosilylation with 1-alkene (1-pentene, 1-hexene, 1-octene, 1-octadecene) derivatives. However, the alkyl-passivated SiQDs proved to be neither photostable nor thermally or long-term stable, although the respective FTIR spectra contained the characteristic vibrational bands of the Si-C bound alkane ligands. As also indicated by the appearance of the Si-O stretching vibration band at 1,070 cm^{-1}, the hydrosilylation reaction occurred only incompletely. The remaining Si–H moieties were oxidized either at higher temperatures (70 °C) or by irradiation at 355 nm. The PL quantum yield of the green and yellow luminescent SiQDs, in particular, decreased with surface oxidation, indicating the incomplete surface saturation with Si–C bound ligands.

Other successful gas-phase methods for the synthesis of SiNPs are based on microwave- or radio-wave-induced generation of Ar/H_2/SiH_4 plasmas. Microwave induced electrical discharging of an Ar/H_2/SiH_4 gas flow generated a hot plasma accomplishing the dissociation of SiH_4. The low-pressure microwave reactor by Giesen et al. [310] con-

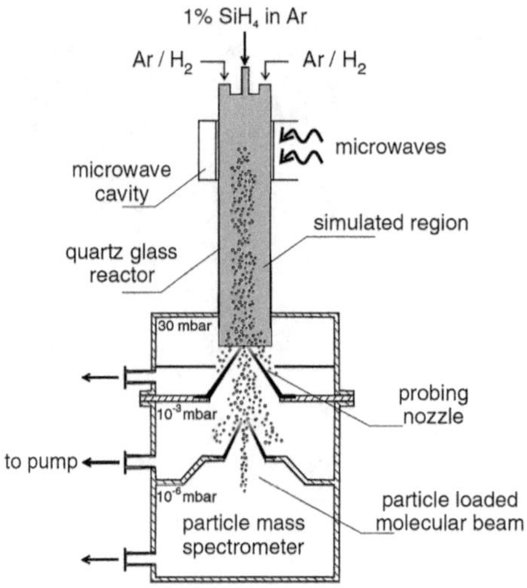

Figure 58: Experimental setup for SiNP production in a microwave reactor [310].

sists of a microwave cavity with an integrated quartz tube and attached analysis chamber (Figure 58). In situ monitoring of SiNP sizes was achieved using a molecular beam-probing system in combination with a particle mass spectrometer. The precursor SiH_4 was injected into the reaction chamber through the central nozzle using argon as carrier gas (99 % Ar and 1 % SiH_4). The plasma gases Ar and H_2 flowed in the reactor through several nozzles around the central one. A thermal $Ar/H_2/SiH_4$ plasma was generated by coupling 2.45 GHz microwaves with a maximum power of 2,000 W into the gas flow forming a flamelike plasma torch. SiH_4 was thermally decomposed in the hot plasma to Si and 2 H_2. A supersaturated silicon aerosol was formed, and ideally, homogeneous nucleation emerged. The subsequent crystal growth by coalescence and coagulation could be controlled through the pressure and temperature gradient in the reactor. Figure 59 shows a TEM image and associated histogram of the size distribution of microwave plasma generated SiNPs. For example, under experimental conditions in the reactor with a microwave power of 200 W, an Ar/H_2 gas pressure of 30 mbar, and a SiH_4 concentration of 1690 ppm, SiNPs were obtained with a log-normal size distribution between 2 and 10 nm and an average size of 5.7 nm. The advantage of this gas-phase synthesis method is the impressively facile easy-and-clean separability of the SiNPs from the gas flow by gravity. Nevertheless, partial oxidation of the SiNPs can not be avoided.

Hence, 5 nm and larger SiNPs were usually synthesized, and the desired sizes of the SiQDs were adjusted via the HF etching process, which must also be followed, if possible, in the same reaction vessel, by surface stabilization due to hydrosilylation with 1-alkene or 1-alkyne (Figure 60, left panel). In contrast to the HF/HNO_3 etching procedure, HF etching of oxidized SiNPs using aqueous/ethanolic 5–10 % HF solutions led to the complete removal of the SiO_x surface layer and complete hydride termination of the SiQDs.

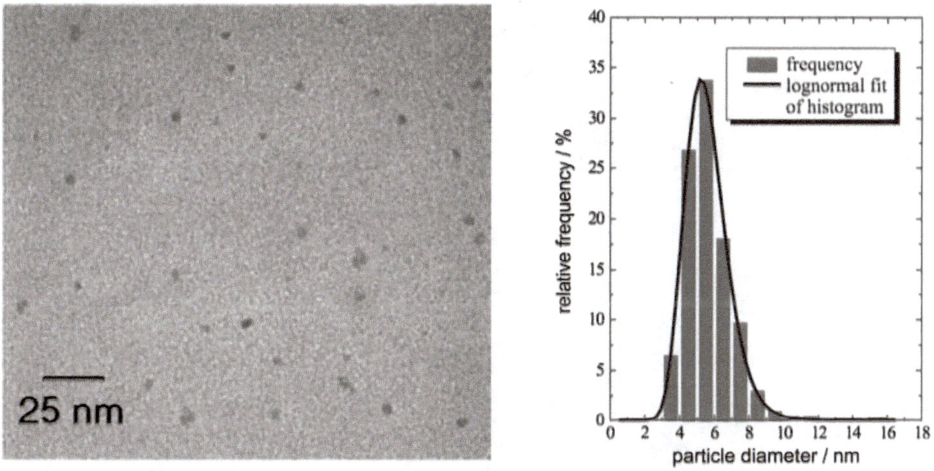

Figure 59: TEM image and histogram of the SiNP size distribution; the experimental parameter are: $p(Ar/H_2)$ = 30 mbar, $c(SiH_4)$ = 1,690 ppm, microwave power = 120 W [310].

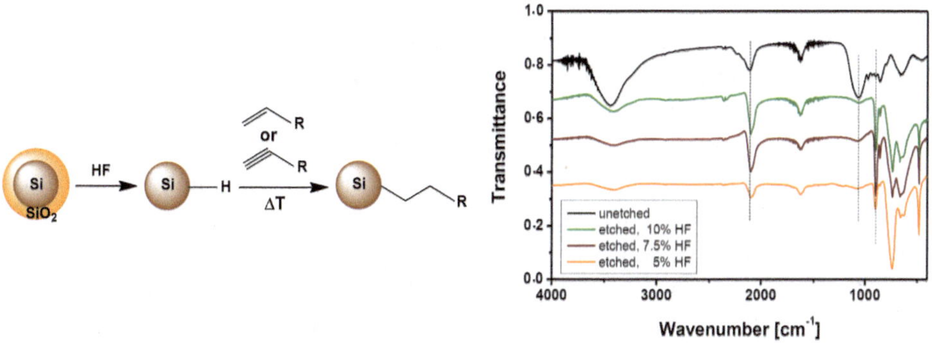

Figure 60: Schematic of the two-step process for the preparation of functionalized SiQDs: HF etching of oxidized SiNPs followed by thermal hydrosilylation of H-terminated SiQDs with 1-alkene or 1-alkyne derivatives (left); FTIR transmission spectra of unetched SiNPs and H-terminated SiQDs after etching with 5 %, 7.5 %, and 10 % HF solutions (right).

The FTIR spectrum of the microwave–plasma synthesized SiNPs displays the symmetric Si-O stretching vibration band at 1,070 cm^{-1} identifying surface oxidation (Figure 60, right panel). The SiNPs were dispersed in ethanol and were etched with a 5 %, 7.5 %, or 10 % HF solution for 10 min. As shown by the respective FTIR spectra, the Si–O stretching band disappeared completely, while at the same time the Si–H stretching vibration band was increasingly formed at 2,088 cm^{-1}.

For the preparation of green and cyan luminescent SiQDs, microwave–plasma synthesized SiNPs with average sizes of 5 nm were used [290–292]. The initially surface oxidized SiNPs were dispersed in ethanol. Etching the SiNPs with a 5 % HF solution for

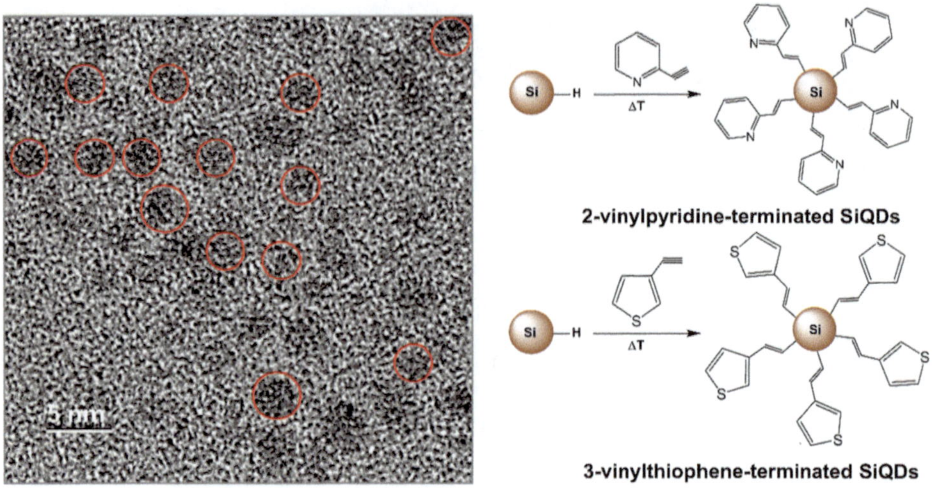

Figure 61: HRTEM image of 3-vinylthiophene-terminated SiQDs (left); schematic of thermal hydrosilylation of H-terminated SiQDs with 2-ethynylpyridine or 3-ethynylthiophene (right).

10 min resulted in H-terminated SiQDs with sizes between 2 and 3 nm (Figure 61, left). The ethanolic HF etching solution was removed by vacuum distillation, and then the SiQDs were passivated by thermal hydrosilylation. The latter proved to be a challenge if the dangling bonds of all Si atoms on the surface are to be saturated by Si–C-bonded ligands, ensuring maximum photostability and thus long-term stability of the SiQDs. This requires, first, complete oxygen exclusion during the HF etching and thermal hydrosilylation process, and, second, linear ("lean") 1-alkene or 1-alkyne derivatives for hydrosilylation. As Groenewegen et al. [291] and Sommer et al. [292] ascertained, 2-ethynylpyridine, 4-ethynylpyridine, or 3-ethynyl-thiophene are particularly suitable hydrosilylation reagents, and o-xylene or p-xylene are appropriate solvents. For the thermal hydrosilylation process, an o-xylene solution of either 2-(or 4-) vinylpyridine or 3-vinylthiophene was degassed by three freeze–pump–thaw cycles and then added to the H-terminated SiQDs in a forming gas atmosphere (Figure 61, right). The reaction suspension was boiled under reflux for 20 hr. The 2-vinylpyridine-terminated SiQDs were purified by washing them over a polyvinylidene fluoride membrane filter with ethanol several times. The FTIR transmission spectrum of 2-vinyl-pyridine-terminated SiQDs in Figure 62 provides evidence of oxide-free SiQD surfaces, apparently achieved by complete hydrosilylation of the surface Si atoms; the Si–O stretching vibration band in the 1,000–1,100 cm^{-1} range is completely absent, while the prominent symmetric Si-C stretching vibration band at 741 cm^{-1} and other vibrational bands characteristic of 2-vinylpyridine indicate covalent binding of the ligand. The FTIR spectra of 4- and 6-month-old SiQD samples did not show any changes, which verifies the long-term stability of the surface chemistry. The PL emission and PL excitation (PLE) spectra of 2-vinylpyridine- and 3-vinylthiophene-terminated SiQDs dispersed in ethanol are shown

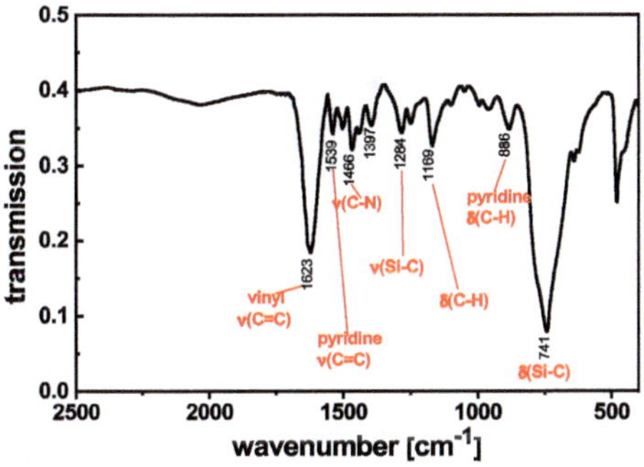

Figure 62: FTIR transmission spectrum of 2-vinylpyridine-terminated SiQDs.

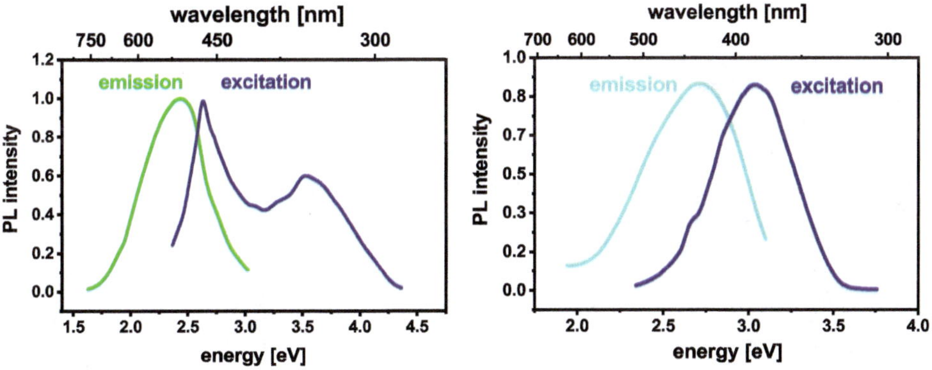

Figure 63: The PL excitation and emission spectra of 2-vinylpyridine- and 4-vinylpyridine-terminated SiQDs dispersed in ethanol.

in Figure 63. The PL emission spectrum of 2-vinylpyridine-terminated SiQDs consists of a Gaussian band at 510 nm, while 3-vinylthiophene-terminated SiQDs exhibited an emission maximum at 460 nm. In comparison, 2-vinylpyridine-terminated SiQDs not only achieved a PL quantum yield of 30 %, but were also water-soluble and biocompatible. The 2-vinylpyridine ligands, which act as Brönsted bases, provided water solubility and positive surface charges of the SiQDs. Hence, these functionalized SiQDs are suitable for biomedical applications and have already been successfully used as a self-tracking transfection tool for siRNA in Caco-2 cells [311]. The positively charged, green luminescent SiQDs were shown to bind siRNA via electrostatic interactions (Figure 64, right). The confocal laser scanning microscopy images proved the cellular uptake of the green luminescent SiQDs, which apparently accumulated solely in the cytoplasm (Figure 64, left).

The plasma reactor engineered by Mangolini et al. [312] consisted of a quartz tube through which a gas mixture of Ar, SiH_4, and H_2 flowed at rates of 50 cm^3/min (Ar), 5–10 cm^3/min (SiH_4), and 1–5 cm^3/min (H_2). Radio waves were generated in the reactor tube by applying radiofrequency (RF) power at 13.56 MHz to two copper-ring electrodes

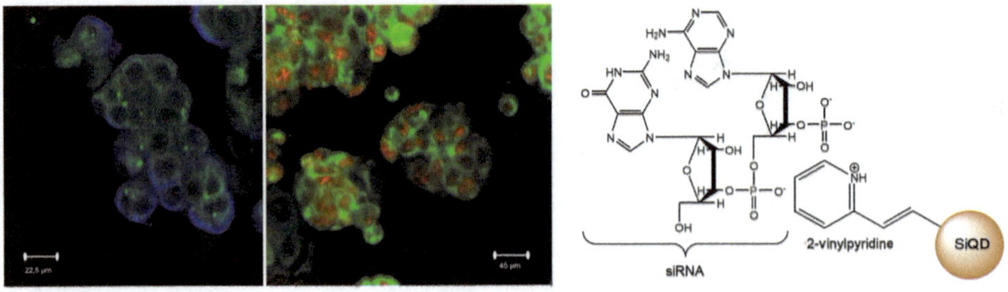

Figure 64: Confocal laser scanning microscopy images of Caco-2 cells with internalized green luminescent SiQDs, immunofluorescence staining of p-glycoprotein (blue), and propidium iodide staining of cell nuclei (left); illustration of electrostatic interactions between siRNA and positively charged 2-vinylpyridine ligands (right) [293].

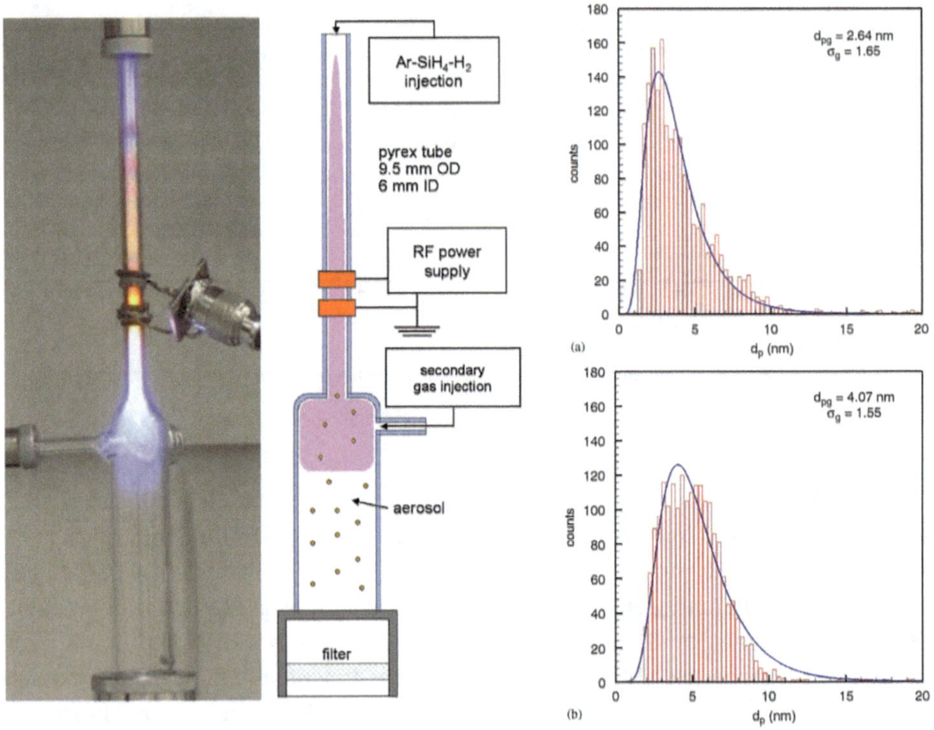

Figure 65: Photograph and schematic of the radio-wave plasma reactor (left) and SiNP size distributions: (a) PL emission at 730 nm, (b) PL emission at 804 nm (right) [312].

arranged around the quartz tube, which generated a nonthermal Ar/SiH_4/H_2 plasma with a pressure of around 190 Pa. SiNPs were formed in the plasma by electron impact dissociation of SiH_4 and subsequent coagulation in the intense plasma zone (Figure 65, left). The plasma generated SiNPs were transported by the gas flow into the expanded quartz tube and collected as a powder on stainless steel mesh filters. This continuous gas-phase procedure allows to synthesize 50 mg/h of luminescent SiQDs. Subsequently, the as-synthesized SiNPs were passivated with 1-dodecene by thermal hydrosilylation in mesitylene (b. p. 165 °C). The size distributions of the red luminescent SiQDs and SiNPs in Figure 65 (right) ranged from 1 nm to 10 nm and thus beyond the quantum-confinement range. The SiQD/SiNP samples were produced under various experimental conditions. The SiQD sample with the average size of 2.6 nm emitted PL with maximum intensity at 730 nm, while the PL spectrum of the SiQDs with a mean size of 4.1 nm peaked at 804 nm. Both samples reached PL quantum yields between 60 and 70 %. Unfortunately, the photostability of the SiQD/SiNP samples was not investigated. In addition, dodecane-terminated SiQDs are water-insoluble and thus are not suitable for biomedical applications. Wet-chemical reduction or oxidation reactions for the preparation of SiNPs and SiQDs are often praised as being much more elegant and cost-effective compared to gas-phase syntheses, and furthermore more suitable for industrial implementation. To date, wet-chemical synthesis methods remain limited by the drawbacks that SiNPs are prepared in a relatively broad-size distribution, are often amorphous, possess at least partially oxidized surfaces, and require a great deal of effort to pu-

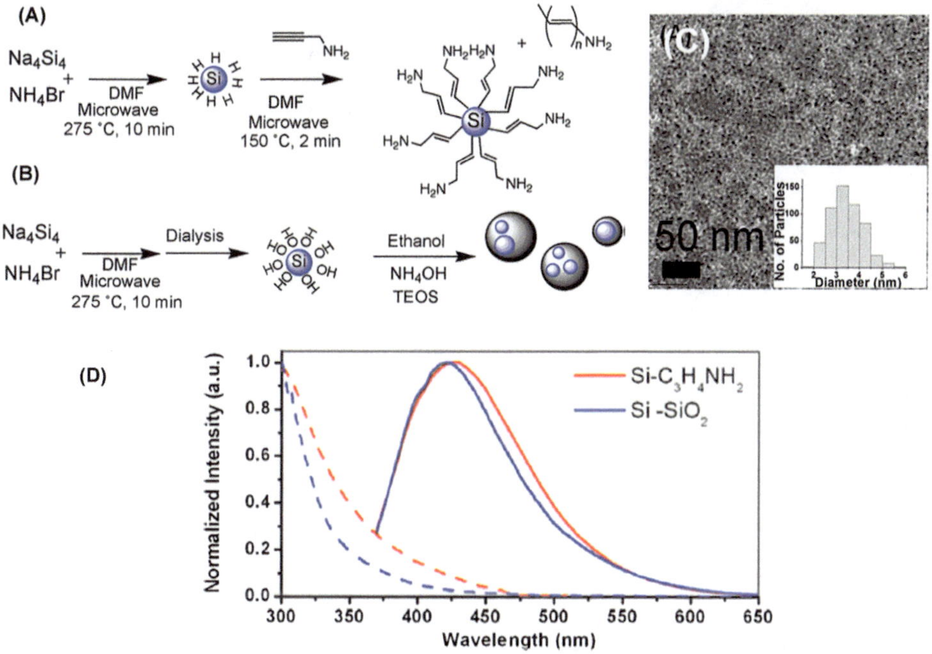

Figure 66: Scheme of the microwave-assisted syntheses of 3-aminopropenyl-terminated SiQDs (A) and silica-encapsulated SiQDs (B); TEM image and histogram of SiQD-size distribution (C); PL emission spectra (solid curves) and absorption spectra (dashed curves) of 3-aminopropyl-terminated SiQDs and silica-encapsulated SiQDs (D) [303].

rify. Rather promising wet-chemical synthesis methods include the reduction of $SiCl_4$ or $SiBr_4$ with $LiAlH_4$ and the oxidation of metal silicides [284, 288, 313]. As already noted, Wilcoxon and Tilley developed the inverse micelle reduction synthesis technique [280, 281]. The inverse micelles are formed by dissolving quaternary ammonium cationic surfactants in toluene or THF by ultrasonication and serve as nanoreactors for the reduction of $SiCl_4$ with $LiAlH_4$ to Si and growth to silicon nanocrystals with size distributions between 1.4 and 10 nm. Tilley could prepare photostable blue luminescent, n-heptyl- and aminopropyl-terminated SiNCs with mean sizes of 1.8 and 1.4 nm, respectively. In contrast, this synthesis technique was less successful for the preparation of monodisperse SiQDs (2–4.2 nm) and SiNPs (>4.2 nm) with oxide-free surfaces. An example of wet-chemical oxidative syntheses of H-terminated SiQDs is the oxidation of metal silicides such as Mg_2Si or Na_4Si_4 with NH_4Br or ethylenediammonium chloride in N,N-dimethylformamide (DMF) [306, 307, 314]. Atkins et al. developed a microwave-assisted synthesis method that enabled the preparation of water-soluble, blue-luminescent 3-aminopropenyl-terminated or silica-encapsulated SiQDs in a two-step process by hydrosilylation of H-terminated SiQDs [303] (Figure 66). As illustrated in Figure 66((A) and (B)), Na_4Si_4 was reacted with NH_4Br in DMF in a microwave-reaction vessel under an

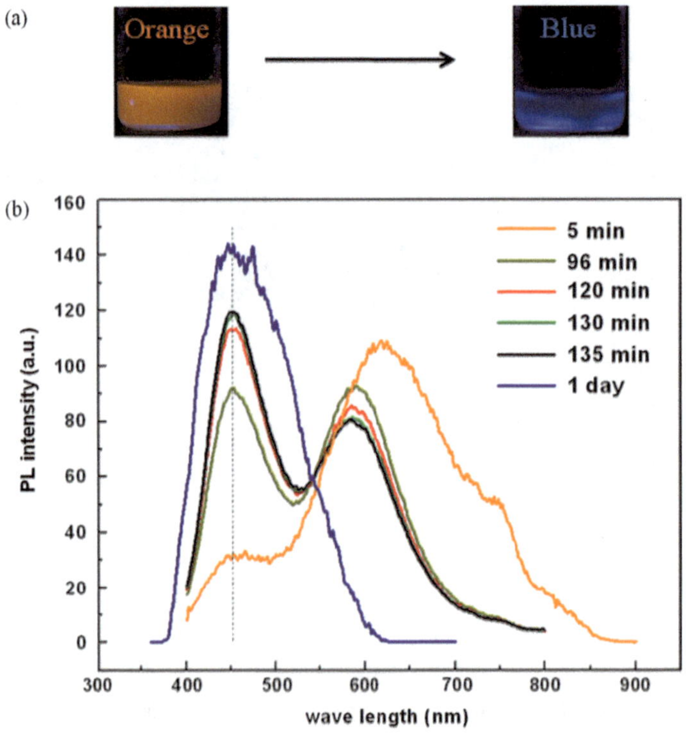

Figure 67: Photographs of orange-luminescent HF-etched and blue-luminescent oxidized SiQDs (a); PL emission spectra of freshly etched SiQDs (5 min) and successively oxidized SiQDs (96 min–24 hr) [302].

inert gas atmosphere. After chromatographic purification of the H-terminated SiQDs, they were hydrosilylated with 3-propargylamine in DMF in the microwave reactor (A). In contrast, in the reaction scheme (B), the H-terminated SiQDs were dialyzed and thus oxidized and then silanized with tetraethoxysilane (TEOS). As can be seen in the TEM image and associated histogram (Figure 66(C)), the SiQDs were synthesized with sizes between 2 and 6 nm and a mean size of 3.5 nm. Surprisingly, the PL emission spectrum of the 3-aminopropenyl terminated SiQDs resembled that of the silica encapsulated oxidized SiQDs with maxima at 415 nm (Figure 67(D)). The PL quantum yields were between 14 and 23 %. The size-independent blue emission was not consistent with the quantum confinement effect. Gupta et al. [302] performed HF etching and reoxidation experiments on gas-phase synthesized SiQDs with similar log-normal-shaped size distributions between 2 and 6 nm. According to their experimental results, freshly etched SiQDs with a mean size of 4.05 nm emit orange PL peaking at 620 nm. After aging or reoxidation under ambient conditions, the orange PL emission of the SiQDs vanished and a size-independent blue emission around 450 nm emerged. The blue PL emission was ascribed to trivalent silicon atoms (P_b resonance centers) in the interface between the amorphous SiO_x layer, and the Si bulk (Figure 67).

3.6 Surface chemistry and functionalization

The fundamental functions of surface coatings are both to control the nucleation and growth processes of nanoparticles and stabilize them in solution. Biomedical applications require surface coatings that enable the formation of stable colloids in physiological media and, in addition, prevents the adsorption of opsonins in blood serum. The key determinants of protein adsorption and protein unfolding are shape, size, and primary surface structures. This implies that suitably sized and functionalized nanoparticles escape removal by the mononuclear phagocytic system (MPS). This prolongs their circulation time needed to reach the targeted organ.

Novel multifunctional targeted nanoplatforms must overcome biological barriers to deliver therapeutic agents directly to diseased tissue. Suitably stabilized and functionalized nanoparticles should act as DNA carriers in gene therapy, CT, and MRI contrast agents, or for targeted transport and delivery of proteins, peptides, or drugs. For specific cancer-cell targeting, the nanoparticle coating should facilitate conjugation with receptor substrates such as folic acid or glucose. Functional surface structures are engineered to specifically bind pharmaceutical agents, fluorescence probes, and tumor-cell targeting small molecules, mRNA, siRNA, aptamers, peptides, and antibodies (Figure 68) [69, 198, 315, 316].

The surfaces of nanoparticles are stabilized during synthesis either by electrostatic interactions with ionic moieties, by chemisorption or physisorption with monodentate or polydentate surfactants, or by coating with polymers. In situ generation of covalently bound or adsorbed surface structures do not only regulate nanoparticle growth in an

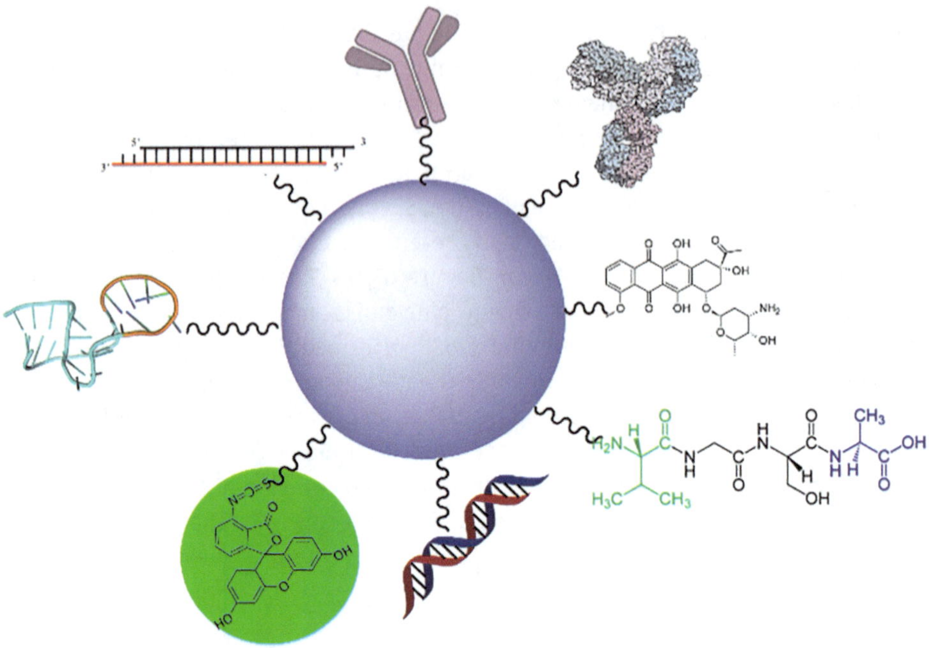

Figure 68: Functionalization of nanoparticles with peptides, aptamer, DNA, siRNA, antibody, drugs, and fluorescent dyes.

isotropic or anisotropic fashion but also makes stable colloids possible and ideally prevents agglomeration. Hydrophilic SPIONs and noble-metal nanoparticles are accessible in aqueous solution by facile co-precipitation and redox reactions. The disadvantages of these low-cost synthesis techniques are poor growth control, only weakly stabilized nanoparticle surfaces and, thus, no long-term colloidal stability. Electrostatically bound ligands, such as citrate anions, ammonium cations, and CTAB bilayer templates, are exchanged for other ions or amphoteric proteins in cellular media [317]. Weakly bound ligands are associated with agglomeration and promote the formation of a protein corona. On the other hand, hydrophobic noble-metal and metal-oxide nanoparticles with defined sizes and shapes are prepared by thermal decomposition of organic metal precursors in nonpolar high-boiling solvents in the presence of unsaturated long-chain carboxylic acids and primary long-chain alkylamines such as oleic acid and oleylamine. These surfactants control the nanoparticle growth through physisorption or chemisorption at the surface and stabilize the nanoparticles through steric (repulsive) interactions by the long alkane chains, but they do not provide biocompatibility.

Biomedical applications of poorly stabilized hydrophilic nanoparticles, as well as initially hydrophobic nanoparticles, require the replacement of native ligands with hydrophilic functional surfactants that make possible specific chemo- or physisorption of drugs, fluorescent dyes, or biomolecules. The post-synthetic surface functionalization can be achieved either by single or multistep ligand exchange or by surface (over)coating with SiO_x or biocompatible polymers [315, 318].

3.6.1 Polymer coating

For nanoparticle coating with polymers, for example, biopolymers such as dextran, alginate, and chitosan or biocompatible polymers such as polyethylene glycol (PEG), polyvinyl alcohol (PVA), poly(vinylamine), polyethyleneimine (PEI), or poly(acrylic acid), PAA are used [319–321]. Water-soluble polymers, like PEI or PAA, are cationic or anionic in nature and are adsorbed onto the surfaces of gold and metal oxide nanoparticles through electrostatic interactions. To prevent desorption of weakly bound polymers, dextran, for example, is cross-linked using epichlorohydrin after adsorption onto SPIONs, which leads to a dramatic increase in the hydrodynamic diameter of the nanoparticles [322]. In addition, epichlorohydrin is cytotoxic [323]. "PEGylation" and other polymer coatings are highly praised in literature [178, 324–326].

PEGylated nanoparticles are understood to prevent surface adsorption of plasma proteins (opsonins), such as IgG or complement fragments, during their passage through the bloodstream [327]. Non-opsonized nanoparticles escape receptor-mediated phagocytosis, increasing their circulation time [328]. These findings, derived essentially from in vitro studies, are in contradiction with results from in vivo experiments [329–331].

PEGylated nanoparticles exhibit a hydrophilic, and uncharged surface that takes on a brush-like or mushroom-like structure, depending on the composition of its molecular weight [327] (Figure 69). These surface structures are stated to provide a hydrophilic

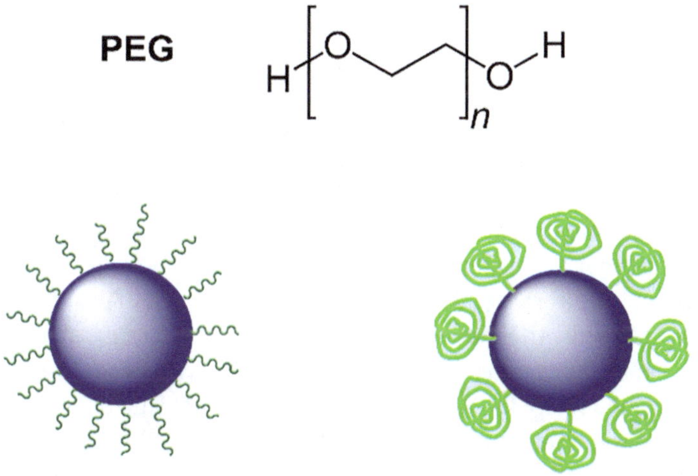

Figure 69: Molecular structure of PEG (top); brush-like and mushroom-like surface structure (bottom).

and steric barrier for plasma protein adsorption from the bloodstream. The so-called "PEG dilemma" is that protein-repelling nanoparticles also do not interact with membrane receptors and thus are not taken up by tumor cells [327]. These findings, obtained essentially from in vitro studies, have now been contradicted by results from in vivo experiments. PEGylated nanoparticles exhibit a hydrophilic and uncharged surface that also may have a brush-like structure depending on the molecular mass of the PEG. The PEG layer should provide a hydrophilic and steric barrier for the adsorption of proteins from blood serum. Such a protein-repellent nanoparticle surface inhibited cellular uptake, making such coated nanoparticles useless for drug delivery. Moreover, the catalytic and optoelectronic properties of nanoparticle surfaces are masked by polymer coatings that increase the size of nanoparticles by tens of nm or more. The topology and surface chemistry of the nanoparticles change in an uncontrolled manner. In addition, multiple nanoparticles may be encapsulated in a polymer layer [332]. Such polymer-stabilized nanoparticle agglomerates (Figure 70) can cause the formation of thrombi in blood vessels [333]. Another disadvantage of polymer coating is that the disordered surface structures are not suitable for controlled functionalization.

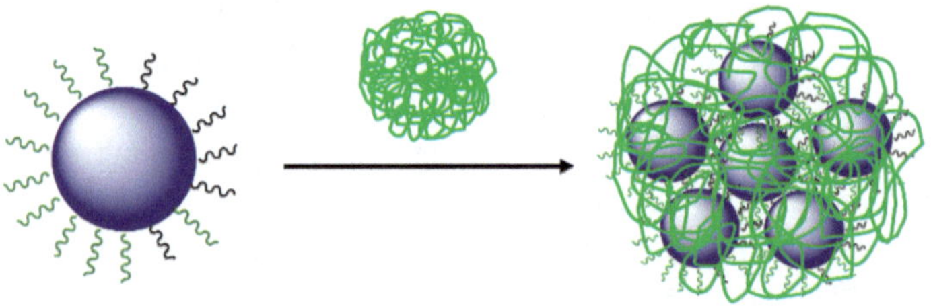

Figure 70: Encapsulation of nanoparticles in a polymer layer.

3.6.2 Silica coating

Silica shells are thin layers of silicon oxide (SiO_2) used to coat or encapsulate nanoparticles, such as SPIONs with a Fe_3O_4 or γ-Fe_2O_3 composition or AuNPs. The silica shell protects the core nanoparticle from degradation in cell media, is biocompatible, and significantly enhances the colloidal stability [334]. On the other hand, it attenuates the surface plasmons of AuNPs and the redox activity of SPIONs. APTES-functionalized silica shells, on the other hand, enable the electrostatic adsorption of biomolecules [335]. Silica shells can be synthesized using a broad variety of methods, including hydrothermal synthesis, sol-gel synthesis, and vapor deposition [336]. The most successful synthesis strategies are based on the Stöber method [174, 336–342]. However, only a few are suited for silica coating with layer thicknesses around 5 nm. More desired are core-silica shell nanoparticles with sizes smaller than 50 nm [343], biocompatibility, good water solubility, and colloidal stability in cell media. In recent years, two different modified Stöber methods have become established for silica coating of magnetite, ferrite, and gold nanoparticles [344]. The direct method, most similar to the Stöber method, is suitable for silica coating of hydrophilic nanoparticles in aqueous ethanolic solutions. The reverse emulsion technique, on the other hand, has been optimized for silica coating of hydrophobic nanoparticles [337, 338, 340, 343]. Moreover, the seeded sol-gel growth method enables precisely adjusting the silica-layer thickness.

In the slightly modified Stöber method, a solution of a silica precursor, usually tetraethyl orthosilicate (TEOS), is mixed with an aqueous dispersion of nanoparticles to be coated [340, 345–347]. The concentration of nanoparticles in dispersion is between 0.1 and 2 %, while the TEOS concentration in ethanolic solution varies between 1 and 10 %. Silica formation is initiated by the addition of an alkaline catalyst NH_3 or KOH, where the pH value is in the range of 11–12. Ideally, the nanoparticles act as nuclei for silica formation so that silica grows as a homogeneous layer on the surface. Suitable nanoparticles are citrate-stabilized SPIONs (with Fe_3O_4 or γ-Fe_2O_3 composition) or AuNPs. The SPIONs are prepared by alkaline co-precipitation with AuNP according to the Turkevich synthesis [135, 136, 205]. The as-synthesized AuNPs are electrostatically stabilized with citrate anions. On the other hand, citrate-terminated SPIONs are obtained after being heated in an alkaline (pH = 10) aqueous citric-acid solution [187, 188]. The negatively charged citrate anions do not only stabilize aqueous AuNP and SPION colloids through repulsive electrostatic interactions but also endow the nanoparticle surface with reactive hydroxyl groups. Under optimal reaction conditions, the silica precursor reacts with the hydroxyl groups of the citrate ligands to form a thin, uniform layer of silica around the nanoparticles (Figure 71).

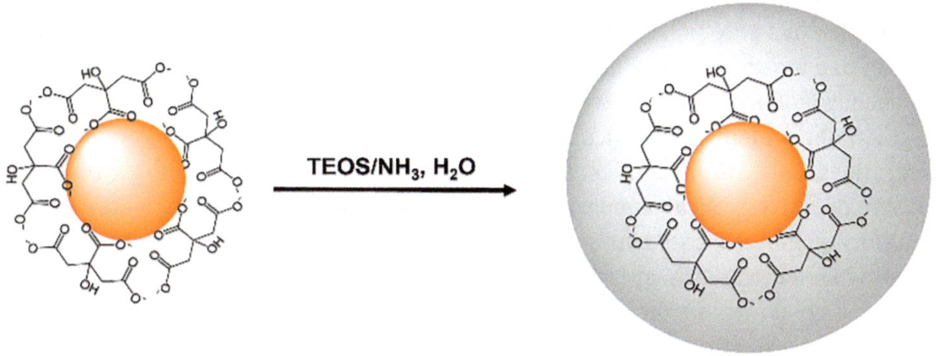

Figure 71: Schematic illustration of the silica shell synthesis around AuNPs.

The silica layer formation occurs via a two-step process (Figure 72). Silicic acid forms through alkali-catalyzed hydrolysis of TEOS and subsequently condenses to silica. The kinetics of the condensation reaction leading to the three-dimensional cross-linking of the Si–O network is determined by the concentration of the base, the temperature, and the TEOS concentration.

$$Si(OC_2H_5)_4 + 4\,H_2O \xrightleftharpoons{OH^-} Si(OH)_4 + 4\,C_2H_5OH$$

$$Si(OH)_4 \rightleftharpoons SiO_2 + 2\,H_2O$$

Figure 72: Reaction scheme of the alkaline hydrolysis of TEOS to silicic acid followed by condensation of silicic acid silica.

The drawbacks of this modified Stöber method arise from nonselective condensation reactions of the silanol groups with the OH groups of the citrate ligands, which competes with silica formation by pure silicic acid polycondensation. In case of AuNPs another problem arises from the insufficient stabilization by citrate ligands which gives rise to the formation of agglomerates. Hence, the as-synthesized silica shell-core nanoparticles are often multinuclear or core-free, have sizes much larger than 100 nm and are polydisperse.

Mine et al. [346] systematically investigated the reaction conditions for the synthesis of silica shells around citrate-stabilized gold nanoparticles with sizes of about 15 nm. Thus, they investigated the influence of the sequence of additions of the ammonia and TEOS solutions as well as the TEOS and ammonia concentrations on the size distribution and the silica shell layer thickness. The authors demonstrated by TEM studies that the interparticle repulsions and the morphology of the silica shell AuNP can be optimized by the timing of the addition of the ammonia solution. The aqueous AuNP colloid was first reacted with TEOS. This was followed by alkaline hydrolysis of TEOS with the ammonia solution. In this way, polycondensation of silica occurred preferentially on the AuNP surface and mostly spherical nanoparticles were formed, which did not show aggregation behavior. Furthermore, it was shown that the thickness of the silica layer depends on the water content in the reaction solution as well as on the TEOS and ammonia concentrations. Thus, the layer thickness increased with the TEOS concentration, while the amount of coreless silica nanoparticles disappeared with increasing ammonia content. The silica shell-AuNP core nanoparticles (SiO_2@AuNPs) obtained by this optimized sol-gel growth method reach layer thicknesses between 30 and 90 nm.

The second method, the reverse microemulsion synthesis, uses reverse micelles as nano-reactors for silica coating of nanoparticles [335–337, 348]. The reverse micelles spatially confine the hydrolysis and condensation reaction of lipophilic silanes (e. g., trialkoxysilanes) and thus the growth of silica shells around nanoparticles. The size of the reverse micelles can be adjusted by the concentrations of ammonium hydroxide, the respective silane monomer, and by the diameter of the nanoparticles. Water-in-oil microemulsion techniques facilitate silica coating of hydrophobic SPIONs, ferrite nanoparticles, and AuNPs with tunable surface layer thicknesses and tailored surface functionalities [343, 346, 348, 349]. The nanoparticles are first synthesized and are preferably stabilized with oleic acid or oleylamine. Silane coupling reagents such as CTAB, sodium bis(2-ethylhexyl) sulfosuccinate (AOT), and poly(4)oxyethylene-4-nonylphenyl ether (Igepal CO-520) are used for the transfer of hydrophobic nanoparticles to the polar phase [337, 350]. Ding et al. [343] and Zhang et al. [342] optimized the reverse emulsion method for silica coating of SPIONs by investigating the influence of the reaction parameters on the growth of the silica shell. They systematically varied the concentrations of the silane coupling agent Igepal CO-520, hydrophobic SPIONs, and ammonium hydroxide (25 %–33 %), the reaction time, and the mixing process. Silica-coated SPIONs with shell thicknesses between 2 and 20 nm were obtained via a two-step synthesis procedure. First, oleic acid-stabilized SPIONs with sizes around 10 nm were synthesized by thermal decomposition of $Fe(acac)_3$ or $Fe(oleate)_3$ under argon atmosphere in the presence of oleic acid and oleylamine at temperatures around 300 °C. In a typical reverse emulsion synthesis, Igepal CO-520 was dispersed in cyclohexane by sonication and mixed with a cyclohexane solution of SPIONs. The amphiphilic Igepal CO-520s interlocked with the alkyl residues of the oleic acid ligands via their nonane tails, so that the hydroxyl-terminated polyoxy residues faced outward (Figure 73). The reverse microemulsion (water-in-oil) was generated by adding an aqueous ammonium hydroxide solution (25–33 %) to the cyclohexane mixture and subsequent mixing in an ultrasonic bath or under vigorous stirring. Inverted micelles were formed, each containing a SPION. The concentration ratio of Igepal CO-520 to ammonium hydroxide was found to determine the size of the micelles. Appropriately, the SPION surfaces consisted of an oleic acid–Igepal CO-520 bilayer with terminal hydroxyl anions and adsorbed water molecules. TEOS was then added. At the $H_2O/^{-}OH$ surface of the micelles, TEOS was hydrolyzed to $Si(OH)_4$, and the silica shell was formed by the subsequent ^{-}OH-catalyzed polycondensation. Ding et al. [343] introduced the so-called equivalent fractionated droplet method for the addition of TEOS to fine-tune the thickness of the silica-shell. The authors were able to show that the thickness of the silica shell can be adjusted by the concentration ratio of ammonia to Igepal CO-520, the ammonia and TEOS concentration, and, in particular, by the number of fractional TEOS additions and the time intervals between them. Figure 74 displays TEM images of SiO_2@SPIONs obtained from

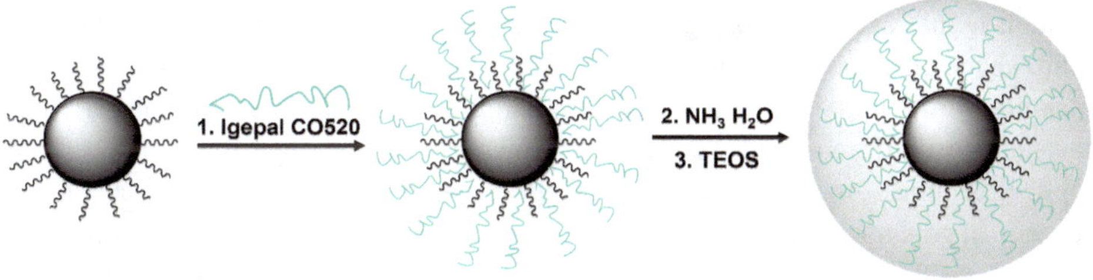

Figure 73: Schematics of the Igepal CO-520-mediated reverse emulsion synthesis of a silica shell around oleic-acid stabilized SPIONs.

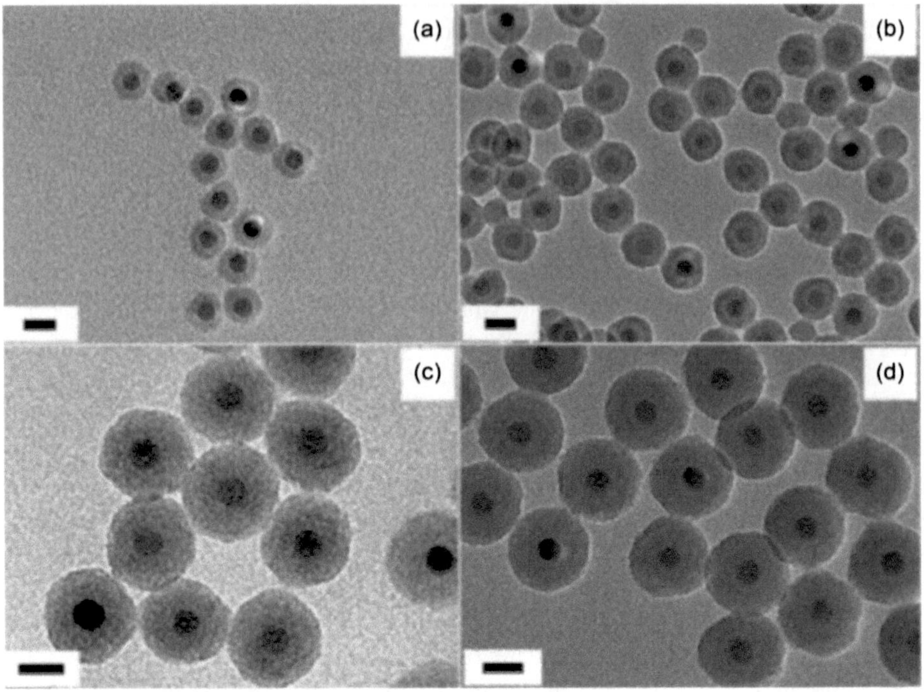

Figure 74: TEM images of SiO_2@SPIONs with various shell thicknesses: (a) 7 nm, 75 µL TEOS (b) 8.7 nm, 150 µL TEOS (c) 13.3 nm, 300 µL TEOS (4 fractionated additions), and (d) 14.9 nm, 600 µL TEOS (8 fractionated additions); scale bar = 20 nm [343].

silica coating of 12.2 nm-sized SPIONs. The reverse emulsion synthesis was performed with a constant concentration ratio of Igepal CO-520 to ammonia. TEOS was added at volumes ranging from 150 to 600 µL either as a single dose (Figures 74(a) and (b)) or divided into four to eight doses of TEOS (Figures 74(c) and (d)).

The reverse microemulsion technique has also proven to be successful for silica coating of AuNPs and AgNPs [336, 337, 351]. Hydrophobic AuNPs are typically synthesized by reduction of gold acid through oleylamine in organic solvents such as chloroform and toluene or in water. However, oleylamine-stabilized AuNPs form unstable colloids. Ligands with reactive hydroxy groups are required for polycondensation of TEOS and other silane alkoxides on nanoparticle surfaces. Ligand exchange for amphiphilic surfactants with terminal hydroxy groups was found to be impossible. Han et al. [351] used the silane coupling agent Igepal CO-520 for the formation of a hydroxy-group containing shell around 12-nm AuNPs. Reverse micelles were formed by mixing the cyclohexane solution of gold nanoparticles coated with Igepal CO-520 with an aqueous ammonia solution. The silica shell was formed by one or two additions of TEOS. The authors could adjust the shell thickness by varying the reaction time, TEOS concentration and number of TEOS additions. A single TEOS addition (20 µL) and a reaction time of 5 h resulted in silica-coated AuNPs with sizes around 24 nm which reached sizes of 35 nm after another 10-h reaction time (Figures 75(a) and (b)). AuNP sizes of 50, 65, and 80 nm were achieved by additions of 50, 100, and 200 µl of TEOS, respectively, and reaction times of 24 h (Figures 75(c)–(f)).

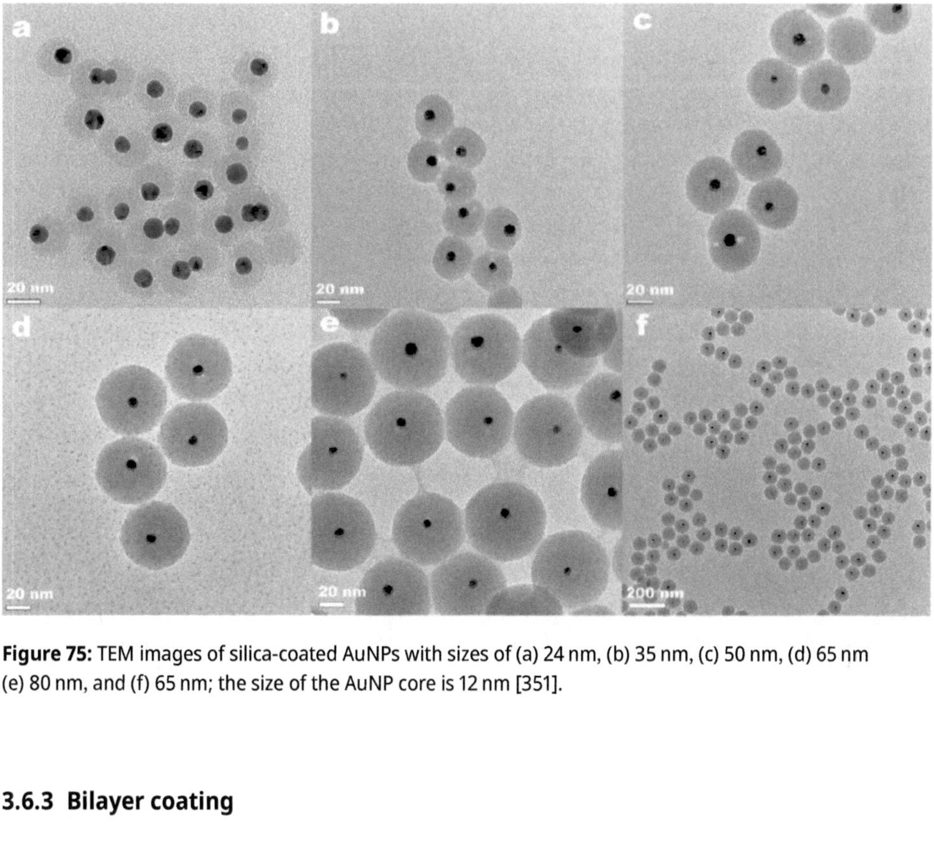

Figure 75: TEM images of silica-coated AuNPs with sizes of (a) 24 nm, (b) 35 nm, (c) 50 nm, (d) 65 nm (e) 80 nm, and (f) 65 nm; the size of the AuNP core is 12 nm [351].

3.6.3 Bilayer coating

In post-synthetic functionalization methods, among ligand exchange, ligand addition of amphiphilic compounds is employed to transfer hydrophobic nanoparticles to the aqueous phase [157, 352–355]. Hydrophobic SPIONs, ferrite, and noble-metal nanoparticles are usually prepared by thermal decomposition of organic metal complexes and are mostly surface stabilized with oleic acid and oleylamine. While oleic acid covalently binds to the SPION or ferrite surface via its carboxy-head group (chemisorption), oleylamine adsorbs most likely electrostatically and more weakly to SPION and AuNP surfaces. In the case of AuNPs, alkyl thiols, such as octadecanethiol, which are covalently bound via Au-S bonds, are better suited as primary ligands [356]. Hydrophobic C18 alkyl chains provide solubility of the nanoparticles in nonpolar solvents. Amphiphilic compounds consisting of a hydrophilic head group and a long alkyl chain tail are dissolved in a polar solvent (e. g., water, $CHCl_3$, acetonitril) that either contains the dispersed hydrophobic nanoparticles or are mixed with an organic nonpolar (*n*-hexane) solution of hydrophobic nanoparticles. The lipophilic alkyl chains of the amphiphilic surfactants act as secondary ligands by interdigitating with the lipophilic alkyl chains of the primary ligands through hydrophobic interactions. This self-assembly process results in an ordered bilayer whose surface consists of the hydrophilic head groups of the secondary ligands. The cell membrane-like bilayer enables the transfer of the nanoparticles into the polar or aqueous phase [246, 332, 353, 357]. Moreover, nanoparticles with highly well-ordered surface structures can penetrate tumor cells by direct permeation [246, 358, 359], which benefits the transport and release of drugs in tumor cells. Another advantage of bilayer coating is the preservation of optical and, for gold nanoparticles, of plasmonic properties. However, bilayer shells increase the hydrodynamic diameter of the nanoparticles and hamper the potentially desirable catalytic activities of the nanoparticle surfaces. In addition, amphiphilic compounds that are weakly bound to nanoparticle surfaces through electrostatic or hydrophobic interactions are steadily exchanged by free amphiphiles in the solvent and are thus in dynamic equilibrium with these free molecules [360]. As a consequence, multiple washes of bilayer-stabilized nanoparticles with pure solvents, as well as dilution of the nanoparticle suspensions, remove the amphiphilic surfactants from the surfaces. The nanoparticles then inevitably agglomerate. Accordingly, nanoparticle suspensions should contain a sufficiently high concentration of free amphiphiles to allow dynamic exchange of amphiphiles in the bilayer with those in the solvent.

Hydrophilic bilayer shells around nanoparticles arise through self-assembly of amphiphilic surfactants at the interface between nonpolar and polar solvent phases or in a polar aprotic solvent. Bilayer formation requires energy input by sonication for 5–60 min or vigorous stirring of the mixture for several hours [246, 352, 355, 357].

Li et al. [357] systematically investigated, and thus optimized, the reaction parameters for the formation of hydrophilic bilayers around 8-nm-sized SPIONs (Figure 76). For this purpose, they mixed aqueous solutions of various amphiphilic compounds and SPION *n*-hexane suspensions by sonication. Surfactant concentrations, sonication amplitude, and time were varied to optimize the bilayer architecture of 13 amphiphilic compounds. These amphiphiles (e. g., ricinoleic acid, lauric acid, oleic acid, sodium dodecylbenzene sulfonate, sodium monododecyl phosphate) consist of lipophilic alkane or alkene chains of various lengths and hydrophilic head groups, which are carboxy, phosphonate or sulfonate groups. The criteria for evaluating the bilayer coating are the hydrodynamic diameter, phase transfer efficiency, and long-term stability of the colloids. The aggregation behavior of the SPIONs coated with the differently composed bilayers was investigated using time-dependent dynamic light scattering. The hydrodynamic diameter, hydrophilicity of the bilayers, and, therewith, the phase-transfer efficiency were found to depend on the alkyl chain length and sonication power. The phosphonate headgroup provides the best water solubility for the SPIONs. Complete bilayer formation was ascribed to hydrodynamic diameters smaller than 30 nm, while larger hydrodynamic

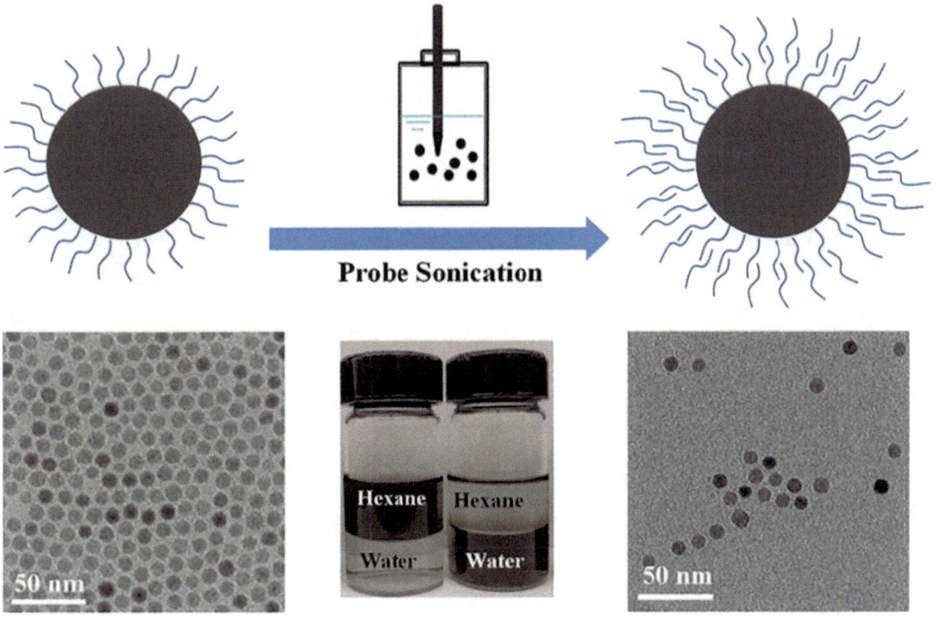

Figure 76: Schematic representation for the phase transfer of 8-nm-sized SPIONs from the *n*-hexane solution to the aqueous solution driven by the bilayer formation process [357].

diameters indicated agglomeration. Long alkyl chains of the secondary ligands (lauric acid, oleic acid) cause efficient hydrophobic and van der Waals interactions between the alkyl chains of the first and second layer.

Prakash et al. [332] prepared fatty-acid bilayer-stabilized SPIONs and CdSe nanocrystals using oleic acid as the bilayer-forming amphiphile and phase-transfer agent. The authors dispersed oleic acid-stabilized 8-nm SPIONs and 4-nm CdSe nanocrystals in *n*-hexane, added oleic acid at different concentrations and mixed the *n*-hexane solution for 1 min. The phase transfer of the nanoparticles from the organic to the aqueous phase was achieved by sonication for 5 min. As shown by dynamic light scattering and small-angle X-ray scattering (SAXS) studies, the optimal oleic acid concentration for the formation of a complete hydrophilic bilayer associated with maximum phase-transfer efficiency was slightly smaller than the critical micelle concentration (0.2 wt %). The thickness of the oleic acid bilayer around the nanoparticles ranged 2.2–2.6 nm. The efficiency of the phase transfer reached a value of 70 %.

An equally successful example of the formation of highly ordered hydrophilic bilayer shell was achieved by 30-min ultrasonic treatment of an aqueous dispersion of oleic acid-stabilized Au–Fe_3O_4 NHDs with sodium 4-dodecylbenzenesulfonate or 1-octadecylpyridinium bromide [246] (Figure 77). Both amphiphiles indented with equal efficiency via their long alkyl tails with the primary oleic acid ligand, as evidenced by the formation of 8.5 nm thick, highly ordered bilayer shells. The surface charges of the bilayer-stabilized Au–Fe_3O_4 NHDs were shown to decide over the cellular entry, intracellular trafficking, and cytotoxicity. The negatively charged (4-dodecylbenzenesulfonate) bilayer-stabilized Au–Fe_3O_4-NHDs were internalized in cancerous MCF-7 cells by direct permeation but spared the healthy MCF-10 cells. In comparison, the positively charged (1-octadecylpyridinium) Au–Fe_3O_4-NHDs were taken up by both cancerous MCF-7 cells and healthy MCF-10A cells via endocytosis and were found to be cytotoxic due to depolarization of the mitochondrial membrane potential.

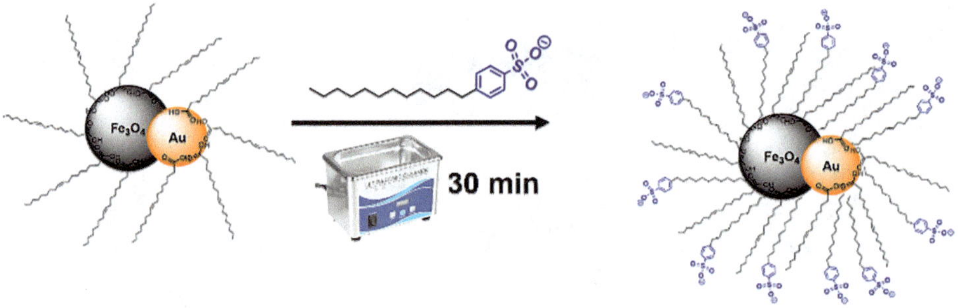

Figure 77: Illustration of the sonication-mediated formation of the 4-dodecylbenzenesulfonate-oleic acid bilayer around Au–Fe_3O_4 NHDs.

3.6.4 Ligand exchange

One of the most commonly used functionalization strategies is single- or multiple-ligand exchange: The primary ligand is replaced by other compounds that have a higher affinity to the nanoparticle surface and/or are added to the colloidal solution in excess concentration [354, 356, 361–365]. Compared to other functionalization processes, ligand-exchange reactions proceed in a very controlled manner, thus enabling a high order of the surface structure and the maintenance of a small hydrodynamic diameter. Ligands consisting of short-chain or aromatic compounds are shorter than 2 nm and thus do not significantly increase the hydrodynamic diameter, but provide sufficient steric repulsion between the nanoparticles for colloidal stabilization. Ligand exchange takes place either in a solvent phase or, in the case of hydrophobic nanoparticles, in a two-phase mixture. The exchange of ligands on the nanoparticle surface requires a high chemical affinity of the anchoring group of the secondary ligand. These reactions can occur through a variety of mechanisms, such as the creation of covalent bonds between AuNPs and thiol-containing compounds, or through coordination chemistry where the new ligands form coordination bonds with the metal cations of the SPION or ferrite nanoparticle surfaces. Ligand-exchange reactions are carried out by stirring or sonicating a one- or two-phase mixture containing the nanoparticles and the secondary ligand in an excess concentration.

Oleic acid binds covalently to the SPION surface via their carboxyl group, while the alkyl chains mediate self-aggregation of the oleic acid ligands on the surface via hydrophilic interactions. The stable binding of the oleic acid ligands makes complete ligand-exchange impossible. Davis et al. [362] studied the exchange of radiolabeled ^{14}C-oleic acid ligands on SPION surfaces for various types of amphiphilic compounds. They analyzed the ligand-exchange reaction using a radiotracer technique, liquid scintillation counting (LSC), and inductively coupled plasma mass spectrometry. The compounds used as secondary ligands included caffeic acid, citric acid, APTES, dopamine sulfonate, phosphonic acid, DMSA, and 2-mercaptoethanol. As the results from liquid scintillation counting experiments showed, the proportion of oleic acid ligands remaining on the SPION surfaces ranged from 18 % (caffeic acid) to 90 % (2-mercaptoethanol). These findings confirm the results of Schechtel et al. [366] obtained from studies of ligand-exchange reactions on TiO_2 nanoparticle surfaces. These authors elucidated the exchange mechanism of the primary oleic acid ligands against various catechol derivatives by means of ^{1}H and ^{13}C NMR spectroscopy. Accordingly, the catecholates displace the oleic acid ligands only partially (approx. 20 %), resulting in a mixed ligand shell. The adsorption of the catecholates on the TiO_2 surface was explained by the Langmuir adsorption model, while the oleic acid desorption exhibits Arrhenius-like behavior.

For oleic acid-stabilized SPIONs, catechol-derivatized compounds (e. g., caffeic acid, quercetin, dopamine), sulfonates, thiols, carboxylic acids, silanes, and phosphonic acid derivatives are suited secondary ligands for surface functionalization [362, 365]. Hydrophilic ligands with catechol anchor groups as allowing bidentate binding geometry, result in highly efficient chemisorption. Ligand-exchange reactions can occur through a variety of mechanisms, such as electrostatic interactions between the nanoparticles and the new ligands, or through coordination chemistry where the new ligands form coordination bonds with the iron atoms on the nanoparticle surface. Korpany et al. [365] thoroughly investigated the influence of the chemical structure of benzoic acid and catechol derivatives on their efficiency in replacing oleic acid and oleylamine ligands on SPION surfaces. The ligands used here possessed a variety of functional groups such as carboxylic acid, hydroxy, catechol, and amine moieties. Using FTIR attenuated reflection spectroscopy and the solubility behavior of the SPIONs, the authors were able to show that the binding mode depends on the positions of the hydroxyl and carboxy groups (Figures 78 and 79). On the one hand, dihydroxy benzoic acid (DHBA) isomers including 2,3-, 2,4-, and 3,5-DHBA bind via their carboxy groups (CA) in the bridging mode, whereas 2,5-DHBA binds by salicylate bridging and carboxylate chelating. On the other hand, in the case of 3,4-didoxybenzoic acid and dopamine, covalent binding to the SPION surface occurs via their catechol group.

The ligand exchange reactions were carried out as follows. The desired secondary ligand was shaken with 5 mg of SPIONs in dichloromethane overnight (~18.5 h) at room temperature using a vortex mixer. The success of the ligand exchange reaction was evaluated by TEM images. The stability of the aqueous colloids of the DHBA functionalized SPIONs was revealed by the absence of aggregates.

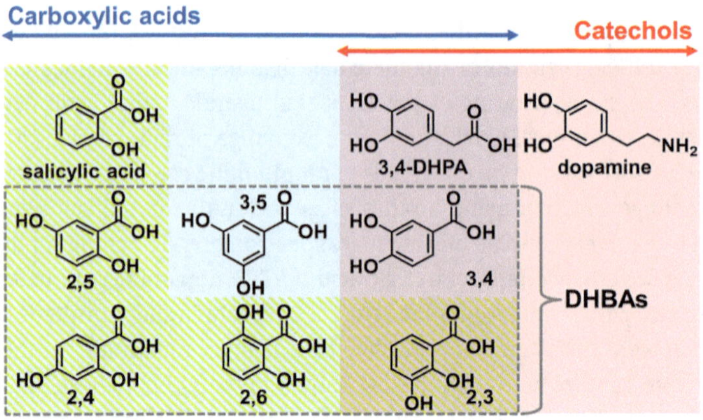

Figure 78: Scheme of positional isomers of dihydroxybenzoic acid (DHBA), salicylic acid and dopamine used for replacement of oleic acid and oleylamine ligands on the SPION surface [365].

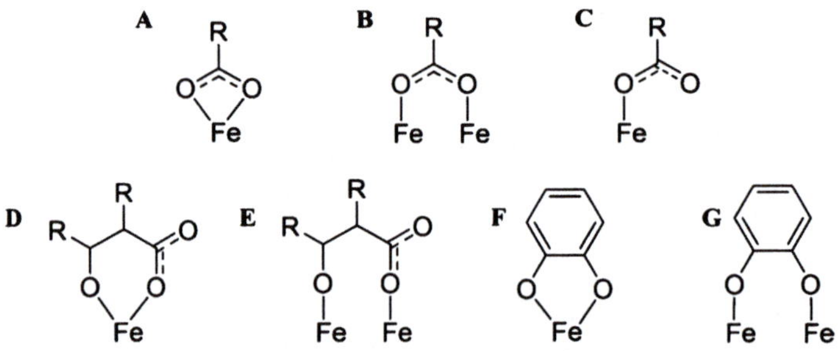

Figure 79: Carboxylate: (A) bidentate chelate, (B) bidentate bridging, (C) monodentate coordination, (D) bidentate mononuclear "salicylate" binding, (E) bidentate binuclear salicylate binding, (F) bidentate mononuclear catechol chelate binding, (G) bidentate binuclear catechol binding [365].

For biomedical applications, oleic acid-stabilized Au–Fe_3O_4 NHDs were transferred into the aqueous phase by performing ligand-exchange reactions with $NOBF_4$ and several catechol derivatives such as 2,3-DHBA, caffeic acid, tiron, and quercetin [233, 234, 244, 247]. The ligand- exchange reactions were carried out by dissolving the hydrophobic Au–Fe_3O_4 NHDs in *n*-hexane and mixing with aqueous solution of the respective catechol derivative or an acetonitrile solution of $NOBF_4$ in an ultrasonic bath for 30–60 minutes followed by stirring for several hr (Figure 80). As shown in the photographs in Figure 80, the secondary ligands act as phase-transfer agents and transfer the Au–Fe_3O_4 NHDs to the aqueous phase. Klein et al. [233] demonstrated by FTIR absorption spectroscopy and zeta potential measurements that oleic acid only partially dissociated from the Au–Fe_3O_4 surface during the ligand exchange process, which confirms the results in the literature [362]. The NO^+ and BF_4^- were rationalized to adsorb electrostatically to the Fe_3O_4 and Au surfaces [364]. Caffeic acid, on the other hand, was preferably attached to

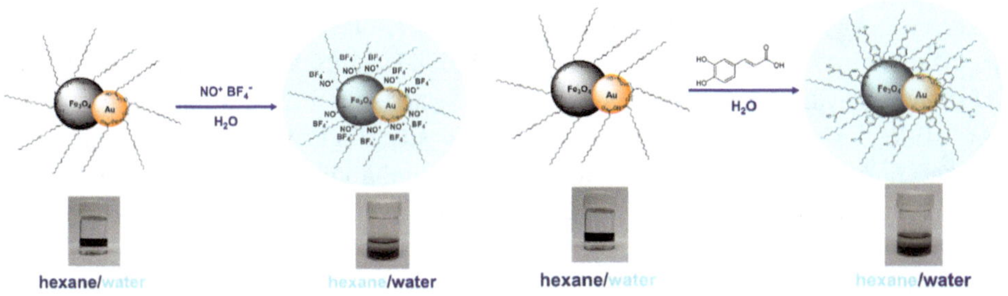

Figure 80: Exchange of oleic acid ligands of Au–Fe_3O_4 NHDs for $NOBF_4$ (left) or caffeic acid (right) by 30 min sonication and stirring overnight.

the Fe_3O_4 surface via the carboxyl group, while the catechol moiety of DHBA, quercetin, and tiron anchored via bidentate bridging to the Fe_3O_4 surface [233, 234, 247].

Depending on the synthesis route, the AuNPs obtained were endowed with hydrophilic surface structures such as citrate anions or hydrophobic ligands such as oleylamine, triphenylphosphine (TPP) or trioctylphosphine oxide (TOPO) [356, 367, 368]. TPP-stabilized AuNPs had sizes smaller than 2 nm. On the other hand, citrate-stabilized AuNPs can be produced with sizes in the range of 10–150 nm and are water-soluble. Functionalization of AuNP surfaces with thiolates is preferred because the sulfide anions form a coordinate covalent bond with the metal surface. However, the exchange with thiol derivatives is driven by the formation of covalent Au-S bonds in aprotic solvents. Therefore, hydrophobic, weakly physisorbed ligands such as oleylamine are best suited for ligand exchange with thiolates. In AuNP synthesis, oleylamine acts not only as a surfactant but also as a solvent and reducing agent at temperatures above 80 °C. The oleylamine-based synthesis strategy enables the production of AuNPs whose sizes can be adjusted in the range of 4 to 22 nm via reaction time and temperature [367]. Yang et al. [368] demonstrated that oleylamine-stabilized AuNPs underwent complete ligand ex-

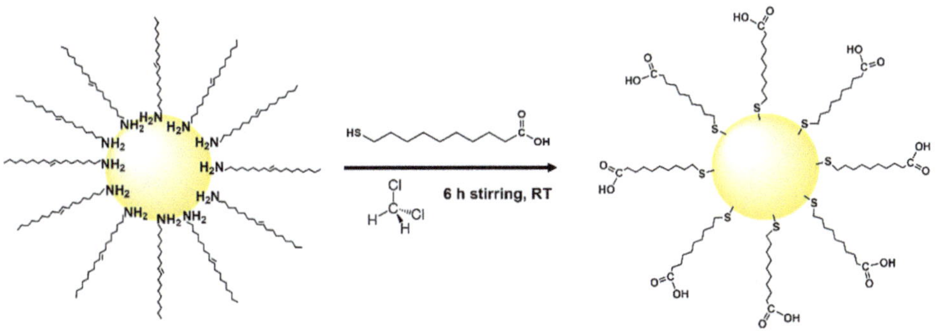

Figure 81: Ligand-exchange reaction of oleylamine-stabilized AuNPs; the primary ligands are replaced by 11-mercaptoundecanoic acid (MUA) on AuNP surfaces by stirring a reaction mixture of oleylamine-stabilized and MUA in DCM.

change against hydrophilic thiol derivatives such as 11-mercaptoundecanoic acid (MUA) and 11-mercapto-1-undecanoic sulfonic acid (MUS). For this, a dichloromethane (DCM) solution of oleylamine-stabilized AuNPs was injected into a dichloromethane solution of the thiol derivative. The reaction mixture was stirred at room temperature for 6 hr. The reaction was terminated by adding ice-cold methanol or acetone (Figure 81). The thiolated AuNPs were separated by repeated centrifugation.

4 Cell response on nanoparticles

4.1 Protein corona

Nanoparticles take several potential routes of entry into the human body including inhalation, ingestion, injection, or absorption through the skin [369]. During their journey through blood vessels, lymph, and cytoplasm, nanoparticles encounter a broad variety of proteins (e. g., albumin, transferrin, gamma globulin, and fibrinogen) and other biomolecules [370–373]. Competitive and cooperative adsorption of different protein species result in the formation of a dynamic protein corona around nanoparticles. "Dynamic" implies that the composition of the protein corona is continuously changing: Initially bound low-affinity, highly abundant proteins are successively exchanged for less abundantly occurring proteins that bind more strongly to nanoparticle surfaces [374, 375]. This implies that the binding affinity dictates the final selection of proteins. First, competitive protein adsorption results in the formation of a hard corona with proteins tightly bound to the nanoparticle surface [375–377]. Weakly bound proteins on top of the hard corona constitutes the soft corona [378, 379] as they are steadily exchanged with those in the immediate environment (Figure 82) [380]. The sequential exchange of different proteins on surfaces over time defines the so-called Vroman effect [380, 381]. Protein adsorption is a kinetic and thermodynamic function of both the individual proteins and nanoparticle properties such as surface modification, composition, and diameter [371, 372, 382–386]. Marichal et al. [384] studied the protein-exchange process on nanoparticle surfaces using Langmuir adsorption isotherms by fitting them to the nanoparticle–protein dissociation process. Protein dissociation from multilayer protein corona on heterogeneous nanoparticle surfaces is better fitted with Langmuir–Freundlich isotherms [385]. Spinozzi et al. [386] studied the association–dissociation

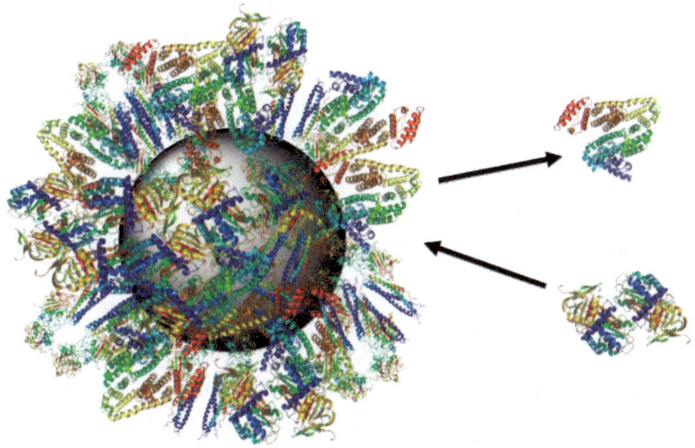

Figure 82: Schematic of nanoparticle–protein corona formation process.

https://doi.org/10.1515/9783110605495-004

equilibrium between nanoparticle-bound proteins and free protein molecules in solution within the framework of the Hill cooperative model. The nanoparticles and proteins under study are citrate-stabilized AuNPs (⌀ = 60 nm) and human serum albumin (HSA), respectively. They proposed that the surface structures (citrate) of the nanoparticles form the first shell, and the second (outer) shell is formed by the binding of proteins and their self-assembly into a monolayer. The Hill equation (Eq. (10)) is based on the following two assumptions: An association–dissociation equilibrium exists between proteins bound to an AuNP and free proteins in solution. The n proteins (L) of the monolayer dissociate simultaneously from n identical sites (S) of the AuNP surface:

$$\frac{1}{n}L_n S_n \leftrightarrow L + \frac{1}{n}S_n. \tag{10}$$

Here, the Hill coefficient n describes the cooperative binding behavior of proteins. For $n > 1$, the surface-bound protein increases the binding affinity of other proteins. This so-called positive adsorption cooperativity results from attractive hydrophobic interactions between proteins, which are manifested in high adsorption efficiency and the formation of a hard protein corona. In contrast, for $0 < n < 1$ (i. e., negative adsorption cooperativity), the binding affinity of the other proteins is reduced by the binding of the first protein, which prevents the formation of a protein corona. When $n = 1$, the proteins bind independently according to the classical Langmuir adsorption model and presumably form the soft protein corona. The dissociation equilibrium constant K_D depends on the concentrations of n protein-bound sites ($C_{L_nS_n}$), the free proteins in solution (C_L) and n free sites (C_{S_n}):

$$K_D = e^{\Delta G^0/k_B T} = C_L C_{S_n}^{1/n} / C_{L_nS_n}^{1/n}. \tag{11}$$

ΔG° denotes the dissociation enthalpy, and k_B and T are the Boltzmann constant and temperature, respectively. The authors investigated the structural and thermodynamic properties of AuNP-HSA conjugates and the dissociation process under physiological conditions using small-angle X-ray scattering (SAXS) and small-angle neutron scattering (SANS), systematically varying the concentrations of nanoparticles and proteins. The dissociation constant K_D, the Hill coefficient n, and the stoichiometry of the AuNP-HSA conjugates were determined by fitting the Hill adsorption isotherm to concentration-dependent X-ray and neutron scattering data simultaneously.

The protein corona of nanoparticles has been characterized using dynamic light scattering (DLS), microelectrophoresis (zeta potential measurements), transmission electron microscopy (TEM), differential centrifugal sedimentation (DCS), isothermal titration calorimetry (ITC), and mass spectrometry-based proteomics [369, 387, 388]. Proteomic analysis provides a deeper understanding of the molecular processes involved in the formation of protein coronas and the cellular responses to nanoparticle exposure [389]. The elucidation of the protein corona composition requires the complete detachment of the proteins from the nanoparticle surface and the separation of the proteins with high spatial resolution. Proteins can be quantified after separation from the nanoparticles using the bicinchoninic assay [373]. Two-dimensional sodium dodecyl sulfate-polyacrylamide gel electrophoresis (2D-SDS-PAGE) [390–393] can separate up to 10,000 different proteins from complex biological samples such as blood plasma and cell lysates in a single gel. Simultaneous quantitative visualization of protein profiles can be achieved by two-dimensional differential fluorescence gel electrophoresis (2D-DIGE). Although these qualitative analytical methods are suitable for fractionating a mixture of proteins based on differences in density, they provide only relative quantification. Precise identification and quantification of the corona protein composition is achieved by SDS-PAGE combined with liquid chromatography-tandem mass spectrometry (LC-MS/MS) [317]. In addition to mass spectrometry, other proteomics techniques, such as protein sequencing and peptide fingerprinting, can be used to provide more detailed information about the proteins in the corona, including their sequence and structure [392].

Blume et al. [394] developed an automated LC-MS/MS-based platform (Proteograph™) enabling the proteomic analysis of protein coronae formed by incubation of SPIONs in human blood plasma. The SPIONs used are 200–300 nm in size. The SPIONs were coated with either a thin layer of silica, poly(dimethylaminopropyl methacrylamide) (PDMAPMA), or poly(ethylene glycol) (PEG). These coatings differed in their surface structures and exhibited negative or positive surface charges (−36.9, +25.8, and −0.4 mV at pH 7.4). The automated Proteograph platform was validated using a panel of 46 differently coated SPIONs surrounded by differently composed protein coronae. In addition, the authors used these validated proteomic analyses to develop a SPION-based blood test for the detection of cancer-associated proteins. For this purpose, the blood test was applied to non-small cell lung cancer (NSCLC) patients and more than 2,000 proteins were quantified in 141 patients.

Figure 83 outlines the workflow of the Proteograph. Incubation of the different SPION samples in human blood serum is performed in 96-well plates (step 1). Subsequent to the protein corona formation, the SPIONs are separated from the rest of the human blood plasma by magnetic separation, washing, and resuspension (step 2). The SPION-bound proteins are then tryptically digested, purified, eluted (step 3), and subjected to quantitative identification by liquid chromatography-tandem mass spectrometry (LC-MS/MS) (step 4). The authors showed that the Proteograph performed the quantification of >700 protein groups in nine samples and more than 500 protein groups for each SPION type. Thus, the multi-nanoparticle protein corona strategy facilitates the identification of a wide range of plasma proteins, especially those of low abundance.

The formation of a protein corona is driven by non-covalent, nonspecific, and hydrophobic interactions between proteins and nanoparticles. Protein exchange on nanoparticles may be accompanied by conformational relaxation and reorientation processes [395–398]. In the worst case, nanoparticle surfaces induce protein unfolding which stimulates inflammatory signaling pathways or, at least, impairs the protein function. The extent of conformational changes depends on the curvature of the nanoparticles: The smaller the nanoparticles are, the larger is their curvature and the

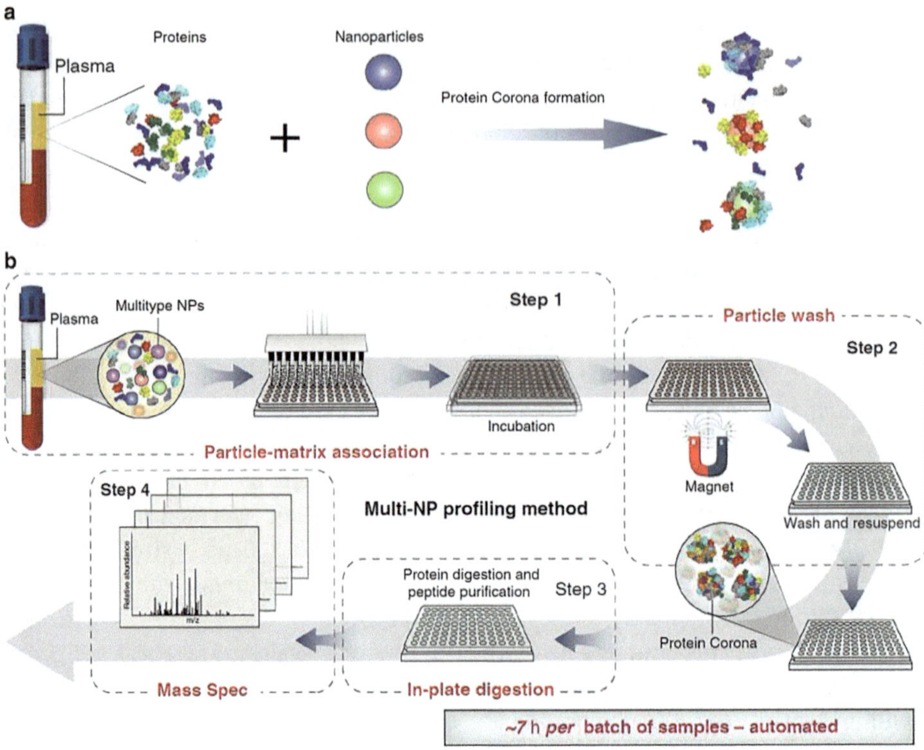

Figure 83: Workflow of the Proteograph: a) formation of a protein corona around three differently coated SPIONs by incubation in a 96 well plate; b) processing of protein coronae of diverse SPIONs followed by mass spectrometry-based plasma proteome analysis; step 1: formation of the protein corona by incubation of SPIONs in human blood plasma, step 2: SPION washing procedure using magnetic separation, step 3: digestion of corona proteins, step 4: protein identification by LC-MS/MS analysis [394].

less is their impact on the protein conformation. Nanoparticles with diameters smaller than 10 nm can only bind proteins such as HSA (4–8 nm) or immunoglobulin G (150 kDa, 10–15 nm) [399] in a ratio of 1:1, whereby the binding affinity is low due to the small contact surface [374]. Nanoparticles with diameters larger than 50 nm not only have a surface area many times larger (>25) but also a much smaller surface curvature and can thus bind 100 or more proteins. As Piella et al. [374] demonstrated, spherical AuNPs with sizes smaller than 12 nm adsorb few proteins, while AuNPs with sizes in the range of 12–80 nm are surrounded by a protein monolayer (Figure 84). AuNPs larger than 80 nm can form protein coronae with multiple layers. The authors investigated protein adsorption on citrate-stabilized AuNPs as a function of nanoparticle size in the range 3.5–150 nm. These AuNPs were incubated in cell culture medium supplemented with fetal bovine serum. AuNP-protein complexes, protein monolayers, and multilayers surrounding AuNPs were characterized and identified by UV-Vis absorption spectroscopy, dynamic light scattering, and zeta potential measurements. AuNPs smaller than 10 nm

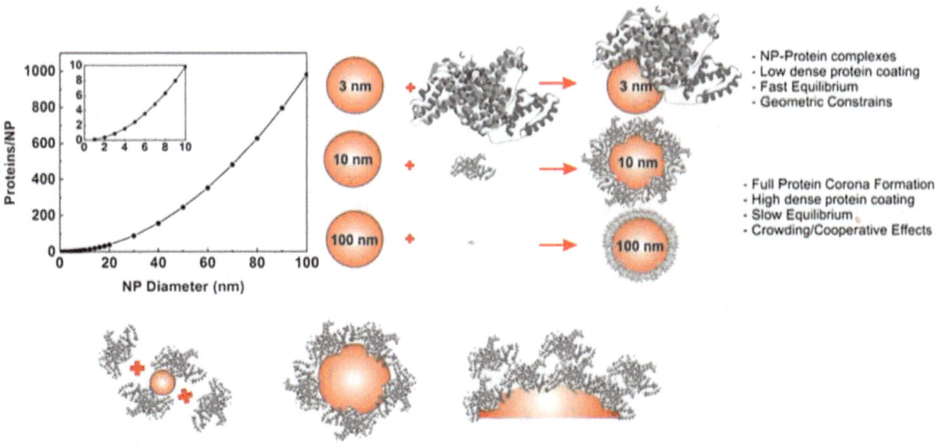

Figure 84: Formation of protein coronae around citrate-stabilized AuNPs with various sizes: (i) incomplete protein corona, (ii) the formation of a protein monolayer, and (iii) the formation of a multilayer protein corona. Graph (left) shows the number of proteins adsorbing on the surface of the AuNP as a monolayer [374].

are smaller than plasma proteins and therefore can bind only a few proteins. No protein corona is formed. Since such small AuNPs have a very large surface curvature, proteins are only weakly bound due to the small contact area. Conformational changes of the adsorbed proteins are therefore unlikely. Similarly, the weakly bound proteins have little influence on the mechanism of cellular uptake, which most likely occurs by direct permeation. Based on these considerations, AuNPs with sizes between 20 and 80 nm allowed the formation of a protein monolayer. Multilayer protein coronae were only detectable for AuNPs larger than 80 nm.

Roach et al. [400] unambiguously demonstrated that the adsorption characteristics and conformational changes of proteins on nanoparticle surfaces can be controlled by their surface chemistry and topography. Hydrophobic and hydrophilic silica nanoparticles (SiO_2 NPs) with sizes ranging from 15 to 165 nm were used for these studies. The hydrophilic SiO_2 NPs carried OH groups and thus negative surface charges in cell culture media, while hydrophobicity was achieved by surface-bound methyl groups. Bovine fibrinogen (BF) and bovine serum albumin (BSA) were used as model proteins. The globular serum albumin is the most abundant protein in the bloodstream and 6.4 nm × 5.5 nm × 5 nm in size. The rod-shaped plasma protein fibrinogen, measuring 4 × 46 nm, is involved in the blood coagulation process. The adsorption behavior and the organization of proteins into layers around SiO_2 NPs, as well as the surface-induced conformational changes of the proteins, were studied using FTIR-ATR and UV-Vis absorption spectroscopy. The amide I and amide II bands in the FTIR spectra of proteins are sensitive to changes in secondary structure and thus act as probes for protein conformational changes. The amide I band, which occurs in the range of 1,600–1,700 cm^{-1} is mainly attributed to the C = O stretching vibration (70–85 %), while the amide II band between

1,510 and 1,580 cm^{-1} is caused by the N–H bending vibration (40–60 %), C–N stretching vibration (18–40 %), and C–C stretching vibration (10 %). On the other hand, the adsorption isotherms of proteins on SiO_2 NPs surfaces can be monitored by measuring the protein absorbance at 280 nm, which is directly proportional to the concentration of free proteins in the solution. The concentration of adsorbed surface proteins corresponds to the difference between the total protein concentration and the concentration of free proteins. The binding affinity of BSA and BF was found to increase with the surface curvature. BSA binds more strongly to hydrophobic than to hydrophilic surfaces. However, the saturation behavior of the BSA adsorption isotherms on hydrophobic surfaces revealed the formation of an incomplete protein monolayer. The size of the nanoparticles, which is reciprocal to the surface curvature, affected the secondary structure of BSA and BF in different ways. Remarkably, BSA underwent progressively larger secondary structure changes by adsorption on SiO_2 NPs larger than 80 nm, leading to the loss of the helical structure. The extent of protein unfolding was enhanced on hydrophobic surfaces. In contrast, BF adsorption was largely controlled by surface curvature, while surface chemistry played a minor role. The conformational disorder of BF increased with decreasing nanoparticle size and was maximal for nanoparticles of 15-nm size.

Nanoparticle morphology, which is directly related to surface curvature, has also been found to influence protein affinity and conformation. For example, Gagner et al. [401] showed that gold nanorods (AuNRs) and spherical gold nanoparticles (AuNPs), which have similar diameters (about 10 nm), differ significantly in their adsorption capacity for the proteins lysozyme and α-chymotrypsin. The AuNPs loosely bind a few proteins, whereas the AuNRs (10 nm × 36 nm) are covered by a protein monolayer. The secondary structure of the proteins is altered, especially that of lysozyme. Furthermore, the surface chemistry of nanoparticles does not only control the selective adsorption of proteins but can also induce changes in the secondary structure of the adsorbed proteins [397]. Thus, the composition of the hard protein corona depends on the size, shape, and surface properties of the nanoparticles, which determine the specificity and affinity of protein binding. The initial surface chemistry of the nanoparticles determines protein adsorption. Positively and negatively charged nanoparticles differ in their affinity for protein species [396]. On the other hand, proteins adsorbed on the surface alter the physical and chemical properties of the nanoparticle, including its size, surface charge, and hydrophobicity.

Although there are several thousand proteins present in human biological fluids, the soft and hard coronas consist of fewer different proteins. The most abundant serum proteins are serum albumin, immunoglobulin G1 (IgG1), alpha-2 macroglobulin (A2M), and apolipoprotein A-1 (apoA1) [398, 402, 403]. The size and composition of the hard protein corona do not only determine the cellular uptake mechanism, but also guide the intracellular fate and control the efficacy of the nanoparticles. Nanoparticles encapsulated in a thick protein corona can be recognized and eliminated by the immune system, while a thinner protein corona, and thus a smaller nanoparticle size, enables the nanoparticles to circulate longer [381]. The composition of the protein corona can influence cellular uptake. Certain proteins interact with cell surface receptors and control nanoparticle internalization [404]. This implies that certain proteins in the corona may enhance cellular uptake by promoting receptor-mediated endocytosis, while other proteins may inhibit cellular uptake [325]. In addition, the protein corona significantly influences the in vivo organ distribution and elimination of nanoparticles.

One method is to functionalize nanoparticles with ligands that minimize nonspecific protein adsorption to preserve the nanomaterial initial surface properties and avoid, therewith, activation of random signaling pathways [324, 405, 406]. PEG, for example, has been broadly employed to this end. Its aptitude to resist protein adsorption is associated with steric repulsion and to the ability to form a strong hydrogen-bonding hydration layer. Nanomaterial functionalization with PEG usually improves the material stability and bioavailability. However, a disadvantage of PEGylation is that, for example, reactive surface properties are completely masked and thus become ineffective. In contrast, Verhoef et al. [331] showed that PEGylated nanoparticles adsorb serum proteins and other specific proteins to a significant extent. Moreover Brandenberger et al. [407] demonstrated that 13-nm PEG coated AuNPs induce acute inflammation and apoptosis in mouse livers.

Recent studies have shown that zwitterions, besides hemocompatible, are able to escape immune-system recognition, providing longer blood circulation times, to resist the formation of bacterial biofilms, and to avoid phagocytosis [408]. Therefore, an alternative method to the PEGylation is the functionalization of the nanoparticles surface with zwitterions, which have been shown to hinder the adsorption of nonspecific proteins [409–411]. Amphoteric nanoparticles were obtained from co-binding of positively and negatively charged ligands at the surface. Since zwitterionic surface chemistry does hinder protein adsorption, amphoteric nanoparticles enable functionalization or bioconjugation with anticancer drugs, small-molecule inhibitors, or miRNA that may be freely released to its intracellular target [412, 413]. However, further development of the fabrication of 'corona-free' nanoparticles is essential for their use as biosensors [408].

4.2 Cellular uptake machinery

The cellular uptake machinery is based on complex molecular mechanisms that enable cells to internalize nutrients, hormones, proteins, and nanoparticles [414]. Cellular uptake processes allow the maintenance of cellular homeostasis, regulation of signaling pathways, and defense against infectious agents. Cellular uptake occurs by recognition, binding, internalization, and sorting of incoming molecules or nanoparticles. Many cellular uptake mechanisms rely on receptor-mediated endocytosis, in which specific receptors on the cell surface recognize and bind the target molecule, resulting in the formation of specialized membrane invaginations called clathrin-coated pits [415–417]. Other mechanisms, such as phagocytosis and macropinocytosis, involve the uptake of large particles and extracellular fluids, respectively.

Nanoparticles interact via their protein corona with plasma membrane receptors responsible for cellular uptake, intracellular trafficking, and cell defense mechanisms [370, 418, 419]. Nanoparticle uptake mechanisms involve multiple processes that depend on the size, shape, and surface chemistry of the nanoparticle. These properties can influence the efficiency and specificity of the nanoparticle uptake, as well as the intracellular distribution and fate of the nanoparticles. Interaction of nanoparticles with cell membranes occurs through various mechanisms such as electrostatic interactions, hydrophobic interactions, and ligand-receptor interactions [404, 420]. Once bound to the cell surface, nanoparticles can be internalized through various endocytic pathways, including clathrin-mediated endocytosis, caveolae-mediated endocytosis, and macropinocytosis [407, 416, 418, 421–424].

Understanding the cellular uptake machinery of nanoparticles is critical for the design and development of effective and safe nanoparticle-based therapeutics. By targeting specific cell types and tissues, nanoparticles can enhance drug delivery, reduce toxicity, and improve therapeutic outcomes. However, the complexity of the cellular uptake machinery and the multitude of factors that influence nanoparticle uptake and distribution pose significant challenges to the development of nanoparticle-based therapeutics.

4.2.1 Cellular uptake pathways

Cellular plasma membranes of mammalian cells are composed of phospholipid bilayers. The amphiphilic properties of phospholipids make their bilayer structure an efficient selective barrier since a "balanced" ratio of hydrophobicity and hydrophilicity is required for a variety of small biomolecules to enter the cell by passive diffusion. Similarly, plasma membranes maintain cell homeostasis and ion concentration gradient, and control the entry and exit of charged small molecules, macromolecules, and nutrients (Figure 85) [425, 426]. Membrane proteins are embedded in the fluid matrix of this bilayer, which controls specific membrane functions. Specialized membrane pro-

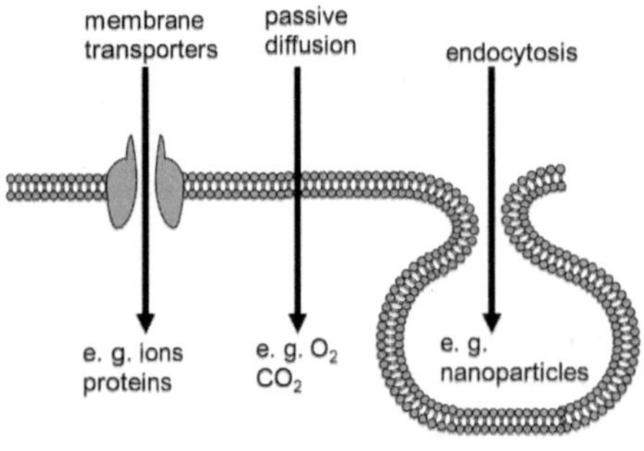

Figure 85: Different mechanisms of cell entry for nonpolar or polar small molecules, macromolecules, or nanoparticles.

tein channels and transporters regulate the uptake of essential ions and nanometer-sized proteins [426].

The topology, morphology, and surface chemistry of nanoparticles determine their interactions with the cell membrane and thus orchestrate the cellular uptake pathway. For example, proteins adsorbed on the nanoparticle surface can trigger plasma membrane receptor proteins that activate the cellular uptake machinery [373]. Cellular internalization may involve active (receptor-mediated) or passive transport across the cell membrane.

Nanoparticles that reach the outer plasma membrane will interact with components of the plasma membrane or extracellular matrix. They enter cells mainly by endocytosis [423, 427, 428] that starts with the entrapment of nanoparticles in membrane invaginations, followed by their budding and pinching off to form endocytic vesicles. These endosomes are then transported to specialized intracellular sorting and trafficking compartments. Depending on the cell type and the proteins, lipids, and other molecules involved in the process, endocytosis can be classified into several types: phagocytosis, clathrin-mediated endocytosis, caveolin-mediated endocytosis, clathrin and caveolae-independent endocytosis, and macropinocytosis. A minority of nanoparticles was found to cross the plasma membrane by non-endocytic pathways like passive diffusion, direct permeation or hole formation [359, 429]. There are also artificially induced mechanisms that enable nanoparticles to enter cells, such as microinjection [430] or electroporation [431].

4.2.1.1 Endocytosis

Endocytosis occurs through an invagination of the cell membrane that makes possible the uptake of nanoparticles. Membrane invaginations are eventually pinched off, form-

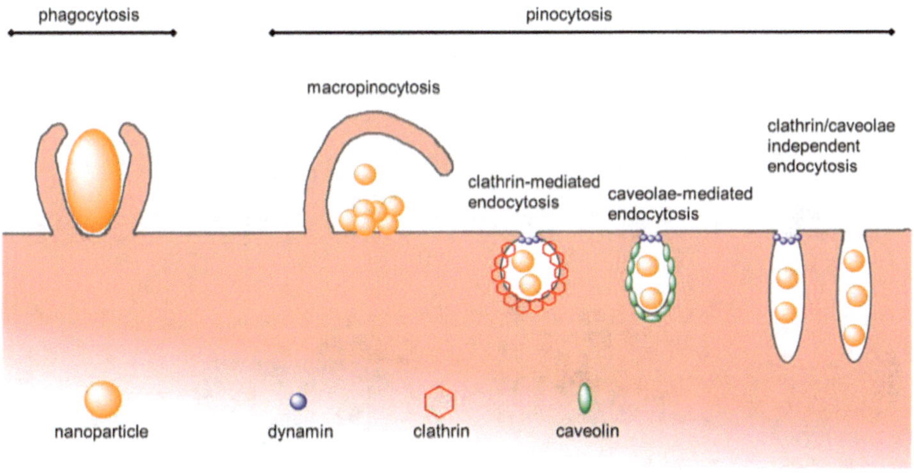

Figure 86: The main categories of endocytosis.

ing the characteristic endocytic vesicles with the nanoparticles inside (Figure 86) [432]. These vesicles participate in intracellular membrane trafficking. Two main categories of endocytosis may be distinguished: phagocytosis and pinocytosis. Phagocytosis is typically restricted to specialized cells (e. g.. macrophages, monocytes, neutrophils), whereas pinocytosis is more widespread and occurs in many cell types [432]. Pinocytosis can be divided into four distinct mechanisms: macropinocytosis, clathrin-mediated endocytosis, caveolae-mediated endocytosis, and clathrin and caveolae-independent endocytosis (Figure 86) [432, 433].

4.2.1.2 Phagocytosis

Particles larger than 500 nm, including microorganisms, foreign nanoparticles, and apoptotic cells, are engulfed by phagocytosis. Phagocytosis is essential for microbial elimination, but also for maintaining tissue homeostasis. Only specialized cells, professional phagocytes, perform phagocytosis with high efficiency. Professional phagocytes include a wide variety of white blood cells such as monocytes, macrophages, neutrophils, dendritic cells, osteoblasts, and eosinophils. In addition, the non-professional phagocytes, namely, fibroblasts, epithelial, and endothelial cells, can perform phagocytosis with low efficiency [434–436]. The main steps of phagocytosis can be described as: 1) recognition of the target particle, 2) activation of particle internalization, 3) phagosome formation, and 4) phagosome maturation (Figure 87).

The first step in phagocytosis is for phagocytes to recognize nanoparticles. Two types of specialized surface receptors help phagocytes recognize foreign nanoparticles. Non-opsonic receptors identify pathogen-associated molecular patterns not found in higher organisms [434, 435, 437]. Opsonic receptors interact with host-driven opsonins that bind to the surface of foreign particles. Opsonins include coagulation factors, im-

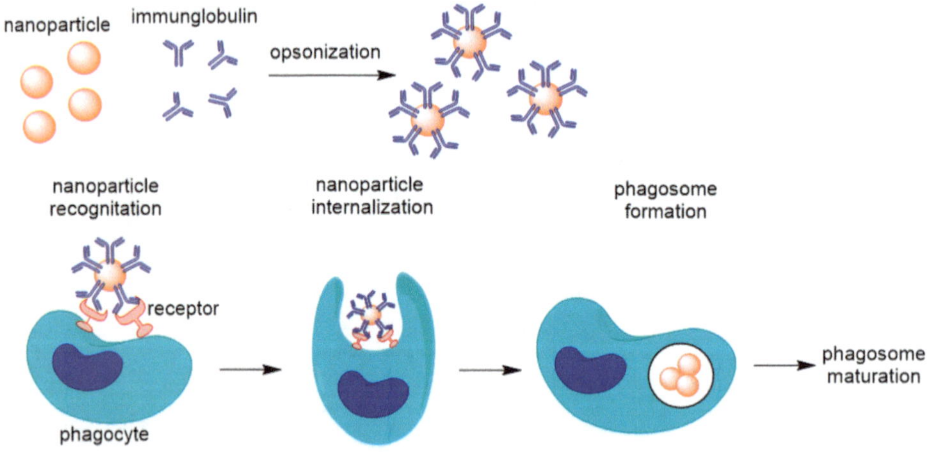

Figure 87: Phagocytosis of nanoparticles.

munoglobulins, apolipoproteins, cell adhesion mediators, and acute phase factors such as antibodies, milk fat globulin (lactadherin), and mannose-binding lectin [438]. Opsonization is a real challenge for the bioavailability of nanoparticles. Upon contact with blood, the binding of opsonins to the nanoparticle surface is followed by rapid recognition by the mononuclear phagocytic system. Nanoparticle characteristics such as size, surface chemistry, and charge influence opsonization. Once a nanoparticle is recognized by phagocyte receptors, multiple signaling pathways are triggered to initiate phagocytosis. This is followed by structural changes in the cell membrane and reorganization of the actin cytoskeleton [436]. A depression—the phagocytic cup—is formed in the area of the cell membrane in contact with the nanoparticle. Pseudopods emerge from the membrane surrounding the nanoparticle and cover the nanoparticle. After a few minutes, the distal ends of the pseudopods close, forming a new phagosome [439].

After internalization of the new phagosome, a process begins in which its membrane composition and contents change. This process is called phagosome maturation, at the end of which the phagosome becomes a new vesicle, the phagolysosome (Figure 88) [434, 435]. Initially, the new phagosome fuses with early endosomes. This fusion is regulated by the small GTPase Rab5 [440]. The early phagosome becomes slightly acidic (pH 6.1–6.5) due to the action of V-ATPase, which accumulates on its membrane. The slightly acidic phagosomes transiently fuse with more acidic vesicles. V-ATPase transports protons (H^+) into the lumen of the phagosome. However, the early phagosome is not acidic enough to activate the cathepsin family of acidic proteases needed to degrade the phagocytosed nanoparticle. Although the phagosome fuses with several early endosomes, the volume of the phagosome does not increase because new vesicles, the recycling endosomes, are simultaneously released [441]. Rab7 appears on the phagosome membrane after Rab5 recruitment to be included in the maturation sequence. As several V-ATPases accumulate on the phagosome membrane and pump protons into the phagosome lumen, the late phagosomes reach pH values in the range of 5.5–6.0. Furthermore, lysosomal-associated membrane proteins and luminal proteases from late endosomes are incorporated [442, 443]. In the final step of phagosome maturation, the late phagosome combines with lysosomes to form the phagolysosome. Phagolysosomes contain many degradative enzymes, including several cathepsins, proteases, lysozymes, and lipases required to degrade microorganisms and cellular debris. Due to the accumulation of several V-ATPases on the membrane, phagolysosomes are now very acidic ($pH < 4.5$). In addition, phagolysosomes carry NADPH oxidase complexes on their membrane that generate reactive oxygen species [444, 445].

In this context, magnetic iron oxide nanoparticles are classified into three types according to their size: (1) ultrasmall superparamagnetic iron oxide nanoparticles (USPIONs) 20–50 nm [446, 447], (2) superparamagnetic iron oxide nanoparticles (SPIONs) 50–150 nm [446, 447], and (3) micron-sized ferrimagnetic iron oxide nanoparticles (MPIONs) with diameters approaching or exceeding 1 µm [447–449]. Metz et al. [446] reported a significantly higher phagocytotic uptake of SPIONs compared to USPIONs. MPIONs are

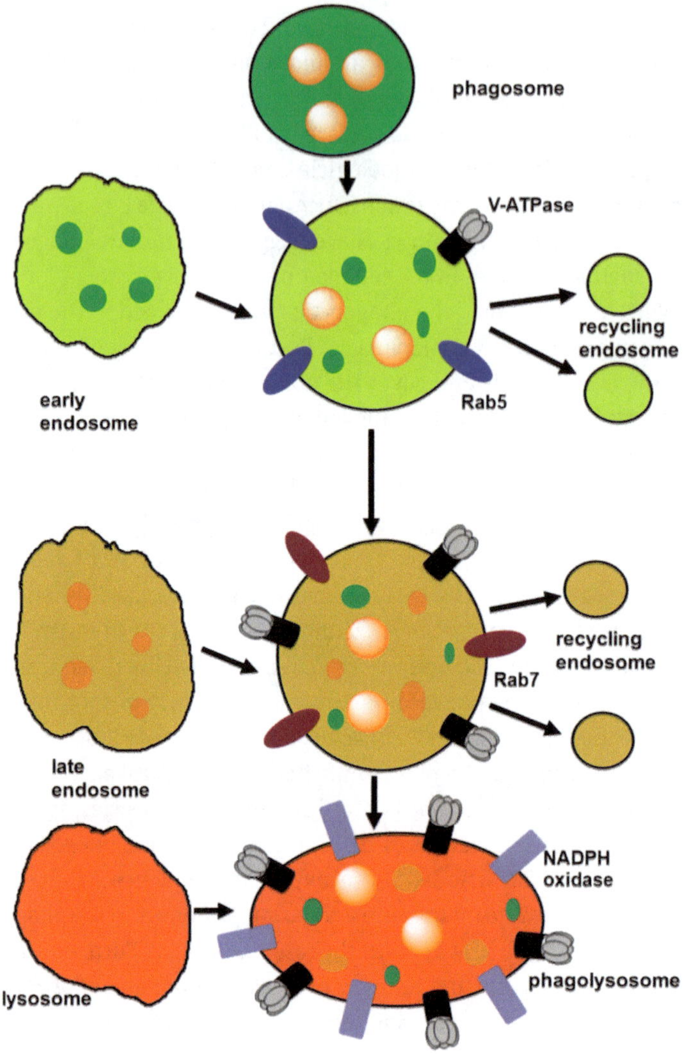

Figure 88: Phagosome maturation.

preferentially taken up by phagocytic cells such as macrophages, dendritic cells, or hepatocytes [447–449], whereas the uptake by non-phagocytic cells is very limited. The enhanced uptake of larger particles by phagocytic cells can be exploited for the treatment of certain diseases such as tuberculosis.

Mycobacterium tuberculosis (M. tuberculosis) survives and replicates in human macrophages by outwitting immune mechanisms of patients. MSNPs of 100-nm size, loaded with antibiotics, were found to effectively target M. tuberculosis-infected phagocytes. PH-sensitive coatings prevented the release of the antibiotic in the blood, while the acidic phagosomes and phagolysosomes triggered the release inside the phagocytes [450]. The size-dependence of AuNPs for phagocytosis was also investigated [451, 452]. Anionic AuNPs of 30-nm size were reported to be internalized by macrophages through a variety of endocytosis pathways, including phagocytosis and clathrin- and caveolin-mediated pathways, while 150 nm AuNPs were taken up preferentially by scavenger receptor-mediated phagocytosis. Tsai et al. [452] compared the cellular uptake efficiency of macrophages for AuNPs of different sizes. According to their results, 4 nm AuNPs showed the highest cellular uptake efficiency compared to AuNPs with sizes of 11 nm, 19 nm, 35 nm, and 45 nm. On the other hand, intravenously administered 2-nm and 40-nm-sized AuNPs were found to be rapidly phagocytized by Kupffer cells in the liver independent of their size [453], whereas intratracheally administered 2-nm, 40-nm and, 100-nm AuNPs mainly accumulated in lung macrophages [454]. Apparently, uptake of gold nanoparticles smaller than 100 nm by macrophages may occur via endocytic pathways other than phagocytosis. In addition to size and surface chemistry, the nanoparticle morphology is another key factor for cell recognition and uptake by phagocytic cells and can also determine circulation time in the bloodstream. To demonstrate the shape effect, PEGylated spherical AuNPs and rod-like AuNRs were injected intravenously into female mice bearing an ovarian tumor [455]. Compared to spherical AuNPs, the internalization of AuNRs by macrophages was significantly lower. In addition, the AuNRs showed a longer circulation time and a higher accumulation in tumors. Hutter et al. [456] investigated the uptake efficiency of spherical AuNPs (∅ = 23 nm), AuNRs (∅ = 12 nm, length = 43 nm), and gold nanourchins (AuNUs) (∅ = 77 nm) by phagocytic microglial cells and non-phagocytic neurons. The relatively slender AuNRs are favored for internalization into neurons due to their morphology. The AuNUs were phagocytosed by the microglial cells, a process that does not occur along the neuritic dendrites and arbors. In conclusion, nonspherical nanostructures were shown to have longer in vivo circulation times and reduced efficiency of phagocytic uptake by macrophages [455, 457]. The in vivo biodistribution of rod-like mesoporous SiNPs (MSNPs) with an aspect ratio of 1.5 (short-rod MSNPs) and 5 (long-rod MSNPs) was investigated by Huang et al. [458]. At 2 h after intravenous injection, no difference was observed in the concentration of the short-rod and long-rod MSNPs in the blood. After 24 h, the concentration of short-rod MSNPs was significantly decreased, while the concentration of long-rod MSNPs remained the same. This indicates that the long-rod MSNPs have a longer circulation time in the blood. The results reported by Park et al. [459] for dextran-coated SPIONs and superparamagnetic iron oxide nanoworms (SPIONWs) are in contrast to these findings. The authors examined the blood circulation times of SPIONs (30.3 nm) and SPIONWs (66.8 nm) in mice that were administered by intravenous injection. Although the SPIONs and SPIONWs differed significantly in shape and size, they exhibited similarly long circulation times (half-live times of 15–18 h) with a first-order elimination rate.

Since nanoparticles interact with cells via their surface structures, their uptake by phagocytosis and phagocytic cells can be controlled by appropriate surface coatings [460]. In addition, surface-adsorbed proteins determine the degree of opsonization by interacting with specific membrane receptors that trigger phagocytosis. For direct uptake of nanoparticles by macrophages, specific molecules and proteins are used that are recognized by the phagocyte Fc receptor or mannose receptor. However, evading the mononuclear phagocytic system is still the greatest challenge for the medical application of nanoparticles. Effective clearance of systemically administered nanoparticles drastically limits their therapeutic efficacy [460]. Therefore, there is a great interest in reducing the nonspecific uptake of opsonins on the nanoparticle surface by making them "bio-invisible". Bio-invisibility can be achieved by using brush layers of hydrophilic polymers (e. g., PEG, dextran), zwitterionic surfaces, self-identifying proteins (e. g., CD47), or coating with cell membranes (e. g., red blood cell membranes) [411, 460, 461].

4.2.1.3 Pinocytosis

During pinocytosis, pinosomes with diameters less than 150 nm are formed, which perform the uptake of extracellular fluid containing suspended nanoparticles into the cytosol. Pinocytosis is an endocytic process for the uptake of extracellular fluid, small molecules, and nanoparticles smaller than 200 nm and can occur as macropinocytosis or receptor-mediated micropinocytosis (Figure 89). During macropinocytosis, dissolved macromolecules with a diameter of more than 0.2 μm are not taken up selectively [462]. Nanoparticles ranging in size from 10 nm to 200 nm are internalized by all cell types via micropinocytosis including clathrin-mediated endocytosis, caveolae/lipid raft-mediated endocytosis, and clathrin/caveolae-independent endocytosis [463].

Recent research has shown that the size, shape [415], and surface charge [447, 464] of nanoparticles determine the cellular uptake mechanism and subsequent cytosolic access of nanoparticles into living cells. Cationic nanoparticles have a higher adhesion efficiency on the cell membrane than anionic ones [465]. Nanoparticles smaller than 200 nm in diameter are catured in clathrin-coated pits, whereas the efficiency of caveolae-mediated endocytosis increases with nanoparticle size [466]. Nanoparticle surface chemistry, size, and shape also influence the exocytosis efficiency [427] and are key determinants of cellular response [467].

Macropinocytosis mediates the uptake of nutrients and other molecules in the fluid phase through the formation of intracellular vesicles called macropinosomes that are 0.2 to 5 μm in diameter. Compared to other endocytic mechanisms, macropinocytosis allows the uptake of extracellular fluid and particles, including those up to 1 μm in size. Macropinocytotic processes are triggered by extracellular stimuli such as the platelet-derived growth factor (PDGF) or the granulocyte-macrophage colony-stimulating factor (GM-CSF) [468]. The resulting signaling cascade leads to an increase in actin polymerization at the plasma membrane via activation of RhoGTPases [469]. Actin-mediated ruffling provides the formation of macropinosomes. For example, macropinocytosis is used by the immune system to enable the transfer of antigens to class I and class II cells. In addition, macropinocytosis promotes increased metabolism of cancer cells.

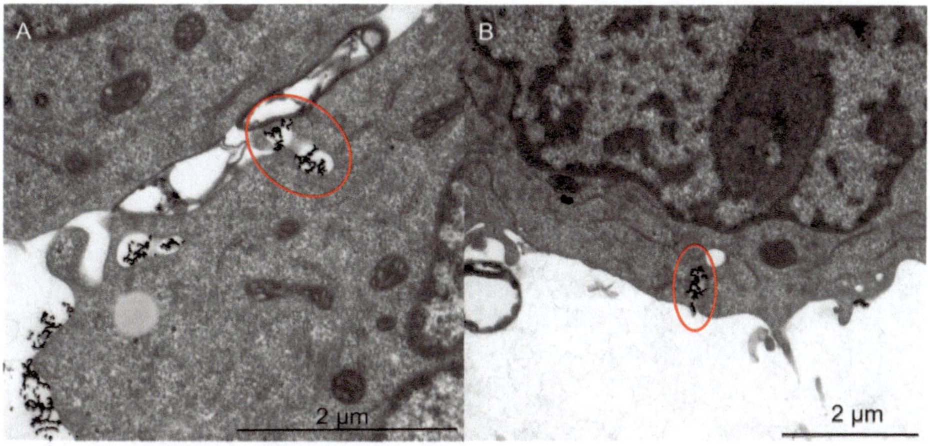

Figure 89: Pinocytosis of citrate-coated AuNPs into (A) A549 lung cancer cells and (B) MCF-7 breast cancer cells.

4.2.1.3.1 Clathrin-mediated endocytosis

Clathrin-mediated endocytosis is the major endocytic pathway in mammalian cells [470, 471]. This pathway plays a key role in many processes in the cell such as intercellular and intracellular signal transduction, membrane recycling and membrane, uptake of transmembrane receptors, and regulation of cell surface signaling [472, 473]. Furthermore, clathrin-mediated endocytosis mediates the internalization for membrane pumps responsible for transporting small molecules and ions across the plasma membrane. In addition, CME contributes to cell and serum homeostasis and to the recovery of small serum proteins after filtration by the kidney [432]. Clathrin-mediated endocytosis is initiated by the binding of cargo molecules or nanoparticles to specific receptors, such as cholesterol-loaded low-density lipoprotein (LDL) to the LDL receptor or iron-loaded transferrin to the transferrin receptor [470]. Clathrin-mediated endocytosis proceeds through six main stages: (1) cargo molecules bind to specific cell surface receptors, (2) adaptor proteins recruit clathrin to form a coated pit, (3) the pit invaginates and dynamin facilitates vesicle scission, releasing a clathrin-coated vesicle into the cytoplasm, (4) the clathrin coat is disassembled by chaperone proteins, (5) the uncoated vesicle fuses with early endosomes, and (6) the internalized cargo is sorted for recycling or degradation (Figure 90).

Clathrin-mediated endocytosis is initiated when endocytic coat proteins from the cytosol begin to accumulate on the inner surface of the plasma membrane. The major coat protein clathrin has a three-legged structure [432] (Figure 90(A)). The so-called triskelion consists of three clathrin heavy chains (ca. 190 kDA), each tightly linked to a clathrin light chain (*ca.* 25 kDa). Low salt and high calcium concentrations can induce spontaneous aggregation of clathrin triskelions into polygonal cages [471]. Several endocytic accessory proteins (EAPs) support the polymerization of clathrin triskelions

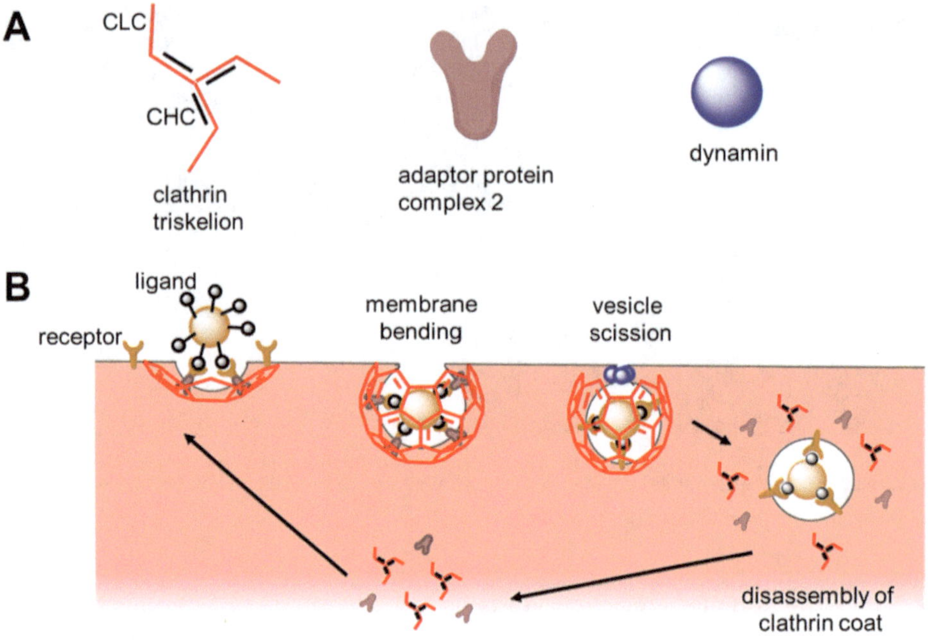

Figure 90: A) Main components of the clathrin-mediated endocytosis: clathrin triskelion consisting of three clathrin light chains (CLC) and three clathrin heavy chains (CHC), adaptor protein complex 2 and dynamin; B) the five stages of the clathrin-mediated endocytosis.

into polygonal lattices that drive membrane invagination and stabilize membrane curvature. In addition, the actin cytoskeleton can contribute to membrane bending (Figure 90(B)). As membrane invagination progresses, a clathrin-coated vesicle is formed due to the activity of heterodimer adaptor proteins (AP 1–4) and monomeric adaptors such as AP180 or epsins. The major adaptor protein to which clathrin binds at the plasma membrane is the adaptor protein 2 (AP2). AP2 does not only bind to clathrin, but also interacts with internalization receptors and cargo molecules. The GTPase dynamin controls the scission of clathrin-coated vesicles from the donor membrane. This enzyme is recruited by BAR domain-containing proteins that attach preferentially to the vesicle neck. Dynamin oligomers form helices around the neck. Hydrolysis of GTP initiates further constriction of the dynamin helices, leading to membrane rupture. After clathrin-coated vesicles are detached, the clathrin coat is disassembled, several homotypic fusions take place, and the cargo is finally released into early endosomes [474]. Several studies have been devoted to the uptake of nanoparticles by clathrin-mediated endocytosis. To distinguish between clathrin-mediated, caveolae-mediated, and caveolae/clathrin-independent endocytosis pathways as the preferred uptake mechanism for AuNPs, Ng et al. [475] used specific endocytosis inhibitors, which are concanavalin A, chlorpromazine, and nystatin. While concanavalin A and chlorpromazine were used to block clathrin-mediated endocytosis, nystatin selectively inhibit caveolae-mediated endocytosis. Using these inhibitors, the authors showed that 20-nm AuNPs, previously incubated in fetal bovine serum, were preferentially internalized by MRC-5 lung fibroblasts and Chang liver cells via the clathrin-mediated endocytosis pathway. By analog, Engelberg et al. [422] demonstrated clathrin-mediated endocytosis as a cellular uptake mechanism for aptamer-functionalized quantum dots (S15-APT QDs) in human lung cancer cells (A549). For this purpose, A549 cells were treated, for example, with cytochalasin D, which acts as an actin polymerization inhibitor to disrupt clathrin-mediated endocytosis. A549 cells modified in this way showed a significant decrease in the uptake of S15-APT QDs, which was taken as evidence for endocytic uptake. In addition, A549 cells were pretreated with amiloride, which inhibits macropinocytosis, or with filipin, an inhibitor of caveolae-mediated endocytosis, and no effect on the cellular uptake efficiency of the S15-APT QDs was observed. These results also confirmed clathrin-mediated endocytosis as the major uptake mechanism for the quantum dots. Chithrani et al. [415] examined the impact of the nanoparticle morphology on the uptake mechanism and kinetics. They demonstrated that spherical and rod-shaped transferrin-coated AuNPs were internalized into STO, HeLa, and SNB19 cells via the clathrin-mediated endocytosis pathway. As expected, the uptake rate decreases with AuNP size and is larger for spherical AuNPs than for AuNRs of similar size. Harush–Frenkel et al. [476] investigated the influence of surface charge of approximately 90–96-nm polymeric nanoparticles (PNPs) on the endocytic pathway in HeLa cells. While negatively charged PNPs exhibit low endocytosis efficiency, positively charged PNPs are rapidly endocytosed by chlatrin-mediated endocytosis and to a lesser extent by caveolae-mediated endocytosis.

4.2.1.3.2 Caveolae-mediated endocytosis

Caveolae, which appear as 60–80 nm invaginations in the cell membrane, are flask-shaped and uncoated. However, they may be equipped with a stromatal diaphragm at the neck of the flask [477–479]. Caveolae are abundant in adipocytes, fibroblasts, and endothelial and muscle cells, but are undetectable in other cell types such as lymphocytes and many neuronal cells [424, 480]. Caveolae-mediated endocytosis is involved in many biological processes such as cell signaling, transcytosis, lipid regulation, and membrane tension. It is also possible that caveolae-mediated endocytosis plays a role in diseases such as cancer, diabetes, and viral infections [481–484]. Caveolae are membrane microdomains rich in cholesterol and sphingolipids that are stabilized by the protein caveolin [485]. These lipid-based microdomains form lipid rafts (Figure 91(A)) of highly saturated lipids and cholesterol. The straight chains of the saturated, polar lipids are tightly packed with the cholesterol. These highly ordered lipid rafts are surrounded by a "lipid disordered" bilayer of unsaturated phospholipids. Because of their curved hydrocarbon chains, the phospholipids cannot be packed as tightly together. Therefore, the biochemical properties of raft and non-raft domains are different [480]. The development of caveolae depends on the expression of caveolin-1 (Cav-1) in non-muscle cells and caveolin-3 (Cav-3) in muscle cells. The central hydrophobic domain of caveolins forms a

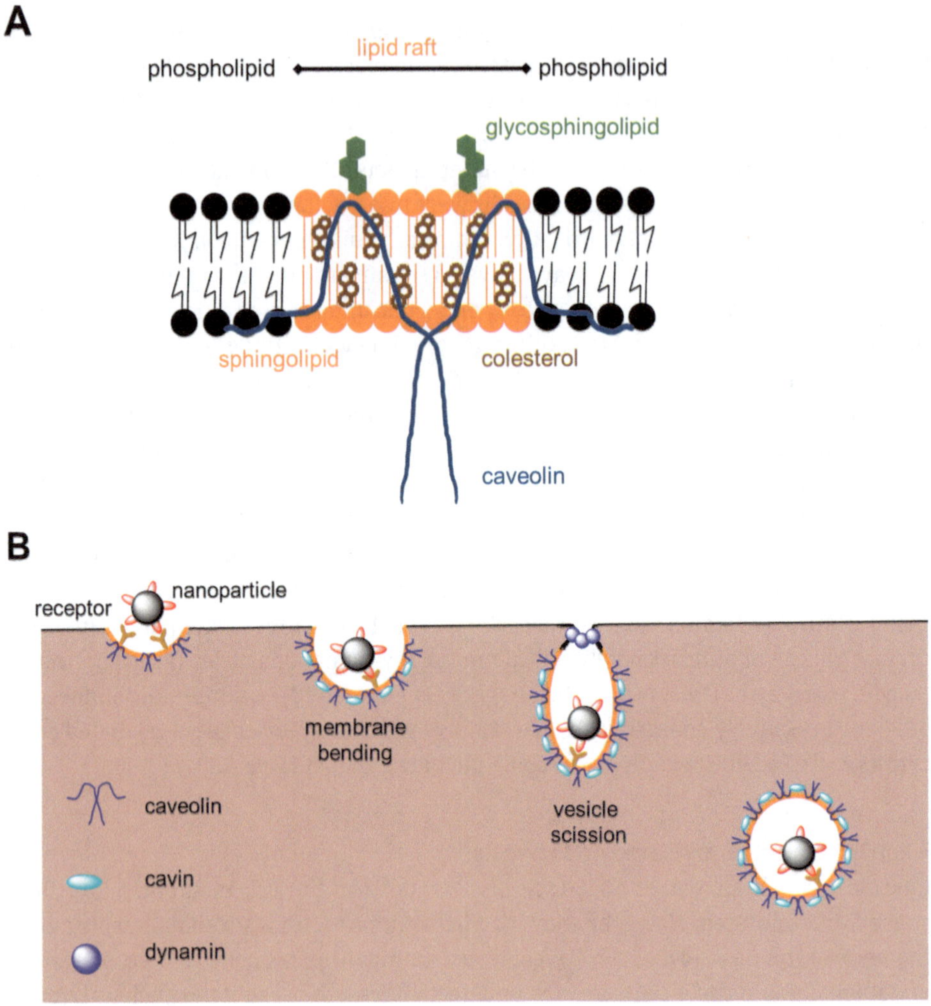

Figure 91: Lipid-raft (A) and caveolae-mediated endocytosis (B).

hairpin loop in the lipid bilayer, while the N- and C-terminal regions of the protein are cytoplasmic. Caveolins interact with cholesterol to form dimers and oligomers [478]. Caveolae budding is highly regulated by phosphatases and kinases [482]. As described for the CME, the fission of the caveolar vesicle is dependent on the GTPase dynamin [424] (Figure 91(B)).

To selectively inhibit caveolae-mediated endocytosis, cholesterol-depleting drugs are used. Either cholesterol is removed from the membrane (e. g., methyl-β-cyclodextrin (MβCD)) or cholesterol synthesis is inhibited (e. g., nystatin, progesterone) [423]. Cholesterol depletion as a marker for caveolae-mediated endocytosis clearly distinguishes this pathway from CME. However, there are other lipid-raft endocytosis pathways that are independent of caveolae. There is evidence that certain proteins, pathogens, or macromolecules can escape lysosomal degradation after caveolae-mediated endocytosis [416]. Based on these observations, there is increasing motivation to coat nanoparticles with proteins known to be internalized by caveolae-mediated endocytosis, such as albumin [416, 479].

4.2.1.3.3 Clathrin/caveolae independent endocytosis

Clathrin-mediated endocytosis and caveolae-mediated endocytosis involve the formation of specific coated vesicles. However, there is evidence that cargoes are internalized by a clathrin- and caveolin-independent endocytic pathway. In a study of cells expressing a temperature-sensitive dynamin, clathrin-mediated endocytosis and probably also caveolae-mediated endocytosis are inhibited by an increase in temperature because of their dynamin dependence. After 30 min, fluid-phase uptake recovers while receptor-mediated endocytosis remains blocked [486]. This uptake mechanism is also present in cells with lipid rafts without caveolae (e. g., lymphocytes, some neuronal cells) [487], suggesting a cholesterol-dependent but caveolae-independent endocytosis. Unfortunately, the underlying mechanisms of this type of endocytosis are still poorly understood. Glycosylphosphatidyllinositol (GPI)-anchored proteins were the first cargoes identified for internalization in a clathrin/caveolae independent pathway [488]. Therefore, GPI-anchored proteins are classical markers for the study of this endocytosis. After internalization, GPI-anchored proteins are found in tubular or ring-shaped structures called GEECs (GPI-enriched endosomal compartment). This uptake mechanism is also referred to as the clathrin-independent carrier (CLIC)-GEEC pathway [489]. Key factors in this pathway are the small GTP-binding protein CDC42 and ADP-ribosylation factor 1 (ARF1), the GTPase regulator associated with focal adhesion kinase (GRAF1), cholesterol, actin, and the cell-surface glycoprotein CD44 [463, 490]. A second pathway has been defined by major histocompatibility class I (MHC I) uptake and regulation by the GTP-binding protein ARF6 [491, 492]. This process may overlap with the CLIC-GEEC pathway. Other proteins and processes have been implicated in clathrin/caveolae-independent endocytosis, but these pathways are less well characterized. Since there is no consensus in the literature on the categorization of clathrin/caveolae-independent endocytosis, all the various pathways are summarized in this chapter. Another feature of clathrin/caveolae-independent endocytosis is its dependence on dynamin. Normally, this lipid-raft non-caveolar endocytosis is independent of the protein dynamin. However, a dynamin-dependent, lipid raft-mediated uptake involving non-caveolar carriers has also been documented for the interleukin (IL)-2 and γc cytokine receptor [493, 494]. The uptake of nanoparticles by clathrin/caveolae independent endocytosis has been reported. Bohmer et al. [421] showed that SPIONs and silica-iron oxide core-shell nanoparticles with negative surface charge were internalized into HeLa cells via CDC42 (CLIC-GEEC) and caveolae-mediated pathway. In another study, polydopamine-coated MSNPs were taken up by HeLa cells through caveolae-mediated endocytosis, ARF6-dependent endocytosis and micropinocytosis [495]. This observation suggests that nanoparticles are internalized by more than one endocytic pathway simultaneously. It is still difficult to separate clathrin-mediated endocytosis, caveolae-mediated, and clathrin/caveolae-independent pathways. The lack of knowledge about the underlying mechanisms also hinders a detailed understanding of nanoparticle uptake by endocytosis.

4.2.1.3.4 Macropinocytosis

Macropinocytosis is involved in many biological processes including nutrient uptake and sensing, signaling, antigen presentation, and cell migration. A wide range of cell types can utilize macropinocytosis: dendritic, B and T cells, macrophages, epithelial and endothelial cells, fibroblasts, neurons, and microglia and cancer cells [468, 496, 497].

Macropinocytosis is usually regulated by cell surface receptors such as receptor tyrosine kinases, cell-surface proteoglycans or G-protein coupled receptors. These receptors activate a signaling cascade, which triggers the actin polymerization at the plasma membrane to generate membrane ruffling [468]. Membrane ruffles exist in several forms depending on the cell type: planar (lamellipodia), circular cup-shaped, or large plasma-membrane extrusions (blebs) (Figure 92) [496]. After the plasma membrane protrusion has engulfed sufficient extracellular fluid, molecules, and nanoparticles, the edges of the protrusion fuse to form a closed bubble, the macropinosome [497]. The macropinosome contains the ingested extracellular fluid with dispersed nanoparticles or dissolved molecules. Actin polymerization also plays an important role in the contraction of the macropinocytic bleb. Actin filaments located along the bleb contract and gradually constrict the bleb. Actin-dependent constriction leads to complete separation of the macropinosome from the plasma membrane. The formation of circular membrane ruffles occurs by a mechanism similar to phagocytosis and is partially dependent on the GTPase dynamin, whereas membrane scission of lamellipodial and bleb-like membrane extensions is independent of dynamin.

The macropinosome, now a self-contained intracellular vesicle, is transported by the cytoskeleton and molecular motors along the cytoskeleton to the interior of the cell. This transport process can be controlled by motor proteins such as dynein and kinesin that move along microtubules or actin filaments. The macropinosome can fuse with lysosomes, which contain digestive enzymes to break down ingested particles or molecules. This step allows the cell to extract nutrients from the extracellular fluid and recover recyclable materials. The indigestible remains can later be excreted from the cell by exocytosis. The precise molecular mechanism of macropinocytosis is complex and involves a number of signaling molecules and regulatory proteins. Some of the key molecules are GTPases, such as Rac1 and Cdc42, which reorganize the actin cytoskeleton and promote the formation of ruffling. Phosphoinositide lipids, such as phosphatidylinositol 3-kinases (PI3K), also play important roles in signal transduction and regulation of membrane envelope formation during micropinocytosis [462].

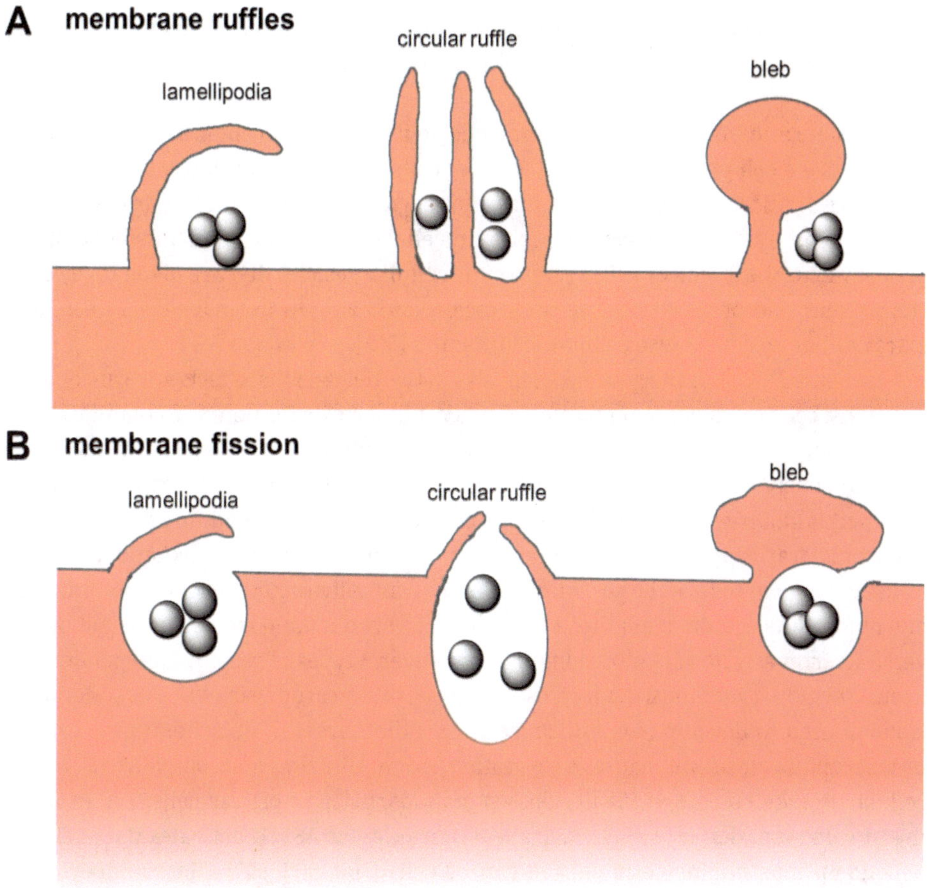

Figure 92: Membrane ruffles (A) and membrane fission (B).

Macropinocytosis plays an important role in tumor progression in certain types of cancer. When cancer cells are deficient of nutrients, they can internalize extracellular proteins via macropinocytosis [498]. These proteins are degraded into amino acids such as alanine and glutamine. Alanine participates in the tricarboxylic acid cycle and the production of glycolytic intermediates. In this process, alanine is deaminated to pyruvate as a significant carbon source, which sustain the cellular metabolism even under glucose deficiency [499]. Next to glucose, glutamine is an important nutrient for cancer cell growth and proliferation. Glutamine is an essential source of carbon and nitrogen. Therefore, it participates in the biosynthesis of amino acids, nucleic acids, glutathione or fatty acids, and the tricarboxylic acid cycle [33]. Macropinocytosis also delivers other nutrients such as carbohydrates, fats, nucleotides, and extracellular ATP [500]. Macropinocytosis is also an efficient way to internalize some surface receptors to maintain cell survival. For example, cancer cells can downregulate cell-surface death receptors to evade TNF-related apoptosis-inducing ligands (TRAILs) and ultimately apoptosis [501].

Prostate and breast cancer cells translocate the growth receptor ErbB3 into the nucleus to drive cell proliferation [502]. Cancer cells take up extracellular vesicles, called exosomes or microvesicles, through macropinocytosis. These extracellular vesicles contain proteins, lipids, or nucleic acids for extracellular communication and metabolic replenishment. Many cancer cells secrete extracellular vesicles that are taken up by other cancer cells and promote disease progression [503, 504]. In normal cells, internalized extracellular vesicles cause cellular dysfunction [505].

Increased macropinocytosis could also have deleterious effects on cancer cells. Normally, cells can sustain high rates of macropinocytosis for hours. To maintain equilibrium, small recycling vesicles return membrane components to the cell surface while simultaneously excluding water [462]. Excessive stimulation of macropinocytosis can disrupt this balance and eventually lead to cell death, known as methuosis. In methuosis, endosomal trafficking does not function. Vacuoles formed during macropinocytosis fuse in the cytoplasm to form larger vacuoles that eventually occupy most of the cytoplasm and deform the plasma membrane. Cytoplasmic vacuolization may be transient or irreversible. Irreversible vacuolization results in several types of caspase-independent cell death, including methuosis, paraptosis, oncosis, and necroptosis [497, 500]. Methuosis is being used to develop new anticancer drugs that target the biochemical regulation of macropinocytosis and cause its overstimulation. Utilizing macropinocytosis for the cellular uptake could specifically deliver nanoparticles to cancer cells. For example, the functionalization of nanoparticles with the aptamer AS1411 activates the nucleolin-dependent macropinocytosis only in cancer cells [500, 506, 507]. Macropinocytosis is

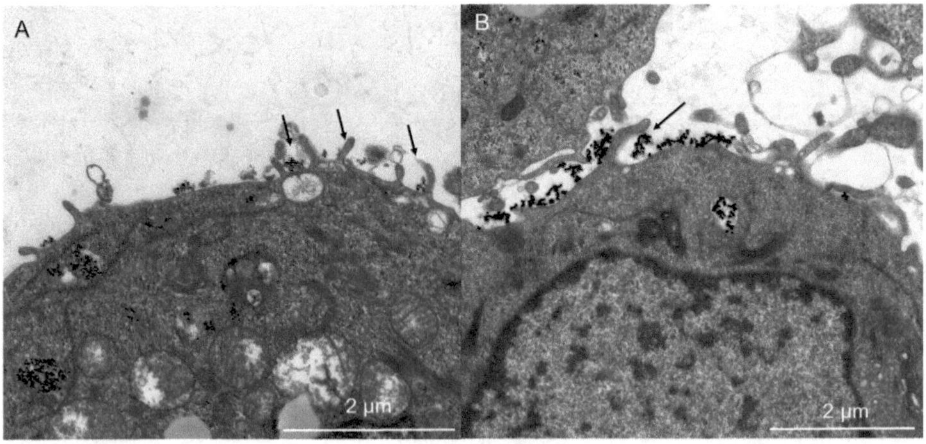

Figure 93: Macropinocytosis (black arrow) of citrate-coated SPIONs (A) and citrate-coated AuNPs (B) internalized by MCF-7 breast cancer cells.

initiated as a nonselective process by the mere contact of nanoparticles with the plasma membrane. This is illustrated by two TEM images illustrating citrate-coated SPIONs (A) and citrate-coated AuNPs (B) that were internalized by MCF-7 breast cancer cells via the process of macropinocytosis (Figure 93). Cellular uptake by macropinocytosis is only slightly influenced by the morphological properties of the nanoparticles. Larger nanoparticles are more likely to be taken up by macropinocytosis than small nanoparticles.

4.2.1.3.5 Direct permeation

In addition to endocytosis as the major uptake pathway for nanoparticles, non-endocytic translocation pathways are also involved in nanoparticle cell entry. A major disadvantage of endocytosis is the entrapment of nanoparticles in vesicles of the endosomal pathway, which eventually leads to their degradation inside lysosomes. In contrast, direct permeation as a non-endocytic process facilitates rapid and efficient cellular uptake, prolonged cytosolic distribution, and insignificant cellular trafficking along the endosomal pathway and therefore low lysosomal degradation [508, 509]. Direct permeation of nanoparticles requires a translocation across the cell membrane, which can be induced by positive surface charges or by ordered surface structures [358]. As the example of cell-penetrating peptides shows, not only the functional groups but also their conformation determine the interaction process with the cell membrane. Cell-penetrating peptides must adopt an amphiphatic α-helix structure in order to either fuse with or penetrate the cell membrane [510].

Nanoparticle size is a critical factor in determining the mechanism of cellular entry. Internalization via receptor-mediated endocytosis is initiated by the interaction of nanoparticles with cell surface receptors. Two competing processes are involved in the subsequent coating process during endocytosis: (a) membrane-nanoparticle adhesion, which facilitates coating, and (b) membrane deformation and membrane tension, which inhibit further coating due to the energetic costs associated with bending the membrane. The bending energy required for membrane wrapping increases with decreasing nanoparticle size because smaller nanoparticles necessitate a higher degree of membrane curvature. Thus, there is a minimum nanoparticle size below which the energy cost for the endocytic pathway becomes unfavorable. While nanoparticles smaller than this threshold may not be efficiently internalized via traditional endocytic pathways, they are not primarily taken up by direct permeation. Instead, they may utilize alternative endocytic mechanisms or specialized uptake processes to enter the cell.

L-histidine-terminated silver nanocluster (AgNC) with an average size of 1.6 nm [511] enter colon adenocarcinoma cells (Caco-2) by direct permeation. As shown in the TEM image in Figure 94(A), AgNC are located close to the cell membrane immediately after cellular uptake, while no endosome formation is visible. Analogous results were obtained for aminosilanized silicon nanocluster (SiNC) with an average size of 1 nm. The positively charged SiNCs are taken up by MCF-7 breast cancer cells by direct permeation.

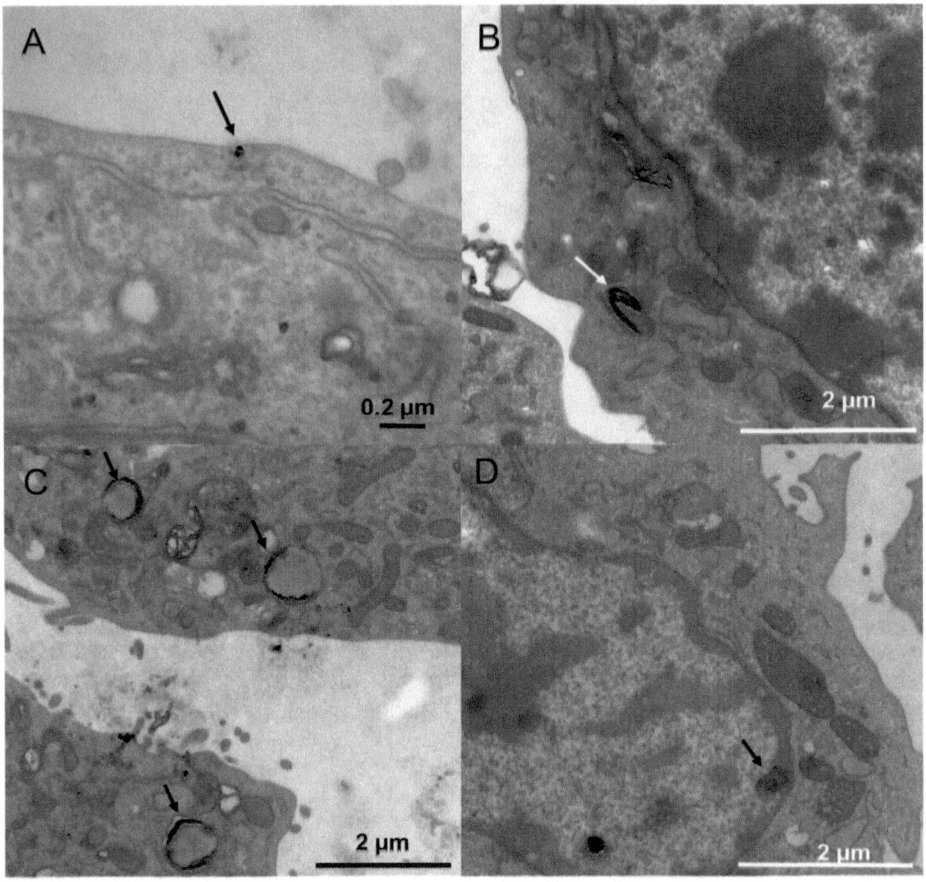

Figure 94: (A) Direct permeation of L-histidine-terminated AgNCs (black arrow) into colon cancer cells Caco-2; after direct permeation: (B) ultrasmall amino-silanized SiNPs are found membranes of cell organelles (white arrow), (C) ultrasmall uncoated SiNPs agglomerate in cytoplasm and cell organelle membranes (black arrow), (D) the SiNPs reach the cell nucleus (black arrow).

As shown in the TEM image in Figure 94(B), the positively charged SiNC preferentially adhere to the membranes of cell organelles such as the endoplasmic reticulum or mitochondria. By comparison, uncoated oxidized SiNC in the size range of 1 to 2 nm were also detected in the cytosol and nucleus (Figures 94(C) and (D)) [277]. These clear experimental results indicate that the preferred cellular entry pathway for AgNC [509, 512], AuNPs [513–516], and SiNC [277, 517, 518] is direct permeation. Therefore, nanoparticle size plays a very important role in determining the entry modality.

Direct permeation has been observed for positively or negatively charged nanoparticles and neutral or zwitterionic nanoparticles. The chemistry and polarity of the ligands and their structural arrangement on the nanoparticle surface determine the cellular uptake mechanism, which can occur either by direct permeation or by the com-

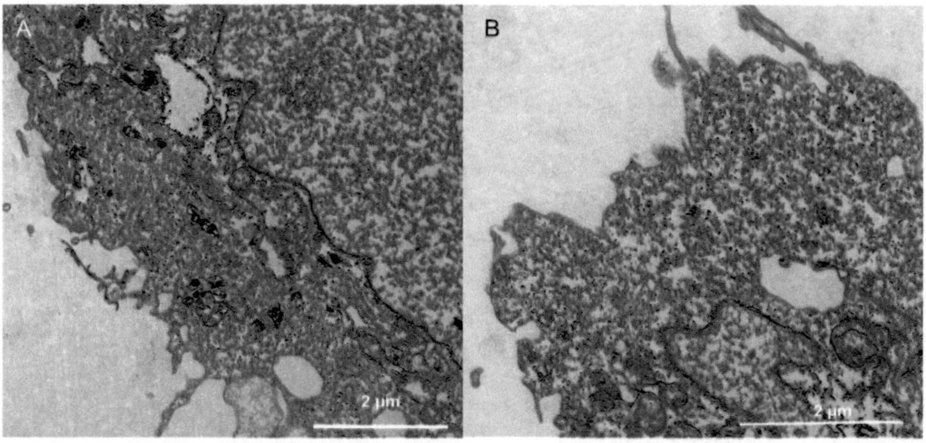

Figure 95: Direct permeation and intracellular distribution of bilayer shell-coated Au–Fe_3O_4 NHDs: (A) positively charged bilayer shell-coated Au–Fe_3O_4 NHDs and (B) negatively charged bilayer shell- coated Au–Fe_3O_4 NHDs in breast cancer MCF-7 cells.

petitive process of endocytosis [358, 515, 516]. As reported by Verma et al. [359], ordered amphipathic surface structures support direct permeation of nanoparticles into cells. The authors coated AuNPs with binary mixtures of hydrophobic and hydrophilic organic molecules (i. e., 11-mercapto-1-undecanesulfate and 1-octanethiol), which self-assembled into ordered structures at suitable concentration ratios. By contrast, AuNPs with a disordered arrangement of ligands were entrapped in endosomes. Klein et al. obtained comparable results for oleic acid terminated Au–Fe_3O_4-NHDs mixed with anionic or cationic amphiphilic compounds (sodium 4-dodecylbenzenesulfonate or 1-octadecylpyridinium bromide) by ultrasonication. The amphiphilic compounds formed a hydrophilic, highly ordered bilayer by interlocking with the oleic acid ligands, thus allowing direct permeation (Figure 95) [246].

4.2.1.4 Physical methods for direct permeation

Direct permeation of nanoparticles can also be achieved by physical methods, but only in vitro or ex vivo [429, 519]. By using microinjection, electroporation, ultrasound, or optoporation, to name the main physical methods, transient defects or pores are created in the cell membrane that allow the nanoparticles to permeate directly into the cytosol. This prevents the endosomal pathway for cellular uptake of the nanoparticles. However, the application of microinjection or an external electric or electromagnetic force field leads to cellular stress and provokes cell mortality. On the other hand, the direct permeation of nanoparticles into the cell interior induced by these physical methods can be used for targeted therapeutic delivery and intracellular imaging.

Microinjection provides a simple mechanical technique for cell membrane poration [520]. Fine-tipped glass microcapillaries (0.5–5 μm) are used in combination with an optical microscope. Nanoparticles dispersed in a cell medium are introduced directly into the cytoplasm or cell nucleus using a fine-tipped glass microcapillary under the guidance of an optical microscope. This approach provides high payload efficiency and cell survival. Derfus et al. [521] demonstrated the applicability of microinjection for targeted delivery of PEGylated CdS/ZnS QDs into HeLa tumor cells using epifluorescence microscopy. They observed active transport of PEGylated CdS/ZnS QDs into the cytosol. On the other hand, microinjected PEGylated CdS/ZnS QDs conjugated to either a 23mer nuclear localization sequence peptide (NLS) peptide or a 28mer mitochondrial localization sequence peptide (MLS) peptide were found to be localized in the nucleus and mitochondria, respectively. Medintz et al. [522] also used microinjection to introduce fluorescent CdS/ZnS QD-protein conjugates into two different cell lines, HEK293/T17 and COS-1 cells. They demonstrated a homogeneous distribution of CdS/ZnS QDs in the cytoplasm by fluorescence microscopy. Candeloro et al. [430] used the microinjection technique in combination with Raman spectroscopy to study the biochemical response of HeLa tumor cells to Ag and Fe_3O_4 nanoparticles. They were able to detect a reduction in glycogen and fructose-6-phosphate levels, which they attributed to oxidative stress caused by the nanoparticles.

Another well-established physical technique enabling direct permeation of nanoparticles is electroporation, consisting of the application of short high-voltage electrical pulses to cells [523]. These electrical pulses affect the plasma membrane, causing phospholipid molecules to change orientation and conformation. The electrophoretic movement of charged lipid headgroups and subsequent rearrangement of lipid molecules lead to the formation of transient nanopores that increase the permeability of the cell membrane for nanoparticle entry. The advantage of electrotransfection is its high efficiency and cell-type independence [524]. Electroporation, also known as electrotransfection, has been and continues to be successfully used to introduce polar and charged substances such as DNA, RNA, proteins, and amino acids into cells [523–525]. In addition, electroporation is used for the controlled permeation of nanoparticles into cells [526–529]. Electropermeation enables rapid cellular uptake and, moreover, penetration of nanoparticles into the cell nucleus while maintaining cell viability. The entry into the cell nucleus can be achieved by an appropriate adjustment of the parameters (i. e., voltage, frequency, pulse duration) of the applied electrical pulses. Targeted delivery of drugs into the cell nucleus offers a promising strategy for cancer therapy. Nanoparticles electropermeabilized into the nucleus could thus be used as a delivery vehicle for anticancer drugs. Phonesouk et al. [527] loaded human colorectal carcinoma (HCT-116) and RL (B-lymphoma) cells with fluorescent dye-labeled SiO_2 NPs with a size distribution of 28–30 nm by electropermeation. The authors demonstrated by fluorescence microscopy that electroporation is the method of choice for rapid and efficient cellular uptake and nuclear penetration of SiO_2 NPs.

Cationic lipids and cationic polymers are used as nonviral vectors for gene transfection. Cationic polymers interact electrostatically with negatively charged DNA and RNA to form complexes. These polyplexes protect the DNA or RNA from degradation and facilitate its cellular uptake and intracellular transport into the nucleus [530]. However, polyplexes tend to aggregate or disintegrate in physiological conditions. This can be avoided by the use of AuNPs as a polyplex carrier [526]. Gene transfection was performed by electroporation, which allows for rapid and direct cytosolic transfer. In this hybrid approach, cationic polymer molecules protect DNA, while AuNP carriers not only stabilize polyplexes but also enhance transfection efficiency by electroporation. For this purpose, AuNPs of various sizes were coated with positively charged polyethyleneimine (PEI) and were then conjugated with DNA plasmids or siRNA molecules to form an AuNPs polyplex. The AuNP-polyplex conjugates were delivered into the cytosol of, e. g., NIH 3T3 cells by electroporation. As shown by the authors, the transfection efficiency was significantly increased by electroporation, the release of the polyplex was rapid, and the toxicity was unchanged [526].

Highly conductive AuNPs in the electroporation buffer solution of a commercial batch electroporation system (625 V/cm, single 10-ms pulse) can significantly increase transfection efficiency for DNA and cell viability. Zu et al. [529] added uncoated AuNPs and transferrin-functionalized AuNPs (Tf-AuNPs) to the electroporation buffer containing plasmid DNA and mammalian cells (K562 lymphoblast cells). The uncoated AuNPs increased the transfection efficiency of the plasmid DNA sondes by reducing the resistance of the electroporation solution. On the other hand, the Tf-AuNPs specifically bound to the transferrin receptor on the plasma membrane. Transferrin receptors are present on nearly all mammalian cells. The receptor is involved in the cellular uptake of iron from transferrin by receptor-mediated endocytosis. Therefore, transferrin ligands are used to achieve targeted cellular uptake of nanoparticles via receptor-mediated endocytosis.

Due to their high electrical conductivity (4.5×10^6 S/m), the Tf-AuNPs acted as nanoelectrodes by boosting the field strength of the applied electrical pulses in the plasma membrane. The emerging nanoporotation system led to the formation of transient pores that facilitate the passage of plasmid DNA into the cytosol. The authors further found that transfection efficiency increases with AuNP size in the range of 5 to 20 nm.

Sonoporation, derived from medical ultrasound imaging, is another technique for controlled delivery of drug-loaded nanoparticles into cells [531, 532]. This technique exploits the ultrasound-induced cavitation effect to transiently increase the permeability of the plasma membrane, which facilitates cellular uptake of nanoparticles. Medical ultrasound imaging has become a useful diagnostic tool due to its real-time capability, safety, and imaging functionality. Pulses of ultrasound waves generated by an ultrasonic transducer can be used in combination with microbubbles that act as contrast agents [533]. Commercially available microbubbles consist of lipid, protein, or polymer shells filled with perfluorocarbon gas, air, or nitrogen (Figure 96) [534, 535]. Microbubbles with diameters of 1–7 µm have a resonant frequency in the range of 2–15 MHz, which corresponds to the ultrasound frequency range in clinical diagnosis [536]. Drug or DNA-loaded nanoparticles can be either embedded inside the microbubble shell or attached to its surface [535].

Figure 96: Schematic illustration of microbubbles consisting of a (a) lipid shell filled with perfluorocarbon gas, (b) protein shell filled with air, and (c) polymer shell with N_2.

Ultrasonic waves, as being longitudinal waves, cause oscillatory changes in pressure and density during their passage through biological liquids and tissues [533]. Accordingly, microbubbles are periodically compressed and expanded by interaction with ultrasonic waves [537]. During expansion they undergo rarefaction. When the acoustic pressure exceeds a certain threshold, the microbubble bursts. The collapse of the microbubbles is known as cavitation collapse or implosion. During this collapse, the rapid reduction in bubble size generates intense local forces and energy, producing shock waves, high temperatures, and even small jets of liquid. The interaction between the microbubbles and the sound waves results in either stable or inert cavitation, depending on the intensity of the ultrasound and the degree of deformation [532]. Stable cavitation induces pushing and pulling of the cell membrane, resulting in permeability of the plasma membrane. Inertial cavitation, on the other hand, is equivalent to a collapse of the microbubble. Shock waves and microjet streams of liquid are generated, perforating adjacent cell membranes. Transient pores are then formed in the cell membrane. The increased membrane permeability allows nanoparticles, whether encapsulated within the microbubbles or adhered to their surface, to enter the cells. The nanoparticles can then be released from the microbubbles or internalized by the cells through endocytosis or other cellular uptake mechanisms. Zhou et al. [538] reported the creation of plasma membrane pores with sizes of approximately 0.10–0.17 µm using ultrasound pulses with pressure amplitudes of 1.5–1.7 MPa. These results demonstrate that microbubble-mediated pore formation can be performed in a controlled manner. This highly efficient method allows transient opening of the plasma membrane for targeted cellular uptake of drugs and nanoparticles.

Yang et al. [539] prepared polymer microbubbles (3–5 µm) with encapsulated 12-nm-sized Fe_3O_4 nanoparticles and introduced them into tumor cells by sonoporation. They showed that the sonoporation process can be adjusted by the intensity of 1 MHz ultrasound pulses in the range of 0.1–0.75 $W\cdot cm^{-2}$. Apparently, sonoporation under these experimental conditions was caused by transient cavitation characterized by high local temperatures, pressures, and free radicals, which caused increased permeability of the adjacent plasma membrane. In particular, these results demonstrated that the release of Fe_3O_4 nanoparticles from the microbubbles could be controlled by the intensity of the ultrasound pulse. On the other hand, Fe_3O_4 nanoparticles inside microbubbles were found to reduce the acoustic cavitation effects and thus their sonoporation effects on the plasma membrane. This is explained by the fact that the encapsulated Fe_3O_4 nanoparticles increase the viscous damping and stiffness of the microbubble shell.

Wang et al. [540] reported of sonoporation-mediated internalization of AuNRs in MDA-MB231 cells. They produced 1–3 μm albumin microbubbles with embedded AuNRs by sonication of a solution containing human serum albumin, perfluoropropane, and AuNRs. Inertial cavitation doses sufficient for transient pore formation were achieved using a 1 MHz transducer generating ultrasonic pulses at a repetition frequency of 100 Hz and acoustic pressures of approximately 1.82 MPa. The resulting cellular uptake of AuNRs was quantified by determining the gold content of MDA-MB231 cells using ICP-MS measurements.

Optoporation uses ultrashort laser pulses in the visible and NIR to temporarily perforate the plasma membrane of single cells [541–545]. To enhance the pore forming effect of the laser pulses in the membrane and to position the pulses more precisely, AuNPs are attached at the plasma membrane. AuNPs larger than 100 nm are advantageous because they cannot enter the cell by direct permeation or via the endocytic pathway [546]. In addition, appropriate surface functionalities may promote the localization of AuNPs on the plasma membrane. Nonresonant and resonant plasmonic nanostructures were shown to amplify the incident laser pulse radiation many times by their electromagnetic near field [547], so that a strongly localized plasma is formed in the biological media. Rapid decay of the plasma (about 100 fs–1 ps) gives rise to localized phase transformation and cavitation bubble.

Patskovsky et al. [545] developed a laser–microscope system for precise laser pulse-induced nanoparticle-mediated single-point optoporation of single cells. They demonstrated that plasmonic AuNPs deposited on the plasma membrane focus due to near-field amplification the laser pulses, resulting in localization and amplification of the laser pulses on the plasma membrane and thus efficient membrane perforation. In addition, using the precisely controllable laser–microscopy system, the authors were able to control the number and size of optoporated holes on the plasma membrane as well as their exact spatial positions at the subcellular level. Further evidence for the performance of nanoparticle-enhanced fs laser transfection was provided by Baumgart et al. [544], who consistently demonstrated that AuNP localization has a significant impact on transfection efficiency and cell viability after laser irradiation. AuNP-enhanced laser pulse optoporation were performed with 800-nm fs laser pulses at a repetition rate of 1 kHz to internalize the fluorescent dye Lucifer Yellow (LY) or DNA plasmids inside human cancer melanoma cells. The internalized LY allows quantification of transfection efficiency by fluorescence microscopy. Plasmonic-enhanced laser-pulse treatment of the cells results in a very high perforation rate of up to 70 % with 80 % viability of the treated cells. The maximum perforation rate was achieved at a fluence of 100–120 mJ/cm^2.

Du et al. [541] also used 532-nm laser pulses with 6-ns pulse duration in combination with 60-nm AuNPs for laser pulse-induced AuNP-mediated sonoporation of HeLa cells. They detected the formation of transient pores using fluorescence microscopy and fluorescence-activated cell sorting. They also investigated the healing process of the plasma membrane as a function of irradiation fluence. For this purpose, they used potential-sensitive fluorescent dyes that diffuse into the phospholipid bilayer and act as potential probes. The recovery of the membrane structure scaled with the membrane potential, corresponding to the return of the membrane potential to the resting potential. Using time-resolved detection of the fluorescence intensity of membrane potential probes, the authors followed the temporal evolution of the healing process of the plasma membrane previously perforated by the synergistic effect of laser pulses and AuNPs attached to the membrane surface. They determined healing times ranging from 28.6–163.8 s for laser fluences between 0.064 and 1.6 J/cm^2.

4.2.2 Effect of physicochemical properties on the cellular uptake

As delineated in detail in Chapter 3, inorganic nanoparticles can be synthesized with different physicochemical properties. The chemical composition, size, and shape of nanoparticles, as well as their surface structures and functionalities, directly and indirectly influence their biological behavior which includes cellular uptake mechanism and efficiency, intracellular fate, and cytotoxicity [548]. On the other hand, surface-adsorbed nanoparticles may affect cell morphogenesis. Wang et al. [549] demonstrated the influence of 50 nm-sized PEG-coated AuNPs on the temporal evolution of the morphology of different cells by tracking nanoparticle-induced changes in membrane tension by AFM and detecting size changes by fluorescence microscopy.

The surface chemistry of nanoparticles not only determines their interactions with membranes, ions, and organelles, but also dynamically changes or adapts to their environment in biological fluids and tissues [548]. For example, the hydrophobicity and hydrophilicity of nanoparticle surfaces determine the dispersion state in extracellular media, the probability of protein corona formation in blood, and the agglomeration behavior of nanoparticles. Hydrophobic nanoparticles are poorly dispersed in biological fluids. They can adsorb proteins via their hydrophobic residues, but this does not prevent them from forming agglomerates, which at best are taken up by macrophages [387]. Hence, therapeutic and diagnostic applications of nanoparticles require suitable surface structures that either electrostatically bind drugs or biomolecules (DNA, RNA, or proteins) or contain specific functional moieties such as membrane receptor-targeting peptides and aptamers [404, 550]. For instance, cationic nanoparticles are suitable for delivering negatively charged drugs, genes, DNA, and RNA into the cell. Moreover, compared to neutral and anionic nanoparticles, cationic nanoparticles bind efficiently to the predominantly negatively charged cell surface, which favors their uptake into the cell [551]. Xia et al. [552] reported that the cellular uptake of cationic polyethyleneimine-coated mesoporous silica nanoparticles (PEI-MSiO$_2$ NPs) is significantly enhanced compared to the negatively charged phosphonate and PEG-MSiO$_2$ NPs. The frequency and efficiency of cellular uptake of PEI-MSiO$_2$ NPs scale with the chain length of the PEI ligands. However, the cytotoxicity of PEI-MSiO$_2$NPs also increased with the polymer chain length because the plasma membrane was damaged by long cationic polymer chains penetrating the membrane. Minimal cytotoxicity with efficient cellular uptake was achieved by coating the $MSiO_2NPs$ with PEI with molecular masses less than 10 kD.

4.2.2.1 Effect of size and shape

The size and shape of nanoparticles are other key determinants of the interaction mechanisms with the plasma membrane and co-determine the pathway of cellular uptake. Nanoparticles smaller than 10 nm enter cells mainly by direct permeation. Larger nanoparticles are taken up by endocytosis-mediated internalization. However, the type of endocytosis pathway also depends on the size of the nanoparticle. As discussed in detail in Section 4.2, there are five different endocytic pathways available to nanoparticles in mammalian cells: phagocytosis, macropinocytosis, clathrin-mediated, caveolin-mediated, and clathrin/caveolin-mediated, and clathrin/caveolin-independent endocytosis. The size of the nanoparticles must be adapted to the endocytotic portal. In receptor-mediated endocytosis, nanoparticles, depending on their size and shape, bind to one or more receptors on the cell surface, leading to internalization of the nanoparticle upon invagination of the plasma membrane (Figure 97). Optimal sizes for spherical nanoparticles, typically around 50 nm for non-phagocytic cells, are more likely to be taken up by receptor-mediated endocytosis. The shape of nanoparticles also influences their binding efficiency and internalization rate. For example, rod-shaped nanoparticles may interact differently with the cell membrane than spherical nanoparticles due to their larger contact area. For example, rod-shaped nanoparticles have a larger contact area with membrane receptors than spherical nanoparticles when oriented longitudinally. This increases the efficiency of binding to the receptors. Nanoparticles larger than

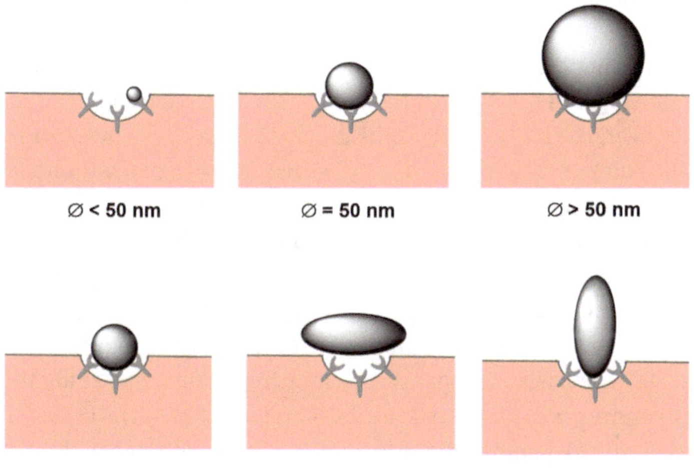

Figure 97: Size- and shape-dependence of the nanoparticle-receptor interactions during invagination of the plasma membrane.

500 nm are internalized by cells via phagocytosis and/or macropinocytosis, depending on the cell type. Cellular uptake of nanoparticles with sizes between 100 and 200 nm occurs preferentially by clathrin-mediated endocytosis, whereas caveolae-mediated endocytosis has been shown to be the uptake mechanism for nanoparticles smaller than 80 nm.

Tree-Udom et al. [553] investigated the effect of nanoparticle shape on the deformation of lipid bilayer membranes prepared from cell-size engineered liposomes. The $NaYF_4$ nanoparticles used were spherical, oblong spherical, and hexagonal prismatic with sizes ranging from 26 nm to 100 nm. Their surface structures consisted of negatively charged COOH–PEO–COOH ligands that were labeled with 6-aminofluorescein for fluorescence microscopy studies. The fluorescence microscopy studies focused on determining the contact area between the nanoparticles and the membrane, which is considered a measure of nanoparticle-membrane association. Adhesion of nanoparticles to membrane surfaces results in indentation of the membrane, which then bends to enclose the nanoparticles. According to the fluorescence microscopy results, the adhesion of the elongated spherical nanoparticles on the bilayer membrane leads to the most efficient association formation, followed by the association formation with the spherical nanoparticles, while the hexagonal prismatic nanoparticles occupy the smallest contact area on the membrane due to their geometry.

4.2.2.2 Effect of surface charge and hydrophobicity

Hydrophobic and electrostatic interactions between nanoparticles and the cell membrane are decisive for the cellular-uptake mechanisms and intracellular trafficking of nanoparticles. Hydrophobic interactions decide on the adsorption geometry of nanoparticles on the plasma membrane surface or inside lipid bilayers. Li et al. [554] performed molecular dynamics calculations to simulate the different localization of hydrophobic and semi-hydrophilic nanoparticles in electroneutral phospholipid bilayers. Charge-neutral PAMAM dendrimer nanoparticles, with a size of 10 nm larger than the thickness of lipid bilayers, do not only adsorb on the lipid bilayer surface but can also be encapsulated within the lipid bilayer. The entrapment in the lipid bilayer disrupts the parallel alignment of the lipid molecules around the nanoparticles, which is called hydrophobic mismatch. The semi-hydrophilic nanoparticles preferentially adsorb to the membrane surface. Wrapping of the semi-hydrophilic nanoparticles by the membrane does not occur spontaneously but requires an energy input. This result confirms the experimental observation of energy-dependent endocytosis of semi-hydrophilic nanoparticles in living cells. Lipid membranes such as phospholipid vesicles [555–557], polymersomes [558], or liposomes [559, 560] are used as model systems for lipid bilayers to study the response of cell plasma membranes to nanoparticles. As shown by Roiter et al. [561], the size of hydrophobic nanoparticles significantly influences their interactions with liposomes. For this study, SiO_2 NPs of various sizes between 5 and 140 nm were deposited on a silicon wafer and lipid bilayers prepared from L-α-dimyristoylphosphatidylcholine were added. SiO_2 NPs larger than 22 nm were completely covered by the lipid bilayer, while the smaller SiO_2 NPs (5–22 nm) partially damaged the lipid bilayer by forming pores (SiO_2 NPs with a diameter >5 nm) smaller than the thickness of the lipid bilayer were embedded in the hydrophobic interior of the lipid membranes (Figure 98).

Dodecanethiol-coated AuNPs of 2-nm size were extruded with phosphatidylcholine (PC) lipid vesicles [562]. The incorporation of AuNPs into the bilayer vesicles was visualized by cryoTEM. The TEM images demonstrated that the AuNPs intercalate between the PC lipid monolayer, deforming but not damaging the lipid bilayer (Figure 98(A)). This intercalation was explained by a strong thermodynamic driving force, which can be attributed to the hydrophobic interactions between the PC lipid molecules and the dodecanethiol ligands of the AuNPs.

Paasonen et al. [563] investigated the interaction of ultrasmall hydrophobic hexanethiol (C6HS)-terminated AuNPs (2.5 nm), hydrophilic negatively charged mercaptosuccinic acid (MSA)-terminated AuNPs (2.8 nm) and dipalmitoylphosphatidylethanolamine (DPPE)-functionalized Au nanocluster (AuNC) (1.4 nm) with dipalmitoylphosphatidylcholine (DPPC) liposomes [564]. While the C_6HS-AuNPs were inserted into the DPPC lipid bilayer, the MSA-AuNPs were encapsulated inside the DPPC liposomes. The DPPE-AuNPs were inserted between the DPPC lipid molecules on the inner and outer surfaces of the liposomes (Figure 98).

Similar results were reported by Amstad et al. [319] for 5-nm palmityl-nitroDOPA-terminated SPIONs and oleic acid-stabilized SPIONs with sizes between 3 and 15 nm. For this study, palmityl-nitro-DOPA- and oleic acid-terminated SPIONs were dispersed in chloroform and mixed with 2-distearoyl-sn-glycero-3-phosphocholine (DSPC) lipids. Agglomeration of oleic acid-stabilized SPIONs induced the formation of a surrounding DSPC micelle (Figure 98(C)), whereas palmityl-nitro-DOPA-terminated SPIONs were embedded in liposome bilayers. This different behavior of DSPC lipid aggregation was explained by the unequal binding affinity of the ligands [565, 566]. Oleic acid adsorbs to the SPION surface via its carboxyl group through electrostatic interactions, which does not only allow partial ligand exchange but also promotes agglomeration by reorganizing the surface structure. These observations have been confirmed by Namiki et al. [567] They mixed agglomerates of oleic acid stabilized SPIONs with phospholipids in organic solvents and then dispersed them in aqueous solution. This led to the formation of stable micelles consisting of SPION agglomerates surrounded by a phospholipid monolayer. The palmityl-nitro-DOPA ligand, on the other hand, binds covalently to the SPION surface via its catechol moiety, resulting in stable hydrophobic surface structures and steric repulsion between the SPIONs.

Lee et al. [568] studied the interactions between 6-nm amphiphilic AuNPs and vesicles formed from 30:70 mixtures of cationic cetyltrimethylammonium tosylate and anionic sodium dodecyl benzene sulfonate surfactants. Amphiphilic surfaces were achieved by functionalizing the AuNPs with hydrophilic deprotonated mercaptoundecanoic acid (MUA) and hydrophobic octadecanethiol (ODT). This mixed surface monolayer is expected to reorganize dynamically in response to variations in the local envi-

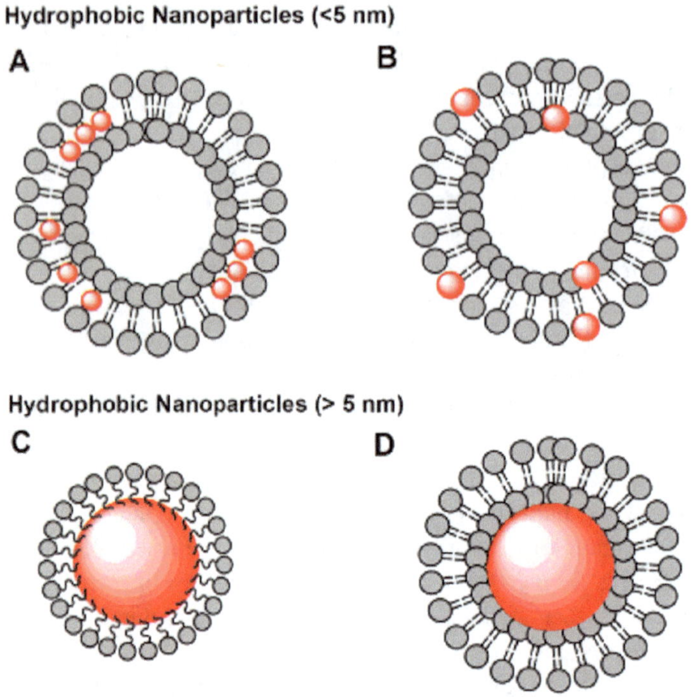

Figure 98: Various interactions between hydrophobic nanoparticles and liposomes: hydrophobic nanoparticles (<5 nm) intercalate within the lipid bilayer (A), lipid-functionalized nanoparticles are incorporated in the inner and outside of the lipid bilayer surface (B); nanoparticles are embedded in a lipid monolayer (C), or in the lipid bilayer (D).

ronment. The AuNPs interact with the preformed vesicles through their hydrophobic ODT ligands, which guide their embedding within the vesicle bilayer (Figure 99).

The hydrophobicity of the nanoparticle surface influences the interaction of nanoparticles with plasma proteins in biological fluids and thus the formation of the protein corona, which determines the cellular uptake mechanism and affects the biodistribution and excretion of the nanoparticles [414]. In this context, Zhu et al. [548] investigated the cellular uptake efficiency of AuNPs with various hydrophobic surface structures in the presence and absence of serum proteins. AuNPs of 2-nm size were functionalized with quaternary amines consisting of the same anchor group but with terminal alkyl moieties of different chain lengths and structures, imparting different degrees of hydrophobicity to the AuNPs. The positive charge of the quaternary amine ligands provides solubility and cellular interactions. Surprisingly, the cellular uptake of highly hydrophobic AuNPs by HeLa cells was inhibited in medium containing serum proteins, which was explained by the formation of AuNP-protein complexes. In the absence of proteins, the cellular uptake efficiency showed no dependence on the degree of hydrophobicity. To confirm these findings, cellular uptake studies were performed with similar bovine serum albu-

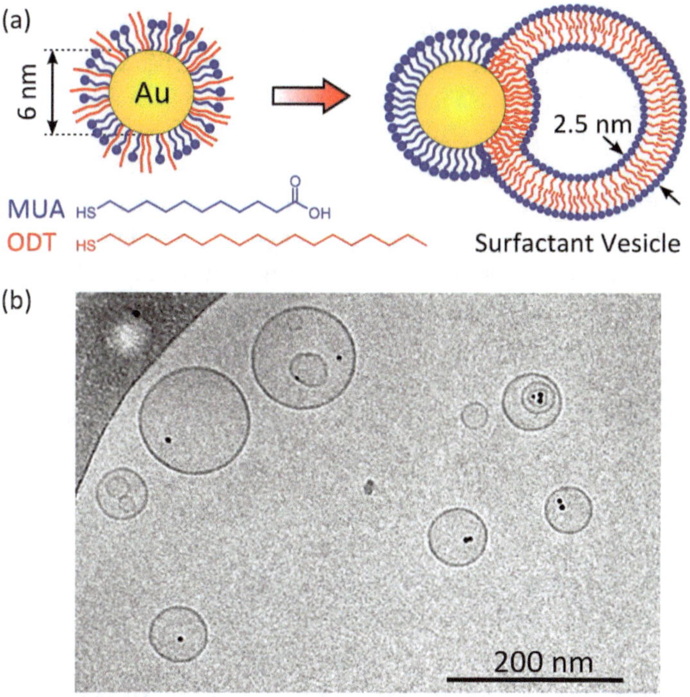

Figure 99: (a) AuNPs functionalized with ODT and deprotonated MUA are embedded within vesicles formed from 30:70 mixtures of cationic cetyltrimethylammonium tosylate and anionic sodium dodecyl benzenesulfonate (b) the Cryo-TEM image illustrates the co-localization of the ODT/MUA terminated ANPs in surfactant vesicles in aqueous solution [568].

min (BSA), immunoglobulin G (IgG), and transferrin. While IgG and transferrin did not affect the cellular uptake of hydrophobic AuNPs, the experiments performed with BSA show similar results to those performed with serum proteins.

Serum albumins are known to possess hydrophobic pockets [220] that bind to hydrophobic AuNPs. The more hydrophobic AuNPs had a higher affinity for BSA. Due to repulsive interactions of the anionic BSA with the negatively charged cell membrane, the cellular uptake efficiency was greater for the more hydrophobic AuNPs. Therefore, other proteins such as apolipoprotein [387] with hydrophobic pockets may also bind to the surface of hydrophobic nanoparticles and affect nanoparticle internalization.

Lin et al. [569] examined the interactions between differently charged AuNPs and cell membrane models by means of coarse-grained molecular dynamic (CG-MD) simulations. The AuNPs used in this study had a core size of 2.2 nm and were terminated with hydrophobic and hydrophilic alkylthiols. The hydrophilic alkylthiol groups were functionalized with either negatively charged carboxylate or positively charged ammonium moieties. In order to determine the influence of the surface charge on the electrostatic interaction with the respective cell membrane model, either uncharged AuNPs (100 %

alkylthiol ligands) or AuNPs with 70 % (i. e., 76 of 108 ligands) cationic or anionic ligands were initially used. Mammalian cell membranes were modelled as an uncharged lipid bilayer consisting of dipalmitoylphosphatidylcholine (DPPC) and as a negatively charged bilayer (DPPC/DPPG) composed of a 3:1 mixture of DPPC with dipalmitoylphosphatidylglycol (DPPG).

The CG-MD simulation results show that different types and densities of surface charge on AuNPs resulted in either repulsion, adhesion, or penetration into the negatively charged or hydrophobic lipid bilayers. Adhesion of the AuNPs to the bilayers was shown to be supported by electrostatic interactions between the ammonium or carboxylate functionalities of the AuNPs and the bilayer head groups. Cationic AuNPs were found to disrupt the lipid bilayer by forming a hydrophilic pore. To elucidate this process, the authors performed CG-MD simulations of the interactions between cationic AuNPs and negatively charged DPPC bilayers. The positive surface charge density of the AuNPs was varied between 10 and 100 %. The time evolution of the simulation results in Figure 100 shows that the penetration of the cationic AuNPs into the anionic bilayer and the resulting membrane destruction depend crucially on the AuNP charge density, with the penetration depth being significantly greater at low charge densities. The degree of membrane destruction increases rapidly at higher charge densities. The latter is attributed to the formation of hydrophilic pores, which allow the transport of water molecules to the upper leaflet of the membrane and thus its destruction. Thus, the cytotoxicity of cationic AuNPs should correlate with the charge density.

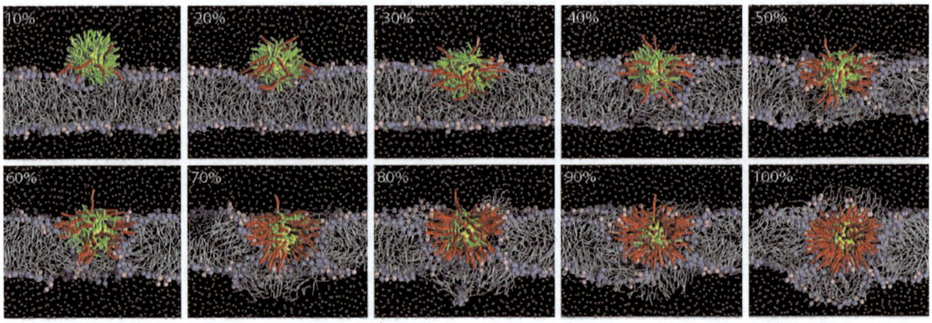

Figure 100: CG-MD simulations of a cationic AuNP interacting with an anionic DPPC/DPPG bilayer, where the surface charge density of the alkylthiol/ammonium alkylthiol terminated AuNP is varied between 10 and 100 %. Snapshots are taken at the end of each 40-ns CG-MD simulation. The gold core is shown in yellow, hydrophobic ligands in green, cationic ligands in red, DPPC headgroups in ice blue, DPPG headgroups in pink, and lipid tails in silver [569].

These simulation results confirm the AFM studies performed by Leroueil et al. [570] on supported DMPC bilayers exposed to dispersions of alkylamine-terminated AuNPs (2 nm). The AFM results prove the hole formation and erosion of the membrane in lipid bilayers triggered by the adsorption of AuNPs.

Wang et al. [571] investigated the influence of positively and negatively charged, 20-nm-sized polystyrene nanoparticles (PSNPs) on the structure of dilaurylphosphocholine (DPPC) bilayers using isothermal titration calorimetry and fluorescence correlation spectroscopy. The adsorbed nanoparticles changed the phase state of the lipid bilayers by interacting with the phosphocholine headgroups. Thus, negatively charged PSNPs induced local gelation in liquid crystalline lipid bilayers, whereas cationic nanoparticles locally liquefied gelled bilayers.

In short, cationic nanoparticles are more likely to be taken up by cells. Ultrasmall cationic nanoparticles, unlike ultrasmall anionic nanoparticles, can be internalized into the cell by direct permeation. However, due to their large radius of curvature and the associated strong local electrostatic interactions with only a few negatively charged lipid head groups in the cell membrane, cationic nanoparticles have been found to induce the formation of hydrophilic holes and other defect structures that can lead to severe cell-membrane damage.

To deepen the understanding of the complex interplay between nanoparticle surface chemistry and cellular response, we present recent research results obtained from different 2D cell cultures incubated with differently functionalized Au–Fe_3O_4 NHDs. Differently charged Au–Fe_3O_4 NHDs were developed for use in radiotherapy as radiation dose enhancers [233, 364]. Snowman-shaped oleylamine/oleic acid-stabilized Au–Fe_3O_4 NHDs were synthesized by thermal decomposition of iron oleate on presynthesized Au nanoparticles. The average size of the Au–Fe_3O_4 NHDs was 16.7 ± 2.3 nm. Subsequent ligand exchange reactions with 4,5-dihydroxybenzene-1,3-disulfonate (Tiron), caffeic acid (CA) or 2,3-dihydroxybenzoic acid (DHBA) ensured the water solubility and biocompatibility of the Au–Fe_3O_4 NHDs (Figure 101). The oleic acid and oleylamine ligands were almost completely or only slightly replaced, depending on the structure of the catechol derivative. Accordingly, the surfaces of the Au–Fe_3O_4 NHDs were proportionally terminated by hydrophobic and hydrophilic ligands. The hydrodynamic diameter and surface charge of the Au–Fe_3O_4 NHDs were quantified by dynamic light scattering (DLS) and zeta potential measurements (Table 1). In the case of DHBA, only a small fraction of the oleic acid and oleylamine ligands were replaced by DHBA, resulting in a very small negative surface charge value (−3.8 mV) and a broad size distribution with a mean size of 55.5 nm, indicating agglomeration. Ligand exchange against Tiron proved to be more efficient, as indicated by the comparatively higher negative surface charge (−18.6 mV) and smaller hydrodynamic diameter at 33.6 nm. The ligand exchange against CA was almost complete. The surface charge of CA@Au–Fe_3O_4 NHDs reached a value of −47.7 mV, and a small hydrodynamic diameter of 28.1 nm. Both results indicate very good water solubility and complete solvation of CA@Au–Fe_3O_4 NHDs. Compared to ultrapure water, biological fluids are characterized by a high ionic strength and contain proteins and other biomolecules. These differences are reflected in larger hydrodynamic diameters of dissolved or dispersed nanoparticles and altered surface charges due to adsorption of amphoteric proteins and ions. The DHBA@Au–Fe_3O_4 NHDs, CA@Au–Fe_3O_4 NHDs and Tiron@Au–Fe_3O_4 NHDs were dissolved in cell culture medium (Dulbecco's

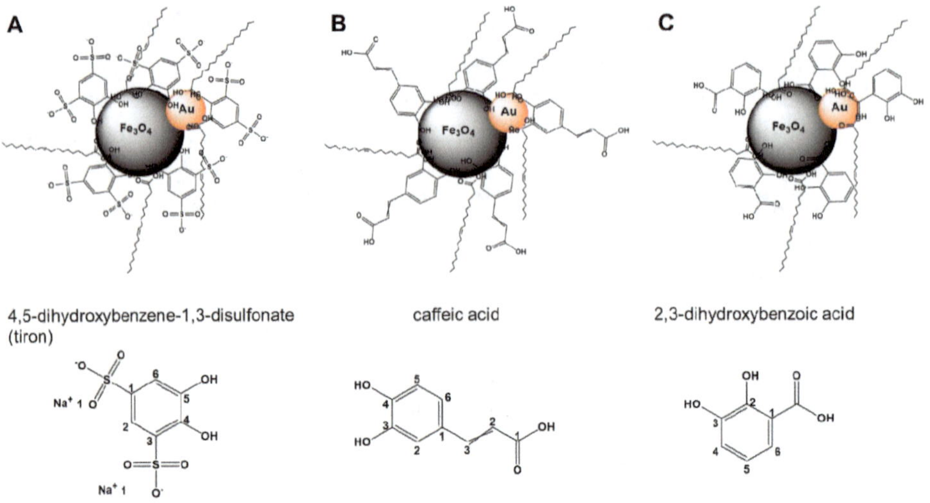

Figure 101: Illustration of snowman-shaped Au–Fe_3O_4 NHDs functionalized with tiron (A) caffeic acid (B) or 2,3-dihydroxybenzoic acid (C).

Table 1: Surface charge and hydrodynamic diameter of the Tiron@Au–Fe_3O_4 NHDs, CA@Au–Fe_3O_4 NHDs and DHBA@Au–Fe_3O_4 NHDs in water or in medium with 10 % serum.

	Hydrodynamic size (water)	Zeta potential (water)	Hydrodynamic size (serum)	Zeta potential (serum)
Tiron@Au–Fe_3O_4	33.6 ± 2.5 nm	−18.6 ± 1.46 mV	60.2 ± 6.9 nm	−8.02 ± 0.76 mV
CA@Au–Fe_3O_4	28.1 ± 3.3 nm	−47.7 ± 3.74 mV	51.2 ± 6.1 nm	−9.78 ± 0.63 mV
DHBA@Au–Fe_3O_4	55.5 ± 7.1 nm	−3.81 ± 0.54 mV	81.7 ± 9.5 nm	+7.44 ± 0.58 mV

modified Eagle's medium (DMEM)) containing 10 % fetal calf serum (FCS). Results from DLS and zeta potential measurements confirmed both, protein and ion adsorption, but not the formation of a complete protein corona (Table 1). All kinds of the Au–Fe_3O_4 NHDs showed increased hydrodynamic diameters. The zeta potential values of anionic CA@Au–Fe_3O_4 NHDs and Tiron@Au–Fe_3O_4 became obviously smaller due to the adsorption of amphoteric proteins or cations. Surprisingly, the zeta potential of the initially anionic DHBA@Au–Fe_3O_4 NHDs, changed to a positive surface charge.

The influence of hydrodynamic size and zeta potential on the efficiency of cellular uptake into breast cancer cells (MCF-7) and non-tumorigenic breast epithelial cells (MCF-10A) was investigated by TEM. Obviously, the differently charged and sized Au–Fe_3O_4 NHDs were taken up intracellularly by endocytosis (black arrow) into both MCF-7 and MCF-10A (Figure 102(A)). This TEM analysis shows that the differences in surface charge and hydrodynamic diameter do not play a role in the mechanism of cellular internalization, which can be explained by the desorption of the presumably loosely attached proteins during the endocytic pathway.

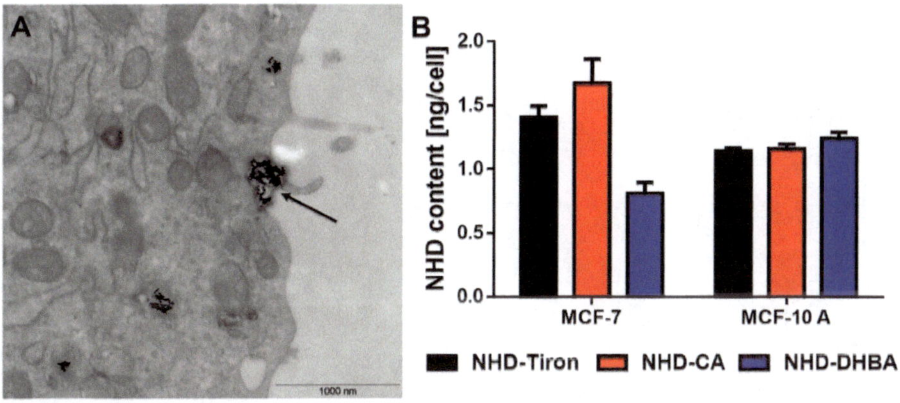

Figure 102: TEM image (A): cellular uptake of the Au–Fe_3O_4 NHDs into MCF-7 cells by endocytosis, (B) the intracellular Au–Fe_3O_4 NHD content (ng/cell) in cancerous MCF-7 cells and healthy MCF-10A cells.

By contrast, the uptake efficiency in MCF-7 cells was found to depend on the surface chemistry of the Au–Fe_3O_4 NHDs. The CA@Au–Fe_3O_4 NHDs with the largest negative surface charge and the smallest hydrodynamic diameter were taken up with the highest efficiency by MCF-7 cells, while the weakly positively charged DHBA@Au–Fe_3O_4 NHDs were internalized with the lowest concentration in MCF-7 cells. On the other hand, the differently functionalized Au–Fe_3O_4-NHDs were taken up by MCF-10A cells to approximately the same extent (Figure 102(B)).

Another example for the impact of surface chemistry and structure on cellular internalization and intracellular fate are Au–Fe_3O_4 NHDs that are encapsulated in hydrophilic membrane-like bilayers [246]. As described in detail in Section 3.6.3 initially oleic acid/oleylamine-terminated Au–Fe_3O_4 NHDs were coated with sodium 4-dodecylbenzenesulfonate or 1-octadecylpyridinium bromide by ultrasonic treatment (Figure 103). These amphiphilic surfactants organized themselves into a highly ordered layer by interlocking their lipophilic alkyl chains with those of the oleylamine and oleic acid ligands. The resulting ordered positively or negatively charged bilayers had a thickness of 8.5 nm. The positively charged (1-octadecylpyridinium) bilayer coated Au–Fe_3O_4 NHDs (pos.shell@Au–Fe_3O_4 NHDs) exhibited a rather small hydrodynamic diameter of 28.2 nm and a zetapotential value of 56.2 mV, when dissolved in ultrapure water (Table 2). Both values were changed by dissolving the pos.shell@Au–Fe_3O_4 NHDs in cell culture medium supplemented with 10 % FCS. The hydrodynamic diameter increased to 105.7 nm and the zeta potential decreased to 6.5 mV. These changes are indicative of protein adsorption on the surface of pos.shell@Au–Fe_3O_4 NHDs, which may cause agglomeration. In contrast to pos.shell@Au–Fe_3O_4 NHDs, dissolution of negatively charged (4-dodecylbenzenesulfonate) bilayer coated Au–Fe_3O_4 NHDs (neg.shell@Au–Fe_3O_4 NHDs) in cell medium increased their hydrodynamic diameter to 50.7 nm, while the zeta potential value rose to −24.4 mV, only. These smaller changes indicate lower protein adsorption.

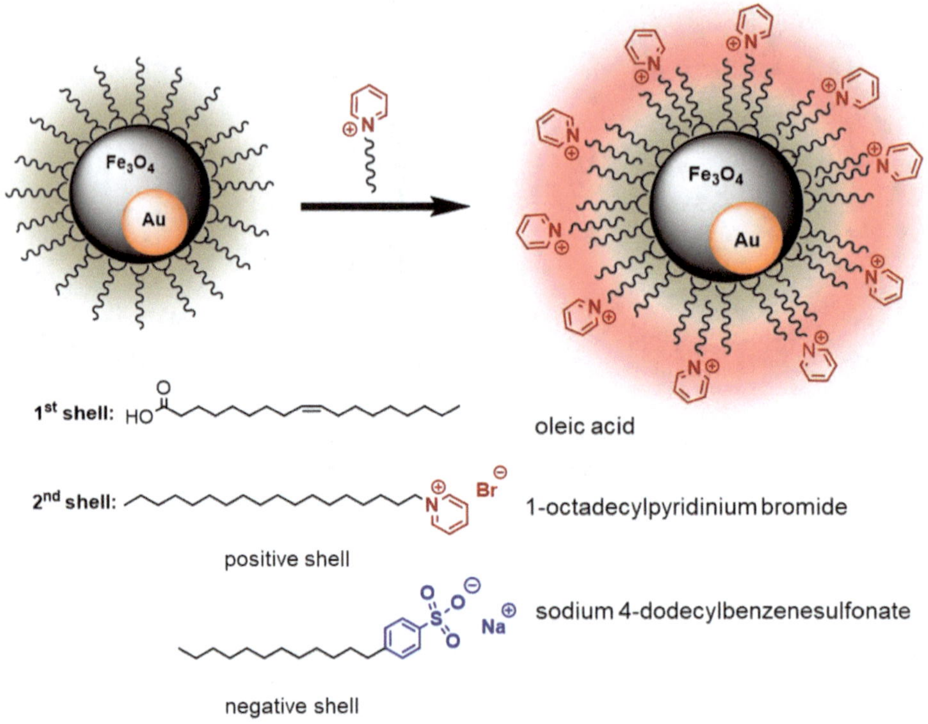

Figure 103: Coating of oleic acid/oleylamine stabilized Au–Fe_3O_4 NHDs with 1-octadecylpyridinium bromide or sodium 4-dodecylbenzenesulfonate.

Table 2: Surface charge and hydrodynamic diameter of the bilayer-coated Au–Fe_3O_4 NHDs in ultrapure water or in medium enriched with 10 % serum.

	Hydrodynamic size (water)	**Zeta potential (water)**	**Hydrodynamic size (serum)**	**Zeta potential (serum)**
pos.shell@Au–Fe_3O_4 NHD[a]	28.21 ± 3.3 nm	+56.2 ± 0.231 mV	105.7 ± 12.7 nm	+6.54 ± 0.27 mV
neg.shell@Au–Fe_3O_4 NHD[b]	28.21 ± 3.4 nm	−47.6 ± 0.723 mV	50.75 ± 4.0 nm	−24.4 ± 1.26 mV

[a]Initially terminated with oleic acid/oleylamine ligands, subsequently coated with 1-octadecylpyridinium.
[b]Initially terminated with oleic acid/oleylamine ligands, subsequently coated with 4-dodecylbenzenesulfonate.

Both the hydrodynamic diameter and the surface charge of the bilayer-stabilized Au–Fe_3O_4 NHDs were shown to determine the cellular entry route, intracellular trafficking, and cytotoxicity. The TEM images in Figure 104 show MCF-7 cells with internalized pos.shell@Au–Fe_3O_4 NHDs (A) and neg.shell@Au–Fe_3O_4 NHDs (B). The Au–Fe_3O_4 NHDs were observed to be cumulatively distributed in the cytoplasm, whereas endosomes with internalized Au–Fe_3O_4 NHDs were not detectable. This implies that the Au–Fe_3O_4

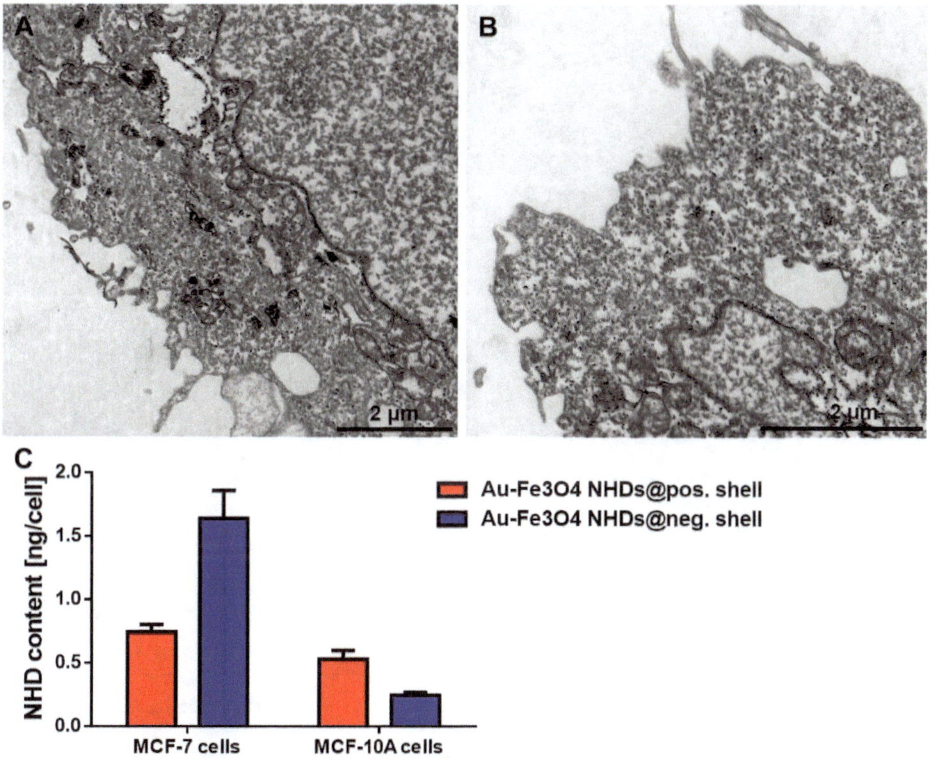

Figure 104: TEM images of pos.shell@Au–Fe_3O_4 NHDs (A) and neg.shell@Au–Fe_3O_4 NHDs (B) internalized into MCF-7 cells, (C) intracellular content of Au–Fe_3O_4 NHDs (ng/cell) MCF-7 and MCF-10A cells.

NHDs encapsulated in highly ordered bilayers, being membrane-like, were taken up by MCF-7 cells through direct permeation, which is consistent with the research results of Verma et al. [358]. Compared with the pos.shell@Au–Fe_3O_4 NHDs, the neg.shell@Au–Fe_3O_4 NHDs were taken up by the MCF-7 cells in significantly larger amounts (Figure 104(C)), which was attributed to the differences in the hydrodynamic diameters and surface charge state. MCF-10A cells contained pos.shell@Au–Fe_3O_4 NHDs at a slightly lower concentration. The difference between the intracellular content of neg.shell@Au–Fe_3O_4 NHDs in MCF-7 cells and the intracellular content in MCF-10A cells is much more pronounced. The observation that neg.shell@Au–Fe_3O_4 NHDs are taken up with high efficiency by direct permeation of tumor cells and sparing of healthy cells shows their benefit for applicability in cancer therapy. On the other hand, the pos.shell@Au–Fe_3O_4-NHDs are completely unsuitable for medical use, as these Au–Fe_3O_4-NHDs adhere to the mitochondria in the cells, depolarize them and are therefore cytotoxic.

Another example of membrane-like ordered bilayer shells as a coating of nanoparticles was published by Stiegler et al. [572]. Al_2O_3 nanoparticles (Al_2O_3 NPs) of 14-nm size were terminated with two types of alkyl chain-bearing phosphonic acids, one with a

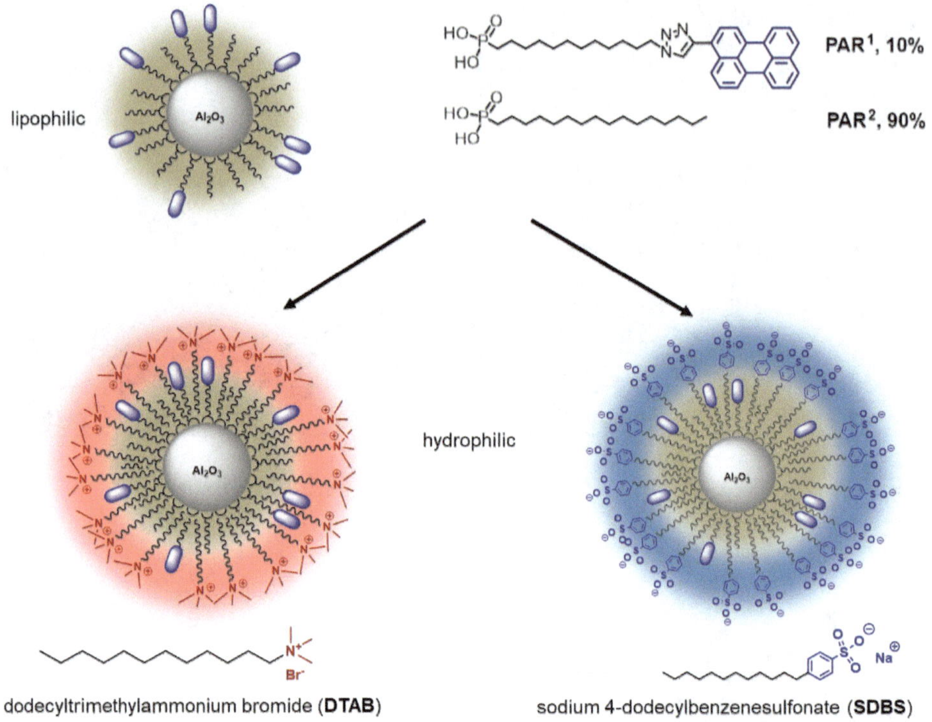

Figure 105: Bilayer architectures around the Al_2O_3 NPs: First layer consists of fluorescent PAR^1 and PAR^2; second layer formed by positively charged dodecyltrimethylammonium (DTAB) or negatively charged 4-dodecylbenzenzesulfonate (SDBS).

Table 3: Surface charge and hydrodynamic diameter of DTAB-PAR^1/PAR^2@Al_2O_3 NPs and SDBS-PAR^1/PAR^2@Al_2O_3 NPs in water or in medium enriched with 10 % serum.

	Hydrodynamic size (water)	Zeta potential (water)	Hydrodynamic size (serum)	Zeta potential (serum)
DTAB-PAR^1/PAR^2@Al_2O_3	100.89 ± 6.80 nm	+33.8 ± 2.46 mV	149.27 ± 12.93 nm	−32.8 ± 0.651 mV
SDBS-PAR^1/PAR^2@Al_2O_3	100.89 ± 6.80 nm	−58.0 ± 2.87 mV	46.13 ± 4.00 nm	−26.4 ± 1.89 mV

fluorescent perylene end group (PAR^1) perylene, and the other an analogous surfactant without perylene fluorophore (PAR^2). These amphiphilic surfactants composed of 10 % PAR^1 and 90 % PAR^2 form a lipophilic layer around the Al_2O_3 NPs (Figure 105). To create positively and negatively charged hydrophilic bilayers, the PAR^1/PAR^2 stabilized Al_2O_3 NPs were dissolved in an aqueous solution of either dodecyltrimethylammonium (DTAB) or 4-dodecylbenzenesulfonate (SDBS) by ultrasound treatment. The results from the DLS and zeta potential experiments indicate agglomeration of the DTAB-PAR^1/PAR^2@Al_2O_3 NPs in water (100.9 nm) and in a cell medium (149 nm) (Table 3). These bilayer-stabilized nanoparticles are positively charged in water as expected (+33.8 mV), but negatively charged in the cell medium (–32.8 mV). The latter result can only be explained by damage to the bilayer through interaction with the serum proteins.

The hydrodynamic diameters of negatively charged SDBS-PAR[1]/PAR[2]@Al_2O_3 NPs in water and cell medium were determined to be 100.9 nm and 46.1 nm, respectively, while the zeta potential value changed from –58 mV to –26.4 mV due to protein adsorption (Table 3). Of note is the decrease in hydrodynamic diameter in the cell medium due to deagglomeration by interaction with ions and proteins in the serum. The influence of different sized and charged bilayer-stabilized Al_2O_3 NPs on the uptake mechanism in lung adenocarcinoma (A549) cells was investigated by TEM (Figures 106(A)–(C)). The TEM images in Figures 106(A) and (B) illustrate the endocytic pathway of DTAB-PAR[1]/PAR[2]@Al_2O_3 NPs by showing the formation of the endosome (Figure 106(A)) and the trafficking of the endosome into the cytosol (Figure 106(B)). By contrast, the negatively charged SDBS-PAR[1]/PAR[2]@Al_2O_3 NPs entered the cell directly (Figure 106(C)). This difference in the cellular uptake mechanism is mainly due to the significantly different hydrodynamic diameters of the two kinds of bilayer-stabilized Al_2O_3 NPs in the cell medium. The dif-

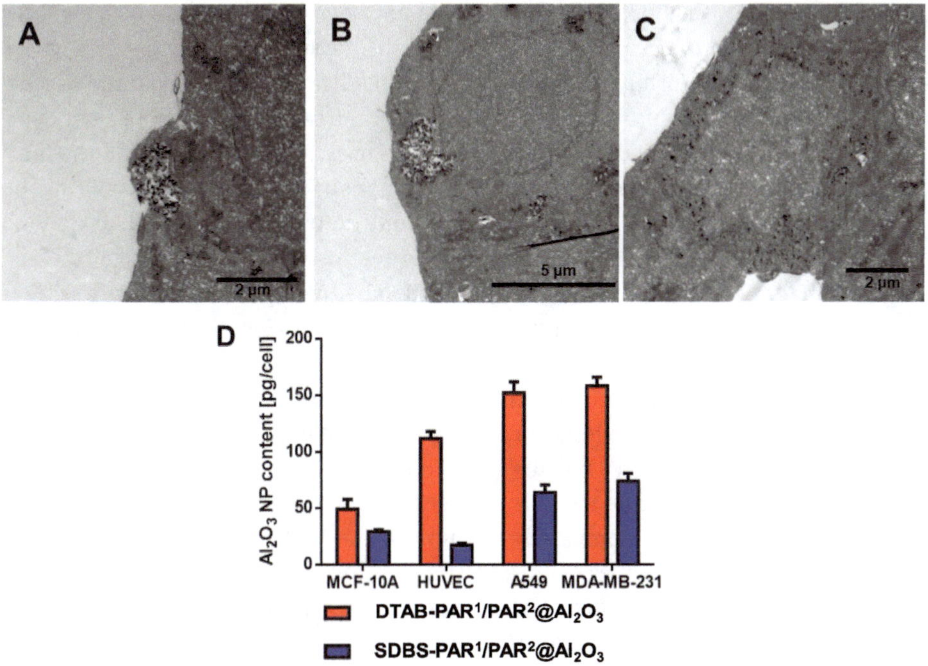

Figure 106: TEM images of DTAB-PAR[1]/PAR[2]@Al_2O_3 NPs internalized by endocytosis (A) and (B) and SDBS-PAR[1]/PAR[2]@Al_2O_3 NPs taken up by direct permeation (C) into lung cancer A549 cells; determination of the intracellular NP content for various cell lines: non-tumorigenic breast epithelial MCF-10A cells and human umbilical-vein endothelial cells (HUVEC) or cancer cell line A549 (lung cancer) and MDA-MB-231 (breast cancer).

ferent uptake mechanisms and, especially, the different hydrodynamic diameters obviously led to different intracellular contents of DTAB-PAR[1]/PAR[2]@Al_2O_3 NPs and SDBS-PAR[1]/PAR[2]@Al_2O_3 NPs in the non-tumorigenic MCF-10A and HUVEC cells as well as in the cancerogenic A549 and MDA-MD-231 cells (Figure 106(D)). Obviously, direct permeation of SDBS-PAR1/PAR2@Al_2O_3 NPs into the four cell lines occurs with significantly lower efficiency than the endocytosis of DTAB-PAR[1]/PAR[2]@Al_2O_3 NPs. This can be explained by the large hydrodynamic diameter of SDBS-PAR[1]/PAR[2]@Al_2O_3 NPs, which is unfavorable for direct permeation.

Gold nanoparticles (AuNPs)

Chithrani et al. [415] investigated the influence of the geometry of transferrin-coated AuNPs on their endocytic and exocytic processes in three different cell lines: STO cells, HeLa cells and SNB19 brain tumor cells. For this purpose, they synthesized AuNPs with sizes between 14 and 100 nm and AuNRs with aspect ratios in the range of 1.5–6.4. As their results unambiguously show, the cellular uptake of both, transferrin-coated AuNPs and AuNRs, occurs mainly via the process of clathrin-mediated endocytosis. Following the results from theoretical modeling of receptor-mediated endocytosis by Gao et al. [573], the uptake process of nanoparticles depends on how a cell membrane, which contains diffuse mobile receptors, wraps around a cylindrical or spherical nanoparticle whose surfaces consist of uniformly distributed, compatible ligands. Derived parameters determine the so-called wrapping time that the membrane requires to enclose a nanoparticle. According to these model calculations, the shortest wrapping time and the most efficient receptor-ligand interaction occurred at a nanoparticle size of about 55 nm.

Chithrani et al. [415] proved this theoretical study for transferrin-coated AuNPs with sizes of 14 nm, 50 nm, and 74 nm. They observed the highest cellular uptake rate and amount for 50-nm AuNPs. By comparison, much smaller nanoparticles can only achieve a sufficiently large contact area on the plasma membrane by agglomeration. This is consistent with the observation of endosomes containing clusters of six 14-nm AuNPs. Since the formation of AuNP clusters on a plasma membrane is not very likely, the cellular uptake efficiency of 14 nm AuNPs is much lower than that of 50-nm AuNPs. On the other hand, 74-nm and larger AuNPs are also taken up more slowly and in smaller numbers because they bind to multiple membrane receptors, which significantly increases the membrane wrapping time. Compared to spherical AuNPs, AuNRs with the widths and lengths 20 nm × 30 nm, 14 nm × 50 nm, and 7 nm × 42 nm show lower cellular uptake efficiency, which is partly attributed to insufficient transferrin coating. Moreover, the rate of cellular uptake decreases with the aspect ratio.

Yue et al. [574] optimized AuNPs as transfection agents for siRNA transport and investigated the effect of size and shape on internalization in U87 glioblastoma cells. Using various formulations, they prepared siRNA-AuNP conjugates containing either spherical 13-nm and 40-nm AuNPs or 40-nm Au nanostars. The siRNA-AuNP conjugates differed not only in geometry but also in surface chemistry and siRNA density. The siRNA of these conjugates is used to transiently downregulate the expression of isocitrate dehydrogenase in the target cells. Transmission and confocal fluorescence microscopy studies showed that all Au nanoconstructs were internalized by U87 glioblastoma cells via receptor-mediated endocytosis. The larger nanoparticles, the spherical 50-nm AuNPs and the 40-nm Au nanostars, showed a significantly higher cellular uptake efficiency than the 13-nm Au nanospheres as remaining longer in the endosomes. These results confirm the findings reported by Chithrany et al. [415] and Gao et al. [573].

Nambara et al. [575] examined the cellular uptake efficiency of 30 nm thick triangular gold nanoparticles (AuNTs) with different edge lengths (46 nm, 55 nm, 72 nm, and 94 nm) in comparison to the cellular uptake efficiency of spherical nanoparticles (AuNSs) with different diameters (22 nm, 39 nm, and 66 nm). The nanoparticle surfaces were functionalized with an anionic nitrilotriacetic acid-terminated alkanethiol. HeLa cells and RAW264.7 mouse macrophages were used for cell experiments to investigate the influence of nanoparticle morphology on different uptake mechanisms, receptor-mediated endocytosis in HeLa cells and phagocytosis in RAW264.7 cells. The ICP-AES results showed that AuNTs with longer sides exhibited higher uptake efficiency in RAW264.7 and HeLa cells. In particular, in HeLa cells, a 20-fold more efficient internalization was observed for AuNTs with a side length of 72 nm compared to 66-nm SNPs that possess a similar surface area. The triangular planar shape of nanoparticles occupies a larger contact area on the plasma membrane than spherical nanoparticles of the same size. The cellular uptake of nanoparticles is a two-step process that starts with the adhesion of the nanoparticles to the plasma membrane and initiates, for example, the formation of endosomes. The AuNTs exert a stronger adhesive force on the membrane as they increase in size. In addition, the edges and vertices of the AuNTs accelerate the membrane wrapping process and thus the internalization through large local curvature.

Nanoparticles tend to aggregate in cell culture media and biological fluids due to the ions and proteins they contain. Therefore, it is essential to elucidate the influence of nanoparticle aggregation on internalization by cells, intracellular trafficking, and exocytosis. Albanese et al. [576] have developed a protocol for the preparation of protein-stabilized AuNP aggregates with sizes between 26 and 98 nm. To achieve receptor-mediated endocytosis into tumor cells, single AuNPs and their aggregates were coated with transferrin. The aggregate size was adjusted to 26 nm, 48 nm and 96 nm by exposing 16-nm citrate-stabilized AuNPs to different concentrations of NaCl and transferrin. NaCl dissociates into Na^+ and Cl^- in aqueous solution. The Na^+ cations electrostatically bind to the carboxy groups of the citrate ligands, reducing the repulsive interactions between the AuNPs and thereby inducing aggregate formation. Simultaneous or subsequent addition of transferrin, which also binds to the citrate ligands but via van-der-Waals interactions, can partially or completely prevent aggregation. Transferrin-coated AuNPs show little tendency to aggregate. By adjusting the NaCl and transferrin concentrations, transferrin-coated AuNP aggregates with sizes of 26 nm, 48 nm, and 96 nm were obtained. The cellular uptake of these AuNP aggregates in HeLa, A549 and MDA-MB 435 tumor cells was investigated by transmission electron microscopy and ICP-AES. For comparison, analogous experiments were performed with monodisperse AuNPs ranging in size from 16 nm to 100 nm.

Since tumor cells differ in the expression of the transferrin receptor (CD71), the CD71 expression of HeLa and A549 and MDA-MB 435 tumor cells was determined by flow cytometry. HeLa cells were found to have the highest CD71 expression, while MDA-MB-435 cells had the lowest number of CD71 receptors in the plasma membrane. Compared to the 16-nm AuNPs, the transferrin-coated AuNPs aggregates were observed to hamper uptake by receptor-mediated endocytosis in HeLa and A549 cells. By contrast, MDA-MB-435 cells took up twice as much of the 98-nm aggregates as of the smaller aggregates and single AuNPs. This is contradictory because MDA-MB-435 cells have the smallest number of transferrin receptors in their plasma membrane. Moreover, large AuNP aggregates need to interact with many more transferrin receptors than small aggregates or single AuNPs to induce the membrane wrapping required for receptor-mediated endocytosis. Therefore, the cellular uptake of large AuNP aggregates in MDA-MB-435 cells should occur through additional, and unfortunately still unclear, uptake pathways.

These results were confirmed by Xie et al. [577] who investigated the cellular internalization of methyl polyethylene glycol-coated Au nanostars (mPEG-AuNSs), nanorods (mPEG-AuNRs), and nanotriangles (mPEG-AuNTs) by RAW264.7 cells. The mPEG-AuNPs all had comparable sizes of 50 nm and hydrodynamic diameters of 60–70 nm as well as carried no surface charges. Thus, the cellular uptake of the similarly sized AuNPs into RAW264.7 cells was determined only by the shape and surface curvature of the nanoparticles. The cellular uptake efficiency increased in the order of mPEG-AuNSs, mPEG-AuNRs and mPEG-AuNTs. This was explained as not only depending on the contact area between the AuNPs and the membrane but also on the impact of the nanoparticle shape on the endocytosis mechanism. As the authors found by inhibiting different endocytic pathways, all three AuNP shapes utilized clathrin-mediated endocytosis. However, cellular uptake of AuNRs also occurred through caveolae/lipid raft-mediated endocytosis. The significantly more efficient cellular uptake of the AuNTs was explained by the increased GTPase activity of dynamin. While clathrin polymerizes during endocytosis and serves as a coating for the vesicles, dynamin causes their cleavage by forming a ring around the neck of the vesicles, which accelerates the internalization of the nanoparticles.

Superparamagnetic iron-oxide nanoparticles

Depending on their size, superparamagnetic iron oxide nanoparticles are divided into three categories: ultrasmall superparamagnetic iron oxide nanoparticles (USPIONs) (10–50 nm); SPIONs (5–150 nm); and micron-sized SPIONs (μSPIONs) [219, 578]. μSPIONs, if they are taken up at all by phagocytosis, remain largely in the extracellular medium or blood vessels. By contrast, properly coated USPIONs and SPIONs are efficiently taken up by a wide variety of cells and can be administered to the patient orally, intravenously, or subcutaneously [319, 579]. Feng et al. [580] systematically investigated the in vitro cell uptake, cytotoxicity, in vivo distribution, exocytosis, and toxicity of 10-nm and 30-nm USPIONs in RAW264.7 macrophages and non-phagocytic SKOV-3 ovarian cancer cells. TEM examination of the commercially available PEI- and PEG-terminated USPIONs revealed monodispersity and spherical shapes, indicating their synthesis by thermal decomposition. PEI-coated 10-nm USPIONs were internalized by both cell lines to a significantly greater extent than PEGylated USPIONs of the same particle size. This was explained by the affinity of the cationic nanoparticles to the negative phospholipid head groups or protein domains. However, PEI-terminated SPIONs form large aggregates in cell culture media and are cytotoxic. In contrast, PEGylation prevents aggregation of USPIONs, while providing biocompatibility. In addition, the 10-nm USPIONs enter cells at higher concentrations than the 30- nm USPIONs. TEM studies revealed that all USPIONs were mainly distributed in endosomes and/or lysosomes in both macrophages and cancer cells. The 10-nm PEI-terminated USPIONs were preferentially internalized by adsorptive endocytosis, whereas the cellular internalization of the PEGylated USPIONs was predominantly mediated by receptor-mediated endocytosis.

Another study on the size dependence of internalization of PEGylated USPIONs in RAW264.7 mouse macrophages led to conflicting results attributed to the different synthesis techniques. The PEGylated USPIONs synthesized by Larsen et al. [581] were obtained by alkaline co-precipitation in the presence of oleic acid. The oleic acid ligand was replaced by commercial silane PEG in a second synthesis step. The PEGylated USPIONs have been obtained in two sizes: 20 nm and 40 nm. The average iron oxide core sizes of the 20-nm and 40-nm USPIONs were reported to be 4.6 and 9.4 nm, respectively. As the TEM images clearly show, the USPIONs are potato-shaped rather than spherical and polydisperse. Using the ferrozine assay, the authors demonstrated that the internalization of 40-nm USPIONs into macrophages was increased 8-fold compared to the 20-nm nanoparticles. Similar results were obtained from *in vivo* experiments in mice with subcutaneously implanted foot tumors. In contrast to the 20 nm USPIONs, only the intravenously administered 40 nm USPIONs were detected in the tumor by MRI. Ge et al. [582] also synthesized USPIONs with various size distributions and indeterminate shapes by alkaline co-precipitation. The USPIONs were coated with chitosan (CS) or meso-2,3-dimercaptosuccinic acid (DMSA). The average sizes of positively charged CS@USPIONs and negatively charged DMSA@USPIONs were determined to be 16.5 ± 6.1 nm and 17.3 ± 4.8 nm, respectively, by TEM analysis. Electrostatic aggregation of CS@USPIONs and DMSA@USPIONs resulted in slightly positively charged CS-DMSA@USPIONs with an average size of 85.7 ± 72.9 nm. The authors investigated the influence of surface chemistry and size of USPIONs on the uptake in oral squamous cell carcinoma (OSCC) cells. The results of quantitative and qualitative studies show that the cellular uptake efficiency is highest for the CS-DMSA@USPIONs, while the concentration of uptaken DMSA@USPIONs is lowest and thus lower than that of DMSA@USPIONs. These results confirm the findings of Feng et al. [580] regarding the influence of surface charge on cellular uptake. On the other hand, the highest uptake efficiency for CS-DMSA@USPIONs can also be explained by the decomposition of the aggregates into the contained

CS@USPIONs and DMSA@USPIONs during the interaction with the negatively charged surface and is thus attributed to a concentration effect rather than a size effect. On the other hand, aggregation of USPIONs with equal surface charges can be driven by proteins and other biomolecules in extracellular fluids. Adsorption of biomolecules onto the SPION surface reduces the electrostatic repulsion between SPIONs, leading to aggregation. Aggregates larger than 100 nm that interact with cells have a low probability of being taken up as aggregates by cells. Since the membrane surface is negatively charged, weakly bound aggregates can dissociate on the membrane. As a result, individual nanoparticles interact with the membrane in higher concentrations, resulting in increased internalization efficiency.

The use of dense, stably bound and highly hydrated polymer shells around USPIONs and SPIONs should make aggregation impossible. Commonly used polymers are dextran, PEG, glycopolymers, and polyoxazolines [169, 583, 584]. The indefinite thickness of the polymer shells and the frequent encapsulation of multiple SPIONs within a polymer shell are disadvantageous. Furthermore, SPIONs with non-polar and uncharged surface coatings can aggregate through magnetic dipole–dipole interactions, which is likely to complicate cellular uptake.

In contrast to AuNPs, the shape diversity of SPIONs is essentially limited to spherical, cubic and undefined (potato) morphologies. Monodisperse SPIONs with spherical and cubic shapes can be reliably prepared by thermal decomposition, while the generally preferred alkaline co-precipitation method yields polydisperse SPIONs with indeterminate shapes. For this reason, the validity of the results of studies on the influence of nanoparticle morphology, when obtained by different synthesis techniques, on cellular uptake is very limited.

Silicon and silica nanostructures

Silicon nanostructures can be prepared wet-chemically, by gas-phase synthesis, by microwave and ultrasound assisted synthesis techniques, by electrochemical etching of silicon wafers, and by laser ablation, as described in detail in Section 3.5. The surface structures and morphologies of wet chemically prepared Si nanostructures differ significantly from spherical SiQDs, such as those prepared by thermal decomposition in the gas phase. In the wet-chemical synthesis described in [517], Si nanoparticles of various sizes were obtained depending on the phase transfer agent, i. e., tetraoctylammonium bromide (TOAB) and tetraoctylammonium fluoride (TBAF), which were subsequently terminated with APTES, i. e., they should have the same positively charged surface structures. The TOAB-assisted synthesis resulted in 3.5 ± 1.5 nm APTES-terminated SiQDs, while TBAF as a phase-transfer agent resulted in a 1.5-nm APTES-terminated Si nanocluster (SiNC). Both, ultrasmall Si nanostructures entered MCF-7 cells by direct permeation (Figure 107). While the 1.5-nm SiNCs preferentially adhered to the membranes of cell organelles, the larger (3.5 nm) SiQDs formed agglomerates in the cytoplasm near the cell

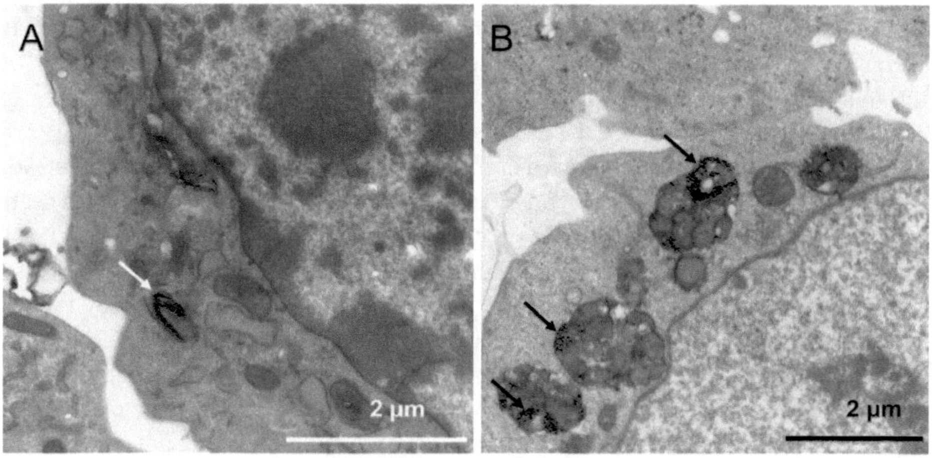

Figure 107: Direct permeation of 3.5-nm SiQDs (A) and 1.5-nm SiNC (B): the 3.5-nm SiQDs form agglomerates in the cytoplasm (black arrows) and the 1.5-nm SiNC adhere to the cell organelles membrane (white arrow) in MCF-7 cells.

organelles. Apparently, the intracellular transport and fate of both Si nanostructures, which have similar surface chemistries, are controlled by the size of the nanoparticles.

Bulky surface structures, such as conjugates with biomolecules, also affect the cellular uptake mechanism of SiQDs that were initially terminated with small molecule ligands. This was observed by Klein et al. [293] for the uptake pathway of SiQD-siRNA conjugates in Caco-2 cells. For these studies, 2.3-nm green luminescent, 2-vinylpyridine-terminated SiQDs were synthesized by microwave-assisted decomposition of SiH_4 in the gas phase followed by thermal hydrosilylation with 2-ethynylpyridine. The water-

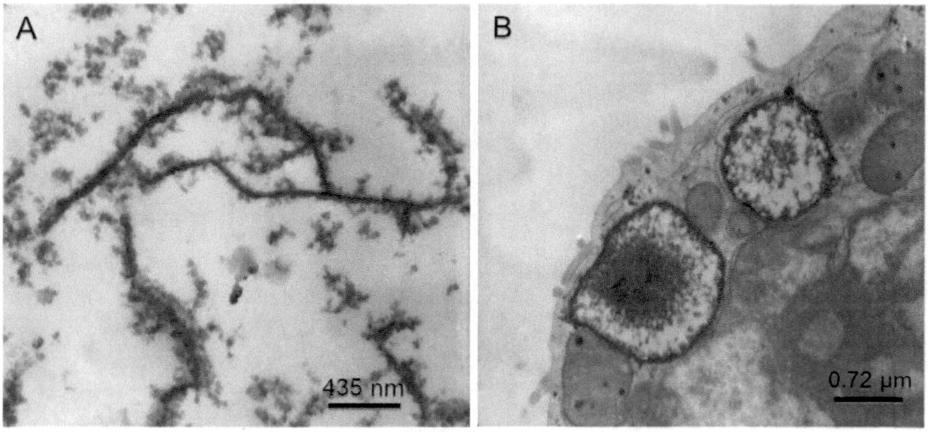

Figure 108: TEM images of the 2-vinylpyridine SiQD-siRNA conjugates (A) and their cellular uptake via endocytosis (B) in Caco-2 cells.

soluble, positively charged SiQDs form stable conjugates with siRNA that are functional for transfection of siRNA into mammalian cells. As illustrated in Figure 108(A), the SiQD–siRNA conjugates, which are at least 10 nm in size, contain multiple SiQDs, explaining their endocytic uptake pathway in Caco-2 cells (Figure 110(B)).

Silica nanoparticles (SiO_2NPs) are prepared with nonporous (NSiO_2NP) and mesoporous surfaces (MSiO_2NP). Hence, another key player, the surface structure, has been suggested for cellular uptake in addition to shape and size [251, 585–587]. Lu et al. [585] developed a simple synthesis procedure that allows the preparation of monodisperse MSiO_2NPs in the size range of 30 and 280 nm and with adjustable charges by tuning the pH. The authors investigated the effect of particle size on cell uptake efficiency in HeLa cells employing confocal laser scanning microscopy (CLSM). Quantitative determination of cellular uptake efficiency was performed by quantifying the total silicon content using inductively coupled plasma mass spectrometry (ICP-MS). Jiang et al. [588] confirmed the size dependence of the cellular uptake efficiency. According to their results, 40–50 nm nanoparticles are optimally suited to activate membrane receptors for the endocytic uptake pathway. On the other hand, Saikia et al. [589] elucidated the influence of SiO_2NP size and porosity on the mechanisms of cellular uptake and toxicity of RAW 264.7 macrophages. To evaluate the effect of porosity on protein adsorption, spherical MSiO_2NP with a diameter of 466 ± 86 nm were compared with NSiO_2NPs (432 ± 18.7 nm), while NSiO_2NPs with a mean size of 46 ± 4.9 nm were used to study the size dependence (Figure 109). The MSiO_2NP surfaces consist of a network of pores with sizes around 3 nm and a surface area of 950 m^2/g that is 140-fold larger than that of the

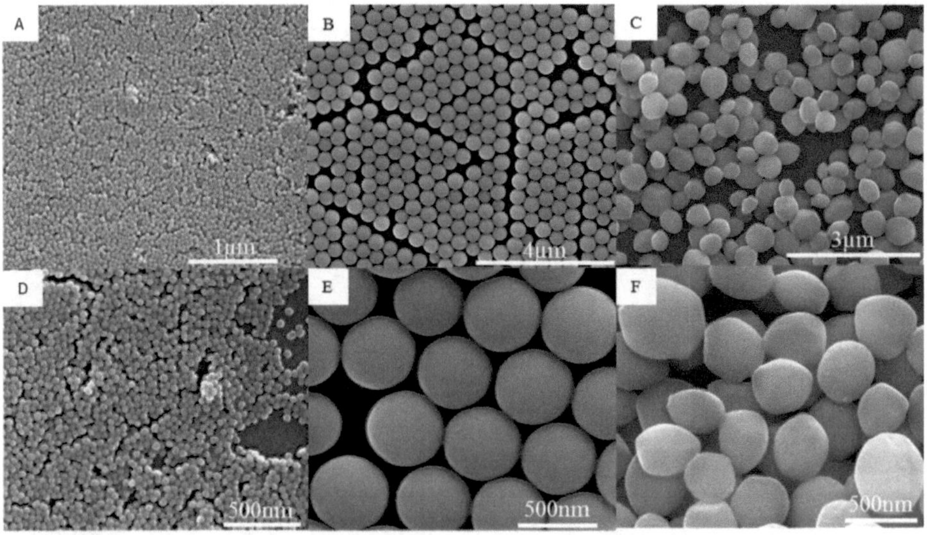

Figure 109: Scanning electron microscopy images of (A, D) NSiO_2NPs with an average diameter of 46 nm, (B, E) NSiO_2NPs with an average diameter of 432 nm, and (C, F) MSiO_2NPs with an average diameter of 466 nm [589].

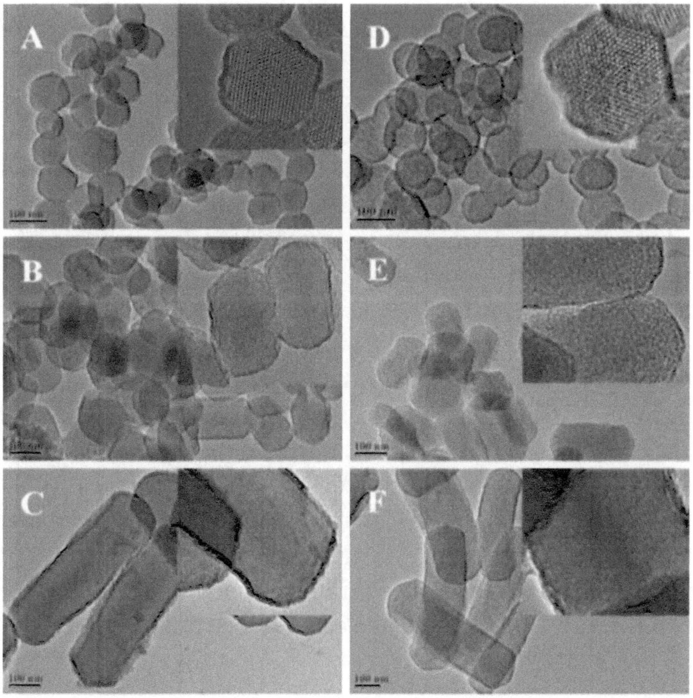

Figure 110: Transmission electron microscopy (TEM) images of naked and PEGylated $FMSiO_2$ NPs. (A) sphere-shaped particles, (B) short-rod-shaped particles, and (C) long rod-shaped particles. (D–F) Indicate the corresponding PEGylated particles of (A–C), respectively. The insets of TEM images with higher magnification show clear mesoporous structures of particles, respectively [587].

equally sized $NSiO_2NPs$ with a surface area 186 m^2/g. The cytotoxicity of SiO_2NPs is due to the surface silanol groups (≡SiOH) that interact with tetraalkylammonium-terminated phospholipids, resulting in cell membrane lysis. In addition, strained three-membered siloxane rings lead to the formation of highly reactive hydroxyl radicals upon reaction with water. Therefore, the adsorption of proteins in biological fluids and ultimately the formation of protein corona reduces the cytotoxicity of SiO_2NPs. The 432-nm $NSiO_2NPs$ were found to bind serum proteins most strongly compared to the 466-nm $MSiO_2NPs$ and 46-nm $NSiO_2NPs$. Consequently, protein adsorption increases with decreasing curvature radius, and smooth SiO_2NP surfaces are preferred over mesoporous surfaces. As expected, the 500-nm $NSiO_2NPs$ showed the lowest cytotoxicity. The uptake mechanism and efficiency were mainly determined by size. In agreement with the literature [258, 585, 588], the amount of intracellular small 46-nm $NSiO_2NPs$ was significantly larger than that of $MSiO_2NPs$ and $MSiO_2NPs$ remained predominantly attached to the outer plasma membrane of macrophages. The 46-nm-sized $NSiO_2NPs$ were localized inside endosomes as well as in the cytosol near the inner membrane, indicating an endocytic mechanism.

It is well known in the literature [433] that the shape of nanoparticles plays a key role in the interaction mechanisms with cells. Hao et al. [587] investigated for the first time the influence of the morphology of mesoporous silica nanoparticles on the cellular uptake pathway and mechanism in Hela cells. For this purpose, spherical and rod-shaped fluorescent (FITC-doped) PEGylated mesoporous silica nanoparticles ($FMSiO_2NPs$) with diameters of 100 nm and aspect ratios (AR) of 1, 2, and 4 were synthesized (Figure 110). The uptake pathway and kinetics were found to be significantly shape dependent. The spherical $FMSiO_2NPs$ were preferentially internalized via clathrin-mediated endocytosis, whereas the cellular uptake mechanism of the long-rod $FMSiO_2NPs$ (AR = 4) was attributed to caveolae-mediated endocytosis. On the other hand, short-rod $FMSiO_2NPs$ (AR = 2) were internalized into Hela cells by both clathrin- and caveolae-mediated endocytosis. The uptake kinetics of the differently shaped $FMSiO_2NPs$ were elucidated by incubating Hela cells with nanoparticles. The intracellular concentrations of the $FMSiO_2NPs$ were then measured by flow cytometry. After 30 min of incubation, the uptake of long-rod $FMSiO_2NPs$ in Hela cells was more than three times higher than that of spherical $FMSiO_2NPs$ and short-rod $FMSiO_2NPs$. The higher uptake rates of long-rod $FMSiO_2NPs$ after 30 min are attributed to the anisotropic contact surfaces inducing membrane wrapping by contacting and activating the endocytotic proteins caveolin.

4.2.2.3 Tumor targeting

The differences between healthy and cancerous tissue are rooted in anatomy, biochemistry, and molecular biology. Rapid and chaotic tumor growth results in incompletely formed, dilated, and thus permeable blood vessels and inadequate lymphatic drainage in tumor tissue. In addition, the plasma membrane of tumor cells is leaky. The hyperpermeability of the neoplastic blood vasculature in the absence of functioning lymphatic vessels allows larger biomolecules and nanoparticles to extravasate and accumulate in the interstitial tumor tissue. Both leaky blood vessels and impaired lymphatic drainage contribute to increased interstitial fluid pressure, which in turn prevents convective transport of nanoparticles. Diffusive transport is hindered by the dense extracellular matrix. This phenomenon is ascribed to enhanced permeation and retention [590].

The enhanced permeation and retention (EPR) effect was first observed in tumor tissue by Maeda in 1986 [591–595]. The accumulation of proteins, in particular of the sample used, which consists of a polymer conjugated to the anticancer protein neocarzinostatin, in tumor tissue was explained by the leaky vasculature and the abnormal tissue structure of the tumor. This implies that EPR-driven tumor targeting not only enables the enrichment of cancerous tissue with nanoparticles but can also be used for the targeted transport of drugs bound to nanoparticle surfaces (nanocarriers) into tumor cells. Healthy tissue is spared from exposure to nanoparticles. However, this advantage is counteracted by nanocarriers smaller than 30 nm because they can easily pass through the leaky capillary wall in the tumor and return to the circulating blood by diffusion [596]. However, the retention of nanoparticles in the tumor can be greatly

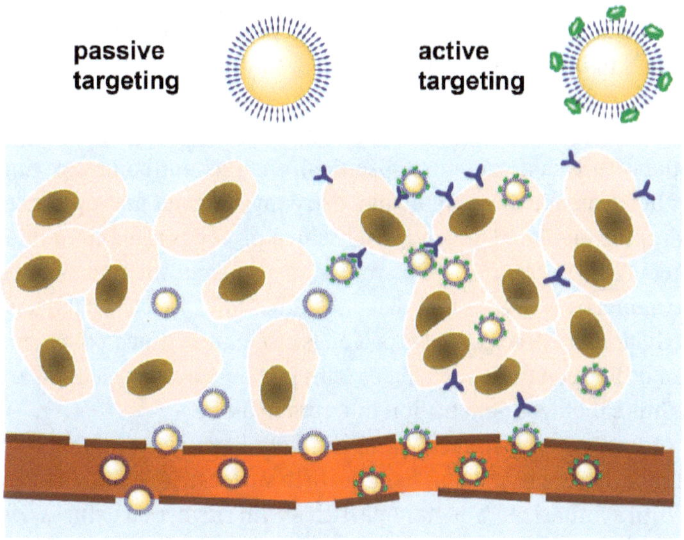

Figure 111: Illustration of the mechanisms for passive (left) and active (right) targeting of PEGylated AuNPs to tumor tissue and PEGylated AuNPs actively targeting the tumor cell receptor.

increased by conjugation with a ligand targeted to an innate tumor-cell receptor [597]. Active tumor targeting uses the specific expression and overexpression of receptors in tumor cells (Figure 111). Nanoparticles conjugated with suitable ligands that specifically bind to tumor cell receptors can be used to transport anticancer drugs into tumor tissue. Endogenous or exogenous triggers allow controlled release of anticancer drugs inside tumor cells [598, 599]. However, targeting ligands being aptamers, small molecules or peptides, can be inactivated by adsorption of serum proteins [600] or destroyed, for example, by enzymatic cleavage [601].

4.2.2.3.1 Passive tumor targeting

Rapid tumor growth is associated with inadequate oxygen and nutrient supply. This leads to irregular neovascularization with structural and functional abnormalities in the tumor blood vessels. These blood vessels are lined with a discontinuous single layer of flattened endothelial cells, have no or leaky basement membranes, and are covered with few or no pericytes [602]. Nanoparticles of suitable size (30–100 nm) are sufficiently small to penetrate the leaky blood vessel membranes into the tumor tissue [603]. They are also large enough that they cannot be removed due to the lack of functional lymphatic vessels [604]. Nanoparticles can only be taken up by tumor cells if they escape recognition by the mononuclear phagocyte system (MPS) and excretion by the kidneys [605]. Depending on the size of the nanoparticles, they are eliminated by hepatobiliary metabolism or via the renal pathway [606]. Nanoparticles with sizes larger than 6 nm are taken up by macrophages, leading to their accumulation in organs of the MPS such as liver, spleen, and lung. Smaller nanoparticles (<6 nm) are below the renal filtration threshold and are too small to be sequestered by the MPS, so they are excreted by the kidney [607].

As discussed in detail in Sections 3.6.1 and 4.1, PEGylated nanoparticles are resistant to surface adsorption of plasma proteins (opsonins), such as antibodies or complement factors during their passage through the bloodstream [327] and non-opsonized nanoparticles escape receptor-mediated phagocytosis, which prolongs their circulation time [328]. These findings, largely from in vitro studies, are in contradiction with results from in vivo experiments [329–331]. Some relevant studies [406, 580, 608, 609] have shown that although PEGylation of nanoparticles prolongs circulation time in the blood, the repulsive interaction of PEGylated nanoparticles with proteins hinders cellular uptake of target cells and thus sabotages penetration into tumor tissue.

Similar repulsive interactions with serum proteins have been reported for zwitterionic self-assembled monolayers (SAMs) on nanoparticle surfaces [411, 610, 611]. For example, Homlin et al. [610] synthesized gold nanoparticles with surface structures consisting of a 1:1 ratio of positively and negatively charged alkanethiol ligands with ammonium and sulfonate head groups. In contrast to uniformly positively or negatively charged SAMs, the zwitterionic SAMs resist to adsorption of fibrinogen and lysozymes.

Zwitterionic structures on nanoparticle surfaces can also consist of ligands that carry both a positive and a negative charge, which cancel each other out. Such zwitterionic compounds can change from the electrically neutral state to a cationic state through protonation in an acidic environment, which affects the cellular uptake efficiency and the toxicity of the nanoparticles. The pH of healthy tissues ranges from 7.35 to 7.45, while the extracellular pH of tumor tissues is weakly acidic at 6.2 to 6.9 [612]

Mizuhara et al. [611] synthesized AuNPs with average sizes of 1.7 nm, whose surfaces they coated with either zwitterionic alkyl sulfonamide (26-mercapto-N-{3-(trimethylamino) propyl-sulfonyl}-3,6,9,12,15-pentaoxahexacosan-1-amide) or acyl-sulfonamide ({4-(23-mercapto-3,6,9,12 tetraoxatricosyl)oxyphenyl}sulfonyl)4(trimethylammonio)butanoyl amide) ligands (Figure 112). While the acyl-sulfonamide groups were shown to undergo a zwitterionic–cationic transition by protonation at pH less than 6.5, the charge-neutral state of the alkyl-sulfonamide ligands was maintained in the pH range of 6–7.5. The authors showed that the uptake of acyl-sulfonamide-terminated AuNPs (AuNP1) in HeLa cells increased several-fold at pH 6 compared to neutral pH. In comparison, the uptake efficiency of alkyl-sulfonamide-terminated AuNPs (AuNP2) showed no pH dependence, consistent with the absence of ligand protonation at pH 6.

Another strategy to optimize cellular uptake, blood circulation, and adaptation to bio-interfaces is based on biomimicking nanoparticle surface structures [613–617]. For example, one such biomimicry approach is to encapsulate nanoparticles in a natural cell membrane, which in turn contains or allows functionalization by, for example, self-identifying proteins. By transferring the cell membrane directly onto the surface of a nanoparticle, the complexity of the membrane with all its lipids, proteins and carbohydrates is preserved. As a result, the resulting membrane-coated nanoparticle takes

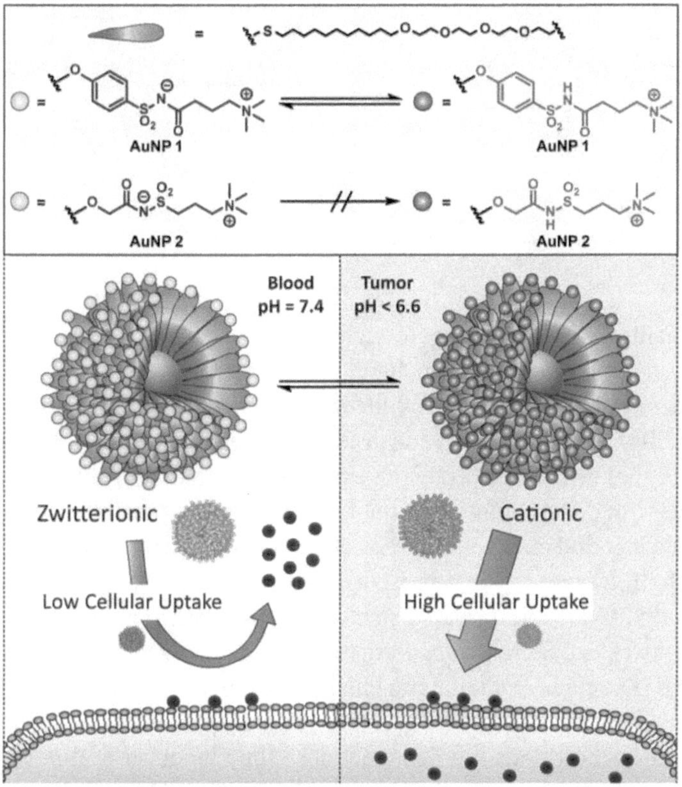

Figure 112: Scheme of protonation of acyl-sulfonamide-terminated AuNPs and non-protonation of alkyl-sulfonamide-terminated AuNPs (top), cellular uptake behavior of zwitterionic, and cationic AuNPs (bottom, left and right panel) [611].

on many of the properties of the original cell. In other words, the nanoparticles camouflage themselves with the cell membrane and use the membrane functions to either remain longer in the bloodstream or/and to penetrate tissues in a targeted manner. Suitable membrane donors were erythrocytes, platelets, bacteria, and leucocytes [614]. The concept of membrane cloaking of nanoparticles was first implemented using red blood cells (RBCs) as membrane donors. Fang et al. [615, 616] isolated the RBC membrane by hypotonic lysis of erythrocytes and subsequent extrusion through porous 100-nm membranes (Figure 113). PLGA nanoparticles are introduced into the RBC membrane vesicles by mechanical extrusion. The authors reported long blood circulation times of the RBC-coated PLGA nanoparticles, reaching elimination half-lives of ca. 40 h after intravenous administration in mouse models. RBCs possess transmembrane proteins (e. g., CD47) that provide blood circulation times of up to 120 days [618].

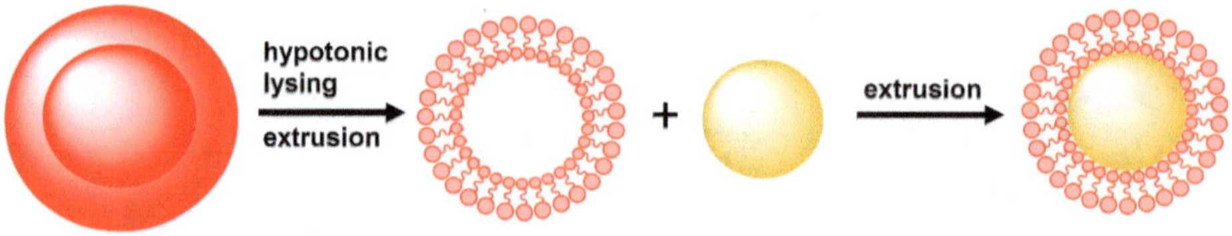

Figure 113: RBC membrane-coated nanoparticles were obtained by hypotonic lysis of erythrocytes and extrusion followed by co-extrusion of RBC membrane with PGLA nanoparticles.

Phagocytes protect organisms by removing foreign bodies such as nanoparticles and foreign cells from the body. On the other hand, they also recognize "self" cells that avoid immune clearance by expressing "self-identifying" membrane proteins like CD47. The glycoprotein CD47 is a transmembrane protein that interacts with integrins, thrombospondins, and signal regulators [619]. The extracellular immunoglobulin-like domain of CD47 interacts with the inhibitory immunoreceptor SIRPα expressed through macrophages of the reticuloendothelial system. The CD47-SIRPα interaction induces inhibitory signaling by the immunoreceptor tyrosine-based inhibition motifs residing in the cytoplasmic tail of SIRPα [620]. A related strategy involves functionalization of nanoparticles with the so-called self-peptides designed as a minimized binding-site structural element of the 117-amino acid immunoglobulin-like domain of the CD47 protein [621]. Rodriguez et al. [622] studied the crystal structure of CD47-SIRPα, selecting the binding site with the highest density of interactions. They modeled a 21-amino acid self-peptide that matched the binding site structure. In vivo tests showed that the self-peptide ligands prolonged the circulation time of the nanobeads in the blood circulation of mice by suppressing clearance. This suggests that self-peptide functionalized nanoparticles are recognized by the SIRPα receptor on the surface of macrophages to signal "self" and inhibit phagocytic uptake.

Passive tumor targeting by the EPR effect enables nanoparticles to penetrate tumor tissue [623]. However, the cellular uptake and distribution of nanoparticles in the tissue is counteracted by the 10–40-fold higher pressure of the interstitial fluid. This leads to a lower nanoparticle concentration and a heterogeneous distribution in the tumor tissue. The necrotic regions inside large tumors or metastases exhibit less angiogenesis compared to tumor growth. As a result, the interstitial pressure increases, making it more difficult for the nanoparticles to penetrate the tumor cells.

4.2.2.3.2 Active tumor targeting

Enhanced uptake of nanoparticles into cancer cells can also be achieved using targeted ligands that recognize and bind to specific receptors on the surface of tumor cells. Active targeting does not only ensure the uptake of nanocarriers into specific cancer cells, but also enables them to cross physiological barriers such as the intestinal mucosa or the blood–brain barrier [597, 624, 625]. The selection of the appropriate targeting

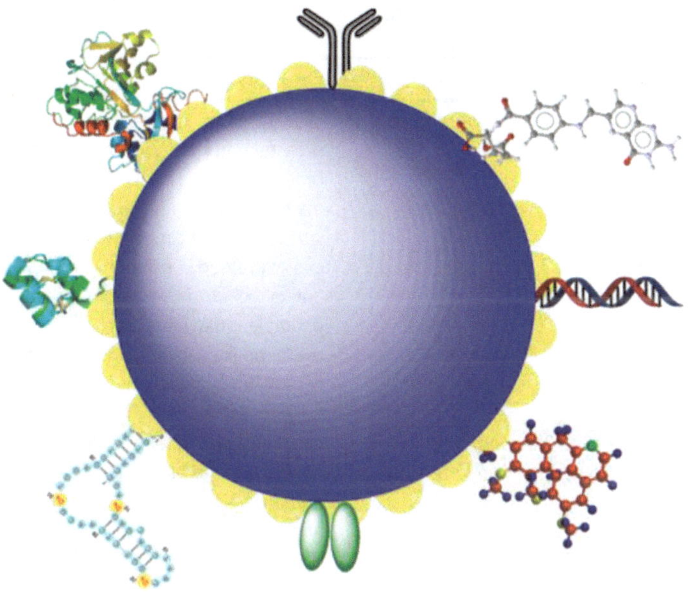

Figure 114: Graphical representation of surface-modified nanoparticles with targeting molecules being antibodies, proteins (e. g. transferrin, integrin), aptamers, peptides, folic acid, lipids and saccharides.

ligand is critical to the efficiency of active targeting. Suitable ligands for active targeting of nanoparticles include antibodies, peptides, aptamers, saccharides, and other small molecules (Figure 114). Exemplary ligands for active targeting of specific cells are listed in Table 4. Targeting ligands conjugated to the surface of nanoparticles facilitate increased cellular uptake through receptor-mediated endocytosis, thereby enhancing drug accumulation in cancer cells. The targeting mechanism relies on the interaction between tumor ligands conjugated on the surface of nanoparticles and cell-surface receptors or antigens on cancer cell surfaces.

Among the various strategies for receptor-mediated internalization of nanocarriers, folate or folic acid receptors (FRs) are used as targets for tumor-specific drug delivery [181, 595]. Folic acid (FA) (Figure 115) is a stable and low immunogenic water-soluble vitamin. It is essential for DNA synthesis and replication, methylation, cell division and growth, and cell survival, especially in rapidly dividing or cancerous cells [630]. Cellular uptake of folic acid occurs through its binding to a FR. A FR-folic acid complex is formed, which can be internalized by endocytosis. The human isoforms FR-α and FR-β, which are glycosylphosphatidylinositol-anchored membrane glycoproteins, have been found to be frequently overexpressed in solid tumors such as ovarian, cervical, lung, breast, renal, colon, and brain. Other FR isoforms are FR-γ and FR-γ0 [625]. These FRs are soluble isoforms of human FR and are found in hematopoietic tissues and lymphoid cells. By contrast, nonproliferating normal cells have very few FRs [631]. By exploiting the

Table 4: Exemplary ligands for active targeting of specific cells.

Ligand type	Targeting ligand	Receptor	Cancer type
Small molecules	folic acid [626]	folate receptor	overexpressed in ovarian, cervical, breast, lung, kidney, colorectal, and brain tumors
Aptamers	AS1411 [507]	nucleolin	overexpressed and localized on the surface of most cancer cells
Peptides	peptides with RGD motif (arginine-glycine-aspartic acid) [627]	integrine $\alpha_v\beta_3$	overexpressed in tumor vasculature and osteo-sarma, neuroblastoma, gliomelanoma, lung cancer
Proteins	transferrin [628]	transferrin receptor 1	overexpressed in ovarian, prostate, breast, lung, liver, colon, brain tumors, and leukemia
Antibodies/ antibody fragments	cetuximab [629]	epidermal growth factor (EGF)	overexpressed in lung, colon, ovary, head, neck and breast cancer, enhancing tumor growth, invasion, and metastasis

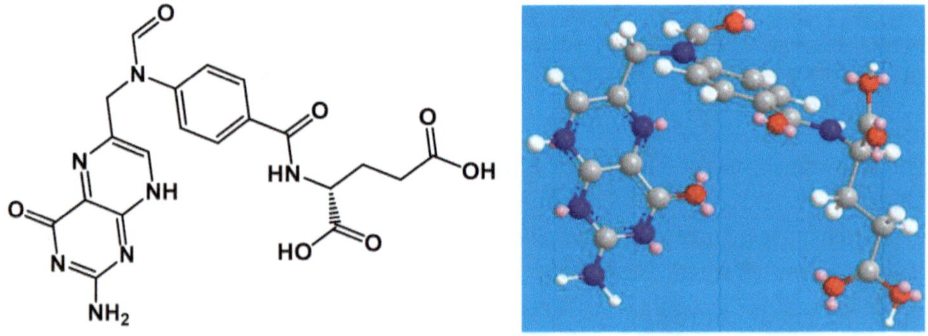

Figure 115: Structure formula of folic acid (left) and its conformation calculated by the MM2 energy minimization method (right, the carbon, nitrogen, oxygen and hydrogen atoms are marked in grey, blue, red, and white, respectively).

overexpression of FRs on certain cancer cells, folic-acid conjugated nanoparticles can selectively target these cancer cells and deliver therapeutic agents to the tumor while sparing normal tissues [626]. For targeting and identification of cancer cells overexpressing plasma membrane-anchored FRs, Nasiri et al. [632] compared the binding affinity of folic acid-conjugated SPIONs to FR-positive versus FR-negative cancer cell lines. Cellular targeting and uptake experiments were performed on FR-negative human skin fibroblasts HSF 1184 and human breast-cancer cells MDA-MB-231, while human breast cancer MDA-MB-231 and human cervical cancer HeLa cells were used as FR-positive cancer cell lines. The authors synthesized citrate-terminated SPIONs (CA-SPIONs) of approximately 13–16 nm by alkaline co-precipitation and conjugated these nanoparticles with folic acid

(FA-CA-SPIONs) to demonstrate molecular recognition of FR-positive cancer cells by the folic-acid ligand. The cellular uptake of these nanoparticle species was quantitatively determined by atomic absorption spectroscopy and Prussian blue staining, while the internalized Prussian blue-stained nanoparticles were visualized by light microscopy. Cellular uptake of FA-CA-SPIONs was shown to occur via FA receptor-mediated endocytosis, whereas CA-SPIONs, which were internalized at much lower concentrations, were presumably shuttled into cells by other receptors. Microscopic images showed that the FA-CA-SPIONs accumulated mainly in the cytoplasm and partially around the nucleus.

Zhang et al. [633] functionalized glutathione-stabilized gold nanoparticles with folic acid by peptide formation between the amino group of folic acid and the carboxyl group of glutathione (GSH). The binding affinity of the FITC-labeled FA-GSH-AuNPs to HeLa and A549 lung cancer cells was investigated by transmission electron microscopy (TEM) and confocal laser scanning microscopy (CLSM). HeLa cells are known for their FR overexpression, while A549 cells completely lack folate receptors. As shown by TEM and CLSM analysis, the FA-GSH-AuNPs were barely detectably taken up by the A549 cells. In contrast, the FA-GSH-AuNPs were found to accumulate in the cytosol and to be entrapped in endosomes in HeLa cells by TEM and CLSM. These results clearly indicate cellular uptake by FR-mediated endocytosis and high uptake efficiency correlating with overexpression on HeLa cell surfaces. TEM studies confirmed the high specificity of FA-GSH-AuNPs for HeLa cells with high folate-receptor expression and supported the confocal imaging data.

Transferrin receptors exist as two subtypes, transferrin receptor 1 (TFR1) and transferrin receptor 2 (TFR2). TFR1 is a homodimeric type II transmembrane glycoprotein that is ubiquitously expressed on the surface of most cells [628]. Another member of the TFR family, TFR2, is expressed primarily on hepatocytes. The molecular weight of TFR as a homodimer is 180 kDa. Each monomer contains a transferrin-binding C-terminal domain, a short N-terminal domain, and a single transmembrane domain. Both TFR1 and TFR2 regulate iron import by binding the glycoprotein transferrin, which has two binding sites for Fe^{3+} ions, providing a source of iron for hemoglobin synthesis. TfR1, which is the major entry point for iron-bound Tf into the cell, circulates to intracellular acidified endosomes in a clathrin/dynamin-dependent manner. Iron is released into the cell and TfR1 returns to the cell surface. Despite its ubiquitous expression, TfR1 is expressed at significantly higher levels in cancer cells than in normal cells, and its expression may correlate with the tumor stage or progression [634]. The high expression of this receptor on cancer cells, its ability to be internalized via endocytosis, and the requirement of iron for cancer cell proliferation make this receptor an attractive target for cancer therapy. Targeting of TfR1 has been achieved by a variety of methods, including strategies using nanocarriers for drug delivery.

Overexpression of the transferrin receptor on the blood–brain barrier (BBB) and glioblastoma cells makes TfR a promising target for nanoparticle-based therapy of glioblastoma. For targeted delivery of chemotherapeutics in gliablastoma multiforme, Luo et al. [635] developed efficient nanocarriers consisting of biocompatible and biodegradable porous silicon nanoparticles (pSiNP). Undecanoic acid-terminated pSiNPs with sizes around 170 nm and pore sizes of 16 nm were obtained by anodic HF etching of boron-doped Si wafers followed by thermal hydrosilylation with undecenyl acid. The carboxylated pSiNPs were covalently conjugated to transferrin by the EDC/NHS coupling reaction, resulting in transferrin-conjugated pSiNPs (tf@pSiNPs) (Figure 116).

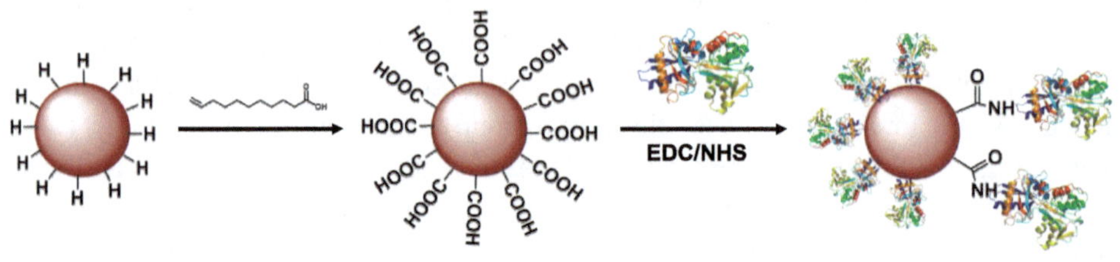

Figure 116: Scheme of the surface modification process for the preparation of Tf@pSiNPs.

To elucidate the mechanism and efficiency of internalization of Tf@pSiNPs into gliablastoma and blood-brain barrier (BBB)-forming cells, the U87-mCherry glioma and brain microvascular endothelial hCMEC/D3 cell lines and HaCaT cells were used as reference. The control cell line, HaCaT, a human cerebral macrovascular endothelial cell line, are keratinocytes representing the non-cancerous cells. To assess the ability of Tf@pSiNPs to cross the BBB, the authors created a transwell BBB model by seeding immortalized hCMEC/D3 cells on transwell inserts. The hCMEC/D3 cells formed a confluent monolayer of tightly packed, elongated cells. The integrity of the monolayer was verified by measuring the transepithelial electrical resistance (30–40 Ω cm^2). Western blotting and immunofluorescence experiments showed that U87 cells achieved the highest TfR expression, followed by hCMEC/D3 and HaCaT cells. According to the Trf receptor expression on the surface of the cancer cell lines and normal cells, the uptake efficiency was highest in U87 cells and lowest in HaCaT cells. The uptake of Tf@pSiNPs into U87 cells was mediated predominantly by clathrin-mediated endocytosis. In hCMEC/D3 cells, both clathrin- and both clathrin- and caveolae-mediated endocytosis are involved in the cellular uptake of Tf@pSiNPs. This also implies that clathrin- and caveolae-mediated endocytosis mechanisms significantly determine the transcytosis of biomolecules into the brain.

Doxorubicin, as the main component of chemotherapy [636], was used as a drug load to investigate the loading efficiency of Tf@pSiNPs for targeted drug delivery. The Tf@pSiNPs were loaded with doxorubicin by incubation in PBS buffer. The release of doxorubicin at the endosomal pH of 5.0 and under physiological conditions (pH = 7.4) was investigated. The loading of doxorubicin most likely occurred by encapsulation into the pores of the Tf@pSiNPs. The authors observed a pH-dependent release kinetics: Only 38 % of the doxorubicin loading amount was released at a physiological pH after 12 h, while doxorubicin release reached 90 % at pH 5. The pH-dependent release profiles were explained by the higher water solubility of doxorubicin in acidic conditions, which favors its intratumoral release.

Integrins are a superfamily of cell adhesion receptors that essentially recognize extracellular matrix ligands and cell surface ligands [637]. Integrins are heterodimeric transmembrane glycoproteins composed of α and β subunits, both of which have a cytoplasmic tail, membrane-spanning helix, and extracellular domain; these subunits interact with the intracellular actin cytoskeleton via adaptor proteins and bind to the surrounding extracellular matrix [638]. In mammals there are 18 distinct α subunits and eight β subunits, which can be combined to form 24 different integrins [639]. Most integrin heterodimers involve the $\beta1$ subunit paired with different α subunits, and the αv subunit can pair with several β subunits to form various integrins. The $\beta1$ subunit can form heterodimers with 12 α subunits, whereas $\beta4$ pairs specifically with $\alpha6$, and $\beta5$, $\beta6$, and $\beta8$ primarily pair with αv. On the other hand, $\alpha4$ and αv interact with different β-subunits, leading to the formation of $\alpha4\beta1$, $\alpha4\beta7$, $\alpha v\beta1$, $\alpha v\beta3$, $\alpha v\beta5$, $\alpha v\beta6$, or $\alpha v\beta8$ integrins [638, 640]. The α subunit largely determines the ligand-binding properties of integrins, but the β subunit also contributes to ligand specificity and intracellular signaling. The main functions of integrins are to attach the cell body to the ECM and to receive signals transmitted by the ECM. The extracellular domain of these integrins can bind to various ECM proteins, such as fibronectin, vitronectin, collagen and laminin. Interactions between integrins and ECM proteins activate signal transduction pathways that control cell growth, survival, cell division, and migration.

Highly malignant cancer cells overexpress certain integrins that provide survival signals to cancer cells, promote tumor angiogenesis, tumor progression, and growth, and drive metastasis [641]. On the other hand, integrins are absent or present in low level in pre-existing endothelial cells and normal tissue. This makes integrins promising targets in cancer therapy. However, some integrins primarily bind to selected ECM ligands; for instance, fibronectin is the principal ligand for the $\alpha5\beta1$ integrin. Other integrins, on the other hand, have broader ligand specificity. For example, the $\alpha v\beta3$ integrin binds to vitronectin, fibronectin, thrombospondin, and fibrinogen [641]. In fact, αv integrins and the $\alpha5\beta1$ integrin were first identified due to their ability to recognize specific amino acid sequences, such as the arginine-glycine-aspartic acid (RGD) peptide of ECM proteins [638, 642]. These integrins are overexpressed on cancerous endothelial cells as well as on some tumor cell types such as glioblastoma, melanoma, ovarian, breast, and prostate cancers [643]. Nanoparticles functionalized with integrin-targeting peptides do not only bind specifically to integrin-expressing cancer cells, but also inhibit integrin-mediated tumor progression and metastasis [642]. In addition, nanoparticles that bind to integrin receptors are mainly internalized by clathrin-mediated endocytosis and then transported to late endosomes or lysosomes.

Wu et al. [627] used $(RGD)_4$ peptide conjugated PEGylated AuNPs (RGD@PEG-AuNPs) to target integrin receptors on various breast cancer cells, including highly invasive MDA-MB-231 and Hs578T cells and minimally invasive SK-BR-3 cell line. PEG-AuNPs with sizes around 40 nm were obtained by surface modification of 20-nm AuNPs with methoxy-PEG-thiol5k and labeled with FITC for confocal immunofluorescence imaging. The PEGylated AuNPs were conjugated with a cysteine-modified $(RGD)_4$ peptide. The authors demonstrated by immunoblotting that the highly invasive breast-cancer cell lines MDA-MB-231 and Hs578T overexpress $\alpha 5$ and αv integrin heterodimers, in contrast to the minimally invasive breast cancer cell line SK-BR-3. Immunofluorescence images showed that RGD/P-AuNPs were efficiently internalized by $\alpha 5$ and αv integrins in MDA-MB-231 and Hs578T cells. After cellular uptake by endocytosis, RGD/P-AuNPs were visualized to accumulate in late endosomes and lysosomes. In contrast, SK-BR-3 cells did not take up RGD/P-AuNPs.

Zhang et al. [644] used (RGD)-peptide conjugated SPIONs (RGD@SPIONs) as MRI T2 contrast agents to monitor and evaluate CCl_4-induced liver fibrosis by molecular MRI. SPIONs with sizes around 14.5 ± 2 nm were conjugated with the (RGD) peptide, which has a high binding affinity to the integrin $\alpha v\beta 3$. In advanced stages of liver fibrosis, the expression of integrin $\alpha v\beta 3$ is upregulated on hepatic stellate cells (HSCs). The activated HSCs can migrate and bind to (RGD) peptide-containing proteins in the extracellular matrix. This makes the integrin $\alpha v\beta 3$ receptor a suitable molecular probe for monitoring activated HSCs during liver fibrogenesis. The protein levels of integrin αv and $\beta 3$ subunits expressed on activated HSCs were found to scale well with the progression of liver fibrosis. The expression level of integrin $\alpha v\beta 3$ in relation to the degree of liver fibrosis was evaluated using Spearman's rank correlation. The RGD@SPIONs were successfully used as MRI T2 contrast agent for MR imaging of activated HSCs during liver fibrosis, allowing MR imaging of the various stages of liver fibrogenesis. The authors not only demonstrated that RGD@SPIONs specifically bind to the integrin $\alpha v\beta 3$ receptor of activated HSCs, but also preferentially accumulate in fibrotic liver samples in contrast to SPIONs and healthy liver tissue. The expression level of integrin $\alpha v\beta 3$ in relation to the degree of liver fibrosis was evaluated using Spearman's rank correlation.

The conjugation of antibodies to nanoparticles is a highly versatile approach for enhancing the specificity and efficacy of nanoparticle-based radiosensitizer, α-emitter and drug delivery systems [629, 645–647]. In cancer therapy, antibodies are used to target cancer cells that overexpress certain receptor proteins or antigens that are hardly or not found on healthy cells. By attaching antibodies to nanoparticles, these targeting ligands specifically direct the nanoparticles to the desired cancer cells, avoiding healthy tissue and minimizing unwanted side effects. Conjugation of antibodies to nanoparticles can be achieved directly through covalent bonds, heterobifunctional cross-linking, or physical adsorption to the surface of the nanoparticle [645]. The antibody should be amenable to conjugation to nanoparticles using a suitable method that preserves its functionality. To date, numerous antibody-conjugated nanoparticles have been developed for cancer therapy, with the type of nanoparticle determining the therapeutic modality. For example, antibody–gold nanoparticle conjugates and antibody–iron oxide nanoparticle conjugates are used for targeted radiotherapy [648], while mesoporous silicon nanoparticles are suitable for targeted delivery of anticancer drugs [649]. Researchers are continually advancing the development of antibody-conjugated nanoparticles, with a focus on improving targeting specificity, pharmacokinetics, and biodistribution. The antibodies predominantly used in targeted cancer therapy are Y-shaped Immunoglobulin G (IgG) antibodies, which consist of an antigen-binding fragment (Fab) region and a fragment crystallizable (Fc) region [650]. The Fab and Fc regions are connected by a flexible hinge region (Figure 117, left). The Fab region is unique to each antibody and consists of two identical arms, each containing a complete light chain (pink) paired with the variable (VH) and constant 1 (CH1) domains of the heavy chain (blue). Each variable domain within the Fab region consists of four framework regions and three hypervariable regions, also known as complementarity-determining regions (CDRs). In each arm, the six hypervariable regions (three from the heavy chain and three from the light chain) come together to form the antigen binding site, which determines the specificity of the antibody. The Fc region is the stem region of the antibody, capable of recruiting complement proteins and interacting with specialized Fc receptors on the surface of leukocytes.

Nanoparticles can be functionalized with antibodies by adsorption, covalent binding, or using adapter molecules [645]. Antibody adsorption includes physisorption and ionic binding. Random immobilization of the antibodies by adsorption may result in four different spatial orientations, which are shown in Figure 117 (right) as end-on, flat-on, head-on, and side-on orientations [645]. Physisorbed antibodies bind to the nanoparticle surface through weak interactions such as electrostatic, hydrogen bonding, hydrophobic, and van der Waals forces.

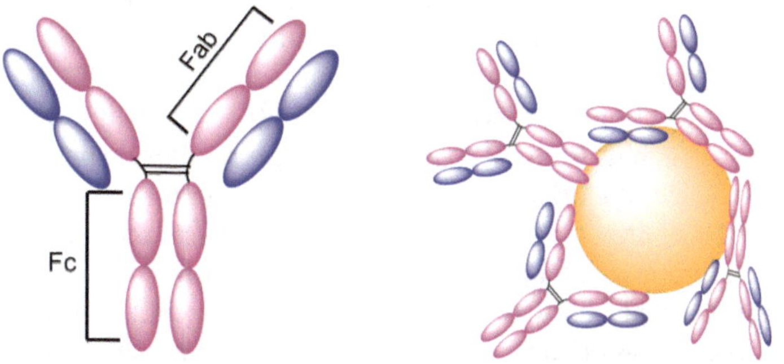

Figure 117: Schematics of the structure of an IgG antibody consisting of heavy chains (purple) and light chains (blue) highlighting the Fab and Fc regions (left) and the adsorption mode of antibody conjugated nanoparticles (right).

On the other hand, ionic binding depends on the charge state of the nanoparticle surface and the antibody, which depends on the ionic strength and pH of the extracellular solutions [648, 651]. In ionic binding, oppositely charged antibodies and nanoparticles are weakly bound via electrostatic interactions. Physisorption also depends on environmental conditions and is more random than reproducible. In addition, hydrophobic interactions can cause significant conformational changes in antibody molecules, leading to denaturation and loss of biological activity. Antibodies have the highest charge density in the central region, which preferentially results in a flattened adsorption geometry. This limits antigen recognition to a volume very close to the nanoparticle surface. While the end-on and flat-on adsorption geometries allow specific binding of the antibody to the matching antigen, the side-on and head-on geometries prevent this due to steric interactions. However, physisorption and ionic binding have less stability compared to covalent linkages and moreover and can cause conformational changes in antibodies, causing denaturation and loss of biological activity. For immobilization of antibodies on nanoparticle surfaces, the end-on orientation is highest desirable. This is achieved by covalent binding the Fc region to the nanoparticle surface. As a result, the antigen binding sites of the Fab region are available for interaction with the matching antigen or surface receptor on the cancer cell. Covalent binding of the Fc region can either be achieved precisely either by suitable surface ligands or by means of suitable linker molecules.

The goal of covalent conjugation of antibodies to nanoparticles is to attach the antibodies—preferably via their Fc region—to the nanoparticle surface, either directly or through linker molecules [652, 653]. This orientation helps to preserve the antigen-binding functionality of the Fab regions, and depending on the reaction chemistry, maintaining the functionality of the Fc region may also be important. The most common covalent coupling reactions are based on carbodiimide chemistry, maleimide chemistry, or “click chemistry”. For example, the amino groups of the antibody can be coupled to carboxyl or phosphate groups on the nanoparticle surface using 1-ethyl-3-(-3-dimethylaminopropyl) carbodiimide (EDC), resulting in an amide bond. Covalent coupling of antibodies to nanoparticles can also be accomplished using the EDC/NHS method. In this two-step reaction, N-hydroxysuccinimide (NHS) or its water-soluble analogue N-hydroxysulfosuccinimide (sulfo-NHS) is added in excess. Unfortunately, the EDC/NHS method cannot control the orientation of the antibodies on the NP surface. Random immobilization of the antibodies may result in loss of biological activity. Optimization of pH conditions is critical to ensure effective coupling and to minimize side reactions such as hydrolysis. The maleimide coupling method utilizes the sulfhydryl (-SH) groups present in cysteine residues of antibodies (which may need to be generated by mild reduction or chemical modification) for covalent binding. The nanoparticle surface is first functionalized with maleimide groups by reacting primary amines with sulfosuccinimidyl trans-4-(N-maleimidomethyl)cyclohexane-1-carboxylate (sulfo-SMCC). The maleimide-activated nanoparticle can then covalently bind to free thiol groups on the antibody via maleimide-thiol coupling. Maleimide chemistry provides greater control over the orientation of the antibodies on nanoparticle surfaces, thereby preserving the antigen-binding activity. It also allows the introduction of spacer arms (such as PEG) to minimize nonspecific binding and improve conjugation efficiency. Click chemistry enables site-specific and oriented conjugation of antibodies to nanoparticles in aqueous solution with high efficiency [645, 651, 653]. A typical reaction is the copper-catalyzed azide-alkyne cycloaddition (CuAAC) between an organic azide and a terminal alkyne to form a stable 1,4-disubstituted 1,2,3-triazole ring. For this purpose, the antibody is equipped with an alkyne in the Fc region and the nanoparticles are endowed with a suitable azide functionality. The azide groups are incorporated into the nanoparticle surface by functionalizing available surface groups with azide-containing linkers using appropriate chemistries, ensuring they are available for the click reaction. Since azides and alkynes do not react with the functional groups of proteins, the click chemistry is highly selective and site-specific.

Another method for oriented immobilization of antibodies via the Fc is non-covalent conjugation via adapter molecules, which ensure the availability of the Fab region [645, 653]. A successful strategy is based on the strong non-covalent interaction between biotin and its binding proteins, avidin or its analogues. Biotin (vitamin B7 or H) binds rapidly and stably to avidin under various conditions, including different pH levels and temperatures. Biotin (vitamin H) is one of the most fascinating cofactors involved in central processes of prokaryotic and eukaryotic cell metabolism. Avidin is a tetrameric basic glycoprotein found in egg whites, with each of its four subunits capable of binding one biotin molecule. However, the glycosylation of avidin can cause non-specific binding to other biomolecules. To reduce this non-specific binding, non-glycosylated and neutral forms of avidin, such as streptavidin or neutravidin (either natural or recombinant), are used. These analogues maintain high-affinity binding to biotin while minimizing non-specific interactions. Other adapter molecules include protein G and protein A, which bind specifically to the Fc region of antibodies. This binding facilitates the oriented immobilization of antibodies, preserving the functionality of the Fab region for antigen recognition.

Exemplary antibodies that have been successfully used in nanoparticle-based cancer therapy are trastuzumab (Herceptin), cetuximab (Erbitux), and rituximab (Rituxan) [650, 654, 655]. Trastuzumab is a recombinant humanized IgG1 monoclonal antibody against human epidermal growth factor receptor 2 (HER2) and was approved in 1998 as the first targeted therapy for HER2-positive breast cancer and later in 2010 for metastatic HER2-positive gastric cancer [656]. Cetuximab is a combined chimeric human immunoglobulin G1 antibody that binds with high affinity to the extracellular domain of the epidermal growth factor receptor (EGFR), which is overexpressed in many colorectal cancers. Binding of cetuximab to EGFR blocks the phosphorylation and activation of receptor-associated kinases. This inhibits receptor activation and leads to receptor internalization and degradation. Rituximab is a chimeric mouse/human monoclonal antibody that targets the transmembrane antigen CD20, a non-glycosylated phosphoprotein found on pre-B and mature B lymphocytes. This antigen is expressed on approximately 95 % of all B-cell non-Hodgkin's lymphomas and on normal B-cells, but not on hematopoietic stem cells or normal/malignant plasma cells [657].

Antibodies are ideal targeting moieties as they are highly specific for a plasma membrane receptor. Since antibodies are taken up by cells via receptor-mediated endocytosis, this provides a target-specific internalization mechanism for antibody-nanoparticle conjugates [658]. However, the pharmacokinetics of antibodies may be altered by conjugation with nanoparticles. For example, Bhattacharyya et al. [659] reported increased endocytosis dynamics of cetuximab-AuNP conjugates compared to free cetuximab in PANC-1, AsPC-1 and MiaPaca-2 pancreatic cancer cells that differ in their EGFR expression level. The investigated nanoconjugates consisted of 5-nm citrate-stabilized Au spheres linked to cetuximab molecules by electrostatic interactions. The altered internalization mechanisms, attributed to clathrin-mediated and clathrin-independent endocytosis, resulted in a different subcellular localization of cetuximab compared to the cetuximab-AuNPs conjugates. Furthermore, the nanoconjugates were found to modulate the involvement of the signal transducing GTPase dynamin-2 in the endocytic processes.

Successful treatment of colorectal cancer with anti-EGFR monoclonal antibodies is often hindered by resistance mechanisms, including the overexpression of competing receptors such as human epidermal growth factor receptor 3 (HER3) and cell adhesion molecules like the epithelial cell adhesion molecule (EpCAM). HER3 contributes to the formation of a heterodimer with EGFR on the surface of CRC cells and activates the intracellular signaling pathway when a ligand binds to the receptor. Overexpression of HER3 in metastatic CRC cells has been associated with resistance to EGFR inhibitors, as HER3 can activate alternative signaling pathways that bypass EGFR blockade. Epithelial cell adhesion molecule (EpCAM), a 40-kDa glycoprotein expressed in 85 % of colorectal cancers, enhances the proliferative and invasive capacity of tumors and can inhibit differentiation and promote proliferation. EpCAM overexpression enhances the proliferative and invasive capacity of tumors and can inhibit differentiation, promoting proliferation. However, in some contexts, loss of EpCAM expression is associated with epithelial-mesenchymal transition (EMT), correlating with CRC progression, metastasis, and poor prognosis. This suggests that EpCAM's role in CRC is complex and may vary depending on the tumor microenvironment and disease stage. Another important surface biomarker, melanoma cell adhesion molecule (MCAM or CD146), has a complex role in colorectal cancer. In some studies, reduced MCAM expression is associated with increased tumorigenesis and cancer stem cell formation, suggesting a tumor suppressor function. Conversely, other research indicates that MCAM overexpression correlates with metastasis and poor prognosis. Therefore, the exact role of MCAM in CRC progression remains subject to ongoing investigation.

Cetuximab-AuNP conjugates were prepared by adsorption of cetuximab onto citrate-stabilized AuNPs with diameters of 25 nm, 40 nm, and 60 nm [629]. These nanoconjugates were used in cytotoxicity studies against BRAF-mutant HT-29 colorectal cancer (CRC) cells. The results indicated that cetuximab-AuNP conjugates with 60 nm AuNPs had the greatest effect on reducing the viability of HT-29 cells, potentially due to enhanced cellular uptake or increased antibody loading on larger nanoparticles. In addition to the cytotoxicity study, the expression levels of cell surface markers EpCAM, MCAM, and HER3 were examined. Treatment with cetuximab-AuNP conjugates led to a decrease in the expression of MCAM and HER3, while the expression of EpCAM showed no significant decrease but exhibited higher heterogeneity among the cells. The inhibition of the EGFR pathway by cetuximab may reduce downstream signaling pathways involving MCAM and HER3, enhancing the therapeutic effect. The decreased expression of MCAM and HER3 could potentially overcome complex mechanisms of resistance to anti-EGFR treatment, although further investigation is needed to fully understand their roles in resistance mechanisms.

One treatment modality for patients with inoperable solid tumors is internal radionuclide therapy [660]. Radionuclides are short-lived radioactive isotopes that emit either electromagnetic radiation or particles (e. g., electrons, positron, helium nuclei (α particles)) when they decay. The study conducted by Dziawer et al. [661] focused on the development and potential therapeutic application of radioactively labeled, antibody conjugated AuNPs as a novel α-radiotherapy agent, specifically for a targeted, highly localized radionuclide therapy, that they called α-nanobrachytherapy. The α-radiotherapy agent was obtained by functionalizing 5 nm, citrate-stabilized AuNPs with PEG carboxylic acid disulfide (i. e., COOH-PEG-SS-PEG-COOH) (Figure 118). The terminal carboxylic acid groups were converted via EDC/NHS chemistry to reactive NHS esters that form amide bonds with amino groups of the HER2-specific monoclonal antibody trastuzumab. The trastuzumab-AuNP conjugates were then labeled with the α-emitter astatine-211 (^{211}At). The range of emitted α particles in the tissue is limited to 5–10 cell diameters, resulting in high localization of the therapeutic dose. The ^{211}At-trastuzumab-

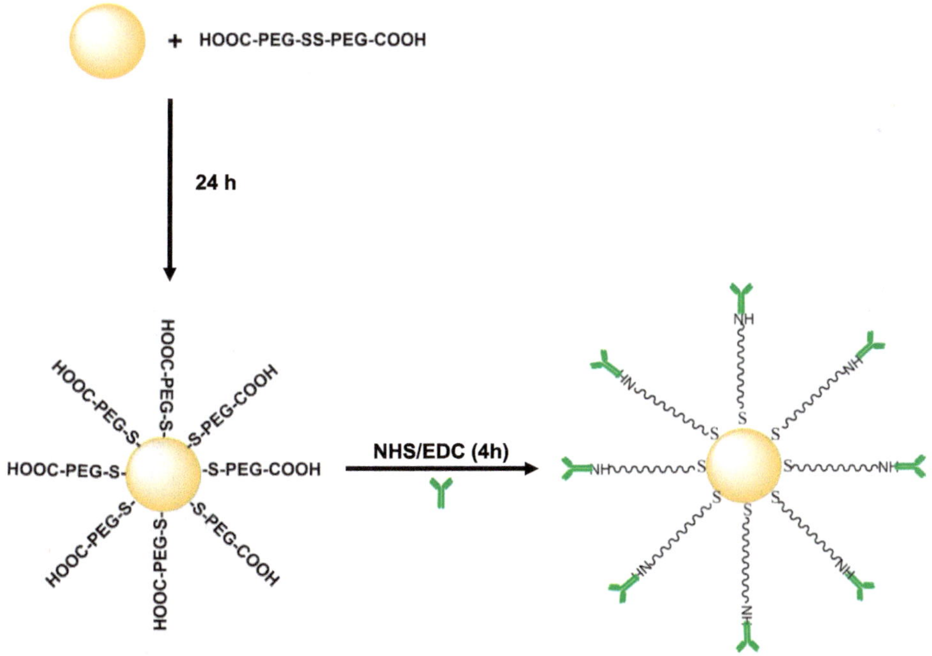

Figure 118: Schematics of PEGylation of AuNPs with subsequent NHS/EDC coupling of trastuzumab.

AuNP conjugate, which binds specifically to HER2 receptors, ensured the targeted administration of the α emitter. To investigate the binding affinity of the ^{211}At-trastuzumab-AuNP conjugates to HER2 receptors and to elucidate the cellular uptake mechanism and the intracellular distribution and cytotoxicity, the HER2-overexpressing human ovarian cell line SKOV-3 cells were incubated with free AuNPs or trastuzumab-AuNP conjugates. Confocal and dark-field microscopy images showed that ^{211}At-trastuzumab-AuNP conjugates were effectively internalized and preferentially localized in the nuclear envelope area of the SKOV-3 cells. The ^{211}At-trastuzumab-AuNP conjugates exhibited a significantly higher cytotoxicity than non-targeted ^{211}At-AuNPs, which was explained by the trastuzumab-induced HER2-mediated endocytosis of the nanoconjugates.

Song et al. [662] conjugated Fe_3O_4 nanoparticles (SPIONs) with the antibody rituximab to establish magnetic nanotheranostics that could not only enhance the therapeutic effect of rituximab on CD20-overexpressing malignant lymphoma cells, but also make the intracellular distribution in the lymphoma detectable by MRI. For this purpose, 2,3-dimercaptosuccinic acid (DMSA)-stabilized Fe_3O_4 nanoparticles with average sizes around 6.5 nm were synthesized via alkaline co-precipitation synthesis. Covalent conjugation of the DMSA@SPIONs with rituximab was performed using the EDC/sulfoNHS coupling reagent, whereby the carboxylic acid groups of the DMSA ligands were converted to reactive sulfoNHS to form amide bonds with amino groups of rituximab. To demonstrate the specific binding affinity of the rituximab-SPION conjugates for the CD20 antigen, human B-lymphoblastoid Raji cells were used as a CD20-positive cell line and CD20-negative lymphoblast K562 cells as a control. TEM and MR imaging experiments for cell targeting and MRI/T_2 measurements showed that the Rituximab@SPION conjugates specifically bind to CD20-positive cells. In comparison to Rituximab and DMSA@SPIONs, the nanoconjugates had a cytotoxic effect on Raji cells, indicating the internalization of the nanoconjugates. The apoptosis of Raji cells induced by the nanoconjugates was correlated with the increase of cytosolic Ca^{2+} cations, Bax/Bcl-2, and reactive oxygen species as well as the drastic decrease of mitochondrial membrane potential.

Similar to antibodies, aptamers have a high specificity and affinity for matching antigens, with dissociation constants in the nanomolar range [663]. In contrast to antibodies, aptamers possess a low molecular weight, can be synthesized easily and reproducibly, and show low toxicity or immunogenicity. Aptamers are single-stranded DNA or RNA oligonucleotides with a length of 25–50 bases and a molecular mass of typically less than 20 kDa. They are obtained by an in vitro selection process called Systematic Evolution of Ligands through Exponential Enrichment (SELEX) [664, 665]. SELEX selects from a large library of randomly synthesized nucleotide sequences those nucleotide base sequences with the highest possible specific binding affinity to a target protein that is either a peptide, enzyme, antibody, or cell-surface receptor. Combinatorial oligonucleotide libraries contain oligonucleotides in the order of 10^{14} to 10^{15} different base sequences per μmole. The selection process begins by incubating the mixture of nucleotide sequences with a target protein. Nucleotide sequences that preferentially bind to the target protein are isolated by washing out the other sequences. The bound sequences are then released from the target protein and amplified by PCR. A new cycle of binding and washing out the weakly bound sequences is started. This process is repeated until one or two oligonucleotides remain. This selection process results in aptamers that bind to their target proteins by folding their 3D structure precisely around the binding partner. The SELEX procedure differs for the selection of RNA and DNA aptamers. In the case of DNA aptamers, the double-stranded DNA from PCR amplification is denatured to obtain single-stranded DNA (ssDNA) for the next selection cycle. For RNA aptamers, the double-stranded DNA serves as a template for in vitro transcription using RNA polymerase to generate RNA aptamers for subsequent selection cycles. DNA aptamers are inherently more stable and less expensive to produce due to the chemical stability of DNA. RNA aptamers can form more diverse three-dimensional conformations and may exhibit higher binding affinities owing to the additional functional groups available for interaction.

Aptamers can be used for early cancer detection. Metastatic tumors spontaneously shed cells into body fluids such as lymph and interstitial fluid and blood. The exfoliated cells are generally outnumbered by healthy cells. The identification of such a small subpopulation of cells requires not only the molecular recognition of surface receptors characteristic of these tumor cells but also the ability to selectively separate these cells, for example, to detect and enrich for rare cancer cells. The ability of the aptamer-conjugated nanoparticles to accurately identify and bind to cancer cells in these different matrices is critical for effective cancer detection and diagnosis. Medley et al. [666] have developed a dual nanoparticle diagnostic assay consisting of superparamagnetic aptamer-conjugated SPIONs (MNPs) and fluorescent aptamer-conjugated SiO_2 nanoparticles (FNPs). While the MNPs ensure the magnetic separation and thus the in-vitro enrichment of the cancer cells, the FNPs provide the unambiguous detection and spatial information of the cancer cells. The influence of nanoparticle size, fluorescent marker and aptamer conjugation scheme of the MNP-FNP diagnostic assay on the efficiency of magnetic cell extraction was systematically investigated by varying the size of the magnetic particles from 60 nm to 8 µm, using FNPs labeled with tetramethylrhodamine, tris(2,2′-bipyridyl) ruthenium(II) dichloride, fluorescein or Cy5-succinimidyl ester, as well as employing different aptamer conjugation schemes. Tetramethylrhodamine (TMR)-labeled FNPs were found to exhibit the brightest fluorescence signature and were therefore used for the FNP-MNP diagnostic assay. All cell studies were performed in culture medium suspensions containing human T-lymphoblast CCRF-CEM cells and B-lymphocyte Ramos cells. A microplate reader and a confocal fluorescence microscope were used for detection of fluorescence intensity and 2D visualization of fluorophore-labeled MNP-FNPs.

The authors carried out different conjugation protocols with the aim of covalently binding the aptamers to the surface of the MNPs and FNPs while maintaining the binding specificity of the aptamers. The surface structures and the surface charge of the nanoparticles were decisive for achieving this objective. Therefore, conjugation schemes utilizing carboxy, phosphonate, and amino groups on the nanoparticle surfaces were used to optimize the selectivity of the aptamer ligands. For this purpose,

FNPs obtained by the reverse micelle microemulsion method from TMR-APTES conjugates and TEOS in a water-cyclohexane emulsion were surface-modified with carboxy- and phosphonate-silane in multiple concentration ratios or with aminosilane.

For conjugation to the nanoparticles, the aptamers were equipped with either a biotin (biotinylated) or a thiol (thiolated) head group. Tricarboxy-silanized FNPs were either conjugated with streptavidin or directly covalently bound to amino-aptamers using EDC/NHS coupling chemistry in both cases. On the other hand, APTES-stabilized FNPs were conjugated to thiolated aptamers using the N-succinimidyl-4-maleimidobutyrate (GMBS) linker chemistry. MNPs were synthesized by alkaline co-precipitation and directly coated with avidin. The conjugation schemes investigated are shown in Figure 119: (A) avidin-coated MNPs were bound to biotinylated aptamers, (B) streptavidin-coated FNPs were conjugated to biotinylated aptamer, (C) tricarboxy-silanized FNPs were conjugated to amino-aptamers via EDC/NHS coupling, and (D) amino-silanized FNPs were conjugated to thiolated aptamers using the GMBS linker chemistry. Medley et al. [666] showed that the concept of conjugation with up to four different aptamer base-pair sequences is very well suited for the recognition of cancer cell receptors, selective binding and magnetic separation of these cancer cells. In comparison with larger streptavidin-coated MNPs ranging from 350 nm to 8030 nm, the 60-nm-sized MNPs increased the cancer cell specific binding affinity and the amount of magnetically separated cancer cells. This result can be explained by the more homogeneous distribution of the smaller MNPs in the cell suspension and the larger number of aptamer ligands relative to the MNP mass. The authors found that the conjugation scheme of avidin-coated MNPs binding to biotinylated aptamers resulted in aptamer ligands with the highest binding affinity for the cancer cell receptors.

Nucleolin is a multifunctional RNA-binding protein that is most abundant in the nucleolus, but it can also be found in the cytoplasm and plasma membrane [667]. Nucleolin binds DNA and RNA, is involved in chromatin remodeling, ribosomal (rRNA) transcription, rRNA maturation, ribosome assembly, nucleocytoplasmic transport, and ribosome biogenesis. It has also been reported to exhibit ATP-dependent helicase activity in vitro. The central region of nucleolin contains four RNA recognition motifs, while the C-terminal region is characterized by an arginine/glycine-rich domain that allows interactions with target mRNAs as well as with other proteins, including ribosomal proteins [668]. Nucleolin is overexpressed in many types of cancer cells. For example, nucleolin expression is increased six-fold in breast- and colon-cancer cells. Interaction with vascular endothelial growth factor (VEGF) and activation of VEGFR signaling pathways are associated with the translocation of nucleolin to the plasma membrane. Nucleolin acts as a membrane-anchored receptor on the cell surface of endothelial and cancer cells, where it regulates cellular localization, adhesion, differentiation, proliferation and, angiogenesis. Furthermore, the nucleolin surface receptor enables the specific binding of nanostructures to cancer cells when they are conjugated with the single-stranded nucleolin-DNA G-quadruplex aptamer AS1411. The nucleolin receptor mediates the endocytosis and transport of such nanoconjugates into the perinuclear region. While

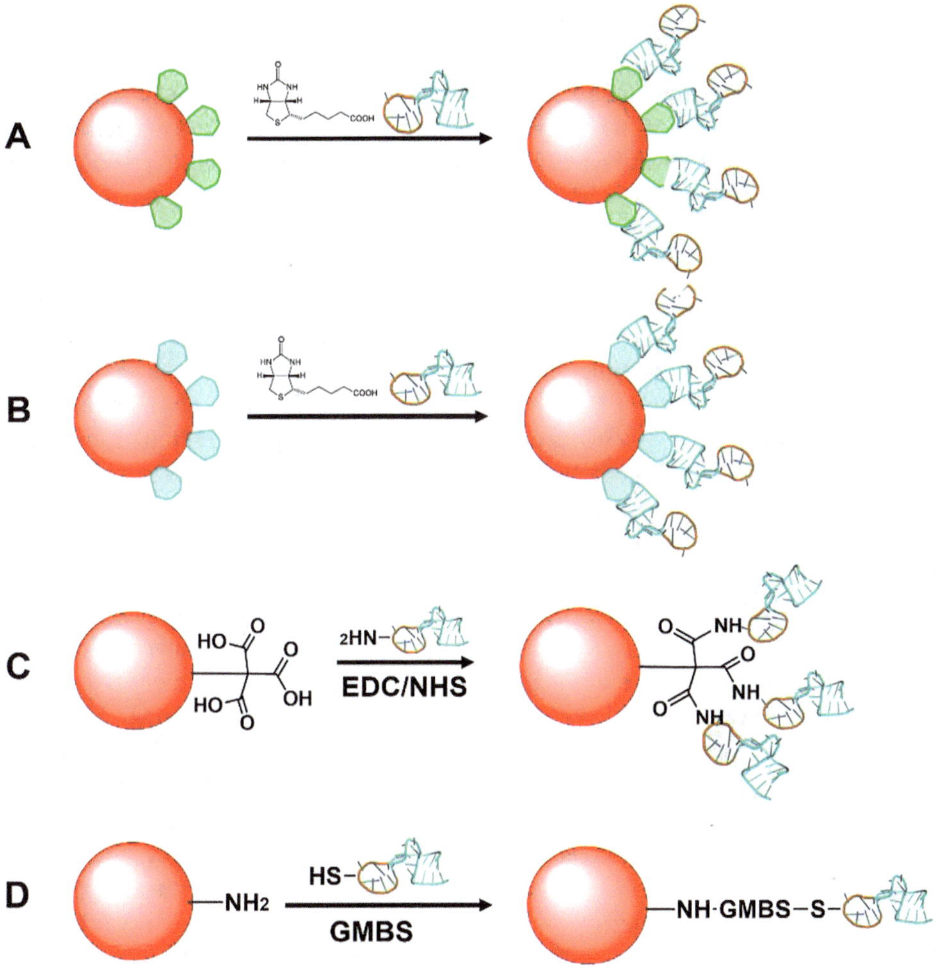

Figure 119: Scheme of the major conjugation chemistries:(A) avidin-coated MNPs conjugated to biotinylated aptamers, (B) streptavidin-coated FNPs conjugated to biotinylated aptamers, (C) tricarboxyl-silanized FNPs conjugated to amine-modified aptamers through EDC chemistry, (D) amino-silanized FNPs conjugated to thiolated aptamers through EDC chemistry with amino-silane.

molecules smaller than approximately 9 nm can passively diffuse through the nuclear pore complex, the nuclear import of nanostructures typically requires active transport mechanisms, even for particles below this size threshold.

The study by Lopes-Nunes et al. [669] focuses on the use of AS1411-functionalized AuNPs as a targeted drug-delivery system for the treatment of cervical and endometrial cancer. Thiolated AS1411 with a regular DNA backbone and a 50-thiol C6-S-S modification was covalently attached to ultrasmall (5-nm) citrate-stabilized AuNPs by overnight co-incubation in phosphate buffered saline. The AS1411-AuNP conjugates were loaded with the green-fluorescent anticancer drugs acridine orange derivative 10-(8-(4-iodobenzamide) octyl)-3,6-bis(dimethylamine) acridinium iodide (C8) [670] and the small-molecule immunomodulator imiquimod (IQ) [671]. IQ has anticancer and antiviral activity, for example, by triggering the capase pathway leading to apoptosis. IQ also activates the TLR7/8 and NF-B pathways, modulating the immune response. The targeted drug-delivery capacity of AS1411-AuNP conjugates was evaluated in the human endometrial adenocarcinoma HEC-1-A and human cervical adenocarcinoma HeLa cell lines. Primary normal human dermal fibroblasts (NHDF) were used as reference cells.

The surface charge and hydrodynamic diameters of AuNPs, AS1411-AuNP conjugates, and their associates with C8 and IQ were examined by zeta potential and DLS measurements. The results show that the negative surface charge of the AS1411-AuNP conjugates decreases from −25.8 mV to about −41 mV when loaded with IQ or C8, which is surprising, especially since IQ is positively charged in aqueous solution and C8 is amphoteric. On the other hand, the 2–5-nm increase in hydrodynamic diameter of the AS1411-AuNP conjugates (16 nm) can be explained by electrostatically bound IQ and C8.

The cellular uptake of free IQ and C8 as well as IQ and C8 bound to AS1411-AuNP conjugates (IQ-AS1411-AuNP and C8-AS1411-AuNP conjugates) was evaluated in HeLa, HEC-1-A, and NHDF cells by confocal fluorescence microscopy and flow cytometry. The impact of free and bound anticancer drugs on cancer cell viability was determined by the MTT assay. The fluorescence microscopy and flow cytometry studies showed that the IQ-AS1411-AuNP and C8-AS1411-AuNP conjugates were preferentially taken up by HeLa and HEC-1-A cells compared to the normal NHDF cells and the cellular internalization of free IQ and C8. This cell specific uptake was ascribed to the nucleolin receptor, which is overexpressed on the surface of the cancerous HeLa and HEC-1-A cells. Moreover, the AS1411-AuNP conjugates were found to accumulate in the lysosomes of the cancer cells, with a higher degree of selectivity observed for C8 when bound to the nanoconjugates. The confocal fluorescence microscopy images in Figures 120 and 121 illustrate the key role of the nucleolin receptor for specific binding and internalization of the nanoconjugates loaded with either C8 (Figure 120) or IQ (Figure 121). To distinguish between plasma membrane, cytoplasm, and nucleus of the HeLa and HEC-1-A cells, nucleolin and the nucleus were stained with the red-fluorescent AlexaFluor 647® dye and the blue-fluorescent Hoechst 33342 dye, respectively (Figures 120 and 121, 1st and 2nd column). Apparently, the C8-labelled nanoconjugates were efficiently internalized by the cancer cells. C8 release took place not only in the cytosol but also in the nucleus (Figure 120, third and fourth column). By comparison, IQ was only released in the cytosol (Figure 121, third and fourth column). The smaller fluorescence intensity of the green-fluorescent IQ is due to the lower fluorescence quantum yield. These results were confirmed by flow cytometric studies for C8-AS1411-AuNP conjugates, which showed higher internalization in HeLa cancer cells compared to normal NHDF cells.

As demonstrated by confocal microscopy, flow cytometry and MTT assays the AS1411-AuNP conjugates were able to enhance the anticancer selectivity such as IQ and C8 in HeLa and HEC-1-A cells The cytotoxicity of IQ-AS1411-AuNP conjugates in

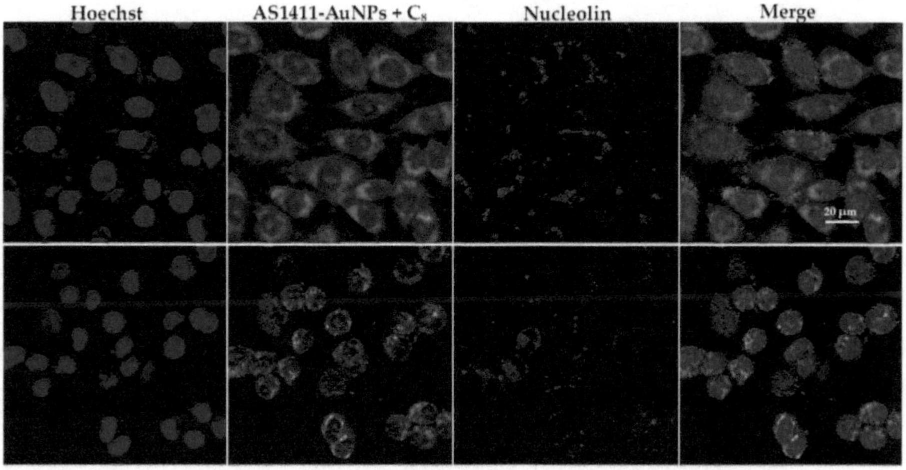

Figure 120: Confocal fluorescence microscopy images of (A) HeLa and (B) HEC-1-A cells incubated with AS1411-AuNP conjugates with C8 (green fluorescence); nucleolin was stained with AlexaFluor 647® (red fluorescence); cell nuclei were stained with the blue fluorescent Hoechst 33342 dye [669].

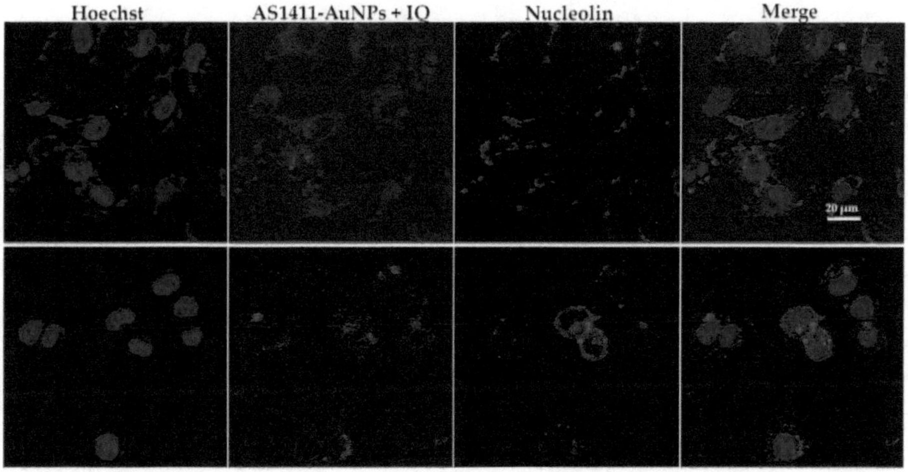

Figure 121: Confocal fluorescence microscopy images of (A) HeLa and (B) HEC-1-A cells incubated with AS1411-AuNP conjugates with IQ (green fluorescence); nucleolin was stained with AlexaFluor 647® (red fluorescence); cell nuclei were stained with the blue fluorescent Hoechst 33342 dye [669].

HeLa and HEC-1-A cells and the therapeutic index compared to normal fibroblasts were both improved as shown by MTT assay. C8-AS1411-AuNP conjugates also showed cancer accumulation of C8 and decrease in cell viability (30.5 % in HeLa and 17.0 % in HEC-1-A cancer cells and 71.3 % in NHDF cells). Similarly, the nanoconjugates are compelling for delivery of IQ, particularly to cervical cancer cells because they enhance the anti-cancer potential and selectivity of IQ itself by significantly reducing the viability of cancer cells (HeLa at 15 % and HEC-1A at 24 %) compared to normal NHDF cells (*ca.* 86 %).

4.3 Intracellular trafficking

Elucidation of the endocytic pathway is crucial for nanoparticle-mediated drug delivery applications. A current challenge is to understand how molecular machines regulate each step of nanoparticle transport and then how these various mechanisms are integrated into the different types of endocytic pathways. For example, macropinocytosis and clathrin-mediated endocytosis ensure the transport of nanoparticles into acidic compartments, and this pathway can be used for pH-controlled drug release [420, 428, 672]. After uptake into the cell, the nanoparticles travel to various intracellular targets to remain there or to be excreted via the process of exocytosis. Once inside the cell, the nanoparticles can enter various intracellular compartments, such as the endosomes or lysosomes, where they can interact with various cellular components, such as membranes, actins, proteins, and enzymes. Intracellular transport and trafficking do not only depend on the mechanism of cellular uptake, but also on the composition, shape and surface structures of the nanoparticles and the cell type. Smaller and more spherical nanoparticles tend to be more efficiently internalized by cells and can be transported to various intracellular compartments more easily. In addition, the surface charge and chemical composition of the nanoparticles can influence their interactions with the plasma membrane and intracellular proteins, which can affect their intracellular trafficking. Understanding the mechanisms of intracellular trafficking is important for developing nanoparticle-based therapies that can effectively target specific cells and tissues in the body. Methods currently used to study the intracellular transport of nanoparticles include optical and electron microscopy techniques such as high-resolution fluorescence microscopy, confocal laser scanning microscopy, dark-field microscopy, transmission electron microscopy, scanning electron microscopy, atomic force microscopy, flow cytometry, mass cytometry, mass cytometry, photoacoustic microscopy, surface-enhanced Raman scattering, inductively coupled plasma laser ablation mass spectrometry, and correlative microscopy [673–675].

4.3.1 Endosomal network

After entering the cell via one of the many endocytic pathways, cargo-laden vesicles assemble within the endosomal network. The endosomal network is a dynamic, interconnected trafficking system that transfers cargo among various different membrane-bound intracellular compartments. Within this network, internalized cargo is sorted and sent to its final destination: (a) recycling back to the cell membrane, (b) degradation in the endosomal–lysosomal pathway, or (c) transfer to the trans-Golgi network [676] (Figure 122).

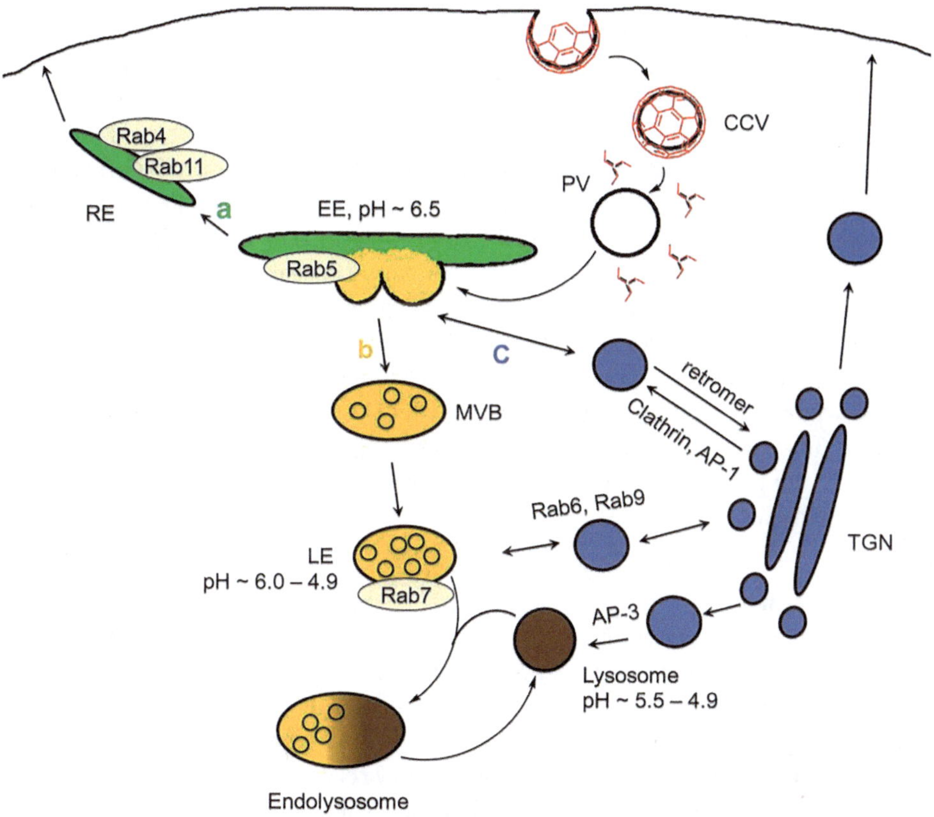

Figure 122: Cargoes internalized via clathrin-mediated endocytosis (CME): clathrin-coated vesicles (CCV) lose their coating and fuse as primary vesicles (PV) with early endosomes (EE); a) recycling pathway: Recycling endosomes (RE) transport cargo back to the plasma membrane; b) Degradation along the endosomal–lysosomal pathway: multivesicular bodies (MVB) mature into late endosomes (LE), which fuse with lysosomes to form endolysosomes; c) Traffic between EE, LE, lysosomes, and the trans-Golgi network (TGN).

4.3.1.1 Internalization via clathrin-mediated endocytosis

Cargoes taken up by clathrin-mediated endocytosis become clathrin-coated vesicles. The clathrin coat rapidly disintegrates upon entry into the cell. These small primary vesicles fuse with early endosomes and the cargo enters the endosomal network [472] (Figure 122).

Early endosomes are highly dynamic and pleiomorphic compartments with a high capacity for homotypic fusion. Thin tubules (60 nm in diameter) and large vesicles (300–400 nm) originate from a cisternal region [677]. Characteristic for early endosomes are a weakly acidic endosomal lumen (pH 6.8–5.9), the small GTPase Ras-related protein 5 (Rab5), and early endosome antigen-1 (EEA-1). Both Rab5 and EEA-1 are required for endosomal homotypic fusion.

Early endosomes

EEA-1 FYVE domains bind preferentially to phosphatidylinositol 3-phosphate (PI(3)P), which is localized only in early endosomal membranes. PI(3)P synthesis is dependent on Rab5 and regulated by the PI3-kinase VPS34, a Rab5 effector [676]. Rab5 can indirectly recruit EEA-1 to early endosomes by stimulating the local production of PI(3)P. Rab5 can also directly bind EEA-1. Thus, Rab5 is required for the recruitment of EEA-1 to the early endosomal membranes. Once bound to the early endosomal membrane, EEA-1 can form homodimeric complexes with EEA-1 from other early endosomes, thereby tethering the different early endosomes together. Membrane fusion of early endosomes requires the formation of a SNARE (soluble N-ethylmaleimide-sensitive factor attachment protein receptor) complex, which forms a bridge between membranes [472]. Early endosomes can receive cargo not only from the plasma membrane, but also from late endosomes or the trans-Golgi network. Early endosomes act as a major sorting station, directing cargo to late endosomes, recycling it back to the plasma membrane, or transporting it to the trans-Golgi network [678].

Recycling pathway

The thin tubules that emerge from the early endosome bud off as recycling endosomes [676, 677]. This event is mediated by Rab4, Rab11 and the SNARE protein Cellubrevin, all of which are found on recycling endosomes and are therefore characteristic of this endosome type [472]. Recycling endosomes are less acidic than early endosomes. Recycling pathways can be divided into a fast and a slow recycling pathway. In the fast recycling pathway, cargo is transported directly from early endosomes back to the plasma membrane [679, 680]. In the slow recycling pathway, cargo is transported from early endosomes to recycling endosomes before being returned to the plasma membrane. Recycling endosomes play a crucial role in sorting and storing cargo before it is recycled to the plasma membrane.

Endosomal–lysosomal pathway

In this pathway, cargoes are transported from early endosomes to late endosomes and finally to lysosomes for degradation. Late endosomes form through the maturation of early endosomes, undergoing significant structural and compositional changes. Mature late endosomes typically have a round or oval shape (250–1,000 nm) and contain a large number of intraluminal vesicles, giving them a multivesicular appearance. They have an acidic pH (around 5.0–6.0) and contain lysosomal proteins, including hydrolases destined for degradation processes [678–680]. Due to these transformation steps, late endosomes bear little resemblance to the early endosomes from which they originated. The major steps of this maturation and transformation process are summarized in Table 5.

Late endosomes eventually fuse with lysosomes to form a hybrid organelle, called the endolysosome. Major cargo hydrolysis occurs in endolysosomes, these organelles mature in lysosomes, which serve as the main degradative compartments containing

Table 5: The endosomal maturation process.

Process	Changes from early to late endosomes
Rab switch	– Rab5 is exchanged for Rab7 – Rab7 is characteristic for late endosomes – Rab7 is involved in the development of the endosome lysosome system – Other Rabs are lost (e. g., Rab4, Rab11) – Other Rabs are added (e. g., Rab9)
Formation of intraluminal vesicles	– Inward budding of endosomal membrane – Fission as intraluminal vesicles inside the late endosomes – Containing membrane proteins and lipids for lysosomal degradation – "backfusion" with endosomal membrane possible – Released as exosomes, when late endosomes fuse with plasma membrane
Acidification	– pH inside the late endosomes decreases – pH range: 5.0–6.0 – V-ATPases (proton pumps) regulate the pH – V-ATPase concentration increases in late endosomes
Size and Morphology	– Thin tubules of the early endosomes are lost – Late endosomes are round or oval – Late endosomes grow in size
Gain of lysosomal proteins	– Lysosomal hydrolases and membrane proteins are transported from the TGN
Fusion specificity	– Late endosomes do not fuse with early endosomes – Homotypic fusion with late endosomes – Fusion with lysosomes
Cytoplasmic motility	– New set of microtubule-dependent motor proteins – Late endosomes move towards the perinuclear region
Luminal ionic environment	– Increase in Cl^- concentration – Change in Ca^{2+}, Na^+ and K^+ concentrations

a high concentration of hydrolases [680]. While lysosomes and late endosomes share some characteristics, lysosomes can be distinguished by their higher concentration of hydrolases, more acidic pH, and unique membrane proteins [677]. Lysosomes can be identified by their electron-dense appearance under the electron microscope, reflecting their high content of hydrolytic enzymes and degraded material. They are characterized by specific lysosomal membrane proteins such as LAMP1 and LAMP2, and the presence of Rab7 [679].

Trans-Golgi network

There is continuous trafficking between endosomes, lysosomes and the trans-Golgi network. During endosome maturation, some endosomal components are removed, and lysosomal proteins are delivered. Sorting and vesicle formation from endosomes to the trans-Golgi network depends on factors such as Rab7, Rab9 and the retromer complex.

The retromer complex is a multiprotein complex with cargo-specific recognition capacity. Sorting nexins in this complex mediate its binding to the endosomal membrane. Recycling endosomes shuttle cargo between the trans-Golgi network and the plasma membrane [680]. Cargoes required for endosomes and lysosomes are transported from the trans-Golgi network to the endocytic pathway. Vesicles coated with clathrin and the adaptor protein AP-1 transport cargoes from the TGN to endosomes. AP-3-coated vesicles transport cargoes directly to lysosomes [681]. From the trans-Golgi network, cargoes are transported in transport vesicles via the secretory pathway to the plasma membrane.

Internalization via caveolae-mediated endocytosis

Cargoes internalized by caveolae/lipid raft-mediated endocytosis are encapsulated by caveolar vesicles, which are enriched in cholesterol, sphingolipids, and caveolin. At the plasma membrane, either single caveolar invaginations or multicaveolar assemblies remain attached to the membrane (Figure 123). Once detached, individual caveolar vesicles can fuse with early endosomes. Furthermore, traffic between caveosomes and early

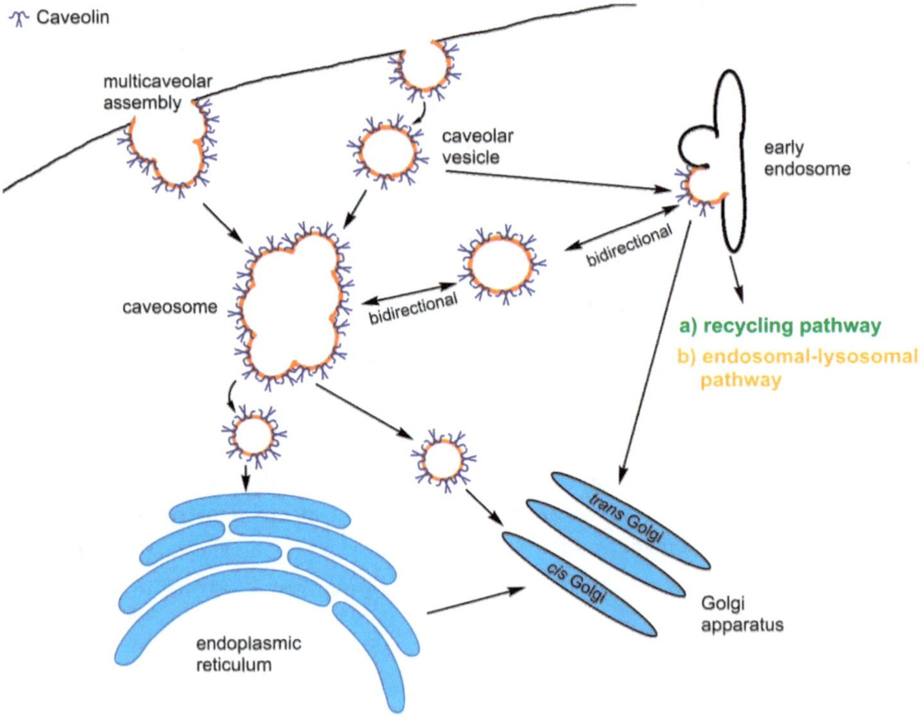

Figure 123: Cargoes internalized by caveolae-mediated endocytosis: either single caveolar vesicles or multicaveolar assemblies from the plasma membrane; multicaveolar assemblies from grape-like caveosomes; from the caveosomes, caveolar vesicles can travel to the early endosomes, the Golgi apparatus, or the endoplasmic reticulum.

endosomes is bidirectional [482, 682]. Fusion of caveolar vesicles with early endosomes is Rab5 dependent [482]. When caveolar vesicles fuse with the endosomal membrane, their membrane domain remains intact and does not mix with the endosomal membrane [485]. After some time, the membrane domains containing caveolin-1 can bud off from the early endosomes as caveolar vesicles. These caveolar vesicles can move along the endosomal pathways: (a) the recycling pathway and (b) the endosomal–lysosomal pathway. Caveosomes are multiple caveolar vesicles with a grape-like structure and an intraluminal neutral pH. Caveosomes never fuse with lysosomes and thus do not reach the degradative end [478]. Caveosomes are intermediate organelles that function as early endosome-equivalent sorting stations in caveolar endocytic pathways. Caveolar vesicles budding from caveosomes are delivered to either the endoplasmic reticulum or the Golgi apparatus (Figure 123). Fusion of caveolar vesicles with the endoplasmic reticulum or the cis-Golgi complex is ADP-ribosylation factor 1 (Arf1)-dependent. During these interactions, the caveolar membrane domains never dissociate. This means that caveolar coats never assemble and disassemble during intracellular trafficking like the clathrin coat. Caveolar coats are assembled during the biosynthesis of new caveolar vesicles. Caveolin is synthesized in the endoplasmic reticulum and then transported to the Golgi complex. At the Golgi complex, caveolin assembles with cholesterol- and glycosphingolipid-rich lipid raft domains that exit the Golgi complex to form new caveolar vesicles [477].

Internalization by clathrin/caveolae-independent endocytosis

Cargoes internalized via the clathrin/caveolae-independent pathway are enclosed in vesicles that do not contain clathrin or caveolin. These vesicles fuse with early endosomes and travel along the classical endosomal pathway [675, 683].

Internalization by macropinocytosis

Membrane ruffles engulf large amounts of cargo and extracellular fluid via macropinocytosis. Cargoes and extracellular fluid end up in large organelles called macropinosomes. Macropinosomes have no apparent coat structure and are larger than 0.2 μm in diameter [684, 685]. In fluorescence microscopy, macropinosomes can be labeled with fluid phase markers such as Lucifer Yellow, horseradish peroxidase, or fluorescently labeled dextran [468]. The final fate of macropinosomes depends on the cell type. In macrophages or kidney cells, macropinosomes fuse with lysosomes in a degradative pathway. In certain cell types such as A-431 (human epidermoid carcinoma cells) or NIH-3T3 (mouse NIH/Swiss embryo fibroblast cells), macropinosomes fuse with each other and eventually recycle back to the plasma membrane without fusing with lysosomes [686]. Macropinosomes in these cells contain EEA-1, which mediates homotypic fusion.

Before fusing with lysosomes, macropinosomes undergo a maturation process (Figure 124). A first step of this maturation is the tubulation and shrinkage of the very large

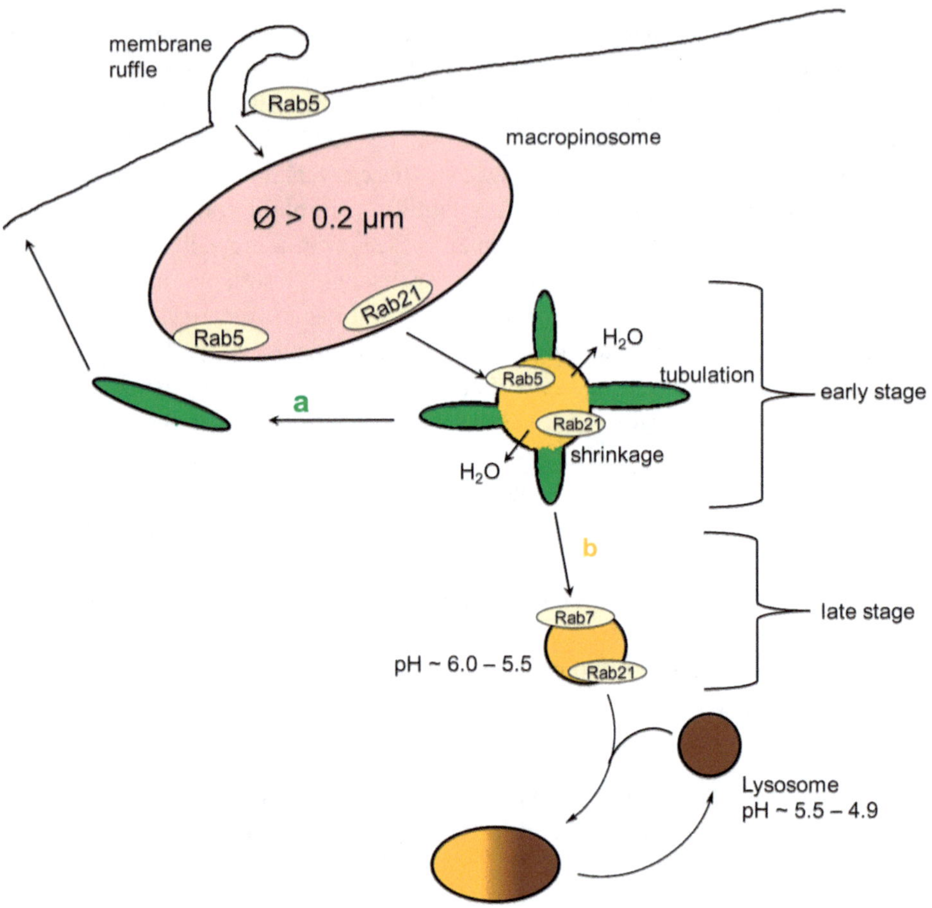

Figure 124: Cargoes internalized by macropinocytosis are found in large macropinosomes; shrinkage of these macropinosomes occurs by intraluminal water loss and tabulation of the micropinosome membrane: a) recycling pathway: tubules recycle back to the plasma membrane; b) degradation pathway: macropinosomes mature via rab5 to rab7 conversion and intraluminal acidification into late endosome-like structures and finally fuse with lysosomes.

macropinosome [500, 686]. Proteins of the sorting nexin (SNX) family oligomerize and induce membrane curvature to form tubular protrusions. As the macropinosome surface area decreases through tubule formation and membrane recycling, osmotic balance and solute concentration gradients within the macropinosome must be adjusted. The tubules are recycled back to the plasma membrane. The reduction of the macropinosome membrane increases the hydrostatic pressure inside the macropinosome. To relieve the hydrostatic pressure and the resulting membrane tension, macropinosomes should lose water. Along with the various cargoes, large amounts of extracellular fluid are internalized by macropinocytosis. The concentrations of inorganic cations and anions, especially

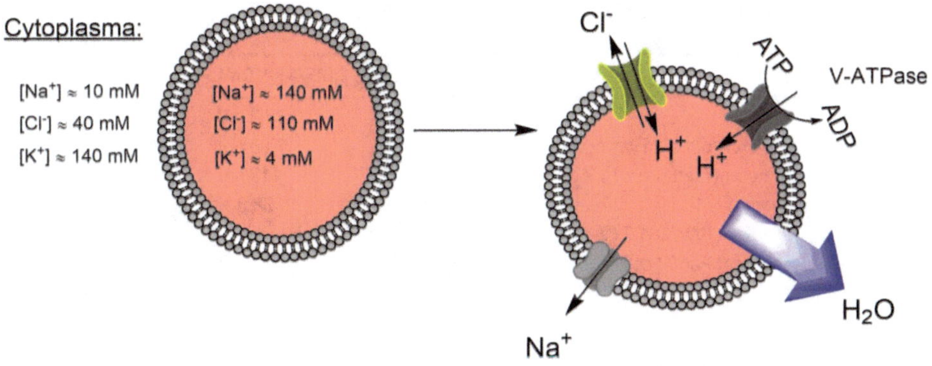

Figure 125: Water efflux and luminal acidification in macropinosomes: The intraluminal ion concentration in macropinosomes is equal to the extracellular ion concentration. Osmotically driven ion exchange causes water efflux into the cytoplasm. Cl^-/H^+ ion exchange and V-ATPases pumping H^+ into the macropinosomes lower the luminal pH.

Na^+, K^+ and Cl^-, are different in the cytoplasm and in the extracellular fluid and therefore for the macropinosomal lumen (Figure 125). The macropinosomal concentrations of Na^+ and Cl^- (110–140 mM) are higher than in the cytoplasm [687]. In contrast, the cytoplasmic K^+ concentration exceeds that of the macrospinosomal lumen [688]. Na^+ ions are transported into the cytoplasm through two pore channels. The Na^+ efflux creates a membrane potential that would prevent further Na^+ transport. This activates Cl^- ion channels such as TMEM206 or ClC-5. These ion exchangers transport two Cl^- ions into the cytoplasm and one H^+ into the macropinosome, helping to regulate the electrochemical gradients. Together with V-ATPases, the Cl^- channels are responsible for the acidification of the macropinosomes. The efflux of Na^+ and Cl^- ions causes an osmotic water loss and shrinkage of the macropinosomes.

In the late stage of macropinosome maturation, Rab5 is replaced by Rab7, similar to late endosomes (Figure 125). Finally, macropinosomes fuse with lysosomes and acquire lysosomal enzymes and proteins.

4.3.2 Transport of nanoparticles

Portilla et al. [689] surveyed the cellular internalization, intracellular transport and degradation of SPIONs with γ-maghemite composition (Fe_2O_3) in two different mouse cell lines, RAW264.7 macrophages and Pan02 mouse pancreatic ductal adenocarcinoma tumor cells, using confocal fluorescence microscopy, TEM, ICP-OES, and high-throughput mass spectrometry. Spherical SPIONs with a size of 12.7 nm were produced by alkaline co-precipitation and subsequently functionalized with aminopropyltriethoxysilane (APS) or dimercaptosuccinic acid (DMSA) to endow them with opposite surface charges and different hydrodynamic diameters (Table 6). APS-SPIONs and

Table 6: Physicochemical features of APS-SPIONs and DMSA-SPIONs.

	APS-SPIONs	DMSA-SPIONs
Surface charge	+23 mV	−34.3 mV
Hydrodynamic diameter (distilled water)	122 nm	83 nm
Hydrodynamic diameter (medium +10 % serum)	1293 nm	117 nm
Relative protein coating	18.3 %	38.3 %
Intracellular iron content RAW264.7 cells	48 pg/cell	58 pg/cell
Intracellular iron content Pan02 cells	55 pg/cell	4 pg/cell

DMSA-SPIONs exhibited surface charges of +23 mV and −34.3 nm, respectively, in distilled water, while the corresponding hydrodynamic diameters were 122 nm and 83 nm. The authors demonstrated that the different nanoparticle coatings activated different macrophage receptors, which obviously affected the endocytic pathway. The APS-SPIONs were internalized by RAW264.7 cells through clathrin-dependent, caveolae-dependent and endocytic pathways [690]. In contrast, DMSA-SPIONs were internalized by RAW264.7 cells through receptor-mediated endocytosis. As a result, APS-MNPs and DMSA-MNPs exhibited different internalization kinetics in RAW264.7 cells. However, in Pan02 cells, APS-MNPs are internalized by macropinocytosis, whereas DMSA-SPIONs utilize all endocytic pathways. To analyze the influence of APS- and DMSA-SPIONs on endosome formation, the subcellular localization of SPIONs in RAW264.7 cells was followed as a function of time by TEM. However, RAW264.7 cells exposed to DMSA-SPIONs produced more SPION-containing endolysosomes than those incubated with APS-SPIONs. Thus, internalization of APS-SPIONs in RAW264.7 cells resulted in the formation of larger endolysosomes (33.9 μm^2) than the cellular uptake of DMSA-SPIONs (15.4 μm^2). Results from both ICP-OES and TEM analysis show that APS- and DSMA-SPIONs have different intracellular transport kinetics. The intracellular trafficking of APS- and DMSA-SPIONs was investigated using confocal fluorescence microscopy, endolysosomal markers, and high throughput analysis of the associated proteome. RAW264.7 and Pan02 cells were incubated with Rab5, Rab7, Rab2a, EEA1, and Lamp1 antibodies to monitor the formation of early endosomes (EEA1+), late endosome-lysosome fusions (Rab7+/Lamp1+), and autophagolysosomes (Rab2A+) around the SPION. Each of these markers defines a specific step in endolysosomal trafficking, including the putative formation of autophagolysosomes resulting from the fusion of autophagic vesicles, late endosomes, and lysosomes. The SPIONs pass through the vesicles with different kinetics. However, DMSA-SPIONs were found to migrate slower than APS-SPIONs in macrophages compared to SPION accumulation in Rab7+ endolysosomes.

Macrophage endolysosomes loaded with DMSA-SPIONs showed higher levels of lytic enzymes and catalytic proteins than endolysosomes loaded with APS-SPIONs. These endolysosomes are acidic due to their increased levels of V-type ATPase. DMSA-SPIONs induce more autophagic vesicles in macrophages, which may trigger the expression of genes and proteins related to iron metabolism, likely due to iron release from degraded

SPIONs. Unlike in Pan02 cells, the SPION coating significantly affects the intracellular trafficking and nature of endolysosomes in macrophages. Clathrin-dependent endocytosis is a receptor-mediated process in which early endosomes containing SPIONs mature into late endosomes before fusing with lysosomes to form endolysosomes. By contrast, macropinocytosis is a receptor-independent process that can deliver cargo to late endosomes or lysosomes, depending on the cell type and cargo. The APS-SPIONs rapidly accumulate in late endosomes or autophagosomes, while DMSA-SPIONs are internalized more slowly and primarily enter mature endolysosomes.

The intra-endolysosomal degradation of the APS- and DMSA-SPIONs in RAW264.7 macrophages and Pan02 tumor cells was investigated: the diameter of the SPIONs inside the endolysosomes decreased by about 25 % after 24 h exposure. The APS- and DMSA-SPIONs, which have opposite surface charges, differ in the mechanism of their cellular uptake, the kinetics of their intracellular transport, their degradation in endolysosomes, and their influence on cell autophagy. For example, DMSA-SPIONs localized in terminally mature endolysosomes have a strong proteolytic profile. Enrichment with the V-type ATPase subunit ensures acidification of the intraluminal space, promoting the activation of lysosomal proteolytic enzymes such as cathepsins. The degradation of DMSA-SPIONs triggers autophagy, leading to the formation of autophagosomes containing damaged organelles such as mitochondria, Golgi, and endoplasmic proteins. Endolysosomes containing DMSA-SPIONs also reflect the strong antioxidant response of cells against ROS production involving various redox enzymes such as CAT, SODM, and those of the thioredoxin system. The intracellular degradation of DMSA-SPIONs was shown to induce autophagy [691]. In comparison, APS-SPIONs migrated to less degradable compartments. In Pan02 tumor cells, both APS- and DMSA-SPIONs were transported to poorly degradative endolysosomes. Induction of iron metabolism correlated with the rate of SPION degradation. DMSA SPIONs increased ferroportin-1 (Slc40a1) transcription in RAW264.7 cells more than APS SPIONs. In contrast, both types of SPIONs caused only weak transcription of the Slc40a1 gene in Pan02 cells, consistent with endolysosomal degradation of SPIONs associated with Fe metabolism: Slc48a1, encoding a heme transporter, and Slc11a2, encoding DMT1. The increased Slc11a2 (DMT1) gene transcription in RAW264.7 cells exposed to DMSA-SPIONs was particularly significant as it correlated with the accumulation of SNX3 enrichment, a known mediator of DMT1 translocation to the endosomal membrane. Furthermore, RAW264.7 cells incubated with DMSA-SPIONs showed a significantly greater accumulation of ferritin protein compared to APS-SPIONs. This indicates the degradation of DMSA-MNPs. By comparison, the ferritin content of Pan02 cells is not affected by the internalization of DMSA and APS-SPIONs.

Van der Zwaag et al. [673] used a combination of super-resolution microscopy and single-particle data analysis to track the cellular uptake and intracellular translocation of polystyrene nanoparticles (PSNPs) in HeLa and dendritic cells. Carboxy acid- and amine-terminated PSNPs were labeled with either the red fluorescent Cy5 or Alexa647 dyes using EDC/NHS coupling chemistry. They successfully used two-color stochastic optical reconstruction microscopy (STORM) to visualize and analyze the size, positioning, and interaction of PSNPs with cellular structures at a resolution of 20 nm inside dendritic cells. The information obtained on the distribution of PSNPs was compared with results from confocal fluorescence and transmission electron microscopy studies. The PSNPs were observed at the plasma membrane and after internalization, as well as their co-localization with various cellular organelles, such as endosomal vesicles. The colocalization of PSNPs in micropinosomes indicates an endosomal uptake mechanism initiated by membrane engulfment. The study successfully demonstrates the application of STORM to resolve subdiffraction particles and provides insight into the mechanism of nanoparticle entry, intracellular trafficking, and potential endocytic pathways. The cellular uptake mechanism of 80-nm PSNPs in HeLa cells was attributed to endocytosis after membrane binding. To visualize nanoparticle-organelle interactions, actin, plasma membrane and nuclear envelope complex, and macropinosomes were labeled with green-fluorescent dyes such as phalloidin-atto488 for actin, wheat germ agglutinin (WGA) alexa 568 for membrane, and nuclear envelope complex, 70-KDa dextran alexa 568 for macropinosomes. Figure 126 shows the colocalization of 80-nm PSNPs with plasma membrane, macropinosome, nuclear membrane, and actin markers. As the

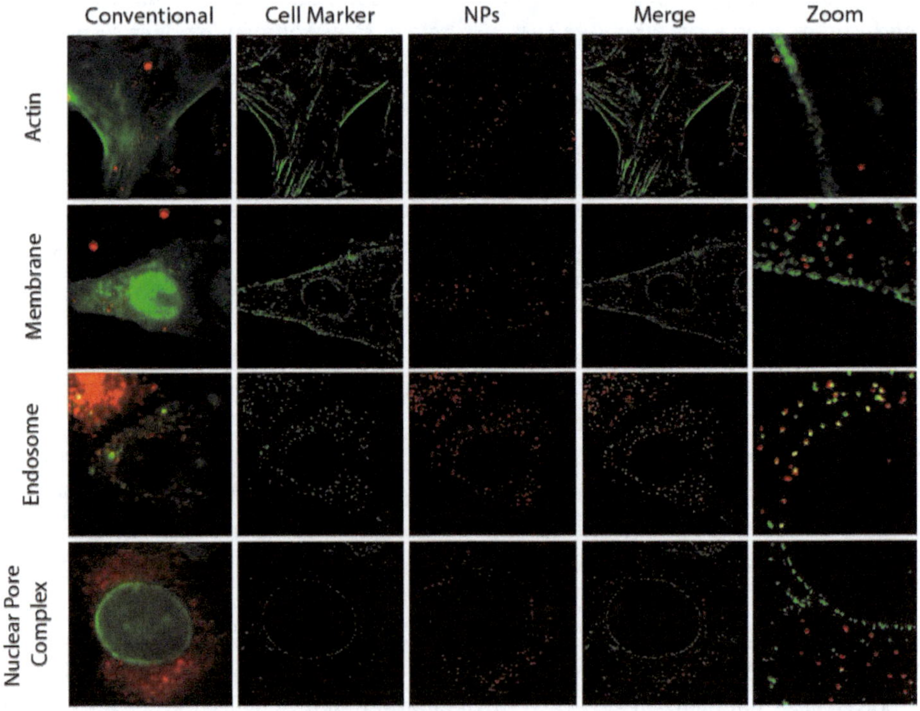

Figure 126: STORM images of colocalization of organelles (green) and internalized 80-nm PSNPs (red). Image size = 40 μm, zoomed image size = 5 μm [673].

merged images show, the PSNPs co-localize with these organelles, illustrating the intracellular transport of the nanoparticles. STORM analysis revealed strong colocalization of PSNPs with dextran-labeled endosomal vesicles, a marker for micropinocytes. Based on these observations of the endosomal pathway, the first steps of internalization could be elucidated.

For the localization and quantification of nanoparticles, Andrian et al. [692] used a correlative fluorescence microscopy (FM) and an electron microscopy method combining direct stochastic optical reconstruction microscopy (dSTORM) and TEM. Poly(lactide-co-glycolide)-poly(ethylene glycol) (PLGA-PEG) coated nanoparticles (PLGA-PEG NPs) were fluorescently labeled with DiI (1,1′-dioctadecyl-3,3,3′,3′-tetramethylindocarbocyanine). HeLa cells were treated with the nanoparticles for specific time points. The cells were then subjected to mild chemical fixation to preserve cellular and membrane structures without quenching the fluorescence signal. The cells were snap frozen to preserve cellular structures, and then processed by freeze substitution, a method where ice is replaced with an organic solvent to better preserve fine ultrastructural details. After further staining and fixation, the cells were embedded in resin at low temperatures and then gradually warmed to room temperature. Ultrathin sections (70–100 nm) of the resin-embedded cells were placed on a TEM grid. This grid was placed on a glass slide containing dSTORM buffer and sealed with a coverslip. The sample was first imaged with dSTORM, then stained and imaged with TEM. Figure 127 shows separate TEM images (Figure 127(A)) and fluorescent images of the PLGA-PEG NPs (Figure 127(B)), along with fiducial markers (Figure 127(C)). Merging these images resulted in correlative FM-TEM images (Figure 127(D)). At higher magnifications (Figures 127(E) and (F)), the dSTORM-TEM images (Figure 127(F)) provided enhanced resolution, allowing the assignment of individual PLGA-PEG NPs to specific cellular compartments identified by TEM. Early endosomes have a diameter of 100–500 nm with an electron-lucent lumen and a few intraluminal vesicles. Late endosomes have a diameter of 250–1000 nm, characterized by densely packed intraluminal vesicles, while lysosomes range from 200 nm to over 1000 nm and are distinguished by their electron-dense lumen and membrane whorls.

One hour after PLGA-PEG NP treatment, most of the PLGA-PEG NPs were still found attached to or surrounding the cell membrane of HeLa cells. However, intermittent intracellular signals indicated that cellular uptake started before 1 h. At 1 h, 48 % of PLGA-PEG NPs were found in early endosomes, 26 % in late endosomes, 20 % in non-endosomal compartments, and 5 % in lysosomes. PLGA-PEG NPs were taken up by endocytosis and transported via the endosomallysosomal pathway. After 8 h, 13 % of PLGA-PEG NPs were in early endosomes, 50 % in late endosomes, and 23 % in lysosomes, indicating further progression of PLGA-PEG NPs along the endosomal–lysosomal pathway. After 24 h, the distribution of PLGA-PEG NPs inside the cells remained the same, but the amount found on the cell membrane decreased. This indicates that the PLGA-PEG NPs were endocytosed at all time points. A few PLGA-PEG NPs were found in non-endosomal compartments. Since PLGA-PEG NPs did not undergo endosomal escape, the authors discussed that either the PLGA-PEG NPs were released from ruptured organelles or the fluorescent

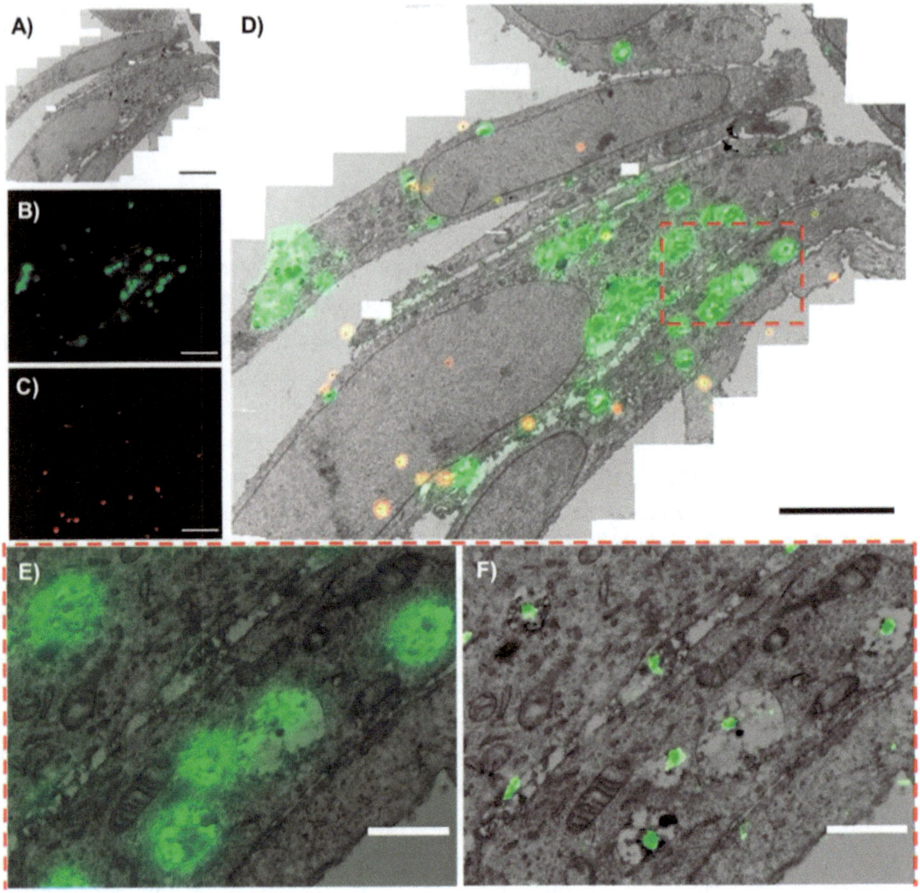

Figure 127: Correlative low-resolution FM/dSTORM and TEM images of PLGA-PEG NPs internalized into HeLa cells. Large images of the region of interest of (A) merged TEM images, (B) PLGA-PEG NPs signal and (C) fiducial markers both in FM; (D) correlative FM-TEM image (scale bars A–D: 5 μm); Magnified views of the boxed region in D: (E) FM-TEM correlated image and (F) dSTORM-TEM correlated image (scale bars E–F: 1 μm) [692].

dye was released from lysosomes after nanoparticle degradation. Between 24 h and 72 h, the number of internalized PLGA-PEG NPs decreased. There could be two reasons for the decrease in PLGA-PEG NP concentration: (a) the PLGA-PEG NPs left the cells via the recycling pathway or other exocytosis pathways or (B) the PLGA-PEG NPs were degraded inside the lysosomes.

Liu et al. [693] used DNA-stabilized AuNPs as dual-emitting fluorescent and plasmonic probes to track their cellular uptake, agglomeration, and intracellular transport in HeLa cervical cancer cells in a time-resolved manner using correlative fluorescence and plasmonic-based dark-field microscopy. The DNA-stabilized AuNPs contained a 50-nm gold core, and their DNA shell was labeled with the red-fluorescent tetramethylindo(di)-carbocyanine dye cy3 (fDNA-AuNP). As shown by real-time correlative fluorescence and plasmonic imaging, the intracellular transport dynamics and mechanism are orchestrated by the size of the fDNA-gold agglomerates. In the early stages of endocytosis, fDNA-AuNPs mostly appear as single nanoparticles, whereas they agglomerate over time during vesicular transport. The transport rate of vesicularly enclosed fDNA-AuNPs was found to depend critically on the size of the agglomerates and not on the nature of the endosomes and lysosomes. The plasmonic resonance band of the AuNPs shifted to longer wavelengths with increasing agglomerate size. While fluorescence microscopy scans the fDNA-AuNPs by detecting cy3 fluorescence, darkfield microscopy records the location and size of the fDNA-AuNP agglomerates by measuring their plasmonic resonance. Both fluorescence and darkfield microscopy images demonstrated that the fDNA-AuNPs were rapidly and efficiently taken up by the HeLa cells. After two h of incubation of HeLa cells with the fDNA-AuNPs, single fDNA-AuNPs (green) were detected near the cell membrane by darkfield microscopy imaging. By contrast, small fDNA-AuNP agglomerates (yellow) were distributed in the cytoplasm, while the large agglomerates (pale yellow) were preferentially located in the perinuclear region (Figure 128).

These results suggest that agglomeration of fDNA-AuNPs occurs during intracellular trafficking. TEM imaging studies revealed that the fDNA AuNPs were mainly located in endosomal vesicles, indicating that both endocytosis and the subsequent intracellular trafficking contribute to their agglomeration. This suggests that the agglomeration of fDNA-AuNPs is induced by the conditions within endocytic vesicles and intracellular trafficking, rather than being caused by self-aggregation of the nanoparticles. The cellular uptake of fDNA-AuNPs can occur via different endocytic pathways, such as caveol- and clathrin-mediated endocytosis and macropinocytosis. The intracellular trafficking of endocytosed fDNA-AuNPs was investigated using co-localization techniques. The fDNA-AuNPs were fused to two specific protein markers of endosomes, the Ras-related proteins 5 & 7 (Rab 5 and Rab 7), which identified the formation of early and late endosomes, respectively (Figures 129(c) and (d)). fDNA-AuNPs co-localized with early endosomes were found near the cell membrane, whereas the late endosomes enclosing fDNA-AuNP agglomerates were located around the nuclei. The fDNA-AuNPs when co-localized with lysosomes that was stained with a lysotracker, were mainly observed in the perinuclear region, demonstrating the transport of fDNA-AuNPs into lysosomes (Figure 129(e)). This result is confirmed by the TEM images in Figure 129(b), which show the early endosome (200 nm), late endosome (500 nm), and lysosome loaded with fDNA-AuNPs. The early endosome contains mainly single fDNA-AuNPs, which gradually agglomerate during transport in the late endosomes and lysosomes. Time-lapse studies revealed the degree of co-localization of fDNA AuNPs with these cellular compartments, which is quantified by the Mander co-localization coefficient threshold (tMr). The Mander co-localization coefficient is proportional to the Cy3 fluorescence intensity that overlaps the GFP fluorescence intensity. The amount of entrapped fDNA-AuNP peaked in early endosomes, late endosomes and lysosomes at 2, 4, and 16 h, respectively

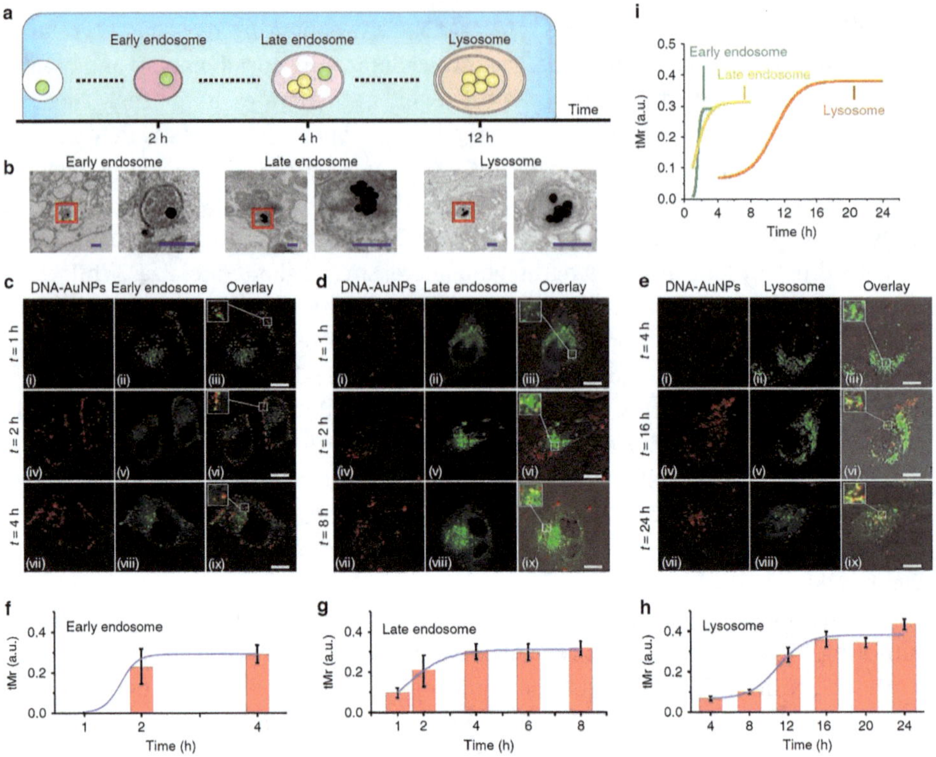

Figure 128: Intracellular trafficking of fDNA-AuNPs: (a) the temporal evolution of intracellular trafficking involving maturation of the early endosome to the lysosome as well as vesicle-induced agglomeration; (b) TEM images of fDNA-AuNPs enclosed in early, late endosomes, and lysosomes (scale bar = 200 nm); (c–e) fluorescence microscopy images showing co-localization of fDNA-AuNPs (red) with early endosomes (green), late endosomes (green) and lysosomes (green) (scale bar = 10 mm); (f–i) plots of thresholded Mander co-localization coefficient (tMr) values over time for early, late endosomes and lysosomes, respectively [693].

(Figures 129(f)–(i)). The temporal evolution of intracellular trafficking involves the maturation of early endosomes into lysosomes, as well as vesicle-induced agglomeration

In addition to determining the localization of nanoparticles within cells, the elucidation of the transport kinetics of nanoparticles from the cell membrane to the various cell organelles along the endosomal pathway is essential for nanoparticle-assisted drug delivery. Vtyurina et al. [674] used live cell imaging to track fluorescent nanoparticles along the endosomal–lysosomal pathway over time. HeLa and HeLa-Kyoto cells were incubated with fluorescent carboxylated polystyrene coated nanoparticles (fPSNPs) of various sizes (40, 100, and 200 nm) by varying the incubation time between 5 and 15 nm. The cells were then imaged at different time points using two-color 3D confocal fluorescence microscopy and analyzed for co-localization. The cells were incubated with LysoTracker to label the acidic compartments of the cells, lysosomes, late endosomes, and

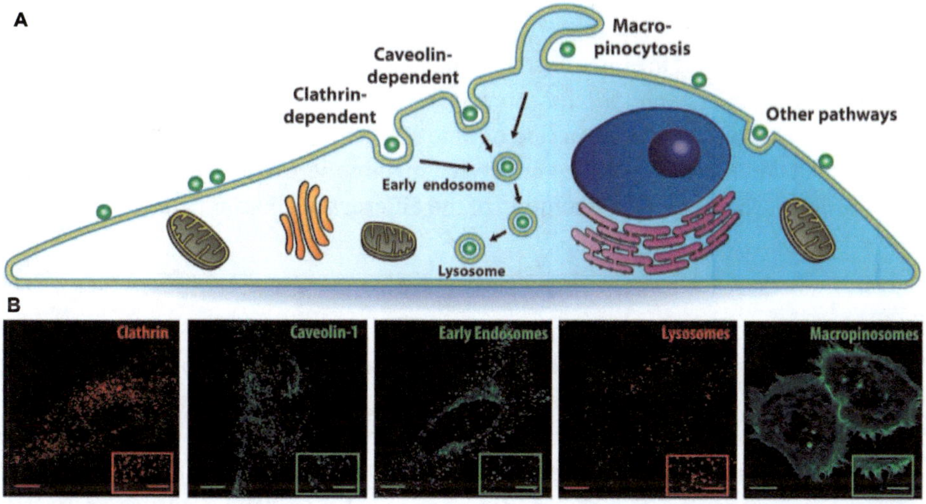

Figure 129: Schematic of a cell exposed to fPSNPs (green spheres) undergoing different endocytic pathways followed by trafficking through an early endosome towards a lysosome (A); fluorescence microscopy mages of HeLa cells labelled with the selected panel of fluorescent markers: mRFP-Clc for clathrin, Cav1-mEGFP for caveolin-1, Rab5 to stain early endosomes (Rab5-GFP), LysoTracker for staining cell lysosomes, pMyrPalm to visualize micropinocytosis (scale bar: 10 μm (main) and 5 μm (inset)) (B) [674].

multivesicular bodies. In addition, HeLa and HeLa Kyoto cells were transfected with a variety of fluorescently labeled endocytic and membrane proteins. This allows the kinetics of intracellular transport of the fPSNPs to lysosomes to be followed (Figure 129). As shown in the fluorescence microscopy images, most nanoparticles are transported to the lysosomes via endosomes by various endocytic mechanisms after cellular uptake. Cells were transfected with a variety of fluorescently labeled endocytic and membrane proteins. At specific time points, two-color 3D confocal fluorescence microscopy images were acquired every 10 to 20 min to determine the kinetics of fPSNPs uptake and trafficking. fPSNPs colocalization with the lysosomes was observed 40 min after exposure. The fraction of fPSNPs that end up in the lysosomes never reached 100 %, but plateaued between 40 % and a maximum of 70 % in case of the 40-nm fPSNPs in HeLa Kyoto cells (Figures 131(A)–(C)).

The trafficking kinetics of nanoparticles from the cell membrane to the lysosomes in single cells was described by the simple differential equation system (Eqs. (12)–(13)) [694]

$$\frac{dN_m}{dt} = -k\,N_m, \tag{12}$$

$$\frac{dN_l}{dt} = k_l\,N_m, \tag{13}$$

where N_m and N_l are the numbers of nanoparticles on the membrane and in the lysosomes, respectively; k is the rate constant of nanoparticles leaving the membrane, and k_l is the rate constant of nanoparticles entering the lysosome. Since the cells were exposed to nanoparticles within a short period of time (5–20 min), the authors assumed that at time t = 0 all nanoparticles resided on the cell membrane, and none were found inside the lysosomes. This leads to the solution of the differential equation system (Eq. (14))

$$\frac{N_{l(t)}}{N_{m(0)}} = f_l(1 - e^{-\frac{t}{\tau}}). \tag{14}$$

The left-hand side of Eq. (14) describes the time-dependent fraction of nanoparticles that have reached the lysosomes (Figure 130(A)), where f_l is the fraction of nanoparticles that reach the lysosomes ($f_l = k_l/k$) and τ denotes the typical departure time required for nanoparticles to leave the membrane ($\tau = k^{-1}$) Fitting this equation to the experimental data showed that this model provides a good mathematical description of the data (Figures 130(A)–(C)). In addition, the values of f_l and τ of differently sized fPSNPs were extracted from these fits (Figures 130(D)–(E)). The accumulation of fPSNPs in lysosomes was calculated from their arrival time ($\tau_\mathrm{l} = k_l^{-1}$). τ_l as a function of the fPSNP size (Figure 130(F)) showed the same trend as the membrane departure time (Figure 130(E)).

In the HeLa cells, the 100-nm fPSNPs exit the membrane and entered the lysosomes much slower than the 40- and 200-nm nanoparticles (Figures 131(E) and (F),

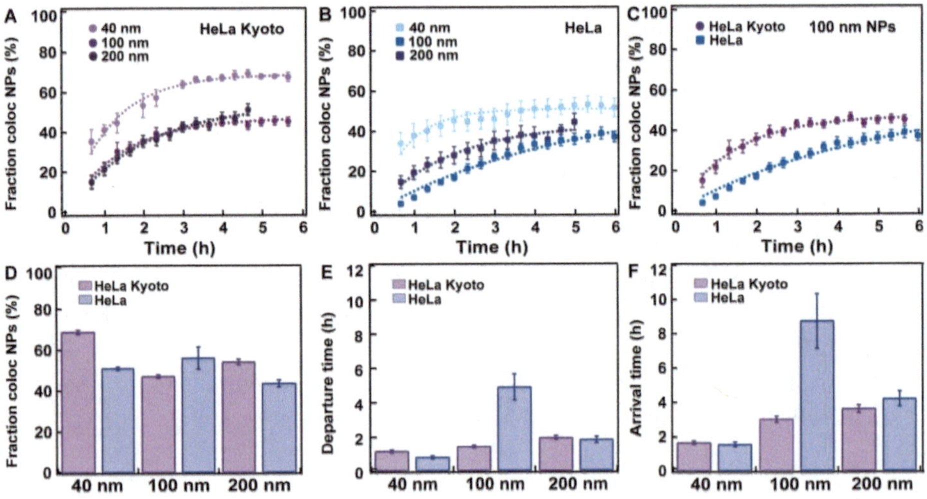

Figure 130: Fitting of the kinetic model to the data for the HeLa and HeLa Kyoto cells; fraction of nanoparticles (f_l) co-localized with lysosomes in HeLa Kyoto cells (A) and HeLa cells (B); comparison of 100-nm fPSNPs in both cell lines; (D) fraction of fPSNPs in the lysosomes (f_l) extracted from the fits in A and B; (E) departure time of the fPSNPs (τ) from the membrane extracted from the fits in A and B; (F) fPSNP arrival time (τ_l) to the lysosomes calculated from D and E [674].

blue columns). This suggests a different uptake pathway and different intracellular trafficking kinetics for the 100-nm fPSNPs compared to the smaller fPSNPs.

As an alternative approach to elucidate intracellular distribution kinetics, Garcia-Romeu et al. [695] used time- and space-resolved flow cytometry of cell organelles. Again, HeLa cells were exposed to fPSNPs (40, 100 and 200 nm) for 30 min. After incubation of HeLa cells with these fPSNPs, cell organelles were extracted by conventional fractionation methods and characterized by high-sensitivity flow cytometry. The organelles were collected at different times (0 to 5 h). The authors hypothesized that the fPSNPs would take the endosomal–lysosomal pathway for endocytosis and subsequently accumulate in the lysosomes. The organelles were labeled with specific antibodies (EEA1 for EE and LAMP1 for lysosomes). Immediately after 30 min incubation with fPSNPs (t = 0 h), 25 % and 5 % of the vesicles surrounding the fPSNPs were early endosomes and lysosomes, respectively. Five hours after incubation, the fraction of fPSNPs in early endosomes decreased to 5 %, while the fraction of fPSNPs in lysosomes increased to 50 %. These observations are in agreement with those obtained from live cell imaging by Vtyurina et al. In addition, Garcia Romeu et al. investigated the effect of nanoparticle size on lysosome arrival kinetics. fPSNPs of 40 nm and 100 nm were observed to enter lysosomes significantly faster than the 200-nm ones. After 5 h, only 20 % of the 200 nm fPSNPs had reached the lysosomes. Subsequently, the measurement time was increased to 24 h. 12 hours after incubation, the co-localization of 40-nm fPSNPs with lysosomes reached a plateau at 60 %, whereas the maximum fraction of entrapped 100-nm fPSNPs was 50 % after 5 h incubation. In the case of 200 nm-fsPSNPs, a steady increase in co-localization was observed even after 24 h, reaching only 35 % for 200 nm. For the 200-nm fPSNPs, the percentage of lysosome-enclosed fPSNPs increased steadily to a maximum of 35 %. The authors concluded that the fPSNPs were endocytosed by moving along the endosomal–lysosomal pathway.

All of these studies demonstrated that differently sized fPSNPs were internalized by endocytosis in HeLa cells and typically trafficked along endosomal–lysosomal pathways. Most of the fPSNPs were shown to reach the lysosomes where they were eventually degraded. However, in all cases, never 100 % of the fPSNPs end up in the lysosomes.

4.3.3 Endosomal escape

The versatile application modalities of nanoparticles in medicine, such as contrast agents, radiation dose-enhancing agents or especially as drug-delivery tools, do not only depend on the efficacy of cellular uptake and, intracellular trafficking routes, but also on the endosomal escape pathway. In the case of nanoparticle-mediated drug delivery, entrapment of drug-loaded nanoparticles in endosomes may be followed by their enzymatic degradation in the lysosome [696]. To ensure cytosolic delivery of therapeutics, effective strategies are needed to facilitate endosomal escape of the nanocarrier and its drug load [697].

Strategies for direct cell membrane penetration

Nanoparticles designed to permeate the cell membrane directly enter the cytoplasm. These nanoparticles bypass endosomal and lysosomal internalization and avoid their degradation. The characteristics of these nanoparticles, including details of the cellular entry mechanism, are described in detail in Section 4.2.1.3.5.

Strategies that use specific trafficking pathways

As described in Section 4.3.1, there are alternative cellular trafficking pathways besides the classical endosomal–lysosomal pathway. An alternative route is the circulation of endosomes between the Golgi apparatus and the endoplasmic reticulum, which are considered to be non-degradative pathways. Nanoparticles can enter cells via caveolin-mediated endocytosis to travel along these trafficking routes. In addition, transport to the trans-Golgi network is also possible after cellular uptake by clathrin-mediated endocytosis or macropinocytosis.

For example, histone-targeted polyplexes have been observed to be transported along the endosome-Golgi apparatus-endoplasmic reticulum pathway by targeting the protein DPY-30 [698]. The regulatory subunit of the DPY-30 histone methyltransferase complex is located in the trans-Golgi network. Alternatively, nanoparticle surfaces can be modified with endoplasmic reticulum membranes or their membrane proteins. Both types of surface modifications result in enhanced transport to the Golgi apparatus and the endoplasmic reticulum. Wang et al. [699] successfully developed an AuNP-peptide conjugate tailored for drug delivery to the Golgi apparatus and endoplasmic reticulum. AuNPs were conjugated to a peptide with a C-terminal Lys-Asp-Glu-Leu (KDEL) motif. The KDEL peptide sequence facilitates retrograde transport of proteins from the Golgi apparatus to the endoplasmic reticulum. Time-dependent fluorescence microscopy measurements showed that the AuNP-KDEL nanoconjugates were internalized via a clathrin-mediated pathway and transported to the endoplasmic reticulum via a COPI retrograde transport pathway. The lysosomal degradation pathway was bypassed. This qualifies the AuNP-KDEL nanoconjugates as nanocarriers for therapeutics and drugs due to their sparing lysosomal degradation.

Strategies for endosomal and lysosomal escape

Most nanoparticles that enter cells via one of the endocytosis pathways are degraded via the endosomal–lysosomal pathway, significantly limiting the therapeutic effectiveness of many drug delivery systems. The cytosolic distribution of therapeutics can only be achieved by the fastest possible ensomal escape. Currently, there are two routes (Figure 131). One is that the nanocarrier-drug system escapes the endosome, or at the latest the lysosome, in time by membrane rupture [700]. The other possibility is that the drug is released from the endosome/lysosome while the nanoparticle carrier system remains trapped in the endosome/lysosome [672].

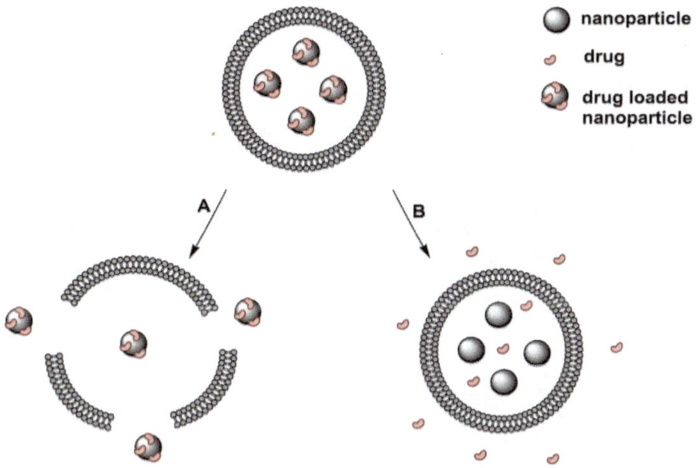

Figure 131: Drug-loaded nanoparticles inside an endosome/lysosome; the nanoparticles are released by two possible mechanisms, following endosomal or lysosomal rupture: The drug-loaded nanocarrier is released into the cytoplasm (A); drug release from the nanocarrier: the drug penetrates the endosomal or lysosomal membrane while the nanocarrier remains entrapped in the endosome or lysosome.

4.3.3.1 Mechanism of endosomal escape

The size, shape, and surface chemistry of nanoparticles can be tuned to promote endosomal escape. To tailor nanoparticles for endosomal escape, the underlying mechanisms must be understood. Unfortunately, there is little consensus in the literature on how nanoparticles or drugs are transported across the endosomal and lysosomal membrane. Several mechanisms have been discussed that may induce endosomal or lysosomal escape [672, 700, 701]. However, it is very likely that more than one mechanism is involved or required for efficient endosomal escape.

Endosomal escape by endosomal/lysosomal membrane rupture

Here we describe several endosomal escape mechanisms that facilitate the release of the nanocarrier drug-delivery system into the cytoplasm. These escape mechanism can involve the following processes: (a) an increase in intra-endosomal pressure until the membrane ruptures, releasing the nanocarrier–drug conjugate; (b) the nanocarrier is equipped with surface structures that perforate the endosomal/lysosomal membrane, allowing the nanocarrier–drug conjugate to escape; (c) photosensitizers in the endosomal membrane cause membrane rupture upon light irradiation; and (d) exogenous triggers induce disruption of the endosomal membrane.

Endosomal escape by pH buffering, osmotic lysis or swelling effect

The pH buffering effect (Figure 132(A)), also called the proton sponge effect, occurs in polymers with a pK_a in the physiologically relevant range. The proton sponge effect oc-

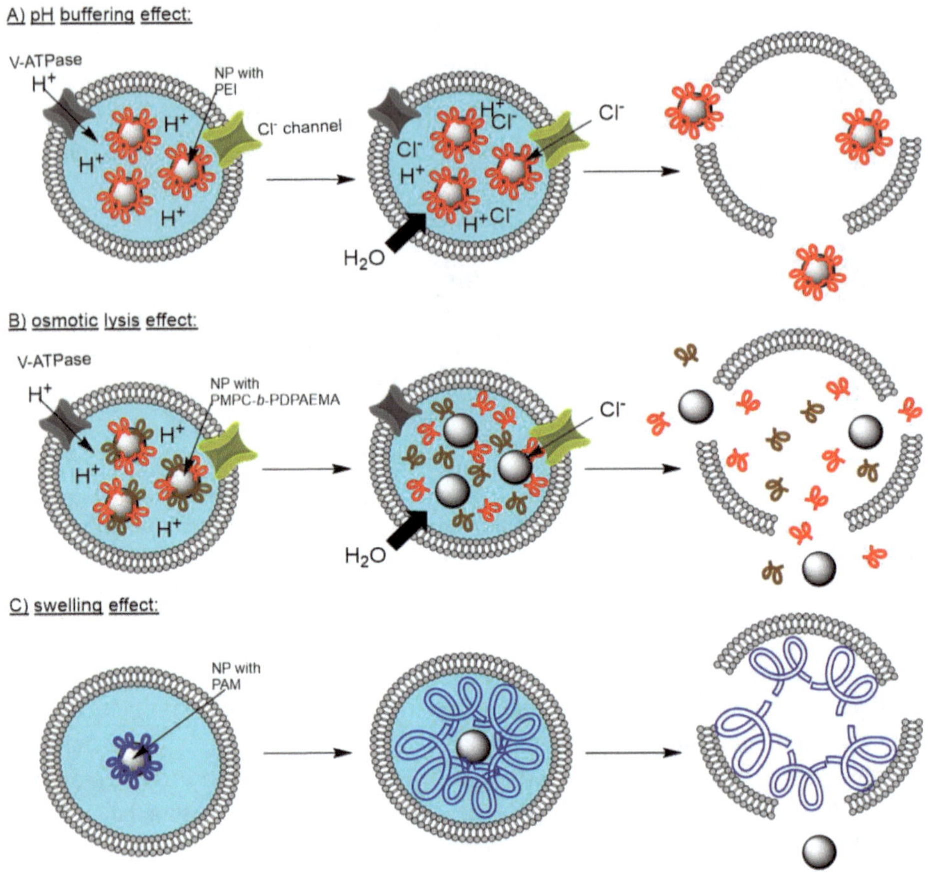

Figure 132: Schematic of endosomal escape involving increased intra-endosomal pressure leading to endosomal membrane rupture: (A) buffering capacity of ligand polymers leads to H^+, Cl^-, and water influx; (B) disassembly of ligand polymers leads to increased osmotic pressure; (C) swelling of ligand polymers increases pressure on endosomal membrane.

curs in Bronsted–Lowry base polymers that act as proton acceptors. These polymers contain amino groups (Table 7). The pH inside endosomes and lysosomes decreases during intracellular trafficking. V-ATPases located in the endosomal membrane pump protons into the endosomes. While the polymer is protonated, protons are pumped into the endosomes and lysosomes. To maintain electroneutrality in the vesicles, chloride anions are transported into the vesicles. The high ionic strength leads to high osmotic pressure. The subsequent influx of water increases the pressure on the endosomal membrane, eventually leading to membrane rupture and facilitating endosomal escape [672, 700]. The pH buffering effect has been extensively discussed in the literature [702–704]. Polymers with increased buffering capacity have been developed that have no significant effect on endosomal escape efficiency. Furthermore, polymers with similar buffering

Table 7: Selected examples of polymeric nanoparticles or nanoparticle coatings for endosomal escape via pH buffering, osmotic lysis, or the swelling effect.

Escape mechanism	Escape agent	Nanoparticles
pH buffering effect	polyethyleneimine (PEI)	magnetic nanoparticles [712] polyplexes [713]
	poly (L-histidine)	polyplexes [714]
	poly(amidoamine) (PAMAM)	polyplexes [715]
	chloroquine	polyplexes [716] lipoplexes
Osmotic lysis effect	poly(2-(methacryloyloxy)ethylphosphorylcholine)-co-poly(2-(diisopropylamino)ethylmethacrylate) (PMPC-PDPA)	polymersome [705]
	poly(2-(methacryloyloxy)ethyl phosphorylcholine)-poly(2-diisopropylaminoethyl methacrylate) (PMPC-PDPA)	polymersome [706]
	calcium phosphate	calcium phosphate nanoparticles [707]
Swelling effect	2-diethylamino ethyl methacrylate (DEAEMA), poly(ethylene glycol) dimethacrylate (PEGDMA)	cross-linked polymer nanoparticles [709]
	2,4,6-Trimethoxy benzaldehyde, 5-methyl-2-(2,4,6-trimethoxy phenyl)-(1,3)-5-dioxanylmethanol	cross-linked polymer nanoparticles [710]
	poly(c-glutamic acid-co-c-distearin glutamate), oly(c-GA-co-c-glutamyl oxysuccinimide)-g-mPEG	gel-caged polymeric vesicles [711]

capacities but different molecular weights show various degrees of endosomal escape. The difficulties in intra-endosomal pH measurements cast doubt on the pH buffering effect as the sole explanation for endosomal escape. The change in pH is certainly an important factor that should be exploited for endosomal escape.

Polymerosomes composed of diblock copolymers (Table 7) rapidly break down into numerous monomers in an acidic environment [703, 704], causing a sudden rise in osmotic pressure. Similarly, inorganic nanoparticles (Table 7) dissolve at low pH, releasing ions inside the endosomes and lysosomes [705, 706]. This increase in ion concentration leads to water influx and builds pressure on the surrounding membrane.

A related mechanism to the proton sponge effect which also involves change in osmotic pressure, is the osmotic lysis effect (Figure 132(B)). pH-responsive nanoparticles disintegrate at decreasing pH, leading to an increased osmotic pressure and subsequent lysis of the endosomal membrane. Polymerosomes composed of diblock copolymers (Table 7) rapidly break down into numerous monomers in an acidic environment [705, 706]. Similarly, inorganic nanoparticles (Table 7) dissolve at low pH, releasing a large number of ions inside the endosomes and lysosomes [707, 708]. The increase in ion concentration leads to water influx and builds pressure on the surrounding membrane.

Another mechanism that increases intraluminal pressure is based on the swelling effect of some polymers (Figure 132(C)) [709–711]. Some polymers are designed to increase their volume at low pH (Table 7). Changes in pH cause changes in electrostatic interactions, solvation, or conformation in the polymer, resulting in an expanding volume. The pH-induced swelling promotes membrane disruption and endosomal escape.

Endosomal escape through pore formation or membrane destabilization

Endosomal escape is promoted by membrane destabilization through the formation of pores or holes. Pore formation occurs when self-assembling peptides create defined pores in the endosomal membrane (Figure 133(A)). Many proteins derived from viral or bacterial peptides typically have a transmembrane domain (Table 8) [716]. As the pH drops along the endosomal–lysosomal pathway, the transmembrane domains undergo conformational changes. The transmembrane domains are inserted into the endosomal membrane and form various types of pores. In the barrel pore model, the peptides aggregate and insert into the endosomal membrane as multimeric channels [717]. The toroidal pore or wormhole model describes pore formation by peptides inserted vertically into the endosomal membrane. The aggregation of these peptides bends the membrane inward, forming a hole in the peptide ring. However, the diameter of these membrane

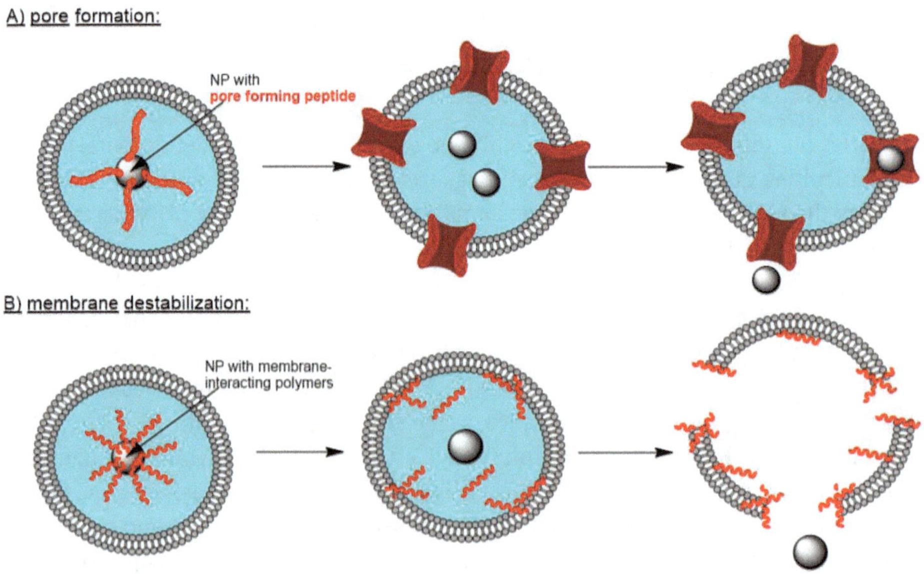

Figure 133: Schematic of endosomal escape involving interaction of peptides and polymers with the endosomal membrane: self-assembling peptides that form pores on the endosomal membrane (A); polymer or peptide coatings that interact directly with the endosomal membrane, causing destabilization and disruption (B).

Table 8: Selected examples of polymeric nanoparticles or nanoparticle coatings for endosomal escape via pore formation or membrane destabilization.

Escape mechanism	Escape agent	Nanoparticles
Pore formation	Listeriolysin O toxin	AuNPs [720]
	HIV gp41 peptide	polyplex [721]
Membrane destabilization	outer membrane protein A (OmpA)	magnetic nanoparticles [722]
	FGF2B-LK15-8R peptide (FLR)	SPION [723]
	Poly(DL-lactide-co-glycolide (PLGA)	PLGA nanoparticles [724]
	Cell penetrating peptides: penetratin, TP10, MPG Δ^{NLS}, EB1	polyplexes [725]

pores is very small with 1–2 nm, which limits the size of nanoparticles or drugs that can pass through these pores.

Another possibility is the use of polymers or peptides that destabilize the endosome through electrostatic interactions (Figure 133(B)). In an acidic environment, the surface charge of these polymers or peptides changes to positive. These positively charged polymers/peptides (Table 8) interact with the negatively charged endosomal membrane resulting in membrane destabilization and damage. Another example is the use of cell-penetrating peptides (Table 8). The decrease in pH triggers the insertion of pH-responsive cell-penetrating peptides into the endosomal membrane, causing membrane leakage and disruption. In addition, ionizable lipids ($pK_a \approx 6.2$) in endosomes are protonated at pH ≤ 6. The then positively charged lipids change their structure from a conical to an inverted hexagonal (HII) shape. This change induces a localized disruption of the endosomal membrane [718, 719].

Endosomal escape through membrane fusion

Fusogenic lipids or amphiphilic molecules (Table 9) allow fusion with the endosomal membrane. Membrane fusion (Figure 134) occurs when lipid, polymer, or peptide nano-assemblies are tightly bound to the endosomal lipid bilayer. The nano-assemblies then fuse with the lipid bilayer of the endosomes/lysosomes, releasing their cargo into the cytoplasm. It is challenging to trigger this fusion process inside the endosomes and lysosomes. pH-responsive polymers or lipids can initiate the fusion process at a specific pH. For example, neutral polymers cover the surface of cationic nano-assemblies at neutral pH and leave the surface at low pH [696].

Another example is the protonation of polymers or lipids at acidic pH, which forms positively charged surfaces on the nano-assemblies [697]. Decreasing pH can also induce conformational changes, shifting from hydrophilic coils to hydrophobic helices, thus facilitating membrane integration (Table 9).

Table 9: Selected examples for fusogenic lipids and amphiphilic molecules for endosomal escape via membrane fusion.

Escape mechanism	Escape agent	Nanoparticles
Membrane fusion	Lipoid S75, D-a-tocopheryl polyethylene glycol succinate	Lipid carriers with SPIONs [726]
	dimethylaminoethane carbamoyl conjugated cholesterol	Liposomes [727]
	N-(1-(2,3-dioleyloxy)propyl)-N,N,N-trimethylammonium chloride (DOTMA)	Lipoplex [728]
	cholesterol-grafted poly(-amidoamine)	Lipid/polymer hybrid nanoparticles [729]

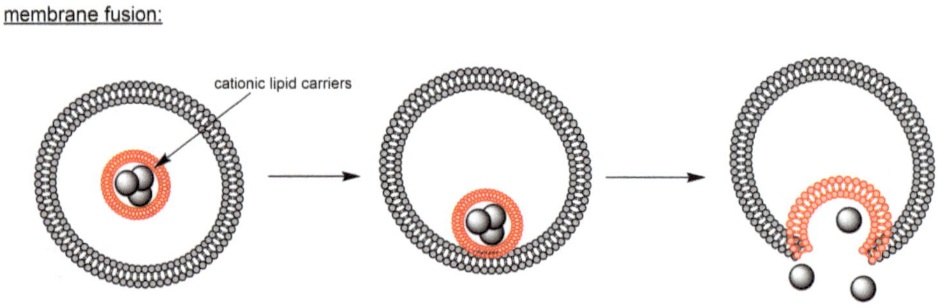

Figure 134: Schematic of endosomal escape involving membrane fusion: Lipid carriers fuse with the endosomal membrane and release their cargo.

Endosomal escape by photochemical disruption

Photochemical internalization uses photosensitizers for effective delivery of nanocarrier drug systems from endosomes into the cytosol [730] (Table 10). In general, photosensitizers have an amphiphilic structure suitable for intercalation into endocytic membranes (Figure 135). This allows them to enter cells via the endocytic machinery, where they become trapped in endosomal vesicles. The most commonly used photosensitizers are disulfonated meso-tetraphenyl-porphine, disulfonated aluminum phthalocyanine, dendrimer phthalocyanine, 5,10,15-tri(4-acetamidophenyl)-20-mono-(4-carboxylphenyl) porphyrin, and tetra(4-sulfonatophenyl) porphyrin (Table 10). Photosensitizers are excited by the absorption of light. The photo-excited state ($^1P^*$) deactivates to the triplet state (3P*) by intersystem crossing, fluorescence, and internal conversion that causes heat dissipation [731]. The $^3P^*$-excited photosensitizer decays by emitting heat or phosphorescence light. When $^3P^*$-excited photosensitizers interact with biomolecules or molecular oxygen, they can transfer electrons and excitation energy, generating radicals or reactive oxygen species. These highly reactive radicals can disrupt the endosomal membrane and promote the release of nanocarriers.

Table 10: Selected examples for polymer nanoparticles or nanoparticles coatings for endosomal escape via photochemical disruption.

Escape mechanism	Photosensitizer	Nanoparticles
Photochemical disruption	Chlorin e6	urchin-like hollow magnetic microspheres [732]
	silicon phthalocyanine	MSNanoparticles [733]
	dendrimer phthalocyanine	Polymeric micelles [734]

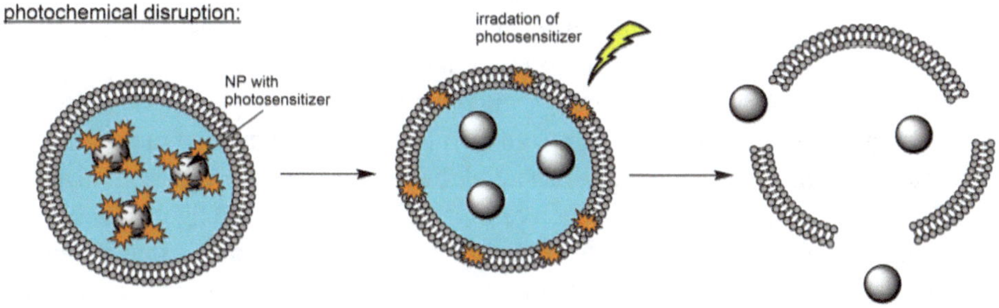

Figure 135: Schematic of endosomal escape involving photochemical disruption: Photosensitizers intercalate into the endosomal membrane and generate reactive oxygen species upon light irradiation.

Escape through membrane disruption by exogenous stimuli

Drug carriers can release drugs not only by exogenous or endogenous stimuli, but also by stimulated modulation of their volume, chemical reaction, or light- or magnetic-field-induced release of heat to disrupt the endosomal membrane (Table 11). Exogenous stimuli are ultrasound, light, heat, magnetic fields, and electric fields, while endogenous stimuli are pH changes (pH 5–6), enzymatic or oxidation, or reduction reactions [730, 735]. A promising exogenous stimulus is ultrasound, which, due to its nonin-

Table 11: Selected examples for polymer nanoparticles or nanoparticles coatings for endosomal membrane disruption by exogenous triggers.

Exogenous trigger	Nanoparticles
Ultrasound	doxorubicin-loaded cationic liposomes [738] poly(ethylene oxide)-block-poly(2-(diethylamino)ethyl methacrylate)-stat-poly(methoxyethyl methacrylate) polymersomes [735]
External heating	thermosensitive pegylated "bubble" liposomes [737]
Near-infrared light	Liposome-tethered AuNPs [739] poly(diallyl dimethylammonium chloride)and hyaluronic acid coated AuNPs [740]
Magnetic field	thermosensitive cationic magnetic liposomes [741] Carboxymethyldextran coated SPIONs [742]

vasive nature and relatively high penetration depth into living tissue, is suitable for the disruption of endosomal membranes and the associated triggered release of drug carriers. Ultrasound-responsive drug carriers include microbubbles, nanobubbles, nanodroplets, liposomes, emulsions, and micelles. However, the clinical application of some of these ultrasound-responsive drug carriers is hampered by their poor colloidal stability. Nevertheless, nano- and microbubbles are already being used as ultrasound contrast agents. They consist of a lipid shell filled with perfluorocarbon gas [736]. Ultrasound waves induce volumetric oscillations of the nanobubbles, which cause periodic pressure changes. In the case of doxorubicin-loaded ultrasound-responsive polymersomes, these periodic pressure oscillations have been shown to disrupt the endosomal membrane and release the anticancer agent into the cytosol [735].

Localized temperature elevation within endosomes, e. g., by plasmonic heaters or thermoresponsive liposomes, can also lead to endosomal membrane rupture and thus promote endosomal release. Suitable plasmonic heaters are gold nanocages, gold nanostars, and AuNRs, whose surface plasmon resonance is excited by NIR light [106]. The absorbed light energy is released to the environment as heat. This implies that plasmonic heaters inside endosomes can destroy the endosomal membrane due to the locally gen-

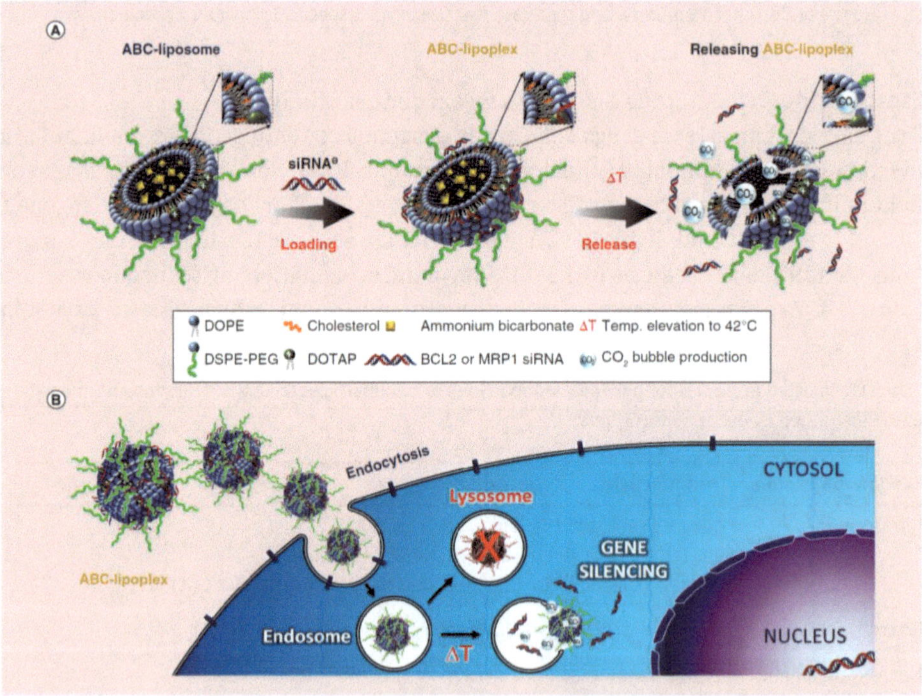

Figure 136: Schematic of loading of the ammonium bicarbonate containing liposome (ABC-liposome) and heating-induced release of CO_2 followed by siRNA release in the cytosol (A); endocytic pathway of the ABC-liposome in the cancer cell followed by the heat-induced release of the siRNA and gene silencing (B) [737].

erated heat. Other so-called photothermal transduction agents that convert NIR light into heat are chalcogenides or carbonaceous nanomaterials, porphyrins, cyanines, or semiconducting polymer nanoparticles [730]. Another encouraging example of thermoresponsive nanocarriers are PEGylated bubble liposome nanovectors tailored for efficient siRNA delivery (Figure 136) [737]. The liposome nanovectors contain the positively charged 1,2-distearoyl-sn-glycero-3-phosphoethanolamine-N-(amino(polyethylene glycol)-2000) as a lipid molecule in their vesicle membrane, which electrostatically binds *MRP1* and *BCL2* siRNA to their surface. The 200-nm liposome nanovectors encapsulating aluminum bicarbonate are taken up by small carcinoma lung cancer cells by endocytosis. By heating the cancer cells to 42 °C, ammonium bicarbonate decomposes and releases CO_2, which destroys not only the liposome but also the endosome.

This allows siRNA to escape lysosomal degradation and to be released into the cytosol. Intracytosomal *MRP1* and *BCL2* siRNA was shown to cause 80–90 % gene silencing by demonstrating a dramatic reduction in *MRP1* and *BCL2* protein production.

Endosomal escape via nondisruptive membrane penetration

As an alternative to destructive endosomal escape mechanisms, amphiphilic nanoparticles, or loaded drugs translocate directly across the membrane without damaging the endosome lysosome.

Endosomal escape through pH-sensitive bond

Drug release within the acidic environment of endosomes/lysosomes was achieved through pH-sensitive binding [672]. Doxorubicin was attached to PEGylated AuNPs via a pH-sensitive hydrazone linker. At low pH, the hydrazone linker was hydrolytically cleaved, releasing doxorubicin from the AuNP surface. Confocal microscopy confirmed that doxorubicin reached the cytoplasm. The authors also demonstrated that endosomal release of doxorubicin can bypass the efflux caused by multi-resistance transport proteins. A higher intracellular drug concentration was found when the AuNPs delivered the doxorubicin into the cells compared to the free doxorubicin [743].

Endosomal escape by enzyme-cleavable bond

For spatial and temporal drug release, Sangtani et al. [744] developed quantum dot-peptide conjugates. Luminescent CdSe@QZnS core-shell quantum dots were conjugated with two different peptides: a cell-penetrating peptide (CPP) and a so-called display peptide. The CPP facilitates the endocytic entry of the quantum dots into cells, while the display peptide is conjugated to the drug, doxorubicin, via three different bonds: ester, disulfide, and hydrazone. As explained in the previous section, hydrazone linkers are pH-sensitive. The ester and disulfide bonds are sensitive to enzymatic cleavage or reducing conditions, respectively. The authors demonstrated that the most effective drug release from the endosome was achieved with doxorubicin via ester linkages. After enzymatic cleavage of the ester bond, the doxorubicin was found in the nucleus. Endosome-specific enzymes can be utilized to ensure drug release within endosomes. One example of endosome-specific enzymes is cathepsin B, the main enzyme responsible for the degradation of insulin-like growth factor 1, which is active in endosomes at a pH 5.5–7.0. Acar et al. [745] linked peptides to liposomes using Cathepsin B cleavable peptide amphiphiles. Cathepsin B, an endosome-specific enzyme, recognizes specific cleavage sequences and releases the loaded peptide.

Endosomal escape by degradation

Proteins and small molecules can also be protected by various shells that promote cellular uptake. After internalization the shell is degraded inside the endosomes and lysosomes allowing the protein or molecules to be released from the endocytic compartments. Prasetyanto et al. [746] encapsulated cytotoxic proteins in organosilica nanoshells consisting of enzyme cleavable disulfide bonds. These nanoshells were internalized by C6 glioma cells and were found in the endosomal and lysosomal compartments. After 24 h, degradation of the organosilica shell was initiated by lysosomal enzyme activity. The degradation, confirmed by TEM images, results in the release of cytotoxic proteins that kill the C6 glioma cell. The experiments were repeated using the green fluorescent protein GFP instead of the cytotoxic proteins. After degradation of the nanoshells, fluorescence emission was observed inside the cells, confirming the migration of GFP from the endosomes into the cytoplasm.

Yang et al. [747] synthesized 2–3-nm AuNPs and stabilized their surfaces with 11-mercaptoundecane sulfonate and 1-octanethiol. These amphiphilic surface structures did not only allow the incorporation of drug molecules into the hydrophobic pockets, but also enable the conjugation with antibodies or antibody fragments. The amphiphilic AuNPs were conjugated with antibodies and single-chain antibody fragments derived from camelids to achieve their targeted uptake into cytotoxic T lymphocytes ($CD8^+$ T cells) via endocytosis. Inside the endosomes and lysosomes, the antibodies or single-chain antibody fragments were shown to be proteolytically degraded inside endosomes and lysosomes, allowing the drug-loaded AuNPs to enter the cytosol of the CD8+ antibodies that were proteolytically degraded.

Intracellular trafficking requires the combination of more than one technique. Therefore, it is necessary to understand the information that can be obtained with each technique and the limitations of each technique. The major detection techniques are described next.

Fluorescent labeling

Fluorescent labeling of nanoparticles or intracellular structures is a common method to identify endosomal escape. Fluorescent nanoparticles entrapped in endosomes appear as a punctate structure under the fluorescence microscope, while escaped fluorescent nanoparticles show a diffuse appearance. For a better understanding of the nanoparticle transportation, co-localization studies of fluorescent nanoparticles and labeled cellular compartments are performed. Acidic vesicles can be labeled with acidotropic dyes such as acridine-orange [748] or lysotracker [749]. These dyes stain specific acidic compartments and provide basic information on cellular trafficking. Alternatively, cell markers can specifically label cell organelles such as LAMP-1 for lysosomes or Rab5 for early endosomes. Fluorescent imaging can provide information for nanoparticle trafficking and endosomal escape capacity, but information on escape mechanisms is limited.

Leakage assays

Leakage assays can provide deeper insights into the mechanism of endosomal escape. Under ex vivo or in vitro conditions, nanostructures that induce leakage in cell membranes are studied. Ex vivo models are artificial liposomes or red blood cells. Artificial liposomes used for leakage assays consist of membranes composed of various phospholipids encapsulating fluorescent dyes. These fluorescent dyes, such as 6-carboxyfluorescein, quench their fluorescence at high concentrations but dramatically increase their fluorescence upon release from the liposomes [750, 751]. These artificial liposomes can be used to test various membrane destabilizing materials. Red blood cells offer a more biologically relevant model capable of simulating the pH gradient along the endosomal–lysosomal pathways [752]. Red blood cells are cultured in buffer solutions that mimic the same the pH of endosomes, lysosomes, cytoplasm, or extracellular environment. When escape agents that are pH-sensitive or induce pH changes cause membrane disruption, the released hemoglobin can be measured in the surrounding buffer [753]. In in vitro leakage assays, cells are treated with fluorescent dyes similar to those used for artificial liposomes. Dyes such as calcein, 6-carboxyfluorescein or fluorescently labeled dextran are taken up through endocytic pathways and their fluorescence remains weak inside endocytic vesicles [751]. When leakage of the dyes occurs, the dyes are released into the cytoplasm, causing an increase in fluorescence intensity.

Transfection assays

Nanoparticle-assisted delivery of siRNA and miRNA into cells relies on endosomal escape. Eukaryotic cells require the process of RNA interference (RNAi) to defend themselves against genetic material introduced into the cells, for example, by viruses [754]. The mechanism of RNAi mainly uses microRNAs (miRNAs) and small interfering RNAs (siRNAs), which can modulate gene expression [755, 756] (Figure 137). miRNAs are transcribed from endogenous genes, and their primary transcripts (pre-miRNA) are converted into a precursor miRNA (pre-miRNA) in the nucleus by the ribonuclease enzymes Drosha and DGCR8. The pre-miRNA is transported into the cytoplasm by exportin 5. The enzyme Dicer removes the hairpin structure from the pre-miRNA, resulting in the formation of miRNA. The siRNA is also formed by the action of Dicer. When double-stranded RNA (dsRNA) enters the cell as exogenous genetic material, it is recognized as foreign

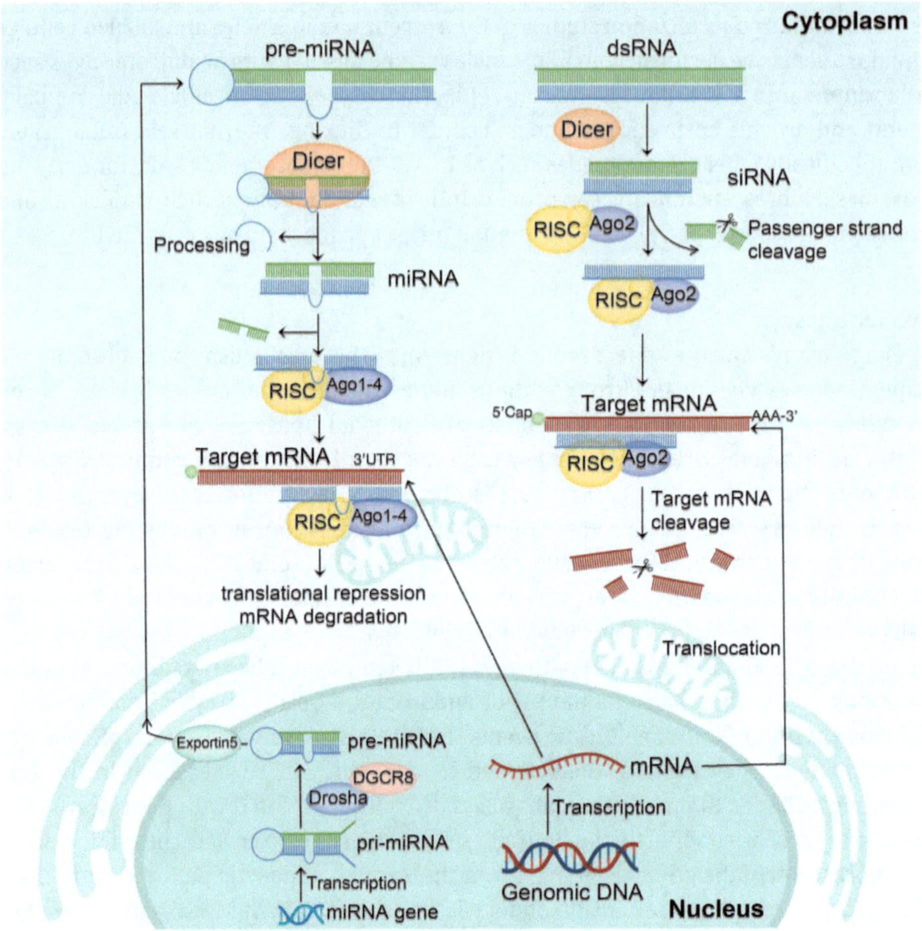

Figure 137: Scheme of RNA interference with microRNA (miRNA) and small interfering RNA (siRNA); miRNA: The miRNA gene is transcribed in the nucleus to pre-mRNA, processed by the ribonuclease enzymes Drosha and DGCR8 to pre-miRNA, pre-miRNA is transported into the cytoplasm by Exportin 5, freed from the stem-loop structure by Dicer, and miRNA is formed; siRNA: Double-stranded RNA (dsRNA) is first recognized by Dicer and cleaved into fragments, the siRNA, which is 21–23 nucleotides long. The siRNA or miRNA is transferred to Argonaute (Ago1-Ago4) of the RNA-induced silencing complex (RISC). The double strands of the bound siRNA or miRNA are cleaved into single strands. The leading RNA strand remains in the RISC, while the other strand leaves the complex and is degraded. The mRNA complementary to the leading strand is incorporated into the RISC by base pairing between the siRNA and mRNA sequences. The Argonaute-2 (Ago2) endonuclease in the RISC cleaves the target mRNA, inhibiting mRNA translation and target-protein expression [756].

RNA and cut into small fragments of 21 to 23 nucleotides in length. These RNA fragments are called siRNA. miRNA and siRNA are bound in the cytosol by the Argonaute proteins (Ago1-Ago4) of the RNA-induced silencing complex (RISC) and cleaved into individual strands. The so-called guide strand remains in the RISC, guides the RISC to the target mRNA, and binds to the complementary sequence of the target mRNA by base pairing. This activates the Argonaute-2 (Ago2) endonuclease, which cleaves the target mRNA. The degradation of the mRNA inhibits translation into protein. Quantification of mRNA levels or protein expression after siRNA delivery provides indirect evidence of endosomal escape. Commonly silenced genes include those for luciferase or β-galactosidase. Another possibility is transfection with plasmid DNA encoding fluorescent proteins such as green fluorescent protein (GFP).

Klein et al. [311] successfully developed green luminescent 2–3 nm-silicon quantum dots (SiQDs) for RNAi-mediated silencing of the ABCB1 gene in Caco-2 cells. Silencing of the ABCB1 gene downregulates the expression of the efflux transport protein P-glycoprotein, which plays a key role in multidrug resistance to cytostatic drugs. The surfaces of the SiQDs were functionalized with covalently linked 2-vinylpyridine ligands. Their positive charges facilitated conjugation with ABCB1-siRNA through electrostatic interactions. The green luminescent SiQDs were shown to act as a self-guiding transfection tool for the ABCB1-siRNA. The cellular uptake and intracellular distribution of the SiQD-siRNA conjugates were imaged by confocal laser scanning microscopy and transmission electron microscopy, while the release of siRNA into the cytoplasm was detected by measuring the decrease in ABCB1-mRNA concentration using real-time PCR. Internalization of the SiQD-siRNA conjugates into Caco-2 cells occurred by endocytosis. As indirectly demonstrated by real-time PCR and transcellular transport studies, the functionalized SiQDs achieved high gene transfection efficiency for ABCB1 siRNA, which they transported into the cytosol of Caco-2 cells. Release and incorporation of siRNA into RISC was followed by detection of ABCB1 mRNA levels and reduced P-glycoprotein transporter efficiency in successfully transfected Caco-2 cells.

4.3.4 Exocytosis

Exocytosis is a cellular process by which unneeded molecules are released from the interior of the cell into the extracellular environment to reduce stress within the cell [757]. Exocytosis starts with the formation and transport of vesicles containing the molecules to be exported. The vesicles dock to the cell membrane with the help of soluble N-ethylmaleimide-sensitive-factor attachment receptor (SNARE) proteins and then fuse with the membrane, releasing their contents. Exocytosis can occur continuously (constitutive) or in response to specific signals (regulated). Once the contents are released, the vesicle membrane can be recycled by endocytosis.

Nanoparticles internalized by the cell are not only be transported into different vesicles, such as endosomes and lysosomes, but can also be found in organelles like the mitochondria, endoplasmic reticulum, Golgi apparatus, nucleus, and cytoplasm and the Golgi apparatus, in the nucleus and in the cytoplasm [758]. Some nanoparticles never leave the cell. Long retention times inside the cell can lead to severe side effects and increased cytotoxicity [759]. For nanoparticle-assisted drug-delivery systems, intracellular retention time is critical for effective drug delivery. Prolonged intracellular retention of drug-loaded nanocarriers in targeted cells increases therapeutic efficacy, while reduced retention time in healthy cells limits serious side effects [760]. Most nanoparticles leave the cell via various exocytosis pathways. Therefore, the elucidation of the exocytosis mechanisms and pathways of nanoparticles is important for biomedical applications.

4.3.4.1 Nanoparticles exocytosis pathways

After endocytosis, nanoparticles are transported in endocytic vesicles. There are several exocytosis pathways for nanoparticles found in endosomes after cellular uptake. Figure 138 schematically illustrates the major exocytosis mechanisms and pathways of nanoparticles. The endosomal recycling pathway, also known as the rapid recycling pathway, involves the separation of recycling endosomes from early endosomes (A). These recycling endosomes return to and fuse with the cell membrane. After endosomal

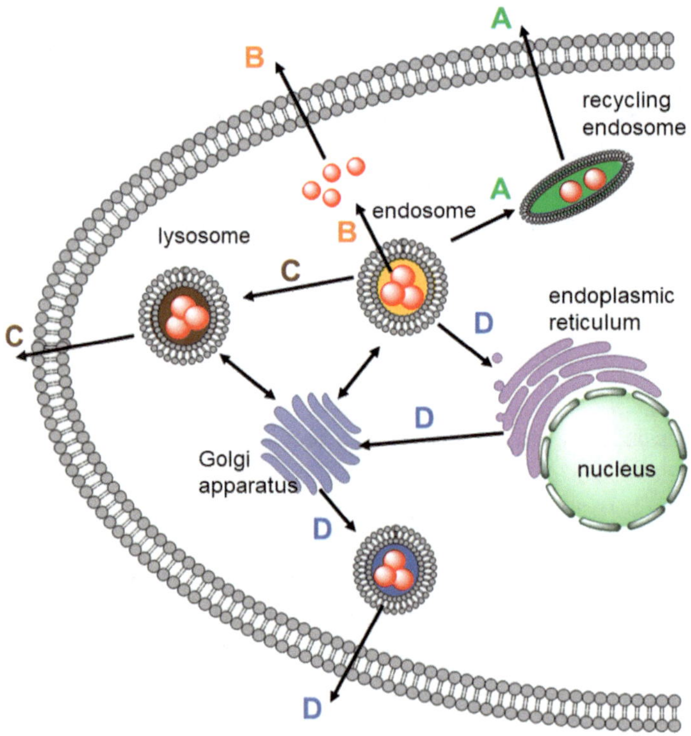

Figure 138: Schematic representation of the major exocytic pathways; A: fast endosomal recycling pathway; B: direct permeation of the cell membrane by diffusion; C: lysosomal pathway; D: endoplasmic reticulum—Golgi apparatus pathway.

escape, nanoparticles accumulate in the cytoplasm. Diffusion allows the nanoparticles to directly and passively penetrate the plasma membrane as an energy-independent exocytosis pathway (B). The diffusion of nanoparticles is controlled by the concentration gradient between the inner and outer environment. Since the diffusion coefficient is inversely proportional to the nanoparticle diameter, diffusion as an exocytosis mechanism is very likely for nanoparticles smaller than 50 nm [757, 761, 762]. The main cellular trafficking route is the endosomal–lysosomal pathway (C). Although this is described as a degradative pathway, lysosomes can also fuse with the cell membrane to release cargos that cannot be digested within lysosomes [763–765]. Nanoparticle-loaded endocytic vesicles can shuttle between the endoplasmic reticulum and the Golgi apparatus (D). This endoplasmic reticulum–Golgi apparatus pathway is also referred to as the slow recycling pathway. Secretory vesicles bud from the endoplasmic reticulum and/or Golgi apparatus and fuse with the plasma membrane. There are alternative exocytic pathways that are found in specific cells and to a lesser extent. For example, cell surface transporters such as P-glycoprotein [766] or $\alpha v\beta 3$ integrin [767], facilitate exocytosis of nanoparticles through these receptors.

4.3.4.2 Methods for detection and quantification of exocytosis

Identification of exocytosis pathways

For the detection of specific endocytosis pathways, inhibitors or accelerators for various endocytic pathways can be used. Table 12 lists the most important inhibitors and activators that distinctly mark the different exocytic pathway. Inhibitors and activators are used to study specific steps in the exocytosis pathway. Inhibitors (e. g., botulinum toxin and tetrodotoxin) block key proteins and signals, allowing the role of these components in exocytosis to be studied. Activators (e. g., forskolin and ionomycin) increase intracellular signaling molecules and cations such as (cyclic Adenosine MonoPhosphate) and Ca^{2+} to stimulate exocytosis and to study the mechanisms involved. The chemical compounds are then added to the cells at specific concentrations and for specific periods

Table 12: Examples of inhibitors and accelerators for the different exocytic pathways.

	Substance	Pathway
Inhibitor	Bafilomycin A1	Lysosomal [764, 768]
	Chloroquine	Lysosomal [769]
	Vacuolin-1	Lysosomal [770]
	Monensin	Golgi apparatus [768]
	Brefeldin A	ER/Golgi apparatus [771]
	Verapamil	P-glycoprotein [768]
Accelerator	Ionomycin	lysosomal [772]
	A23817	Lysosomal [773]

of time. The effects of the inhibitors and activators on exocytosis are examined using fluorescence microscopy, Western blotting, or patch-clamp or biochemical assays [757].

Fluorescence microscopy or confocal laser scanning microscopy can be used to study exocytic pathways by co-localizing intracellular nanoparticles with fluorescent markers for organelles/exocytic pathways. For these microscopic techniques, the nanoparticles are fluorescently labeled or fluorescent nanoparticles are used. The fluorescently labeled nanoparticles have the disadvantage that the fluorescent dye can be detached from the nanoparticle surface resulting in false fluorescence signals. As markers, the same can be used as for cellular trafficking, including early endosome markers (Rab5), late endosome markers (Rab 7), markers for transport from the Golgi apparatus to the cell membrane (Rab3/Rab 26) [774]. In addition, exocytosis pathways are imaged using TEM. This technique allows high-resolution analysis of exocytic pathways. However, it is more difficult to distinguish the various intracellular vesicles from each other.

In the extracellular medium, molecules released by specific endocytosis pathways can be measured. For example, β-hexosaminidase is found specifically in lysosomes. After fusion of lysosomes with the cell membrane, β-hexosaminidase is secreted into the extracellular medium together with the nanoparticles. If β-hexosaminidase is detected along with the nanoparticles, the nanoparticles leave the cells via the lysosomal exocytosis pathway.

Examination of endocytosis

Endocytosis and exocytosis of nanoparticles are directly linked and occur simultaneously. The kinetics of exocytosis depends on the interaction with endocytosis. Nanoparticles with sizes between 10 and 100 nm are most likely to be taken up into cells via the endocytic pathway [775]. As the concentration of intracellular nanoparticles increases, exocytosis becomes more likely and the rate of exocytosis increases until an equilibrium between endocytosis and exocytosis is reached. At this point, the intracellular nanoparticle concentration reaches its maximum, which determines saturation [418]. This is the ideal time to measure exocytosis rates. Pretreated cells saturated with nanoparticles are washed and dispersed in culture medium. The washing step is critical. Any nanoparticles still adhering to the cell surface must be removed since they will interfere with the measurement results [776]. The fresh medium is collected at multiple times for quantitative measurements of the exocytosed nanoparticles. In most cases, the concentration of nanoparticles in the extracellular medium is very low. Therefore, measuring the intracellular and extracellular nanoparticle concentration at the same time provides an accurate result [759, 765]. Cell proliferation is another factor that can affect measurements. Rapidly proliferating cells reduce the intracellular nanoparticle content in individual cells more quickly than slowly proliferating cells. Since the exocytosis rate depends on the concentration gradient between the intracellular and extracellular nanoparticle content, the rate of cell proliferation influences the exocytosis rate. Cell proliferation can be inhibited by certain drugs, though these inhibitors may have cytotoxic effects. To obtain accurate results, the total nanoparticle concentration should be normalized to the number of cells. Fluorescence-based methods (e. g., fluorescence spectroscopy, quantitative flow cytometry) are used for quantification. These methods measure the concentration of either fluorescently labeled or intrinsically fluorescent nanoparticles. Results can also be quantified through elemental analysis, particularly for elements not naturally present in cells, such as gold (Au) or silver (Ag). Nanoparticle concentration is often determined by inductively coupled plasma spectrometry (ICP), including techniques like ICP-AES (atomic emission spectroscopy), ICP-MS (mass spectrometry), and ICP-OES (optical emission spectroscopy).

4.3.4.3 Factors affecting nanoparticle exocytosis

The rate of exocytosis and the intracellular retention time of nanoparticles depend on many factors such as cell type, physicochemical properties of the nanoparticles, and cell culture conditions. Although there are few studies, some factors that strongly influence exocytosis can be identified. However, there is a lack of consistency and no general statements can be made.

4.3.4.3.1 Cell-type effect

The response to nanoparticle treatment varies across different cell types. Various cell lines internalize nanoparticles through distinct endocytosis mechanisms, resulting in different intracellular distributions. This variability also depends on the detoxification pathways that different cells use for same type nanoparticles. However, even when nanoparticles are internalized through the same endocytosis pathways, the exocytosis rate can still vary between cell types.

Chu et al. [777] found that 50-nm SiNPs followed similar endocytosis pathways and exhibited similar intracellular distribution patterns in human lung carcinoma cells (H1299), human esophageal epithelial cells (NE083), and human bronchial epithelial cells (NL20). The NE083 cells exocytosed the SiNPs slower than the H1299 and NL20 cells, showing a clear cell type-dependent exocytosis profile. Slowing et al. [778] treated human umbilical vein endothelial cells (HUVECs) and HeLa cells with FITC-labeled mesoporous silica nanoparticles (FITC-SiO_2 NPs). The exocytosis rate was measured when the intracellular FITC-SiO_2 NPs content reached its saturation concentration. While exocytosis of FITC-SiO_2 NPs started after 40 min in HUVEC cells, no exocytosis was observed in HeLa cells. The ability of released FITC-SiO_2 NPs to be internalized by neighboring cells was tested in cross-cell experiments. In these experiments, two different types of fluorescence-labeled SiO_2 NPs: the FITC-SiO_2 NPs and the tetramethylrhodamine isothiocyanate-labeled mesoporous silica nanoparticles (TRITC-SiO_2 NPs). HeLa cells were treated separately with both types of SiO_2 NPs and were then mixed together. Only about 12 % of the HeLa cells showed fluorescence signals for both FTIC- and TRITC-SiO_2 NPs, whereas a much higher transfer rate was obtained in HUVECs, with

89 % of the cells showing transfer of SiO_2 NPs between them. In another experiment, the transfer rate of SiO_2 NPs between different cell lines was measured. One cell line was labeled with a cell tracer dye, and the other was incubated with FITC-SiO_2 NPs. When the tracer-loaded HUVEC cells were mixed with FITC-SiO_2 NP loaded HeLa cells, only a small amount of SiO_2 NPs was transferred (7 %). In contrast, when FITC-SiO_2 NP-loaded HUVECs were co-cultured with the labeled HeLa cells, a large proportion of the FITC-SiO_2 NPs migrated into the HeLa cells (74 %). When FITC-labeled, phosphonate-modified SiO_2 NPs (FITC-pSiO_2 NPs) with a size of 130 nm were incubated with different human cell lines (Table 13), the effect of cell type was evident by the significantly different exocytosis rates [765]. Nanoparticle exocytosis was correlated with the release of lysosomal enzymes, suggesting the lysosomal exocytosis pathway.

Table 13: Exocytosis rates of phosphonate-modified SiO_2 NPs in various human cell lines.

Cell type	Cell line	Exocytosis rate
Lung cancer	A549	87 %
Breast cancer	MDA-MB-231	81 %
Breast cancer	MCF-7	75 %
Melanoma cancer	MDA-MB-435	61 %
Pancreatic cancer	PANC-1	36 %
Embryonic stem cells	H9	4 %

Not only SiO_2 NPs show a cell type-dependent exocytosis, the same behavior was found for spherical AuNPs [415] or AuNRs [779].

4.3.4.3.2 Effect of physicochemical properties of nanoparticles

Unlike cellular uptake, where the influence of biomolecules on the nanoparticle surface can be minimized by using serum-free medium, it is not possible to prevent the attachment of biomolecules during cellular trafficking. Additionally, changes in nanoparticle properties during intracellular trafficking may affect exocytosis rates. For instance, surface modifications may detach from the nanoparticle surface, or the nanoparticles themselves may degrade, altering their size and number.

The intracellular retention time of nanoparticles is influenced by their physicochemical properties, such as size, shape, and surface modifications. While these properties are as important for exocytosis as they are for endocytosis, their impact on exocytosis is less well-characterized. Unlike cellular uptake, where the influence of biomolecules on the nanoparticle surface can be minimized by using serum-free medium, it is not possible to prevent the attachment of biomolecules during cellular trafficking. In addition, changes in nanoparticle properties during intracellular trafficking may affect exocytosis rates. For instance, surface modifications may detach from the nanoparticle surface, or the nanoparticles may degrade, altering their size and number.

Nanoparticle size

Exocytosis, like endocytosis, is a size-dependent process. Smaller nanoparticles are cleared from cells more easily than larger ones. The exocytosis rate of 14-nm transferrin-coated AuNPs was found to be twice that of 50-nm AuNPs and five times that of 74-nm or 100-nm AuNPs [415]. The authors explained this size dependence with the presence of ligands on the nanoparticle surface. Smaller nanoparticles interact with fewer receptors than larger ones, resulting in a slower cell entry but faster cell efflux. Ligand–receptor interactions were further studied by Ohta et al. [780]. The cellular trafficking of 3-nm allylamine-modified SiQDs was investigated using confocal laser scanning microscopy. The SiQDs co-localized with the dye Lyso-Tracker, indicating their transport along the endosomal lysosomal pathway. The authors proposed a kinetic model to describe their experimental results: the SiQDs bind to surface receptors and are internalized via endocytosis. The acidic conditions inside the lysosomes induce conformational changes in the receptors, causing them to dissociate from the SiQDs and be transported back to the cell membrane for exocytosis. SiQDs still bound to the receptors are recycled to the cell membrane, while free SiQDs remain inside the cells.

Porous silicon microcarriers were loaded with either 15-nm or 30-nm SPIONs and internalized by murine macrophages J774. Excreted irons from these macrophages were determined as an indicator of SPION exocytosis. The release of 15-nm SPIONs from the carrier and macrophages was much faster than that of 30-nm SPIONs at the same initial concentration [761]. A similar size dependence was found for mesoporous SiO_2 NPs in hepatocyte HepG2 cells. The percentage of exocytosed SiO_2 NPs was found to depend on the SiO_2 NP size: 60 nm (63 %), 180 nm (67 %), 370 nm (58 %), and 600 nm (38 %) [781]. Apparently, smaller SiO_2 NPs were exocytosed faster than larger ones.

Complementary studies on poly (D,L-lactide-*co*-glycolide) nanoparticles (PGLA-NPs) were performed to elucidate the mechanism and dynamics of the exocytosis pathway in human vascular smooth muscle cells [782]. It was observed that PGLA-NPs exhibited a size-dependent endosomal recycling pathway. PGLA-NPs of size 100 nm were found in the early endosomes, the Golgi apparatus and the endoplasmic reticulum, which is the intracellular trafficking route for the slow recycling pathway.

Nanoparticle morphology

Using spherical and rod-shaped polystyrene nanoparticles (PSNPs and PSNRs), Zhuang et al. [783] demonstrated that the exocytosis pathway and the retention time of the nanoparticles depend not only on the size but also on the shape of the nanoparticles. The PSNPs had a diameter of 200 nm, while the PSNRs, obtained by stretching the PSNPs, had a length of 400 nm and a thickness of 100 nm. PSNPs and PSNRs exhibited significantly different exocytosis behaviors in a monolayer of Caco-2 cells. Compared to PSNRs, PSNPs favored basolateral exocytosis, whereas PSNRs exited the cells more efficiently via the apical endocytosis pathway than PSNPs. Furthermore, the exocytosis mechanism was shown to depend on the cell type [784]. For example, spherical AuNPs (18 nm) exited mouse breast tumor 4T1 cells faster and more efficiently than AuNRs with dimensions of 14 nm × 58 nm and 16 nm × 110 nm, whereas the exocytosis behavior of the AuNPs and AuNRs in M1/M2-polarized Raw264.7 cells showed the opposite trend. On the other hand, the efflux rate of the AuNPs and AuNRs in STO fibroblast cells did not differ.

Surface modifications of nanoparticles

The surface chemistry and charges of nanoparticles are key determinants of the intracellular distribution, exocytosis pathway, and exocytosis efficacy of the nanoparticles. AuNPs with similar hydrodynamic size and surface charge, differing only in their surface functionality, were studied in HUVEC cells [759]. The AuNP surface was functionalized with a targeting peptide and a non-targeting peptide. The targeting AuNPs were attached to the KATWLPPR peptide, which binds to membrane receptors of endothelial cells and downregulates angiogenic genes. The nontargeting peptide was KPRQPSLP, which does not interact with cell membrane receptors. Tumor-targeted AuNPs enter the cells via specific receptor-mediated endocytosis, the nontargeted AuNPs via nonspecific pathways. The differently functionalized AuNPs differ from each other in their exocytosis profile. Tumor-targeted AuNPs were gradually exocytosed over a period of 6 h. In contrast, the amount of exocytosed nontargeted AuNPs initially increased and then decreased after 4 h, indicating a renewed uptake of the nanoparticles by the cells. After another 2 h, the exocytosis rate increased again.

Kim et al. [785] investigated the specific role of surface chemistry for the exocytosis of differently functionalized 2-nm gold nanoparticles (AuNPs) in MCF-7 cells. Of particular interest was the influence of hydrophobicity or structural functionality of the surface structures on the exocytosis rate. Contrary to expectations, the authors were unable to demonstrate a clear correlation between the amount of exocytosed gold and the hydrophobicity of the ligand for aliphatic head groups. By comparison, the AuNPs with aromatic headgroups showed a significantly higher exocytosis rate. This result underlines the influence of the ligand structure on the exocytosis efficiency.

Ho et al. [786] focused their research activities on demonstrating the influence of the ligand structure on the exocytosis of hydrophilic AuNPs. For this purpose, 13-nm AuNPs were PEGylated, and the PEG surface of the AuNPs was functionalized with structurally different hydrocarbyl groups. The hydrocarbyl groups consisted of either branched or linear alkane and alkene chains or contained cycloalkanes or aromatic hydrocarbons. The differently functionalized hydrophilic PEG-AuNPs were shown to be taken up by RAW264.7 macrophages, C166 endothelial cells and HeLa cells via endocytosis. The PEG-AuNPs functionalized with dodecane ligands showed the highest endocytosis efficiency for all three cell types, which was ten-fold higher compared to the other functionalized PEG-AuNPs. In comparison, the amount of exocytosed PEG-AuNPs was significantly dependent on both cell type and ligand structure. The cyclododecane-terminated PEG-AuNPs were preferentially exocytosed from RAW264.7 and C166 cells but not from HeLa cells, while their exocytosis from HeLa cells was not remarkable. Of note was the release of intraluminal vesicles into the extracellular medium concomitant with exocytosis.

All nanoparticles studied so far had the same surface charge. However, the surface charge may also play a role in nanoparticle exocytosis. Oh et al. [787] observed that the cationic AuNPs had a longer retention time in macrophages compared to PEGylated AuNPs, regardless of their size in the range of 10–40 nm. The cationic AuNPs formed intracellular aggregates that slowed intracellular trafficking and delayed exocytosis. In contrast, PEG-AuNPs migrated through the cells as single nanoparticles, resulting in rapid exocytosis.

4.3.4.3.3 Effect of nanoparticle concentration, incubation time, and localization in cells

The influence of the preincubation time and initial concentration on the exocytosis process remains unclear. Some authors have reported that the initial concentration has no influence on the amount of exocytosed nanoparticles [786], while others have suggested that the preincubation time and the dose affect the exocytosis rate [788].

The localization of nanoparticles in the cell, whether in vesicles or cell organelles, determines the exocytosis pathway, the degradation of nanoparticles and their residence time. The intracellular fate of nanoparticles is also determined by the cellular uptake mechanisms that orchestrate their intracellular transport pathway. For example, molecules or nanoparticles that use the fast recycling pathway have a higher exocytosis rate than nanoparticles that choose the slow recycling pathway [782]. Nanoparticles that escape the endosome and accumulate in the cytoplasm have a longer residence time than those that remain in the endocytic vesicles [789]. Chu et al. [777] observed that SiO_2 NPs retained in lysosomes exited the cells much faster than those located in the cytoplasm. On the other hand, CuO nanoparticles that became trapped in the mitochondria or nucleus in A549 cells cannot leave the cells, which destroys the cells [790].

4.3.4.3.4 Effect of extracellular conditions

The composition of the culture medium in which the nanoparticles are introduced to the cells, or that covers the cells without nanoparticles must be carefully selected. Certain ingredients can alter endocytosis and exocytosis rates. For example, the serum content of the culture medium promotes the exocytosis of nanoparticles. Serum proteins internalized along with the nanoparticles can interact with the endosomal pathways and influence exocytosis. Panyam et al. [782] observed that the exocytosis of PGLA nanoparticles was delayed when the nanoparticles were administered to cells in serum-free medium.

Specific ions in the cell culture medium play a crucial role in the cellular exocytosis. Ca^{2+} cations, in particular, can enormously increase the exocytosis rate of nanoparticles. Nanoparticles migrating through the cells showed an increased rate of exocytosis when treated with Ca^{2+}. When the chelator BAPTA-AM bound the extracellular Ca^{2+}, the exocytosis rate was downregulated. However, once the nanoparticles exited the lysosomal pathway and dispersed into the cytoplasm, Ca^{2+} had no effect on exocytosis efficiency. Ca^{2+} induces conformational changes in the protein synaptotagmin VII [791], which subsequently binds to the SNARE complex at the plasma membrane and promotes the fusion of the lysosome with the cell membrane. An increase in the intracellular fraction can accelerate the fusion of lysosomes with the cell membrane and increase the rate of exocytosis along the endosomal–lysosomal pathway but has no effect on other pathways. Other extracellular cations such as K^+ [792] or Zn^{2+} [793] regulate the release of endocytic vesicles and change the exocytosis dynamics.

4.4 Impact of nanoparticles on cell organelles

4.4.1 Cell organelles and their function

Cell organelles are subcellular structures with specific functions necessary for normal cell survival and function [794]. These organelles can be categorized as membrane-bound or non-membrane-bound (Figure 139). Non-membrane-bound organelles include the plasma membrane, ribosomes, or the cytoskeleton. Single-membrane organelles consist of lysosomes, endosomes, the Golgi apparatus, and the endoplasmic reticulum. Double-membrane organelles are the nucleus and mitochondria. Each of these organelles plays a critical role in maintaining the cellular life and functionality of the cell,

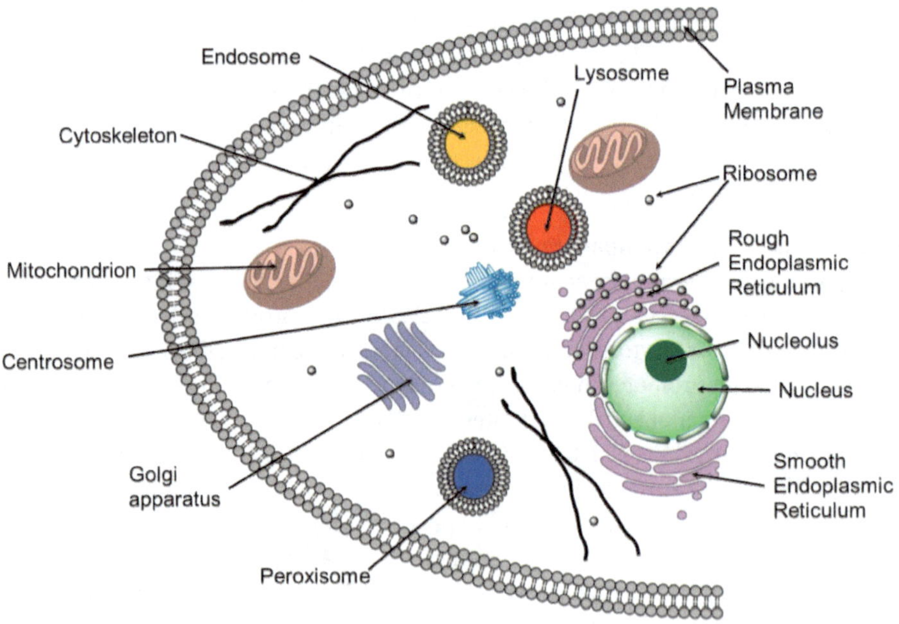

Figure 139: Schematic of cell organelles.

Table 14: Brief summary on cell organelles, their structure and function [794].

Cell organelle	Structure	Function
Plasma membrane	– lipid bilayer embedded with proteins and other biomolecules such as cholesterol	– selectively permeable: controls trafficking of materials in and out of the cell – maintains cellular shape – protects cell interior
Cytoplasm	– gel-like substance between plasma membrane and cell nucleus – consisting of water, organic and inorganic substances	– embedded all cell organelles – maintains metabolic activity – dissolves nutrients and waste products of the cell
Endoplasmic reticulum	– network of membrane canals filled with fluid	– mechanical support and compartmentation – transport of proteins, lipids, enzymes, etc. – detoxification in liver cells – glycogenolysis
Rough endoplasmic reticulum	– ribosomes embedded in the membrane canals	– protein biosynthesis – production of secretory products – glycosylation of proteins and lipids
Smooth endoplasmic reticulum	– membrane canals without ribosomes	– synthesis of hormones – synthesis of lipids for the plasma membrane
Mitochondrion	– elongated cylindrical bodies – consist of an external membrane, intermembrane space, an inner membrane and the matrix – the inner membrane is folded in cristae – contains mitochondrial DNA	– generation of key metabolites – oxidation phosphorylation to produce ATP and H_2O_2 – production and consumption of ROS – regulates apoptosis – Ca^{2+} homeostasis
Ribosomes	– composed of RNA and proteins – consists of 2 subunits – found in the cytoplasm or bound to the endoplasmic reticulum	– protein biosynthesis
Golgi apparatus	– 5 to 8 flattened, disk-shaped membranes (cisternae) – Polarized structure, facing the cell nucleus ("cis site"), facing the cell membrane ("trans site")	– modification, completion and export of secretory proteins and glycoproteins – activation of secretory proteins
Peroxisome	– vesicles of sizes ≈0.2–1.5 µm – spherical shape – surrounded by a single membrane – contain proteins and enzymes	– type II oxidase oxidize organic molecules by using molecular oxygen – produce H_2O_2 – contain catalase, that metabolize H_2O_2 in water and oxygen

Table 14 (continued)

Cell organelle	Structure	Function
Endosome/lyso-somes	- vesicles with a single membrane - contain hydrolases - acidic pH	- intracellular trafficking - degradation of proteins, lipids, nucleic acids
Cytoskeleton	- fibrous network of protein filaments - 4 types: actin filaments, microtubules, intermediate filaments and septins	- maintains shape of the cell - Movement of cell and internal organelles
Centrosomes	- consists of 2 centrioles - c - entriole composed of 9 tubulin fibrils	- organizing microtubules - cell division, mitotic spindle arrangement
Nucleus	- surrounded by a nuclear membrane, called nuclear envelope - contains chromosomes	- contains genetic information - controls cell growth and reproduction - DNA replication - DNA transcription
Nucleolus	- spherical bodies inside the cell nucleus - nucleolus organizing region of the chromosomes	- necessary for ribosomal production - encode the ribosomal RNA

working together in a coordinated manner to ensure that cellular processes are carried out efficiently. Table 14 summarizes the major cell organelles and their functions.

4.4.2 Changes in organelle function

Various cell organelles play important roles in nanoparticle-based therapies. The interaction between nanoparticles and different organelles determines the success of these therapies or the potential for cytotoxicity. Following cellular uptake, endosomes and lysosomes act as the first line of cellular defense, often leading to the degradation of nanoparticles. If nanoparticles successfully escape the endosome and reach the cytoplasm, they can interact with other organelles, enzymes, macromolecules, or small molecules within the cytoplasm.

4.4.2.1 Mitochondria function and dysfunction

Mitochondria are essential organelles found in nearly all eukaryotic cells, often referred to as the "powerhouses" of the cell due to their critical role in energy production [795, 796] (Figure 140). These double-membrane structures are thought to have originated from ancient bacteria through endosymbiosis, retaining some bacterial characteristics such as their own circular DNA and the ability to replicate independently within the cell. The structure of mitochondria is key to their function. They have an outer membrane

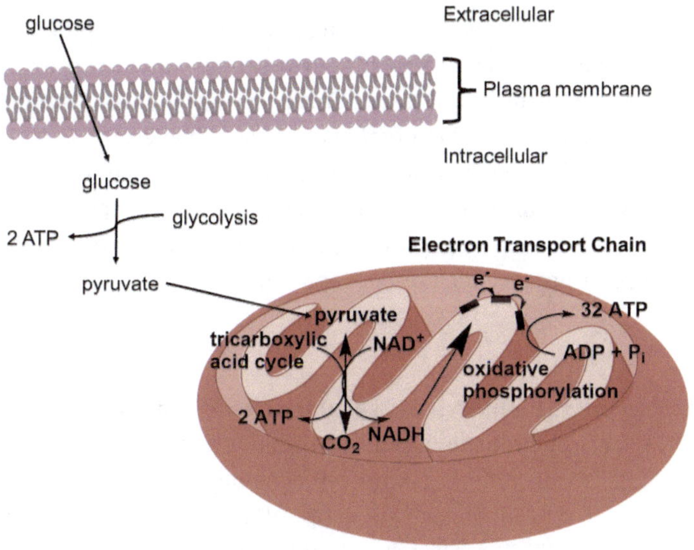

Figure 140: Glucose metabolism and ATP production via oxidative phosphorylation.

that is relatively permeable and an inner membrane that is highly folded into structures called cristae. These cristae greatly increase the surface area of the inner membrane, increasing the organelle capacity for energy production. The space enclosed by the inner membrane is called the matrix, while the space between the outer and inner membranes is called the intermembrane space.

The primary function of mitochondria is to produce cellular energy in the form of adenosine triphosphate (ATP) through the process of oxidative phosphorylation [797]. This process occurs along the electron-transport chain, which is embedded in the inner mitochondrial membrane. The electron transport chain consists of a series of protein complexes that transfer electrons from electron donors to electron acceptors via redox reactions. This electron flow is coupled with the pumping of protons from the matrix into the intermembrane space, creating an electrochemical gradient. The energy stored in this gradient is then used by ATP synthase to convert ADP and inorganic phosphate into ATP.

While ATP production is their best-known role, mitochondria are involved in numerous other cellular processes. They play a critical role in cellular metabolism, participating in the citric acid cycle (also known as the Krebs or TCA cycle), fatty acid oxidation, and the urea cycle. Mitochondria are also central to the biosynthesis of several molecules, including heme, some amino acids, and certain lipids.

Mitochondria are key players in cellular calcium homeostasis. They can rapidly take up calcium from the cytosol and act as a calcium buffer to help regulate intracellular calcium levels. This function is particularly important in neurons and muscle cells, where rapid calcium fluctuations are critical for signaling and contraction. These organelles also play a key role in programmed cell death, or apoptosis. Under certain conditions, mitochondria can release pro-apoptotic factors such as cytochrome c into the cytosol, initiating a cascade of events that leads to cell death. This process is tightly regulated and is essential for normal development and maintenance of tissue homeostasis.

Recent research has also highlighted the role of mitochondria in cellular signaling beyond their metabolic functions. Mitochondria-derived reactive oxygen species (ROS), which are by-products of the electron transport chain, can act as signaling molecules and influence various cellular processes, including adaptation to hypoxia, autophagy, and innate immunity [798].

4.4.2.1.1 Mitochondrial ROS production, RNS and cellular defense system

Reactive oxygen species (ROS) are by-products of cellular oxidative metabolism and play critical roles in cell survival, cell death, differentiation, cell signaling, and the production of inflammatory-factor production [799]. Biologically relevant ROS include singlet oxygen ($^1O_2$), superoxide anion radical ($O_2^{-\bullet}$), hydroxyl radical ($HO^\bullet$), hydroperoxyl radical ($H_2O^\bullet$), and hydrogen peroxide (H_2O_2). Another important class of reactive molecular species involved in both physiological and pathological processes are reactive nitrogen species (RNS) [800]. The primary RNS is nitric oxide ($NO^\bullet$), a gaseous free radical produced by nitric oxide synthases (NOS) through the oxidation of L-arginine. When $NO^\bullet$ reacts with $O_2^{-\bullet}$ or O_2^{2-}, other reactive nitrogen species are formed, namely peroxynitrate (ONOO–) and nitrogen dioxide ($NO_2^\bullet$) (Figure 141). The fact that peroxynitrate is a highly reactive oxidant demonstrates the major impact of the interplay between RNS and ROS on cellular redox biology. Additionally, $NO^\bullet$ can react with hydrogen peroxide (H_2O_2) to produce hydroxyl radical ($HO^\bullet$), further confirming the complex interplay between RNS and ROS.

The biological effects of ROS depend on their concentration and localization within cells and tissues. At low levels, ROS play important regulatory and signaling roles. Hydrogen peroxide, being relatively stable and membrane-permeable, is particularly well suited as a signaling molecule. It can diffuse from its site of generation and interact with specific protein targets, often oxidizing thiol groups on cysteine residues. This can lead to changes in protein structure and function, thereby modulating cellular processes.

At higher concentrations, ROS can have damaging effects on cellular components [801]. They can oxidize lipids, proteins, and nucleic acids, potentially leading to cellular dysfunction. The hydroxyl radical is particularly reactive and can damage virtually any biomolecule in its vicinity. To protect against these harmful effects, organisms have evolved sophisticated antioxidant defense systems. These include enzymes such as superoxide dismutase, catalase, and glutathione peroxidase, as well as nonenzymatic antioxidants such as glutathione, ascorbate, and tocopherol.

The interplay between ROS generation and scavenging is critical for maintaining cellular redox homeostasis. Disruption of this balance, resulting in excessive ROS levels,

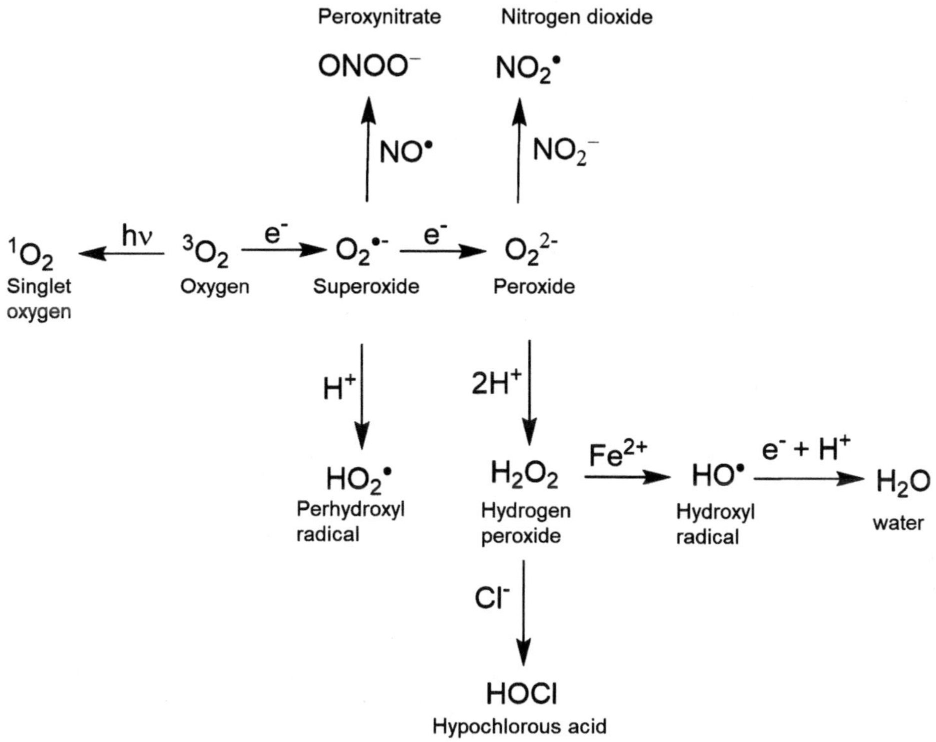

Figure 141: Generation of different types of ROS and RNS.

is termed oxidative stress [802, 803]. This state has been implicated in several pathological conditions and aging processes. However, it is important to note that a certain level of ROS is necessary for normal cellular functions, including cell signaling, immune responses, and regulation of gene expression.

Several fluorescent sensors have been developed to detect different ROS and RNS [804, 805]. For singlet oxygen ($^1O_2$) detection, probes such as DPAX and DMAX rely on the formation of endoperoxides upon reaction with $^1O_2$, leading to increased fluorescence intensity. ATTA-Eu^{3+} is a phosphorescent probe that forms a highly luminescent endoperoxide product upon reaction with $^1O_2$. These probes show high selectivity for $^1O_2$ over other ROS.

Several probes have been developed for the detection of hydrogen peroxide (H_2O_2). Pentafluorobenzenesulfonyl fluoresceins are deprotected by H_2O_2, regenerating fluorescent fluorescein. Chang et al. [806] developed boronate-based probes such as PF1, PR1, and PX1, which rely on H_2O_2-mediated oxidation of arylboronates to phenols, resulting in fluorescence activation. These boronate probes show excellent selectivity for H_2O_2 over other ROS. The Eu^{3+}-tetracycline complex {Eu(tc)} binds H_2O_2 to form a highly fluorescent complex. DPPEA-HC reacts with H_2O_2 to form an oxide and enhances fluorescence through a photoinduced electron transfer mechanism.

For hydroxyl radical ($^{\bullet}OH$) detection, probes such as terephthalic acid and coumarin-3-carboxylic acid undergo hydroxylation to form fluorescent products. 3′-(p-Hydroxyphenyl) fluorescein (HPF) and 3′-(p-aminophenyl) fluorescein (APF) selectively react with highly reactive oxygen species, including $^{\bullet}OH$. TDFT is a FRET-based probe that undergoes DNA strand cleavage by $^{\bullet}OH$, resulting in fluorescence changes. These probes show good selectivity for $^{\bullet}OH$ over other ROS.

For superoxide ($O_2^{-\bullet}$) detection, bis(2,4-dinitrobenzenesulfonyl) fluorescein is deprotected by $O_2^{-\bullet}$ to generate fluorescence. 2-(2-pyridil)-benzothiazoline is oxidized by $O_2^{-\bullet}$ to produce a fluorescent product. These probes exhibit high selectivity for $O_2^{-\bullet}$ over other ROS. The sensitivity of these probes varies, with detection limits ranging from picomolar to micromolar concentrations of the target ROS/RNS. For example, some H_2O_2 probes can detect nanomolar levels, while $O_2^{-\bullet}$ probes have detection limits in the picomolar range. The selectivity and sensitivity of these fluorescent probes make them valuable tools for studying the role of specific ROS and RNS in biological systems, enabling real-time monitoring and imaging in living cells and tissues. However, careful consideration of probe characteristics and potential interferences is necessary when applying them to complex biological environments.

Approximately 90 % of the ROS are produced in mitochondria, among which superoxide is the most abundant [807]. During the oxidative phosphorylation molecular oxygen is reduced to water. However, mitochondria convert 0.2–2.0 % of the molecular oxygen into superoxide. The majority of oxygen consumed by organisms is reduced to water via a four-electron reduction process in the mitochondrial electron transport chain. $O_2^{-\bullet}$ is primarily generated through electron leakage from the electron transport chain. This leakage occurs mainly at complexes I and III, where electrons can prematurely reduce oxygen molecules instead of completing their journey through the electron transport chain. The generation of ROS is influenced by factors such as the strength of the proton gradient, the redox state of the ubiquinone pool, and the availability of ADP. Increased ROS production often occurs under conditions of high membrane potential or during reverse electron transport. Once formed, superoxide is rapidly converted to hydrogen peroxide (H_2O_2) by superoxide dismutase, and H_2O_2 can further react to form other ROS, including the highly reactive $^{\bullet}OH$. While excessive ROS production can lead to oxidative damage of mitochondrial components, including mitochondrial DNA, proteins, and lipids, moderate levels of mitochondrial ROS play important roles in cellular signaling and adaptation. The balance between ROS generation and antioxidant defenses in mitochondria is crucial for maintaining cellular health, with imbalances potentially contributing to various pathological conditions and the aging process.

ATP synthesis occurs primarily through oxidative phosphorylation in mitochondria, a process that begins with the electron-transport chain in the inner mitochondrial membrane. As electrons from NADH and FADH2 move through protein complexes, energy is released and used to pump protons into the intermembrane space, creating a concentration gradient. This gradient creates a proton motive force that ATP synthase utilizes as protons flow back into the matrix, causing parts of the enzyme to rotate and mechanically drive ATP synthesis from ADP and inorganic phosphate. Protons are pumped through complexes I, III, and IV into the intermembrane space (Figure 142), creating a net positive charge in the intermembrane space and a net negative charge in the matrix. This electrochemical potential is also called the mitochondrial membrane potential (MMP, ΔΨ). The flow of protons through the proton channel of complex V back into the matrix induces the generation of ATP [808].

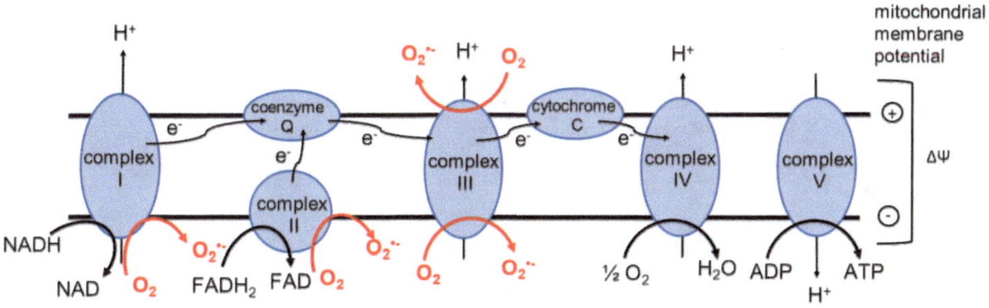

Figure 142: Mitochondrial superoxide generation at complex I to III of the electron transport chain.

Along the electron-transport chain, electrons travel from complex I to complex IV to reduce molecular oxygen to water. Complex I shuttles electrons from NADH to coenzyme Q [809]. During this process, electrons can be transferred to molecular oxygen, producing superoxide. A similar process occurs at complex II. In this case, electrons received from FADH2 generate superoxide. Electrons from complexes I and III are transferred to complex III via coenzyme Q. Complex III converts ubiquinol to ubiquinone (Q cycle) to reduce cytochrome c, which releases an electron to complex IV. Complex III has a pool of ubiquinone that is targeted to the intermembrane space and matrix of the mitochondria. Superoxide is formed at complex III as a result of the auto-oxidation of ubisemiquinone, a Q-cycle intermediate. Superoxide is then released into either the intermembrane space or the matrix, depending on the ubiquinone pool from which it is generated. The energy from the proton gradient enables the formation of the high-energy phosphate bond in ATP, making this endergonic reaction possible. This efficient process can produce approximately 32–34 molecules of ATP per molecule of glucose oxidized, providing the cell with its primary energy currency.

The cellular antioxidant defense system employs a sophisticated network of enzymes to convert harmful reactive oxygen species into less damaging radicals, ultimately producing water and oxygen. This process begins with $O_2^{-\bullet}$, a highly reactive oxygen species that can cause significant cellular damage if not properly regulated (Figure 143). The first line of cellular defense against superoxide radicals involves superoxide dismutases (SODs) [803, 810]. There are two main types of SODs in eukaryotic cells: copper/zinc SOD (Cu/Zn SOD or SOD1) and manganese SOD (MnSOD or SOD2).

Cu/Zn SOD is mainly located in the cytosol and in the intermembrane space of mitochondria, whereas MnSOD resides in the mitochondrial matrix. Both enzymes catalyze the same reaction: the conversion of superoxide radicals into hydrogen peroxide and molecular oxygen. This reaction, known as dismutation, effectively reduces the threat posed by superoxide radicals.

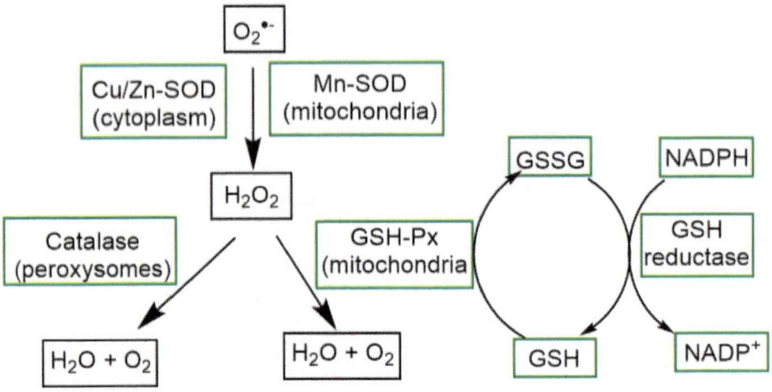

Figure 143: Cellular antioxidant defense system.

However, while hydrogen peroxide is less reactive than superoxide, it can still cause cellular damage and must be further processed. This is where the next line of defense comes into play, involving catalase. Catalase is a highly efficient enzyme found primarily in peroxisomes. It catalyzes the decomposition of hydrogen peroxide into water and oxygen. This reaction is incredibly fast, with each catalase molecule able to convert millions of hydrogen peroxide molecules per second. To complement the action of catalase and provide additional protection, especially in areas where catalase may not be present or abundant, cells also employ the glutathione system. This system involves glutathione peroxidase (GPx) and glutathione reductase (GSH reductase). GPx catalyzes the reduction of hydrogen peroxide to water, while simultaneously oxidizing glutathione (GSH) to its disulfide form (GSSG). GSH reductase plays a crucial role in maintaining the cellular pool of reduced glutathione. It catalyzes the reduction of GSSG back to GSH, using NADPH as an electron donor. This regeneration of GSH is vital for maintaining the cellular antioxidant capacity and overall redox state. The continuous cycling between GSH and GSSG, facilitated by GSH reductase, ensures a constant supply of reduced glutathione to support the ongoing detoxification of hydrogen peroxide and other oxidants.

4.4.2.1.2 Mitochondrial dysfunction and ROS generation caused by nanoparticles

Klein et al. [311, 517] studied the impact of 1–2 nm sized aminosilanized (APTES) silicon nanocluster (NH_2-SiNC) on the mitochondrial function and associated $O_2^{-\bullet}$ generation in MCF-7 and 3T3 cells. As been revealed by TEM images the positively charged NH_2-SiNC

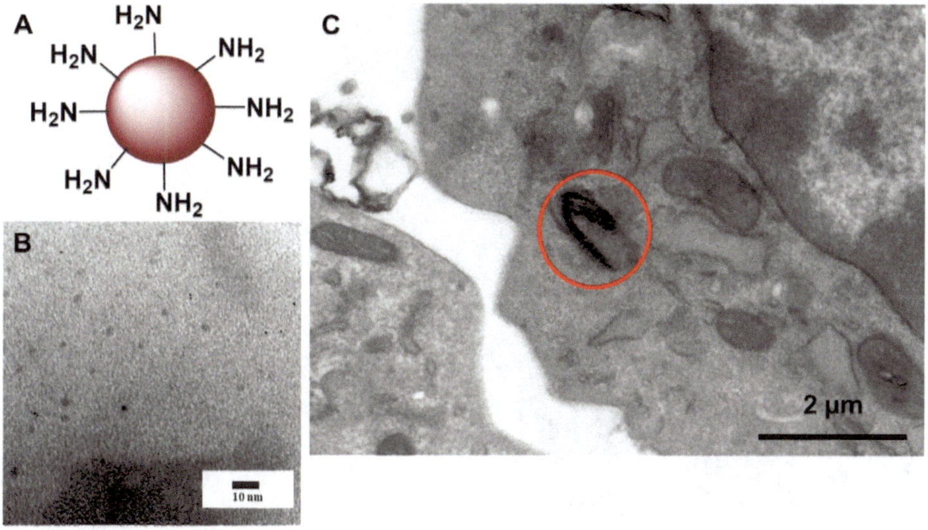

Figure 144: Sketch of a NH_2-SiNC (A); TEM image showing a mean size of 1.5 nm (B); TEM image of a MCF-7 cell with the NH_2-SiNC incorporated into the membranes of organelles; the red circle marks a mitochondrion (C).

directly permeated the plasma membranes of both MCF-7 and 3T3 cells and were predominantly embedded in the mitochondrial membranes (Figure 144). This localization was shown to have a profound effect on mitochondrial function, which was significantly enhanced under x-ray exposure.

The NH_2-SiNC embedded in mitochondrial membranes were proven to enhance the production of ROS, in particular, that of $O_2^{-\bullet}$ (Figure 145). The impact of the membrane-embedded NH_2-SiNC on the mitochondrial function was examined by measuring the mitochondrial membrane potential using the JC-1 dye. The JC-1 dye molecules accumulate in the mitochondrial membrane, where they form red-fluorescent aggregates. As soon as mitochondrial membrane depolarizes or is disrupted, the JC-1 dye leak out of the mitochondria and disintegrate to green-fluorescent monomers. A decrease of the ratio of red-fluorescent JC-1 aggregates: Green-fluorescent JC-1 monomers indicates a mitochondrial membrane dysfunction (Figure 146).

The NH_2-SiNC induced mitochondrial membrane depolarization was rationalized to cause the opening of mitochondrial permeability transition pores that triggered release of cytochrome c into the cytoplasm and subsequent accumulation of ROS. Additionally, inhibition of the respiratory chain was thought to increase ubisemiquinone free radical levels in complex III, further contributing to ROS generation. The release of cytochrome c from mitochondria is a critical event in the activation of the mitochondrial pathway to apoptosis, ultimately leading to DNA fragmentation (Figure 147). This process involves a complex cascade of molecular events:

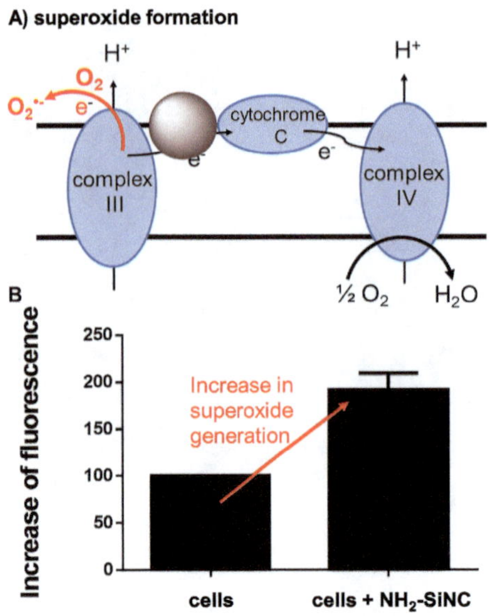

Figure 145: NH_2-SiNP embedded in the mitochondrial membrane hinders the electron transfer between the complexes of the ETC (A), which causes an increase in superoxide production (B).

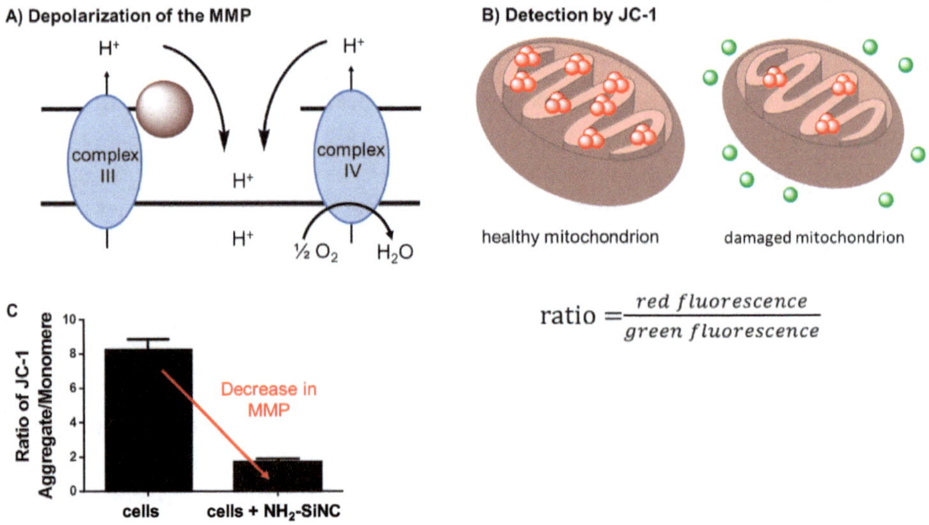

Figure 146: Depolarization of the mitochondrial membrane potential (MMP) caused by an influx of protons (A), depolarization of the MMP detected using fluorescent dye JC-1 (B), NH_2-SiNC caused a decrease in the MMP (C).

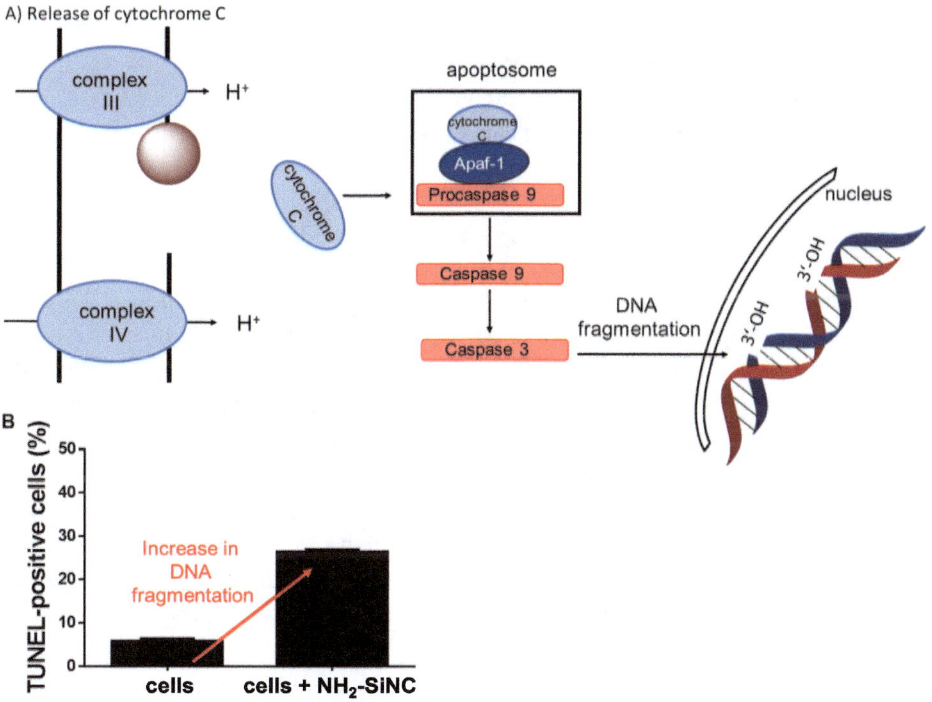

Figure 147: NH_2-SiNC caused release of cytochrome c which activates the mitochondrial pathway of apoptosis followed by DNA fragmentation (A); NH_2-SiNC induced DNA fragmentation.

Cytochrome c binds to the adaptor protein Apaf-1 (apoptotic protease activating factor 1), causing a conformational change that allows Apaf-1 to oligomerize and recruit procaspase-9 [811]. The complex of cytochrome c, Apaf-1, and procaspase-9 forms the apoptosome. Inside the apoptosome, procaspase-9 is activated to form caspase-9, an initiator caspase. Activated caspase-9 then cleaves and activates downstream effector caspases, particularly caspase-3 and caspase-7. These effector caspases are the primary executioners of apoptosis. Activated effector caspases proceed to cleave various cellular substrates, including structural proteins, regulatory proteins, and other enzymes. A critical target of these caspases is an inhibitor of caspase-activated DNase (ICAD). When ICAD is cleaved, it releases and activates caspase-activated DNase (CAD). Once activated, CAD translocates to the nucleus where it begins the process of DNA fragmentation. This endonuclease cleaves genomic DNA between nucleo-somes, producing the characteristic ladder pattern of DNA fragments typically seen in apoptotic cells. This DNA fragmentation is a hallmark of apoptosis and contributes to the condensation of chromatin observed in apoptotic nuclei.

The release of cytochrome c further amplifies the apoptotic signal by disrupting the electron transport chain, leading to increased ROS production. This creates a positive feedback loop, causing further mitochondrial damage and potentially triggering cytochrome c release from additional mitochondria.

Compared to other ROS, H_2O_2 as being the most stable non-radical one exist in the highest concentration in the cytosol of cancer cells. In the presence of Fe^{2+} cations, H_2O_2 is converted to $^{\bullet}OH$ via the Fenton reaction [812, 813]:

$$Fe^{2\oplus} + H_2O_2 \longrightarrow Fe^{3\oplus} + HO^{\bullet} + {}^{\ominus}OH.$$

Iron cations (Fe^{2+} and Fe^{3+}) are essential for many life processes, including cell growth, proliferation, and oxygenation. Iron is bound and transported in the blood stream by transferrin and stored in the cells in two major proteins: ferritin and hemosiderin. A dynamic equilibrium between iron uptake, transport, storage and utilization ensures that iron homeostasis is maintained. Because of its pro-oxidant effects, excess iron is associated with increased toxicity in biological systems. Other metal cations that catalyze the Fenton reaction include Mn^{2+}, Cu^{1+}, V^{2+}, and Cr^{4+}.

Due to their small size, large surface area, and high reactivity, metal oxide nanoparticles are well-suited as Fenton catalysts for generation of hydroxyl radicals via the heterogeneous Fenton reaction [812, 814, 815]. However, SPIONs, with a Fe_3O_4 composition, can only catalyze the intracellular Fenton reaction if they either release Fe^{2+} cations into the cytosol or have Fe^{2+} cations in their surface. This would require the Fe_3O_4 surface to be accessible to H_2O_2 in the cytosol which may be hindered by surface structures that are required to stabilize the surface and prevent them from agglomeration [188]. On the other hand, unprotected Fe_3O_4 surfaces are oxidized to Fe_2O_3 under normal environmental conditions, i. e., in the presence of oxygen. Intracellular Fe^{3+} can be reduced to Fe^{2+} cations by superoxide radicals and then again act as a Fenton catalyst.

$$Fe^{3\oplus} + O_2^{\ominus\bullet} \longrightarrow Fe^{2\oplus} + O_2.$$

To compare the catalytic activity of Fe^{2+} and Fe^{3+} cations and SPIONs in cancer cells, Klein et al. [188] incubated MCF-7 cells with Fe^{2+} and Fe^{3+} cations, uncoated, and citrate-stabilized SPIONs. The relative ROS concentration of MCF-7 cells, as a relative measure of Fenton-catalyzed reactions, was measured using the 2′,7′-dichlorofluorescein diacetate (DCF-DA) assay (Figure 148). The 100 % intensity value of the DCF fluorescence refers to that of MCF cells in culture medium (dashed line in Figure 149). MCF-7 cells with Fe^{2+} or Fe^{3+} cations contained only slightly increased ROS concentrations compared to the reference MCF-7 cells (dashed line at 100 %). In contrast, the increased fluorescence intensity values of 180 % (uncoated SPIONs) and 80 % (citrate-stabilized SPIONs) indicate a high catalytic efficiency of both uncoated SPIONs and citrate-stabilized SPIONs.

The influence of self-assembled monolayers (SAMs) on the biocompatibility and Fenton catalytic activity of superparamagnetic $CoFe_2O_4$ nanoparticles ($CoFe_2O_4$ NPs) was investigated in MCF-7 cancer cells and healthy HUVECs using various cell biological dye assays such as the DCF-DA and neutral red assay. The uncoated $CoFe_2O_4$ NPs with

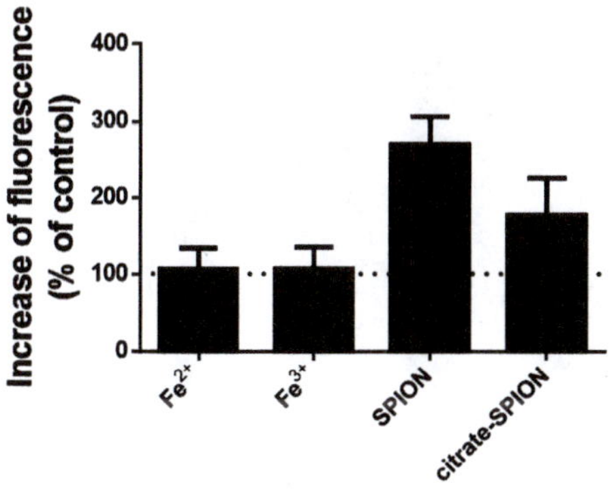

Figure 148: Relative ROS concentration in MCF-7 cells loaded with, Fe^{2+} cations, Fe^{3+} cations, and uncoated and citrate-stabilized SPIONs; the relative ROS concentration is represented as increase of DCF fluorescence intensity (%); $n = 6 / p > 0.05$.

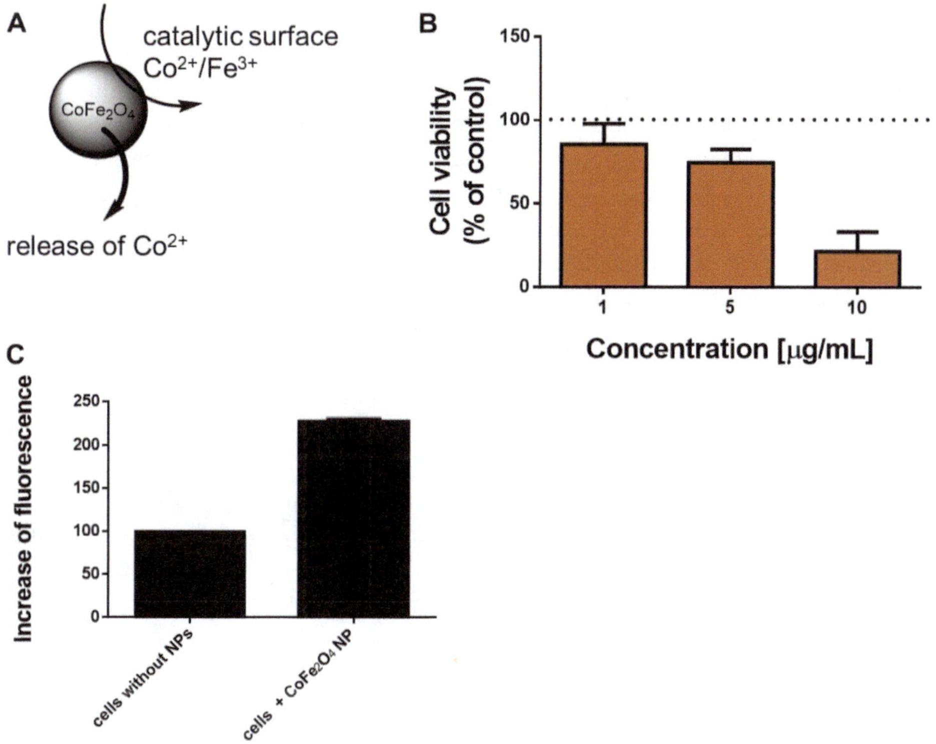

Figure 149: $CoFe_2O_4$ NPs catalyze the Fenton/Haber–Weiss reaction via surface cations or through the release of Co^{2+} (A); decreased cell viability of uncoated $CoFe_2O_4$ NPs (B) and the increased ROS production (C).

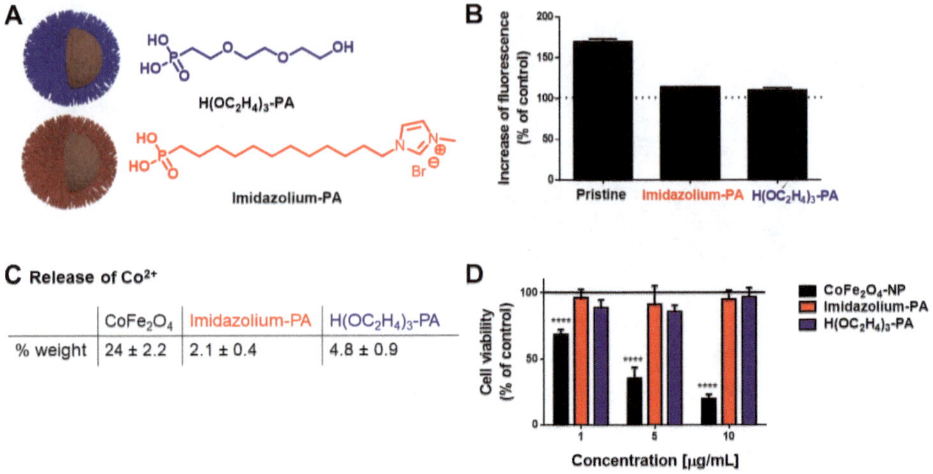

Figure 150: Surface coating of $CoFe_2O_4$ NPs with self-assembling monolayers of $H(OC_2H_4)_3$-PA and Imidazolium-PA (A), DCF fluorescence related ROS concentration (B), release of Co^{2+} from (C) cell viability of $CoFe_2O_4$, Imidazolium-PA-$CoFe_2O_4$, $H(OC_2H_4)_3$-PA-$CoFe_2O_4$ NPs (D).

hydrodynamic diameters between 10 and 30 nm tended to agglomerate in culture media and proved to be cytotoxic as determined by the neutral red assay (Figure 149). The clearly dose-dependent reduction in cell viability was explained by leakage of cytotoxic Co^{2+} cations into the cytoplasm. Not only the Co^{2+} cations released into the cytosol are responsible for the Fenton catalytic activity of the $CoFe_2O_4$ NPs. The $CoFe_2O_4$ NP surfaces, which were shown to be catalytically active, also increased the ROS concentration as detected by the DCF-DA assay. The $CoFe_2O_4$ NPs were coated with the SAM-forming compounds 1-methyl-3-(dodecylphosphonic acid) imidazolium bromide (imidazolium-PA) or (2-{2-(2-hydroxy-ethoxy)-ethoxy}-ethylphosphonic acid ($H(OC_2H_4)_3$-PA) (Figure 150(A)). As expected, the positively charged imidazolium-PA-$CoFe_2O_4$ NPs in culture medium exhibited a narrow distribution of hydrodynamic diameters with 14–16 nm. The SAMs of both imidazolium-PA and ($H(OC_2H_4)_3$-PA almost completely prevent the leakage of cytotoxic Co^{2+} cations into the cytosol (Figure 150(C)), which can be considered not only as a quality feature of optimally dense SAMs, but also ensures very good biocompatibility (Figure 150(D)) and significantly reduced Fenton catalytic activity (Figure 150(B)).

5 Pharmacokinetics, pharmacodynamics, and nanotoxicology

The pharmacokinetics of nanoparticles consists of a sequence of processes in the human body, generally divided into absorption, distribution, metabolism, and excretion, based on fully metabolizable agents [816, 817] (Figure 151). The first process, absorption, involves the uptake of nanoparticles into the human body. Nanoparticles can be administered systemically, such as orally, nasally, dermally, or by intravenous or subcutaneous injection, or locally by injection into tumor tissue. The route of administration influences the second process, the distribution of the nanoparticles in the human organism. This process refers to the reversible transport of nanoparticles between the bloodstream and the extravascular fluids and tissues of the body. The third process, called metabolism, results from the interaction of nanoparticles with cells in tissues and organs. Complete metabolism occurs only with a few nanomaterials, such as SiO_2NPs and SPIONs. The fourth process, excretion, involves the elimination of nanoparticles, which can occur through the kidneys, gastrointestinal tract, or other pathways. Ideally, nanoparticles are

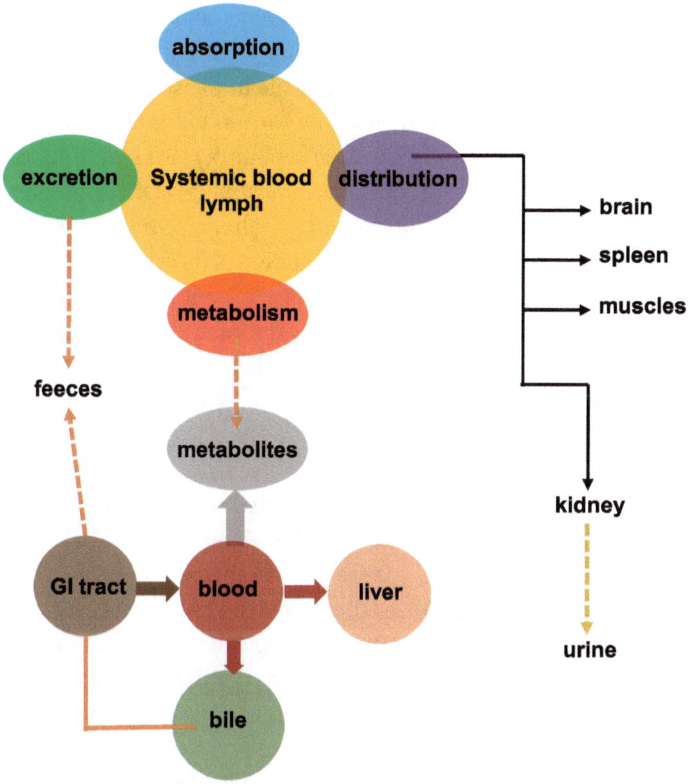

Figure 151: Schematic representation of the four underlying pharmacokinetics processes.

https://doi.org/10.1515/9783110605495-005

absorbed cellularly and exert their effects on the organism, a process governed by pharmacodynamics. This includes the action profile, the dose–response relationship, and the mechanism of action, which largely depend on the surface chemistry of the nanoparticles. Understanding both pharmacokinetics and pharmacodynamics is essential for developing safe, effective, and targeted nanoparticle-based therapies. Pharmacokinetics provides important insights into the optimal method of administration, dosage, and frequency and duration of treatment to ensure efficacy to maximize while minimizing toxicity. Pharmacodynamics, on the other hand, examines the biochemical and physiological effects of nanoparticles on the body.

5.1 Administration routes of nanoparticles

Systemic and local administration of nanoparticles involves the distribution of nanoparticles throughout the patient body so that they can reach the target tissue and specific organs in the body [818]. The systemic administration of drugs or nanoparticles is predominantly carried out either by intravenous, subcutaneous and intramuscular injection, or by oral ingestion [819] (Figure 152). Inhalation, intratumoral injection, and dermal penetration are more localized routes of administration. Intravenous injection allows the nanoparticles to spread rapidly throughout the body, making them particularly effective against cancer cells because they can take advantage of the increased permeability and retention effect in tumor tissue. The subcutaneous and intramuscular routes offer alternatives with varying absorption rates and patient acceptability. Oral administration of nanoparticles leads to their absorption in the gastrointestinal tract. The efficacy of the oral route of administration depends on the chemical stability of the nanoparticles subjected to the harsh gastrointestinal conditions. Inhalation is the preferred route of administration for the treatment of respiratory diseases because it efficiently delivers the nanoparticles to the bronchi and lungs. Pulmonary administration of nanoparticles usually results in localization to the lung surface, avoiding systemic side effects [820]. Intratumoral administration directly targets tumor tissues. Transdermal delivery, in which the nanoparticles are applied to the skin, is advantageous for local treatment [821]. The skin, as the largest organ of the body, acts as a barrier between the environment and the internal organs of the human body. When applied topically, sufficiently small nanoparticles can penetrate the skin and cause systemic damage. Oral and intravenous administration of nanoparticles have an inherently faster systemic effect compared to transdermal administration. Nanoparticles that enter the bloodstream can be transported to the liver. The sequestration of nanoparticles in the liver is often undesirable for therapeutic applications. Nanoparticles that are trapped in liver macrophages are no longer therapeutically available because they cannot reach the target site of the disease [822].

In general, the distribution and clearance of nanoparticles in the body is influenced by photophysical properties of nanoparticles, such as their size, shape, and surface

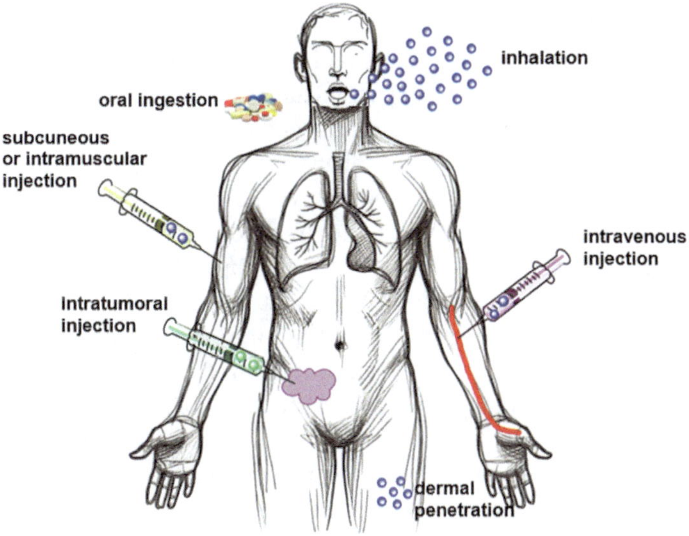

Figure 152: Routes of administration of nanoparticles.

structures, as well as the presence of proteins and other biomolecules in the bloodstream and the cellular composition of the targeted tissue or organ. Understanding the factors that influence the delivery route of nanoparticles is important for the development of safe and effective nanoparticle-based therapies. Efficient systemic delivery of nanoparticles to organs and tumors is hampered by overcoming biological barriers as they enter and travel through the human body [823, 824]. For example, during their passage through the bloodstream, nanoparticles are exposed to opsonization by adsorption of proteins. Nanoparticles in organs can be removed by cells of the mononuclear phagocyte (MNP) or reticuloendothelial (RES) system. In addition, the endothelium of the target organ protects the underlying cells from interaction with the nanoparticles in the bloodstream. On the other hand, the dense extracellular matrix and high interstitial fluid pressure in the abnormal microenvironment of the tumor prevent deep penetration of the nanoparticles into the tissue.

5.2 Pharmacokinetics of administration routes

5.2.1 Intravenous injection

Intravenous administration of nanoparticle suspensions is carried out either as a single dose or as an infusion directly into the bloodstream [825]. The most reproducible method of systemic administration is the peripheral intravenous injection. From the injection site the nanoparticles are transported with the bloodstream via the venous circulation directly into the right ventricle, then into the pulmonary circulation, then into the left atrium and from there into the left ventricle [823, 824]. The nanoparticles are then released into the systemic circulation, which rapidly distributes them to the tissues of the mononuclear phagocytic system (MPS) or reticuloendothelial system (RES), which includes the spleen, liver, kidney, lung, and cervical lymph nodes. The distribution in certain organs of intravenously administered nanoparticles depends on their chemical composition, size, charge, shape, and surface structures [826]. The hydrodynamic diameter of nanoparticles is considered their "true size" in biological fluids because it accounts for the adsorption of serum proteins and the surrounding solvation envelope. Nanoparticles with hydrodynamic diameters in the range of 20–30 nm have the longest blood circulation time, which increases their chance of reaching target organs or cancerous tissue. Larger nanoparticles (>100 nm) are easily internalized by phagocytic cells and accumulate in the liver or spleen. Nanoparticles spend about one eighth of their circulation time in the liver structures. The large blood reservoir in the liver provides more opportunities for the liver cells to interact with and take up the nanoparticles. Nanoparticles move in arteries and veins at velocities between tens and hundreds of cm per second, while, in the liver sinusoid of rats, they slow down to hundreds of μm per second [827]. The slower nanoparticles can diffuse into the vessel walls and thus interact with the liver cells.

The mononuclear phagocyte system (MPS) consists of monocytes and macrophages that circulate in the blood and are found in organs such as the liver, spleen, lymph nodes, bone marrow, lungs, and the brain [828]. Macrophage progenitors develop from hematopoietic stem cells in the bone marrow and are then released into the circulation as monocytes. These circulating monocytes extravasate through the endothelium and then migrate into the tissues, where they differentiate into several major macrophage subsets based on their anatomic location and functional phenotype. The major specialized tissue-resident macrophages are Kupffer cells in the liver, alveolar macrophages in the lung, osteoclasts in the bone marrow, and histiocytes in the interstitial connective tissue. Distinct macrophage populations also exist in secondary lymphoid organs, including marginal zone macrophages in the spleen and subcapsular sinus macrophages in the lymph nodes. Tissue-specific macrophages also patrol highly immune-privileged organs such as the brain (microglia), eyes, and testes. All of these tissue-distributed macrophages cleanse the body of bacteria, viruses, abnormal and old cells, and nanoparticles through phagocytosis.

The clearance of nanoparticles from the bloodstream occurs mainly through the liver and spleen [829]. If the doses of nanoparticles injected are too high, the macrophages in the liver and spleen eliminate only a fraction of the nanoparticles from the bloodstream. The larger proportion of nanoparticles accumulates in other macrophage-rich tissues such as the lungs and fatty tissue.

The liver is the largest internal organ in the body. It is approximately 10 times heavier than the spleen and 4 times heavier than the lungs. The liver macrophages form the largest scavenger cell population in the body and as such have direct access to nanoparticles in the blood [829]. The filtered blood then passes through the endothelial walls of the venous sinusoids and finally returns to the bloodstream. Macrophages in the reticular plexus of the spleen and bone marrow also act as efficient filters for the removal of nanoparticles from the bloodstream. For example, an artery entering the splenic medulla terminates in small, highly porous capillaries that carry blood to the marginal and red pulp zones. Blood is forced through the venules in the red pulp regions into the collecting veins. The macrophages in the marginal zones and the red pulp of the spleen phagocytose the nanoparticles.

The entire macrophage population in the body consists of 80–90 % Kupffer cells, while the red pulp of the spleen contains only 5–15 % [829, 830]. Thus, Kupffer cells play a crucial role in the innate immune response [831]. The remaining macrophages are found in the bone marrow, lungs, pituitary gland, and pancreas. Macrophages in the reticular plexus of the spleen and bone marrow also act as efficient filters for the removal of nanoparticles from the bloodstream. For example, an artery entering the splenic medulla terminates in small, highly porous capillaries that carry blood to the marginal and red pulp zones. Blood is forced through the venules in the red-pulp regions into the collecting veins. The macrophages in the marginal zones and the red pulp of the spleen phagocytose the nanoparticles. The filtered blood then passes through the endothelial walls of the venous sinusoids and finally returns to the bloodstream to absorb the nanoparticles directly from the bloodstream.

Nanoparticles spend about one-eighth of their circulation time in the liver structures. The large number of liver macrophages increases the likelihood that nanoparticles in the liver will be taken up by macrophages. Nanoparticles can enter the liver through both the portal vein and the hepatic artery. The portal vein carries blood containing nutrients and waste from the digestive system to the liver. The hepatic artery supplies the liver with oxygen-rich blood. This blood is mixed in the hepatic sinusoids before leaving the liver through the central vein (Figure 153). The localization of Kupffer cells in the hepatic sinusoids allows them to efficiently phagocytose nanoparticles and other immunoreactive material pathogens that enter the portal circulation from the gastrointestinal tract.

The macrophages near the hepatic artery and portal vein are more phagocytic than the macrophages near the central vein. The macrophages in the periportal vein are larger and have higher lysosomal enzyme activity [822].

The fate of intravenously administered nanoparticles depends on their chemical composition, size, charge, shape, and surface structures. Kupffer cells are the most active phagocytes in the uptake of SPIONs [829]. Before SPIONs can be internalized by macrophages and in particular by Kupffer cells, they are opsonized, which occurs in the bloodstream immediately after intravenous injection [825]. The most common opsonins are immunoglobulins (Ig G and M), plasma proteins such as complement components C3, C4 and C5, and blood serum proteins such as albumins, fibrinogen, C-reactive protein, fibronectin, C-reactive protein, and type I collagen. Coated with these opsonins, SPIONs are recognized by macrophages. The SPIONs induce the process of phagocytosis through receptor–ligand interactions. This initiates the GTPase-mediated cascade that

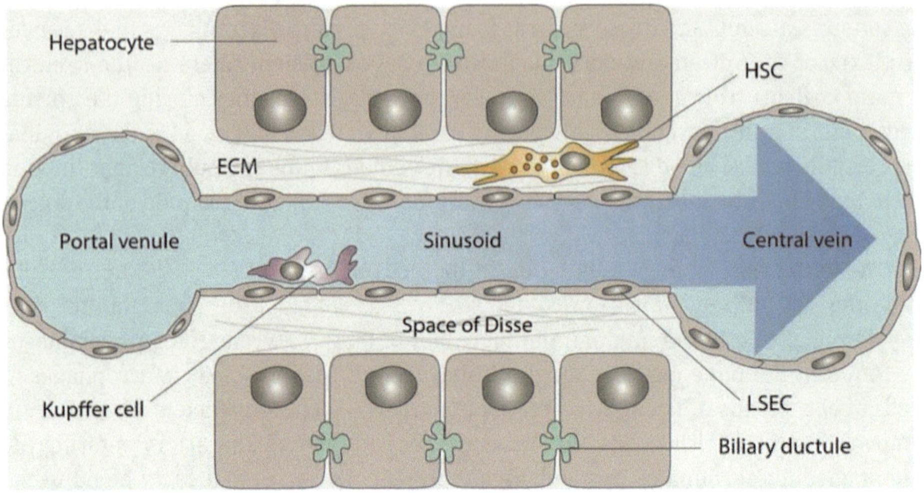

Figure 153: Schematic illustration of a liver sinusoid that transports blood from the hepatic artery and portal vein to the central vein; Kupffer cells and monocyte-derived macrophages are in the lumen of the fenestrated capillaries; hepatic stellate cells (HSCs) are located in the space of disse [832].

triggers actin assembly, followed by invagination and constriction of the macrophage cell membrane. The resulting phagosomes transport the SPIONs through the cytoplasm and make them accessible to lysosomes after actin depolymerization. Fusion with lysosomes to form phagolysosomes occurs through the formation of a continuous bilayer of two separate lipid membranes. Phagolysosomes contain various hydrolytic enzymes such as proteases, nucleases and lipases in an acidic environment (pH 4.5–5) that degrade and metabolize internalized species such as SPIONs.

De Jong et al. [833] investigated the size-dependent uptake of spherical AuNPs in different organs of Wisdar rats. To this end, 10 nm, 50 nm, 100 nm, and 250 nm AuNPs were administered to the rats by intravenous injection. The gold content in the liver, spleen, kidney, testis, thymus, heart, lung, and brain was quantitatively determined using inductively coupled plasma mass spectrometry (ICP-MS). Twenty-four hours after intravenous injection, approximately 46 % of the 10-nm AuNPs, 44 % of the 100-nm AuNPs- and only 31 % of the 250-nm AuNPs were found in the liver, while only 1–2 % accumulated in the spleen. These results were essentially confirmed by other research groups. Fischer et al. [834] studied organ uptake of mercaptoundecanoic acid and BSA stabilized ZnS coated CdS quantum dots with a mean hydrodynamic diameter of 25 nm in Sprague–Dawley rats. Ninety minutes after intravenous injection, 36.4 % and 99.5 % accumulated in the liver, respectively, while the spleen contained a maximum of 2 % of the injected quantum dots. Arnida et al. [455] investigated the influence of the shape of PEGylated AuNPs on the circulation time in the bloodstream and subsequent distribution in the organs of mice with ovarian tumors. For this purpose, they used 50- nm spherical AuNPs and 10 nm × 45 nm AuNRs, which were administered to the mice by intravenous injection.

Both the gold nanorods and the spherical particles were detected in all organs examined, including the brain. The circulation time of the PEGylated nanorods was significantly higher than that of the PEGylated gold nanospheres. Both the spherical and rod-shaped nanoparticles accumulated in the liver for up to 24 h, with the spherical AuNPs being taken up by the liver to a considerably greater extent. After one week, the amount of gold detected in the liver decreased significantly. AuNPs excreted from the liver were detected in the feces. The longer circulation time and lower uptake of AuNRs correlated consistently with the higher accumulation of AuNRs in tumor tissue.

Elucidation of the elimination pathway and process of nanoparticles is crucial for medical applications such as medical devices or nanocarriers for drug delivery. One important reason is to avoid damage to the excretory organs caused by nanoparticles. In this context, Xie et al. [835] investigated the distribution and excretion of intravenously administered SiO_2 NPs in the body of ICR mice and Sprague–Dawley rats. The APTES-modified SiO_2 NPs with sizes of 20 and 80 nm are poorly degradable and were therefore metabolized to a negligible extent in vivo. For this reason, these inert SiNPs should be excreted. The SiO_2 NP surfaces were labeled with radioactive iodine (^{125}I) and fluorescent dyes (CF™680 or FITC) for the quantitative determination of the accumulated excretion rate and the visualization of the biodistribution. The radiolabeled SiO_2 NPs were injected intravenously into ICR mice and Sprague–Dawley rats. The distribution and accumulation of SiO_2 NPs in the lungs, heart, liver, spleen, kidneys, stomach, and intestine was monitored over 8 h by fluorescence microscopy (Figure 154). The accumulated excretion rate for the biliary and renal pathways was determined by measuring the radioactivity of the feces and urine using a γ-counter at various time intervals af-

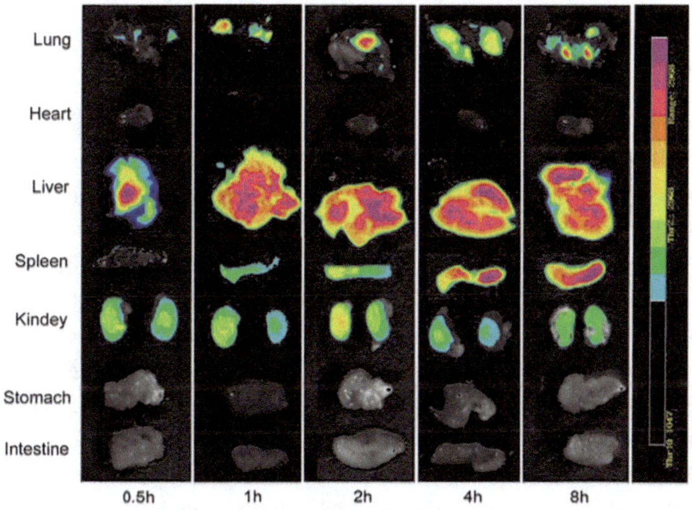

Figure 154: Fluorescence-microscopy study of the distribution of CF™680 labeled SiO_2 NPs in diverse mice organs in dependence on the time: 0.5, 1, 2, 4, and 8 h after intravenous injection [835].

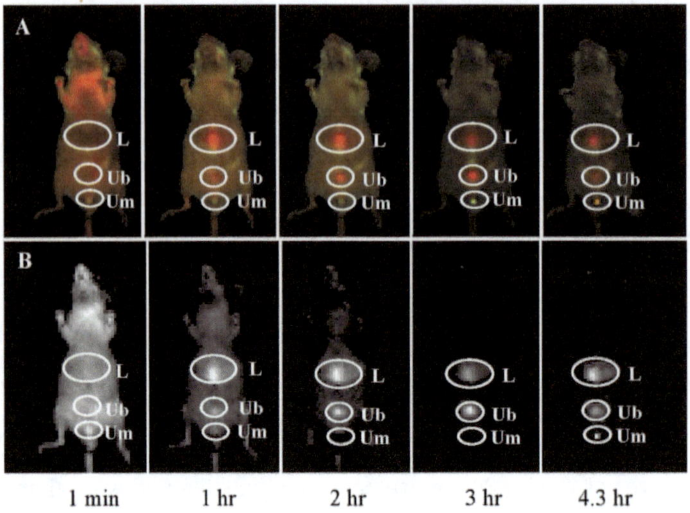

Figure 155: Real-time in vivo fluorescence microscopy imaging at various time intervals after intravenous injection of OH–SiO_2 NPs into mice (A, superimposed image; B, unmixed RuBpy dye doped SiNPs image; liver (L), urinary bladder (Ub), and urinary meatus (Um)) [836].

ter intravenous injection of the ^{125}I-labeled SiO_2 NPs. The accumulated excretion rate is expressed as the percent injected dose (%ID) and is related to the respective hemoconcentration. The fluorescence microscopy images in Figure 155 substantiate the even distribution of SiO_2 NPs in the liver, whereas in the spleen, SiO_2 NPs accumulated predominantly in the white pulp. As being involved in immune defense, the white pulp consists of lymphatic tissue containing B and T lymphocytes. The accumulated excretion rate values obtained showed that the 20-nm and 80-nm SiO_2 NPs were mainly eliminated by the kidneys, with the excretion rate of the 80-nm SiO_2 NPs exceeding that of the 20-nm SiO_2 NPs.

These results confirm those reported by He et al. [836]. The authors investigated the in vivo biodistribution and urinary excretion of SiO_2 NPs terminated with either hydroxy or carboxy moieties or coated with PEG (i. e., OH–SiO_2 NPs, COOH–SiO_2 NPs, and PEG-SiO_2 NPs). The differently surface-modified SiO_2 NPs were administered intravenously to BALB/c mice and Sprague–Dawley rats. The results of the in vivo fluorescence microscopy imaging studies show that the intravenously injected OH–SiO_2 NPs, COOH–SiO_2 NPs, and PEG-SiO_2 NPs with an average size of ca. 45 nm were all cleared from the bloodstream and accumulated in the organs. The blood circulation time (180 ± 40 min) of PEG-SiO_2 NPs exceeds that of OH–SiO_2 NPs (80 ± 30 min) and COOH–SiO_2 NPs (35 ± 10 min), which may be the reason for the lower uptake by the liver than that of OH–SiO_2 NPs and COOH–SiO_2 NPs. Furthermore, fluorescence microscopy images showed that all intravenously injected SiO_2 NPs were partially excreted via the kidneys.

However, a reliable elucidation of the in vivo distribution of SiO_2 NPs in the human organism and their subsequent elimination requires a quantitative analysis of absorption, distribution, excretion, and pharmacokinetics of SiO_2 NPs. The in vivo pharmacokinetics and long-term biodistribution of SiO_2 NPs were quantitatively investigated for the first time as a function of the administered dose using accelerator mass spectrometry (AMS) [837]. AMS detects rare, long-lived isotopes such as ^{14}C in a biological matrix with extremely high sensitivity. Malfatti et al. [837] demonstrated that ^{14}C-labeled nanoparticles in living tissue can be quantitatively determined with an accuracy of 1–3 % over the range from approximately 10 pmol ^{14}C to 1 attomol ^{14}C. In this study, ^{14}C-labeled spherical SiO_2 NPs (^{14}C–SiO_2 NPs) with an average diameter of 33 nm were synthesized via alkaline hydrolysis of TEOS in a microemulsion process, followed by surface modification with ^{14}C-acrylic acid and allyltrimethoxysilane. BALB/c mice were given a single intravenous injection of ^{14}C–SiO_2 NPs through an implanted jugular vein catheter. After intravenous administration, ^{14}C–SiO_2 NPs were preferentially distributed in the spleen, liver, kidney, lung, and cervical lymph nodes. ^{14}C–SiO_2 NPs were also detected in bone tissue, whereas these nanoparticles were hardly detectable in brain, muscle, and adipose tissue. The latter result is due to the poor blood supply to muscle and adipose tissue compared to organs. In addition, ^{14}C–SiO_2 NPs are unable to cross the blood–brain barrier. Quantitative analysis of the tissue distribution of ^{14}C–SiO_2 NPs showed that spleen and liver contained the highest amount of ^{14}C–SiO_2 NPs, followed by kidneys, lymph nodes, and lungs. The ^{14}C–SiO_2 NP concentrations in the different organ tissues reached their maximum concentration value (C_{max}) 2 h after intravenous administration and then decreased at different velocities over a period of 24 h (Figure 156(a)). The spleen contained the highest ^{14}C–SiO_2 NP concentration with approximately 300 µg/g tissue, while the corresponding C_{max} values of the liver, kidney, and cervical lymph nodes were approximately 230 µg/g, 57 µg/g, and 34 µg/g, respectively. Long-term biodistribution measurements showed that a significant amount of ^{14}C–SiO_2 NPs remained in the organ tissues 8 weeks after administration (Figure 156(b)). In the liver and spleen, ^{14}C–SiO_2 NP concentrations decreased by 72 % and 57 %, respectively. The observed increase in the ^{14}C–SiO_2 NP concentration in the organ tissues between 2 and 7 days, and in particular, the steep rise in the ^{14}C–SiO_2 NP concentration in the spleen between 14 and 28 days, were explained by a redistribution of ^{14}C–SiO_2 NPs, probably via the blood circulation and lymphatic drainage. The long-term persistence of ^{14}C–SiO_2 NPs in the organs was attributed to uptake by macrophages. The biliary elimination route takes place via the liver and bile.

The nanoparticles were phagocytosed by the reticuloendothelial system (RES), transported by macrophages to the liver, and then translocated into the bile and subsequently excreted with the feces. Alternatively, renal clearance resulted in the execretion of nanoparticles in the urine. Figure 157 illustrates temporal development of both, biliary and renal elimination of ^{14}C–SiO_2 NPs. Approximately 70 % of the total amount of ^{14}C–SiO_2 NPs excreted within 24 h after intravenous administration was found in the urine, while the residual 30 % was detected in the feces. Urinary excretion occurred with a half-life ($t_{1/2}$) of 1.9 h, while fecal excretion of ^{14}C–SiO_2 NPs was significantly

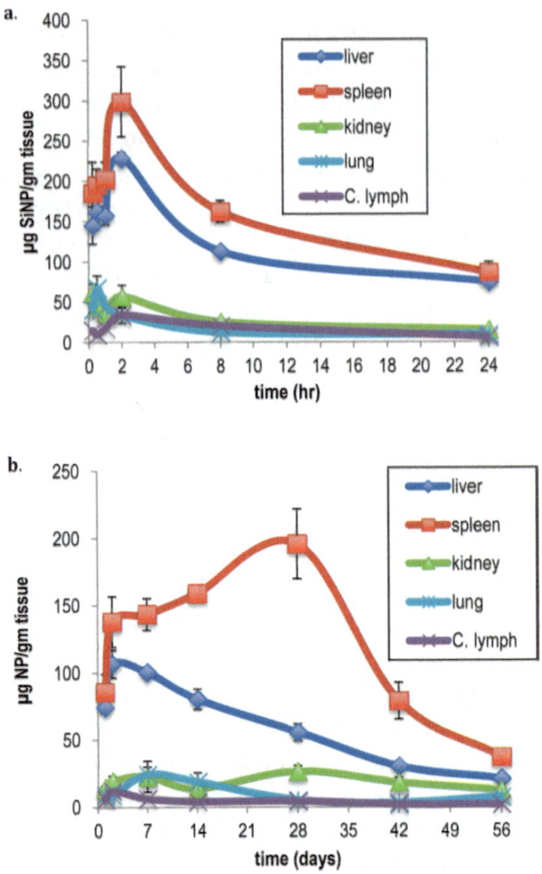

Figure 156: Mean tissue concentration of $^{14}C–SiO_2$ NPs over 24 h after a single intravenous injection into BALB/c male mice (a); mean tissue concentration of $^{14}C–SiO_2$ NPs over 56 days after a single intravenous injection into BALB/c male mice (b) [837].

slower with a $t_{1/2}$ of 21 h. These results indicate that the 33-nm $^{14}C–SiO_2$-NP were mainly excreted via the kidneys which is in contradiction to the size limitation of renal filtration inside the glomerular capillaries. Choi et al. [838], and independently Du et al. [770], found out that the critical hydrodynamic diameter for renal excretion of nanoparticles is 6 nm. The kidney is also responsible for the filtration and excretion of nanoparticles, which considerably reduces the side effects of the therapeutic use of such nanoparticles in distant organs. Therefore, the elucidation of the influence of the kidney on the pharmacokinetics of nanoparticles and the assessment of the nephrotoxic potential is of eminent importance [605, 839]. The kidney contains more than one million nephrons, which are the most important functional and structural units for the homeostasis of body fluids and electrolytes in the intracellular, extracellular, and extravascular compartments. Each nephron consists of the renal corpuscle or glomerulus and the proximal

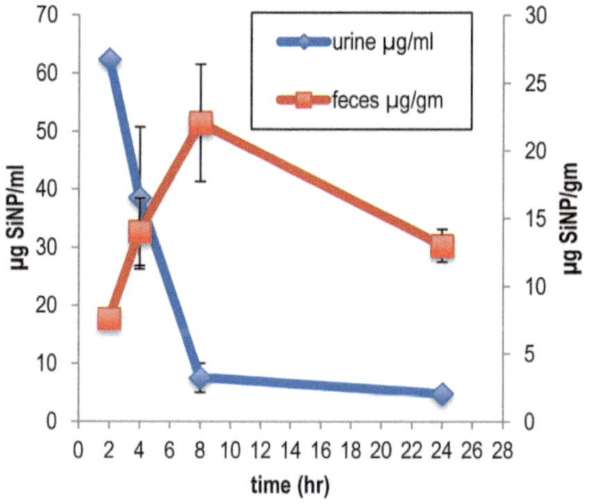

Figure 157: Mean concentration of $^{14}C-SiO_2$ NPs in urine and feces over 24 h after a single intravenous injection into BALB/c male mice [837].

and distal tubules. Blood is transported to the glomeruli by the renal vasculature. The nephrons selectively filter carbohydrates, proteins, and sufficiently small nanoparticles from the blood.

The primary filtrate passes through fenestrated capillaries into Bowman's space, which surrounds the glomerulus and collects the filtrate. The filtrate is then transported to the proximal and distal renal tubules where nutrients, water, and ions are reabsorbed and waste products are excreted. Nanoparticles injected intravenously into a living mouse or rat were transported through the renal hilum into the blood vessels of the nephrons and eventually excreted via the ureter and then the urinary bladder with the urine. Driven by glomerular hydrostatic pressure, the nanoparticles should pass through the glomerular capillaries which are the blood filtration sites in the nephrons. The filtered blood leaves the glomeruli through the efferent arterioles [840]. Renal filtration occurs through the glomerular filtration membrane (GFM), which is located between the glomerular capillary lumen and Bowman's space (Figure 158). The GFM is composed of four layers: the endothelial glycocalyx, the endothelial cells, the glomerular basement membrane (GBM), and the podocytes. The negatively charged GBM consists mainly of collagen type IV, laminin and proteoglycans and has pores of 2–8 nm in size. On the side facing the glomerular capillary lumen, the GBM is covered by a monolayer of endothelial cells with 70–90 nm sized fenestrations. The other side of the GBM, within Bowman's space, is covered with podocytes separated by openings 4–11 nm in size. The podocyte layer is covered by a approximately 200-nm-thick glycocalyx. The nanoparticles filtered from the blood (ultrafiltrates) must pass through the 2–8 nm pores of the GBM, which limits their hydrodynamic diameter to 6 nm [838, 841]. The ultrafiltrates are directed into Bowman's space, leaving the bloodstream and reaching the renal pelvis, where they are directed through the ureters to the bladder. This ultimately implies that

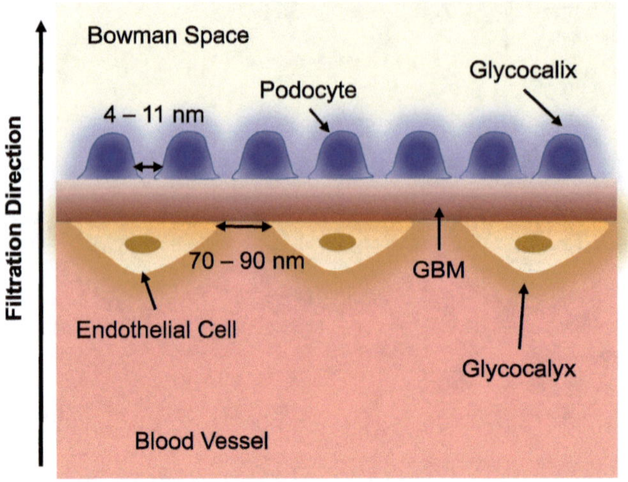

Figure 158: Scheme of the glomerular filtration membrane (GFM) located between the glomerular capillary lumen and Bowman space; the GFM is composed of an endothelial glycocalyx, endothelial cells, the glomerular basement membrane (GBM) and podocytes.

the filtration limit is determined by the size of the GBM pores. Nanoparticles with an hydrodynamic diameters of 6–100 nm can pass through the endothelial fenestrations, but will be trapped between the endothelium and the GBM [841]. However, not only Malfatti et al. but also other research groups have detected nanoparticles in urine that could not have overcome the GFM due to their too-large size [841, 842]. It has been hypothesized that these unfiltered nanoparticles migrate directly through endothelial and epithelial cells into the renal tubules and are excreted in the urine [842]. So far, the mechanisms of uptake of large nanoparticles (>10 nm) from the bloodstream and subsequent transport through epithelial cells into the tubule lumen have not yet been clarified.

Another and more plausible explanation for the renal excretion of large nanoparticles is based on the assumption that the nanoparticles, when in close contact with the GBM, partially damage the membrane.

Renal clearance is influenced not only by the size of nanoparticles but also by their shape and charge, as illustrated in Figure 159 [839]. Nanoparticles with a hydrodynamic diameter of less than 6 nm undergo globular filtration, while nanoparticles with a hydrodynamic diameter larger than 8 nm are not eliminated via the kidneys but accumulate in the liver. On the other hand, glutathione stabilized Au nanoclusters consisting of 10–18 Au atoms exhibited a four- to nine-fold reduction in renal clearance efficiency, because they were initially trapped and retained in the endothelial glycocalyx [605, 770]. The surface charge of nanoparticles is another important determinant of renal filtration behavior. The GFM also acts as a charge selective filter. This is due to the endothelial glycocalyx, which contains anionic heparan sulfate proteoglycan as a building block. Consequently, cationic nanoparticles can more easily surmount the GFM, while the passage of

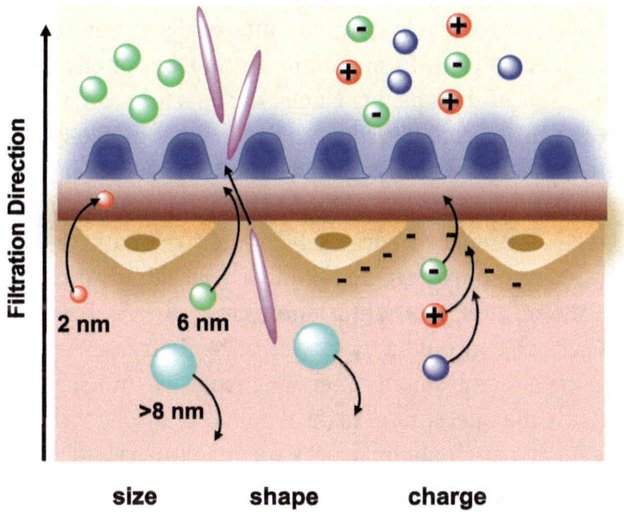

Figure 159: Scheme of the glomerular filtration membrane (GFM) illustrating the filtration of the nanoparticles in the glomerulus in function of size, shape, and charge.

negatively charged nanoparticles through the GBM is hindered by repulsive interaction forces. Experimental evidence for this hypothesis was provided by Choi et al. [838], who prepared anionic, cationic, and uncharged nanoparticles with hydrodynamic diameters smaller than 10 nm by coating ZnS-encapsulated CdSe QDs with dihydrolipoic acid, cysteamine and DHLA-linked PEG, respectively. In fact, the cationic ZnS@CdSe QDs were observed to pass through the negatively charged GBM and podocytes more efficiently than the neutral QDs followed by the anionic QDs. These results were confirmed by Du et al. [605].

The influence of nanoparticle shape on the glomerular filtration mechanism was investigated using single-wall carbon nanotubes (SWCNTs) and graphene oxide layers (GOLs) [843, 844]. The SWCNTs had a diameter of 1 nm and an average length of 200–300 nm, while the GOLs reached a planar dimension of about 1 µm and were 2–3 nm thick. Both, the SWCNTs and GOLs crossed the GBM and were detectably excreted in urine. Apparently, the respective thickness of 1–2 nm allowed the passage of 1D and 2D nanostructures through the 2–8-nm large pores or slits of the GBM, indicating a blood flow-driven transport of the SWCNTs and GOLs along their longitudinal axis or plane.

5.2.2 Subcutaneous and intramuscular injection

In subcutaneous administration, nanoparticles and other therapeutics are injected into the interstitial area under the dermis of the skin [845]. The interstitium or subcutaneous tissue is poorly supplied with blood and consists mainly of negatively charged glycosaminoglycan, salts, and plasma proteins stabilized by a network of collagen fibers

[846]. The reduced blood flow results in a relatively slow but uniform absorption rate of (nano)therapeutics. Conventional subcutaneous formulations are aqueous or oily solutions, suspensions or simple emulsions of therapeutic agents. Nanoparticle-containing subcutaneous formulations can also be injected directly into the lymph. Ultimately, the size of the nanoparticle determines the route of absorption. Nanoparticles smaller than 10 nm can penetrate the blood capillaries and are then transferred to the systemic circulation. By contrast, nanoparticles (10–100 nm) are more likely to enter lymphatic vessels during their passage through the interstitium due to their limited permeability through the vascular endothelium. The rate of interstitial diffusion of nanoparticles is determined by their size, shape, and surface hydrophilicity, as well as by their interactions with endogenous components in the interstitium. In general, nanoparticle transport in the lymph is driven by the fluid from the interstitium to the lymphatic vessels, with the flow velocity being location dependent. Larger nanoparticles are unable to pass through the 100-nm-wide water channel in the interstitium and therefore remain in the vicinity of the injection site.

Lymph is formed from the interstitial fluid and the plasma exudate, initially flowing into the smallest lymphatic vessels, the lymphatic capillaries, which are widely distributed throughout the body in close proximity to the blood capillaries [847]. These lymph capillaries are often blind ending vessels (Figure 160), but they can also be organized as a network of interconnected vessels. Lymph flows, driven by local hydrostatic forces, within an initial network of lymphatic vessels. External tissue pressure favors forward flow. Substances dissolved in the lymph or dispersed nanoparticles diffuse freely within the vessels. The lymph capillaries drain into larger collecting vessels that transport the lymph via the lymph nodes to the thoracic duct. The thoracic duct then ascends through the thoracic cavity and finally empties the systemic circulation at the junction of the left internal jugular vein and the left subclavian vein.

The detection sensitivity of 20-nm dextran-coated SPIONs as MRI contrast agents in axillary lymph nodes of Wistar rats was evaluated using susceptibility-based gradient echo pulse sequences [848]. For this purpose, the SPIONs were administered to rats in ferrofluids with Fe concentrations of 2.2 mg/mL by intravenous and subcutaneous injection. MR lymphography measurements showed that a dose of only 0.028 mg iron per kg was sufficient to visualize axillary lymph nodes in the case of the subcutaneously administered SPIONs and a minimum dose of 0.16 mg Fe/kg for the intravenously injected SPIONs. These results show that the subcutaneously administered SPIONs reach the target organ via the lymphatic system more efficiently than the intravenously administered SPIONs, which reach the target organ via the bloodstream. On the other hand, Johnson et al. [849] used the MRI contrast agent Endorem, which consists of dextran-coated SPIONs with hydrodynamic diameters of 80 and 150 nm consisting of 4–6-nm iron oxide cores, to detect metastases in sentinel lymph nodes of patients. Endorem was injected subcutaneously into the breast of cancer patients. Despite their large hydrodynamic diameter, the SPIONs were transported via the lymphatic vessel system to the sentinel

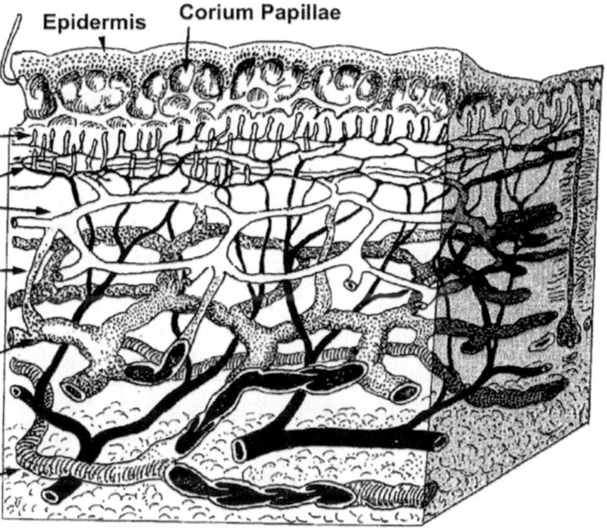

Figure 160: Block diagram of cutaneous lymphatic and vascular networks in human skin [847].

lymph nodes, where they remained. Surprisingly, the SPIONs spared the metastases, which may be due to their large hydrodynamic diameter.

Intramuscular administration of nanoparticles is performed by injecting liquid nanoparticle dispersions into the highly perfused muscle tissue. This provides a moderately rapid onset of action, 5–10 minutes, which should be faster than subcutaneous administration [845]. Furthermore, intramuscular injections are more suitable than intravenous injections when a longer duration of action is required or when oily solutions are to be administered. In addition, by injecting appropriately dosed nanoparticles deep into a large muscle, tissue absorption of nanoparticles is expected to be much faster and more reliable than with oral administration. To complicate matters, the uptake and transport of nanoparticles into muscle tissue is critically dependent on the size, shape and surface chemistry of the dispersed nanoparticles, as well as muscle blood flow and muscle activity. To date, there are no reliable in vivo results on the intramuscular administration of nanoparticles [850, 851]. Leclerc et al. [850] investigated the accumulation and efficacy of negatively charged 70-nm gold-core SiO_2-shell nanoparticles in the testicles of male C57BL/6 mice after intramuscular injection. To track the SiO_2@Au NPs in the testicles, high-resolution lateral methods such as TEM and fluorescence microscopy and the element-specific detection method ICP-AES were used. All these experimental investigations failed to provide evidence of these nanoparticles inside the testicles. The results were consistent with the histological examination of the testicles. Fu et al. [851] obtained similar findings from a study of absorption, distribution, excretion, and toxicity of 110-nm SiO_2 NPs administered via various routes, intravenous, intramuscular, hypodermal injection and oral ingestion, to female ICR mice. In contrast to the intravenous injection and oral ingestion, the intramuscularly injected SiO_2 NPs were unable to overcome the various biological barriers in the liver. Moreover, the intramuscularly and subcutaneously injected SiO_2 NPs induced an inflammatory reaction around the injection sites, while SiO_2 NPs showed good biocompatibility after oral ingestion and intravenous injection.

5.2.3 Dermal administration

The skin consists of the epidermis and dermis and lies above the hypodermis [852]. The human epidermis is a stratified squamous epithelium and is classified as either thin or thick skin [853]. In thick skin, the epidermis consists of five layers of epithelial cells, whereas the epidermis of thin skin has only 4 layers of epithelial cells [854]. The epidermis of thick skin consists of the stratum corneum, stratum lucidum, stratum granulosum, stratum spinosum, and stratum basale (Figure 161). The stratum corneum is approximately 10–20 µm thick and consists of 15–20 sublayers of flattened, denucleated keratinocytes called corneocytes. The corneocytes are filled with keratin filaments. The keratinized envelope of the corneocytes is formed by keratins, filaggrin, loricrin, involucrin, and some other proteins, which are cross-linked by transglutaminases. Intracellular keratins and intercellular lipids adhere to the approximately 10–15-nm thick corneal envelope. The stratum basale consists of a single layer of basal keratinocytes anchored to the basal lamina

The stratum lucidum consists of one or two cell layers of nucleusless keratinocytes that are further flattened and densely packed with eleidin, a clear protein rich in lipids.

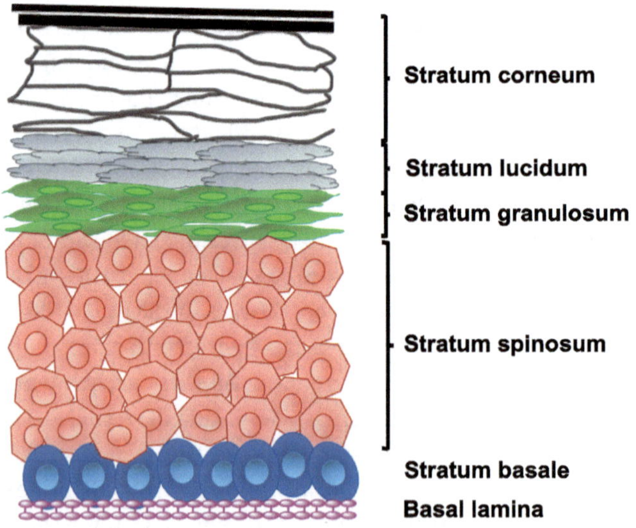

Figure 161: Illustration of the five epithelial layers in the epidermis of thick skin.

The stratum granulosum consists of 3–5 sublayers of keratinocytes. The granular appearance of the stratum corneum is due to the keratins and keratohyalin (lamellar granules) in the keratinocytes. These keratohyalin granules are composed of profilaggrin, a 400–500 kDa protein. Profilaggrin breaks down to ~26–48 kDa filaggrins as keratinocytes move from the stratum granulosum to the stratum corneum. The stratum spinosum consists of 8–10 sublayers of polygonal keratinocytes. It derives its spiky morphology from protrusions formed by desmosomes connecting the keratinocytes. The spiky keratinocytes produce water-repellent glycolipids. In addition to keratinocytes, the stratum spinosum also contains Langerhans cells, which are tissue-resident macrophages capable of recognizing and absorbing antigens and other foreign substances such as nanoparticles in the epidermis. The stratum basale consists of a single layer of basal keratinocytes anchored to the basal lamina. The stratum basale also contains Merkel cells and immunological cells such as melanocytes, which are melanin-producing cells derived from the neural crest. The stratum basale is separated from the dermis by the basal lamina. The dermis supplies nutrients to the epidermis and contains blood and lymph vessels, nerves, and sebaceous glands. The transfer of fluid from the dermis to the epidermis, with the epidermis acting as a very efficient barrier that reduces fluid loss. The dermis is stabilized by an extracellular matrix composed of collagen, elastic microfilaments, fibronectin, and proteoglycans. The dermis also contains endothelial cells, fibroblasts, lymphocytes, and macrophages. The hypodermis as being the subcutaneous tissue consists mainly of collagen and adipose tissue and serves to insulate the body.

In principle, transdermal delivery of nanoparticles is limited by the stratum corneum, which acts as a thin barrier within the epidermis. The shape, size, and surface chemistry of the nanoparticles are the key parameters that determine the efficiency and route of skin permeation. For example, skin permeation can be enhanced by surfactants such as oleic acid [855, 856]. Ultrasmall nanoparticles can penetrate the epidermis through various routes, including the intercellular pathway through corneocytes and the intracellular pathway around corneocytes. The larger ones permeate via the transappendageal route which involves passage through sweat glands and the follicles [857] (Figure 162). It is currently accepted that ultrasmall nanoparticles of 4 nm and 6 nm can penetrate intact skin, slightly larger ones of less than 20 nm can potentially penetrate intact and damaged skin, while nanoparticles between 21 and 45 nm can only penetrate damaged skin [853].

Hsiao et al. [858] investigated the penetration and distribution of ultrasmall gold nanospheres (AuNSs) and gold nanorods (AuNRs) in the skin using an in vivo rat model. Since oleylamine as a lipophilic surfactant promotes skin permeation, the surfaces of 5-nm and 10-nm gold nanospheres were stabilized with poly(ethylene glycol)-oleylamine (PEG-OAm). For confocal fluorescence microscopy studies, the PEG-OAm AuNPs were conjugated with the fluorescent dye FITC. The PEG-OAm-AuNS and -AuNR suspensions were administered topically to Sprague–Dawley (SD) rats by soaking dressings with the respective AuNP suspension and applying them to the skin surface. The permeation and

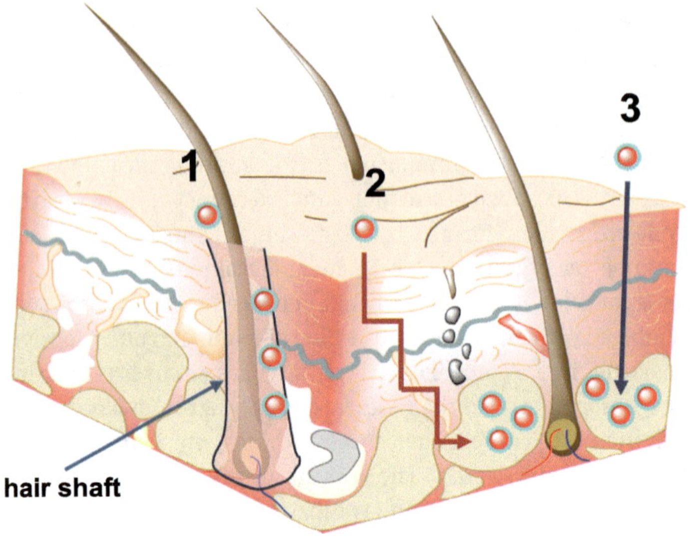

Figure 162: Illustration of the penetration pathways of nanoparticles into the skin: (1) appendageal pathway, (2) intracellular pathway, (3) intercellular pathway.

distribution of the PEG-OAm-AuNSs and AuNRs in the rat skin were examined 1 h after topical application using confocal laser scanning microscopy. The localization of AuNPs in subcutaneous adipose tissue was visualized by transmission electron microscopy. Fluorescence micrographs showed that the permeation properties of ultrasmall AuNPs are indistinguishable for sizes of 5 and 10 nm, but depend significantly on the shape. AuNRs are deposited in the epidermis at concentrations 1.9 times higher than that of AuNSs. The tracking and concentration of AuNSs and AuNRs in the subcutaneous tissue is both evidence and a measure of skin penetration. The TEM images show that the AuNRs have a 1.7 times larger concentration in subcutaneous adipose tissue than the AuNSs. In addition, AuNRs were detected in higher concentrations than AuNSs in hair follicles. Nanoparticles enter the hair follicles via the transappendageal route, which favors nanoparticles with diameters smaller than 20 nm [854]. By comparison, Baroli et al. [855] showed that maghemite and iron nanoparticles with diameters less than 10 nm passively penetrated the skin through the stratum corneum lipid matrix, partially aggregated at the stratum corneum interface, and reached the viable epidermis to a very small extent. In this study, the skin penetration behavior of 6-nm tetramethylammonium hydroxide-stabilized SPIONs (TMAOH-SPIONs) was investigated by exposing human skin specimens to PBS solutions of the TMAOH-SPIONs for 24 h. To analyze the translocation profile of SPIONs in the skin layers, untreated and TMAOH-SPION-treated skin samples were systematically examined using light microscopy, resistivity measurements, ICP-AES, TEM, and EDS-SEM. The measurement results show that the ultrasmall nanoparticles penetrate both the hair follicle and the stratum corneum, which was expected. However, only a few nanoparticles reach the epidermis, which was disappointing. Complete penetration of the nanoparticles into the skin could not be demonstrated, making them unusable as nanocarriers for drugs, for example. This disappointing result can be explained by the efficient formation of μm-sized agglomerates, which could also be detected by DLS measurements. On the other hand, dead skin samples may also be unsuitable for studies of nanoparticle skin permeation because they lack the elasticity and transport properties of living tissue.

Transdermal delivery of nanoparticles can be realized by a simple incision in the skin, by tape stripping, and by low-frequency sonophoresis with the support of chemical penetration enhancers, for example, using amphiphilic substances such as oleic acid [859]. The use of these methods, and especially their combination, significantly increases the permeability of the stratum corneum by partially disrupting its lipid bilayer structure, which is highly questionable for medical applications. Oleic acid alters the surface of the stratum corneum by extracting endogenous membrane components, while tape stripping as a physical-penetration enhancer damages the surface structures of the stratum corneum by physical ablation [859]. Low-frequency sonophoresis uses low-frequency ultrasound to improve the permeability of the skin to drugs using a low-frequency ultrasound frequency (20–100 kHz). The operating principle of the low-frequency sonophoresis is based on the mechanical interaction of sound waves with matter, resulting in the rapid growth and subsequent collapse of symmetric and asymmetric oscillating bubbles. The symmetric bubble collapse produces shock waves, while the asymmetric bubble collapse generates high-velocity jets. Both shock waves and high velocity jets can damage the stratum corneum through pore formation [860]. Lee et al. [859] used 2–10 nm sized $LaNO_3$ NPs. Hairless mice were treated with low frequency ultrasound and the 1–2 μm pores induced in the surface of the stratum corneum were visualized by scanning electron microscopy (SEM) and TEM. Despite their very small size, $LaNO_3$ NPs did not penetrate intact skin but agglomerated on the surface of the stratum corneum. The authors investigated the effect of oleylamine, tape stripping, and low-frequency ultrasound (25 kHz), first separately and then in combination, on the penetration behavior of $LaNO_3$ NPs into the skin of nude mice. Treatment of mouse skin with oleylamine allowed $LaNO_3$ NPs to invade the intercellular spaces of the stratum corneum and the stratum corneostratum granulosum interface (Figure 163(a)–(c)). Tape stripping removes the top layer of the stratum corneum, resulting in a similar but less efficient penetration behavior of the $LaNO_3$ NPs. The $LaNO_3$ NPs were detected in the uppermost stratum corneum layer in the corneocytes as well as in the intercellular domains in the stratum granulosum and stratum corneum. When the physically or chemically damaged mouse skin was then exposed to low-frequency ultrasound, the $LaNO_3$ NPs penetrated the skin significantly more efficiently and deeply. The $LaNO_3$ NPs were homogeneously distributed throughout the epidermis (Figure 163(d)–(f)).

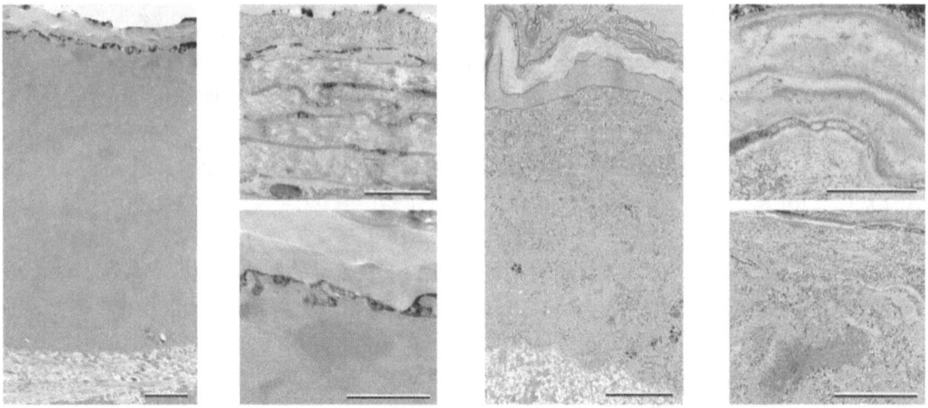

Figure 163: TEM images of mouse skin sections after treatment with oleylamine (a–c) and after treatment with low-frequency ultrasound and oleylamine (d–f) in the presence of $LaNO_3$ NPs; (a) skin treated with oleylamine and $LaNO_3$ NPs; (b) magnified section of TEM image (a): $LaNO_3$ NPs in the intracellular domains of corneocytes in the upper stratum corneum; (c) magnified view of the stratum corneum–stratum granulosum interface with $LaNO_3$ NPs; (d) after low-frequency ultrasound treatment followed by oleylamine treatment; (e) enlarged section of the stratum corneum with $LaNO_3$ NPs in the intracellular areas of the corneocytes in the upper stratum corneum; (f) enlarged section of the underlying epidermis with diffusely distributed $LaNO_3$ NPs (scale bars = 2 µm (a, d), 1 µm (b, c, e, f)) [859].

5.2.4 Inhalation

The respiratory system is categorized into the upper and lower respiratory tracts. The upper respiratory tract includes the nasal cavity, pharynx, and larynx, whereas the lower respiratory tract includes the trachea, bronchi, and lungs (Figure 164) [818]. In the human lung, the airways are structured as a dichotomous tree with approximately 8-million terminal branches, the alveolar ducts and sacs. Each bronchus branches into bronchioles, which are connected to the surrounding area with 4.8×10^8 alveoli with a total surface area of 130 m^2 [861]. The alveoli are responsible for gas exchange. The airways consist of a gradually thinning columnar epithelium, with the bronchial epithelium and bronchiolar epithelium having thicknesses of 3–5 µm and 0.5–1 µm, respectively [862]. The upper respiratory tract allows air to pass through and protects the lower respiratory tract. Air, including airborne particles, enters the trachea via the nasal and oral cavities, passes through the tracheobronchial tree and enters the alveoli, where the size of the particles largely determines their distribution throughout the respiratory system. Micrometer-sized particles or nanoparticle aggregates with sizes between 5 and 30 µm remain in the nasopharyngeal region or enter the tracheobronchiolar region, which is composed of secretory and ciliated epithelial cells [863]. Secretory epithelial cells produce the mucus, while the ciliated epithelial cells remove the particles and pathogens trapped in the mucus for excretion by the mucociliary escalator. This primary defense system functions for microparticles (>6 µm). 20-nm nanoparticles and smaller nanopar-

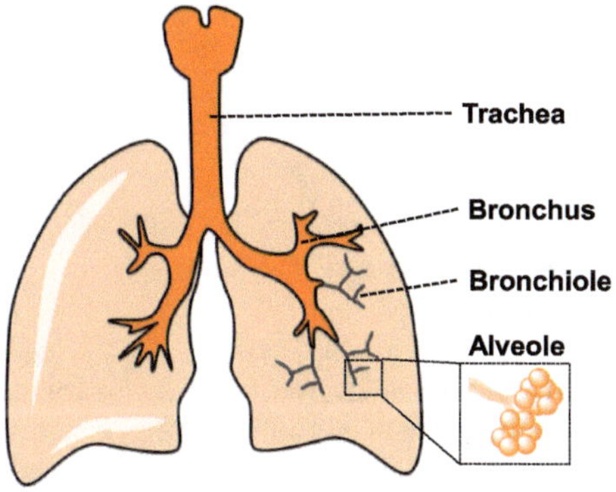

Figure 164: Schematics of the lower respiratory tract.

ticles arrive in the alveolar region. The distance between the air in the alveolar lumen and the capillary blood flow is less than 400 nm. Ultrasmall nanoparticles may efficiently penetrate the alveolar epithelium and capillary endothelial cells and cross through the extremely thin air–blood barrier to reach the cardiovascular system and other internal organs. In the conducting airways, airborne nanoparticles encounter relevant epithelial barriers by mucus-producing goblet cells (GC) in the bronchus and Clara cells (CC) in the bronchiolus, as well as by alveolar epithelial cells type II in the aveoli (Figure 165) [864]. The alveolar epithelium consists of alveolar type I (AT-I) and type II (AT-II) types that are integrated into the epithelium [865]. The AT-I cells are specialized for gas exchange with the underlying endothelium at the air–blood barrier. The surfactant-coated AT-II cells are able to proliferate and differentiate into AT-I cells. On the inner surface of the alveoli and bronchioles reside the macrophages (MA), which are also part of the lung defense system. They are reinforced by dentritic cells, which act as antigen-presenting cells in the conducting airways and in the distal lung. The epithelial basement membrane (MB) of the endothelial cell layer lies close to the endothelial cell layer in the thinnest areas of the gas exchange surface, increasing the efficiency of gas exchange between epithelial and endothelial cells. The surfactant-coated AT-II cells regulate the penetration of foreign bodies into the body by exerting their innate immune response and contribute to alveolar epithelial repair and regeneration. In the healthy lung, ATII cells orchestrate host defense mechanisms by secreting surfactant proteins, which are responsible for protecting the lung from pathogens. AT-II cells contain a high density of mitochondria and are equipped with specialized apical microvilli. They represent approximately 15 % of all cells, but only about 5 % of the alveolar surface area of the healthy adult human lung. The pulmonary surfactant is a lipid-rich film that spreads at the alveolar air-liquid interface [863]. It consists of about 80 % phospholipids, 10 % neutral lipids, and 10 % proteins. The role of their surfactant proteins is to maintain homeostasis in the distal lung and to reduce surface tension at the air–liquid interface

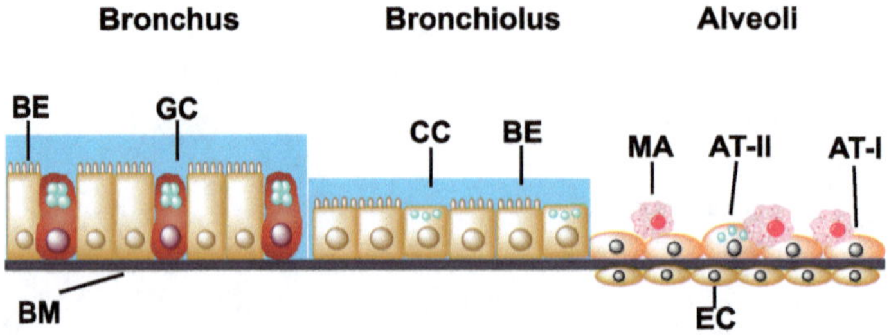

Figure 165: Schematic representation of the surfaces of the conducting airways with basement membrane (BM) and endothelial cells (EC) at the bottom; bronchus: bronchial epithelial cells with cilia (BE) and mucus-producing goblet cells (blue); bronchiolus: bronchial epithelial cells (BE) and mucus-producing Clara cells (CC); aveoli: alveolar epithelial cells type I (AT-I) and surfactant producing type II (AT-II) cells with alveolar macrophages (M).

of the lung. The surfactant barrier is the most permeable barrier in the human body with a thickness of only 0.1–0.2 μm.

Nanoparticle sizes smaller than 20 nm facilitate uptake into cells and transcytosis via epithelial and endothelial cells into the blood and lymphatic circulation [866]. This allows them to reach target sites such as bone marrow, lymph nodes, spleen, and heart. Ultrasmall nanoparticles can also enter the central nervous system and ganglia via translocation along the axons and dendrites of neurons. Surprisingly, 90 % of inhaled 1-nm nanoparticles are deposited in the nasopharyngeal compartment, only about 10 % in the tracheobronchial area and virtually none in the alveolar area. By comparison, 5-nm nanoparticles are equally likely to be deposited in all three regions. On the other hand, 20-nm nanoparticles accumulate with 50 % efficiency in the alveolar region (~50 %), while their deposition efficiency in the tracheobronchial and nasopharyngeal regions is only about 15 %. These results show that the transfer pathways of nanoparticles to the extrapulmonary organs are very sensitive to nanoparticle size. Thus, nanoparticles can be distributed in the body by transcytosis across the epithelia of the respiratory tract into the interstitium and through the air–blood barrier directly into the bloodstream or via the lymphatic vessels. Nanoparticles that have diffused into the alveoli can be eliminated by alveolar macrophages through phagocytosis. The retention time of nanoparticles in the alveolar region that are cleared by macrophages is about 70 days in rats and up to 700 days in humans. Approximately 20 % of 15–20-nm and 80-nm nanoparticles are internalized and cleared by macrophages. This means that 80 % of these nanoparticles remain in the lungs.

Miller et al. [867] investigated the influence of inhaled AuNPs on the development of cardiovascular disease in humans. For this purpose, young healthy volunteers were exposed to aerosols of AuNPs by inhalation for 2 h. The size effect of AuNPs on their transport from the lung to the bloodstream and renal clearance was considered by using nanoparticles with a primary size of either 4 nm or 34 nm. The AuNPs were found to agglomerate in the aerosol to a mean aerodynamic diameter of 18 or 52 nm. The translocation kinetics of AuNPs in the volunteers was monitored by quantifying the AuNP content in repeated blood and urine samples using high-resolution inductively coupled mass spectrometry and Raman microscopy. AuNPs were detected in blood and urine within 15 min to 24 h after inhalation and were still detectable after 3 months. The concentration of 5-nm AuNPs in blood and urine was significantly higher than that of 3-nm AuNPs. These results show that the inhaled AuNPs rapidly enter the bloodstream, which is attributed to a passive transport mechanism. In a complementary study, AuNPs in the size range of 2–200 nm were administered by inhalation via the lungs of a well-validated mouse model of atherosclerosis. Translocation of AuNPs was detected in blood and liver. In particular, AuNPs less than 10 nm in size were found in high concentrations in the liver, confirming that the liver is the primary route for clearance after translocation. In addition, AuNPs were studied after inhalation in surgical specimens from patients at risk for stroke and carotid artery disease. Translocation and accumulation of AuNPs in atherosclerotic plaques of these patients were confirmed. Importantly, the nanoparticles also accumulated in inflamed vascular sites, but not in non-atherosclerotic blood vessels.

Kreyling et al. [868] investigated pulmonary deposition and subsequent distribution of AuNPs to other organs in variously aged rats ranging from newborn to adulthood. The AuNPs used in this study are 20 nm in size and have been radioactively labeled with ^{195}Au. Wistar–Kyoto rats were intratracheally inhaled with ^{195}Au-AuNPs by negative pressure ventilation for 2 h. Immediately thereafter, the lungs were excised, inflated, and dried, and the deposition pattern of ^{195}Au-AuNPs was analyzed using single-photon emission computed tomography (SPECT), and autoradiography. In addition, the distribution of ^{195}Au-AuNPs in major organs and body tissues was quantified at intervals ranging 1–28 days. As shown by the SPECT images, intratracheal inhalation resulted in accumulation of ^{195}Au-AuNPs predominantly in the caudal lung, independent of age. Approximately 30 % of the ^{195}Au-AuNPs were deposited on the epithelia of the respiratory tract, resulting in rapid mucociliary clearance and subsequent transport to the gastrointestinal tract. The majority of ^{195}Au-AuNPs diffused into the alveoli, where they were translocated from the epithelium to the interstitium within 24 h. Long-term clearance of ^{195}Au-AuNPs in the alveolar region was mainly carried out by alveolar macrophages (LT-MC). By comparison, translocation of ^{195}Au-AuNPs across the air–blood barrier or via the lymphatic system was found to be less efficient. The ^{195}Au-AuNPs left the bloodstream to enter secondary organs and tissues, resulting in persistent retention. As expected, the highest ^{195}Au-AuNPs content was detected in the liver. Surprisingly, the second highest level was found in soft tissues. The age-independent inhomogeneous ^{195}Au-AuNPs deposition was probably caused by negative pressure ventilation. In the context of these results, Tsuda et al. [869] hypothesized that the age-dependent permeability of the air-blood barrier in rodents and humans to inhaled AuNPs should be detectable in the time window of lung maturation. Structural alveolarization of the human lung, which has only 15 % of the alveoli of the adult lung immediately after birth, occurs mainly postnatally. In their study, the authors used a rat model to represent postnatal lung development. To systematically evaluate the permeability of the air–blood barrier for polyvinylpyrrolidone (PVP)-coated AuNPs of two sizes, 5 or 100 nm, in rats at different postnatal developmental stages; 3-day-old, 7-day-old, and 21-day-old neonatal rats were selected. The 3-day-old neonatal rat corresponds to a premature human baby at the stage of beginning alveolus, while the 7-day-old and 21-day-old rats have reached the developmental stage of human neonates and human infants, respectively. 5-nm and 100-nm PVP-coated AuNPs were administered to neonatal rats by intratracheal instillation. The content of AuNPs in pulmonary and extrapulmonary tissues was quantified using inductively coupled plasma mass spectrometry. As their research showed, the permeability of the air–blood barrier in the lungs of neonates for AuNPs was independent of their size, while in the lungs of adult animals, the translocation of smaller AuNPs across the air–blood barrier occurred more efficiently than for larger AuNPs. In addition, the translocation efficiency of AuNPs in 3-day-old neonates was shown to be significantly higher than that in older animals. Obviously, the systemic translocation efficiency of AuNPs in infant rats exhibited both, age, and size dependency. The difference in air–blood barrier permeability between neonatal and adult animals suggests that translocation of AuNPs in the immature lung may occur through different mechanisms compared to the mature lung. The authors attributed these results to functional differences between the developing and adult lungs. These include the epithelial barrier and its components, such as tight junctions, which suggests the possibility of paracellular transport of nanoparticles in infants. However, this also implies that inhaled nanoparticles can pose a significant toxic risk to infants, potentially damaging lung development and long-term lung function. Since a significant amount of inhaled nanoparticles can cross the tissue barrier of the infant lungs, these nanoparticles can also reach and damage other organs and tissues in the body.

Sadhukha et al. [724] investigated the efficacy of inhalable SPIONs for lung-cancer therapy using magnetic hyperthermia. SPIONs were synthesized by alkaline co-precipitation, then coated with Pluronic F127 and conjugated with EGFR-targeting peptides. The EGFR-targeting SPIONs were developed for the treatment of non-small-cell lung cancer (NSCLC) by means of magnetic hyperthermia. NSCLC is a lung tumor of epithelial nature and therefore accessible to inhaled nanoparticles. The EGFR-targeted SPIONs were administered to orthotopic mouse models of NSCLC either by inhalation or tracheal instillation. The maximum SPION content achieved in the mouse lung by inhalation was 73 μg, while the administration of SPIONs by tracheal instillation resulted in a 5-fold higher lung content. Finally, the results showed that pulmonary administration, either by tracheal instillation or inhalation, resulted in therapeutically effective concentrations of SPIONs in the lung with minimal exposure to other organs. For example, one hour after tracheal instillation of SPIONs, the average lung concentration of iron oxide was 1.2 mg/g tissue. By comparison, the average liver and blood concentrations were 130 μg/g and 27 μg/g tissue, respectively. The other organs, including spleen, kidney, and heart, contained less than 2 % of the administered SPION dose, demonstrating that inhalation and tracheal instillation are efficient routes of administration for increasing the local concentration of nanotherapeutics in the lung.

5.2.5 Oral ingestion

The human gastrointestinal tract (GIT) is divided into the upper and lower sections [818, 870]. The upper GIT consists of the esophagus, stomach, and duodenum, while the small intestine and large intestine form the lower GIT (Figure 166). The digestive system is responsible for the ingestion and digestion of food, absorption of nutrients and excretion of waste products. It also facilitates the breakdown and absorption of nutrients such as carbohydrates, peptides, and fats. The oral cavity, pharynx, and oesophagus are lined by stratified squamous epithelia, while the intestinal epithelium consists of a single layer of cells [870]. The intestinal lamina propria contains more antibody-producing B cells than any other organ in the body and is therefore immunologically active. Ma-

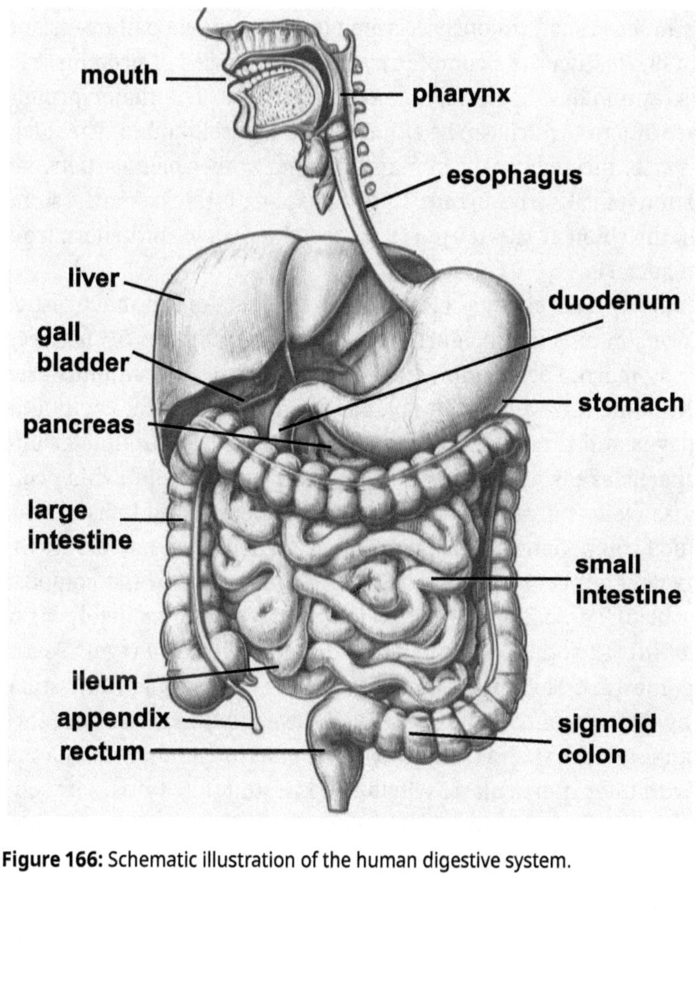

Figure 166: Schematic illustration of the human digestive system.

ture and differentiated intestinal epithelium is continuously replaced by epithelial invaginations (i. e., Lieberkühn crypts). The progeny cells that ascend from the base of the crypt to the villus undergo 4–6 rounds of cell division followed by maturation and differentiation. The final differentiated intestinal cells are absorptive enterocytes, goblet cells, enteroendocrine cells, and Paneth cells. Absorptive enterocytes are found only in the small intestine and are responsible for sodium, chloride, and nutrient absorption. Goblet cells secrete mucus into the lumen of the small and large intestine. Enteroendocrine cells are smaller and secrete various gut hormones. Paneth cells in the small intestine contain large apical secretory granules and express specific proteins, including lysozyme, tumor necrosis factor, and defensins. Enterocytes are 20- µm-high prismatic epithelial cells with rigid, densely packed microvilli on their apical surface. This gives them a greatly increased absorptive surface. The tips of the microvilli contain large negatively charged carbohydrate side chains of glycoproteins that form a continuous, filamentous brush border glycocalyx. The enterocytes are interconnected by tight junctions to form a barrier to the intestinal lumen.

The gastrointestinal epithelium is lined with a layer of mucus that protects the epithelial surfaces by efficiently trapping and rapidly removing pathogens and foreign nanoparticles [871]. The mucus is continuously secreted to both remove pathogens and lubricate the epithelium [872]. Mucus is a complex hydrogel composed of proteins, carbohydrates, lipids, salts, antibodies, bacteria, and cellular debris. The major protein components of mucus are mucins, which can be either secreted or cell-bound. The thickness of the mucus layer varies throughout the human GIT to maintain a balance between protective function and nutrient absorption rate. The mucus layer is thickest in the stomach at 40–450 µm and in the colon at 110–160 µm to protect the gastric epithelium from acid and the colon from bacteria.

In the gastrointestinal tract, pH changes along the various sections of the digestive system [818]. Gastric juice contains hydrochloric acid, which maintains a pH between 1.2 and 2.0 in an empty stomach. For example, when nanoparticles are administered to a patient orally at the same time as food, a bolus is formed in the oral cavity that travels down the esophagus and is transported through the esophageal sphincter into the stomach. The nanoparticles disperse in the highly acidic gastric fluid, which contains hydrochloric acid as well as digestive enzymes (gastric pepsin and lipase), salts, and partially digested food components. Transport of nanoparticles into and across the gastrointestinal mucosa depends on peristalsis, pH in the stomach, and the composition of the nanoparticle [870]. For example, iron oxide nanoparticles completely disintegrate in the stomach at pH < 2 and SiO_2 NPs partially dissolve, while inert AuNPs can pass through the entire intestine. The uptake of nanoparticles into the gastrointestinal mucosa is determined by the segmental heterogeneity of the epithelium. The residence time of acid-resistant nanoparticles such as AuNPs in the stomach depends on the nature of the food consumed with them, particularly whether it is liquid (fast transit) or solid (slow transit). Nanoparticles dispersed in gastric fluid pass through the pyloric sphincter at the bottom of the stomach into the small intestine, which consists of the duodenum,

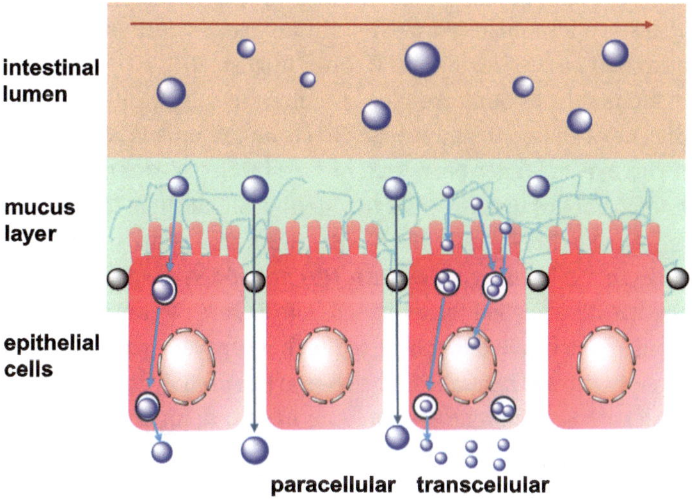

Figure 167: Schematic of paracellular and transcellular transport of nanoparticles of various sizes across the intestinal epithelium; paracellular transport occurs via gap junctions, whereas transcellular transport is associated with endocytic uptake and clearance of the nanoparticles.

jejunum, and ileum. The small intestine contains digestive enzymes (pancreatic amylase, lipase, and proteases), bile salts, phospholipids, and mineral ions at a pH of 5–7. Digestion and absorption take place in the small intestine. In addition to pH, peristalsis can also influence the physicochemical properties of nanoparticles as the pressure can reach up to 150 mm Hg. Most mucins in the mucus layer are glycosylated and therefore negatively charged. This means that positively charged nanoparticles can adhere to the negatively charged mucus layer through electrostatic interactions and thus do not reach the epithelium. Because the mucus layer has pores of 400 nm, it is easily penetrated by smaller uncharged or negatively charged nanoparticles. Certain food enzymes, such as papain and bromelain, which break through the mucus layer, facilitate the transport of nanoparticles to the underlying epithelium. The uptake mechanisms of nanoparticles in the lumen can be paracellular or transcellular by enterocytes, whereby nanoparticles are mainly transported via the transcellular route (Figure 167) [873]. Large nanoparticles and microparticles can also damage the intestinal epithelium through transcytosis, uptake, and persorption by microfold cells. In addition, nanoparticles can cross the villi through the gaps in their apical zone caused by a nanoparticle-induced dysfunctional process that alters the morphology of the epithelium. In contrast, paracellular uptake of nanoparticles is rare. Disease may have altered cell junctions, which may facilitate the passage of nanoparticles. When they pass through the epithelium, they enter the bloodstream and can therefore reach any organ. Nanoparticles that are not dissolved or taken up by the epithelial cells in the upper regions of the GIT reach the large intestine and are then excreted in the feces.

Inorganic nanoparticles are administered orally to humans as additives in food, medicines, toothpaste, and food packaging [871, 873]. Until August 2022, TiO_2 nanoparticles (E171) were used in foods such as mozzarella to enhance the color white. Because in vivo and in vitro studies have shown negative side effects on the mucus layer in the GIT [874], TiO_2 nanoparticles are no longer permitted in food [875], but are still used in medicines, toothpaste, and cosmetics despite emerging doubts about their safety. SiO_2 NPs (E551), Ag NPs (E174), AuNPs (E175), and SPIONs (E172) are still approved and increasingly used as additives in food. For instance, SiO_2 NPs are added to spices as anticaking agents. SPIONs and AuNPs are used to color food. AgNPs serve as antimicrobial agent in food packaging and are also added to food, despite their proven cytotoxic effect [864]. Undesirable and already proven harmful side effects are caused by the release of metal cations (e. g., Ag^+ and Fe^{2+}/Fe^{3+}) or aggregation. Another cause for concern is the penetration and adsorption of nanoparticles in or on the intestinal mucosa, which can cause inflammatory reactions. The microbiota can also be damaged by interaction with nanoparticles.

Askri et al. [876] examined the effect of 30 nm SPIONs on lethality, cognitive performance, and organ health in young male Wistar rats orally administered 100 mg/kg or 200 mg/kg SPIONs per kg body weight. The uncoated SPIONs synthesized by thermal decomposition are negatively charged and unfortunately tend to agglomerate. This study shows that orally administered SPIONs reduced the relative mass of organs such as liver, kidney, brain, lung, and stomach. The decrease in relative organ mass after oral administration of 100 mg/kg (D1) or 200 mg/kg (D2) doses of SPIONs is explained by the hypotrophy of these organs, which apparently correlates with the systemic effect of SPIONs. Ten days after oral administration of SPIONs, the animals were sacrificed. Liver sections were prepared and stained with hematoxylin-eosin (H&E). The histopathologic examination of the liver revealed iron deposition in hepatocytes and Kupffer cells, leading to inflammatory reactions and necrosis, which also manifests itself as shrinkage of the cell nucleus (i. e., nuclear pyknosis).

Correlative instrumental neutron activation analysis (INAA) and transmission electron microscopy (TEM) were used to quantitatively and qualitatively investigate the gastrointestinal uptake and subsequent organ and tissue distribution of AuNPs ranging in size of 4–58 nm after oral administration in mice [877]. Spherical AuNPs with diameters of 4 nm, 10 nm, 28 nm, and 58 nm were prepared using the Turkevich method. BALB/c mice were used as the animal model. The mice received a suspension of the different AuNPs at an approximate concentration of 2×10^5 parts per billion ad libitum in their drinking water for 7 days. INAA studies were used to quantify the translocation of AuNPs into the bloodstream and their subsequent distribution in the organs. As expected, a size-dependent uptake of AuNPs into the organism was observed: AuNPs with sizes between 4 and 28 nm passed through the gastrointestinal tract with significantly higher efficiency than 58-nm AuNPs. The 4-nm AuNPs were taken up into the bloodstream and subsequently into the brain, lungs, heart, spleen, and especially into the kidneys with significantly higher concentrations compared to the larger AuNPs, while the concentrations of the 28-nm AuNPs in the small intestine and stomach were 6-fold and 1.5-fold higher, respectively. In vivo TEM observations showed that intestinal uptake of 4-nm and 10-nm AuNPs occurred by persorption into dying enterocytes as they extruded from a villus. Persorption of the AuNPs was observed at the tips and sides of villi in the proximal ileum, distal ileum, and Peyer patches of the small intestine. Single nanoparticles and small agglomerates were detected in the gastrointestinal lumen. Figure 168 illustrates the persorption of 4-nm AuNPs. The TEM images in Figures 168(A) and B show an extruding enterocyte with a degrading nucleus and the lacteal basal in the lateral side of a villus in mouse ileum and a magnified section, respectively. The TEM image in Figure 168(B) depicts weakly visible AuNPs. The apical part of an extruding enterocyte and a magnified section are shown in Figures 168(C) and (D), respectively. The area marked with a square has been further magnified and shows two AuNPs (Figure 168(D), inset). These AuNPs were found in the cytosol at the apical part of the extruding enterocyte. The TE micrograph of the lateral side of the extruding enterocyte and a magnified section, indicated by a square, are presented in Figures 168(E) and (F). Here, the AuNPs resided in the cytosol, but in the lateral part of the extruding enterocyte. As suggested by these TE micrographs, the AuNPs were translocated from the apical side of the dead enterocyte to the lacteal walls and presumably from there into the bloodstream. Apparently, the AuNPs were found only in the cytosol of the disintegrating enterocyte, which they easily reached through membrane ruptures.

The number of AuNPs inside the dying, extruding enterocytes showed a significant size dependence. Especially in the basal part of the extruding enterocytes and near the lacteal walls, mainly 4-nm AuNPs were found. In contrast, intestinal uptake of 28-nm and 58-nm AuNPs could not be observed microscopically. These AuNPs were detected only in the gastrointestinal lumen and not in the intestinal epithelium. Persorption of small AuNPs was not observed either in the crypts or by the M cells in the Peyer patches of the ileum.

Chronic inflammatory bowel diseases (IBD) such as ulcerative colitis and Crohn's disease are on the rise as countries become more industrialized, increasing the need for anti-inflammatory drugs [878]. Depending on their size, shape, and surface chemistry, orally administered AuNPs have an anti-inflammatory effect and show particularly promising potential for the treatment of chronic inflammatory bowel diseases. AuNPs are not degraded in the gastrointestinal tract and may therefore alter the intestinal microbiota. Zhu et al. [879] investigated the anti-inflammatory potential of AuNPs of different sizes in ulcerative colitis. AuNPs with sizes ranging from 5 nm to 60 nm were used, which were stabilized with either citrate (CA), tannic acid (TA) or PVP. In vivo experiments were performed in C57BL/6 mice induced with ulcerative colitis by oral administration of dextran sodium sulfate (DSS) for 5 days. During this period, various surface-stabilized AuNPs with sizes between 5 and 60 nm were orally administered to these mice for 8 days.

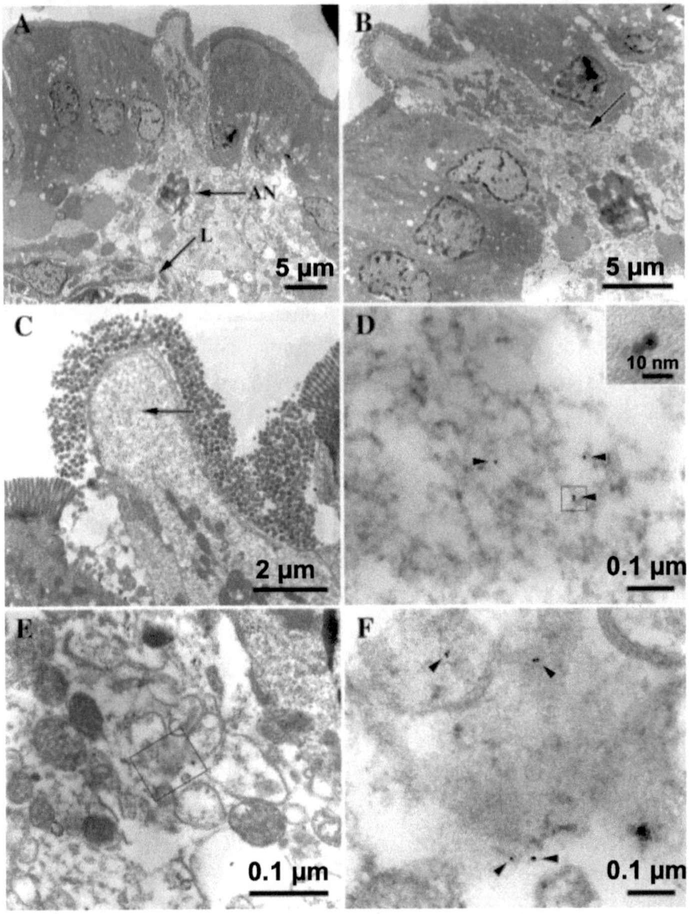

Figure 168: Persorption of 4-nm AuNPs through the intestine: (A) TEM image of an extruding enterocyte with the degrading nucleus (AN) and the lacteal (L) basal; (B) magnified section of (A); (C) apical part of an extruding enterocyte and (D) a magnified section, the area marked by a square is further magnified (inset) showing two AuNPs; (E) TEM image of a lateral side of an extruding enterocyte and (F) a magnified section, the area marked with a square in (E), the arrows in (F) point to AuNPs (Scale bars: (A, B) = 5 µm; (C) = 2 µm; (D, F, H, I, J, K, L) = 100 nm; (D, inset; H, inset; K, inset) = 10 nm; (E) = 1 µm; (G) = 2 µm) [877].

From day 3, the DSS-induced colitis mice treated with PVP- and CA-stabilized AuNP lost significantly less weight than the reference DSS-induced colitis mice. Histological analysis of colon tissue stained with hematoxylin and eosin (H&E) showed that oral administration of AuNPs partially healed the DSS-induced thinning and disrupted mucosal structure and reduced the area of inflammatory cell infiltration. The 5-nm-sized PVP- and CA-stabilized AuNPs showed significantly greater anti-colitis and anti-inflammatory efficacy in this mouse model of ulcerative colitis compared to the TA-stabilized AuNPs from day 6. These results were confirmed by determination of colonic and peripheral interleukin-6 and tumor necrosis factor-alpha levels, as well as peripheral leukocyte and lymphocyte counts. The differential effect of surface coating was explained by the higher bioavailability of PVP-stabilized AuNPs and the peroxynitrite detoxifying activity of citrate AuNPs. The higher bioavailability of the PVP-coated AuNPs is explained by the higher hydrophobicity of the PVP coating, which facilitates more efficient cellular uptake in mammalian cells. The CA-stabilized AuNPs, on the other hand, showed a higher peroxynitrite degrading effect in in vitro experiments, as they catalyzed the disproportionation of peroxynitrite to O_2 and nitrite. With a half-life of more than 1 s, peroxynitrite is a potent oxidant that can cause a variety of diseases, including neurodegenerative disorders, atherogenesis, and the cardiovascular disease diabetes.

To investigate the size dependence of AuNP bioavailability, TA-stabilized AuNPs ranging in size from 5 to 60 nm were orally administered to C57BL/6 mice. The 15-nm TA-stabilized AuNPs, for which a hydrodynamic diameter of 58 nm was determined, showed a peak accumulation in the kidneys. However, no size effect on anti-colitis and anti-inflammatory activities was observed for the TA-stabilized AuNPs.

High-throughput sequencing of fecal 16S rRNA indicates that TA-stabilized AuNPs can induce intestinal dysbiosis in mice by significantly reducing the biodiversity of the gut microbiome. These AuNPs were observed to decrease the Firmicutes/Bacteroidetes ratio, the abundance of short-chain fatty acid-producing bacteria, and the probiotic lactobacilli. These effects were attributed to catalytic-mimetic activities of the TA-stabilized AuNPs: The AuNPs catalyze the degradation of intraluminal hydrogen peroxide produced by probiotic bacteria to maintain a normal and homeostatic microbiota.

5.3 Nanotoxicology

Nanotoxicology plays a central role in elucidating and assessing the potential health and environmental risks of nanomaterials. The aim of toxicological assessment of nanomaterials is to ensure the safe and sustainable development of nanotechnology [880–882]. Nanotoxicology is a multidisciplinary field that focuses on the study and assessment of the interactions and potential adverse effects of nanomaterials on human and environment. To this end, fundamental knowledge from nanoscience, toxicology, chemistry, biology, and materials science is used to identify and elucidate the toxicological properties of engineered nanomaterials. An essential prerequisite for nanotoxicological studies is the characterization of the physicochemical properties of nanomaterials, including size, shape, surface area, surface charge, and chemical composition. These properties have a significant impact on their behavior in the biological environment and subsequent interactions with cellular components. Biomedical applications of nanoparticles as drugs, drug carriers, radiation enhancers, or imaging contrast agents require the intentional, direct ingestion or injection of nanoparticles into the body. It is essential to understand the biological degradation and elimination processes of nanoparticles in the human organism. The cellular uptake of nanoparticles, which may induce oxidative stress, inflam-

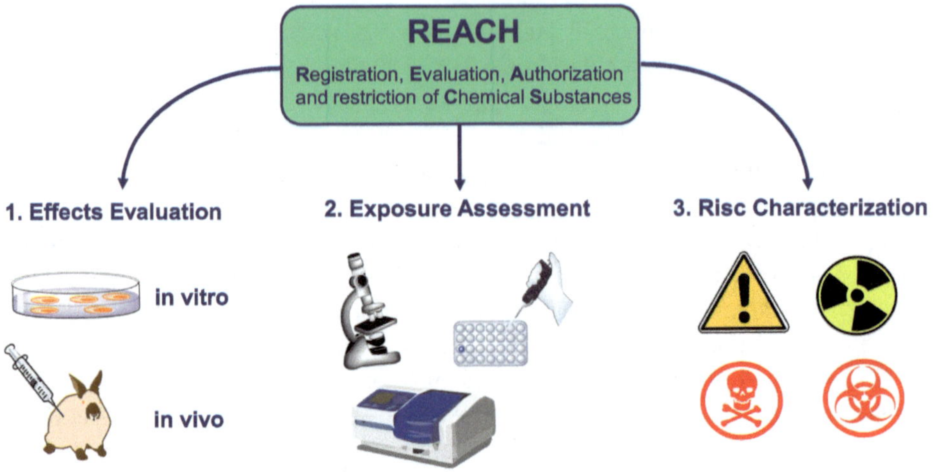

Figure 169: Risk assessment of nanomaterials following the European Chemical Regulation on REACH.

mation and cause damage to organelle integrity or genetic alterations. These undesirably toxic effects of nanoparticles ultimately causing damage to the entire organism. Understanding the processes underlying nanoparticle-induced toxicity is crucial for the development of safer nanomaterials and risk assessment strategies.

According to the European Chemicals Regulation on Registration, Evaluation, Authorization, and Restriction of Chemicals (REACH), the safety assessment of nanoparticles should follow the risk assessment methodology for conventional chemicals [883, 884]. This is based on (1) hazard assessment, (2) exposure assessment, and (3) risk characterization (Figure 169). Impact is assessed using the risk quotient. The risk quotient is considered being acceptable if the estimated exposure value is less than the safe nanoparticle concentration. This concentration is determined in in vitro and/or in vivo studies, which for nanoparticles also provide information on the main physicochemical parameters that may influence toxicity, such as size distribution, state, morphology, surface chemistry, reactivity, water solubility. and long-term stability. The exposure assessment must identify all potential sources of exposure to protect humans from unintentional exposure throughout the nanoparticle manufacturing process. The estimated exposure level is lower than the concentration of nanoparticles that does not cause adverse effects. Risk characterization is based on evidence of toxic tissue and organ damage at nanoparticle concentrations above the estimated exposure level.

As described in detail in Section 5.1, biomedical applications of nanoparticles to the human organism may use diverse routes of application including ingestion, inhalation, transdermal exposure, intravenous, subcutaneous, and intraperitoneal injection. In the case of oral administration, nanoparticles are absorbed throughout the gastrointestinal tract by passive transport mechanism such as transcytosis or endocytosis [885]. The translocation of nanoparticles into the systemic circulation depends on their resis-

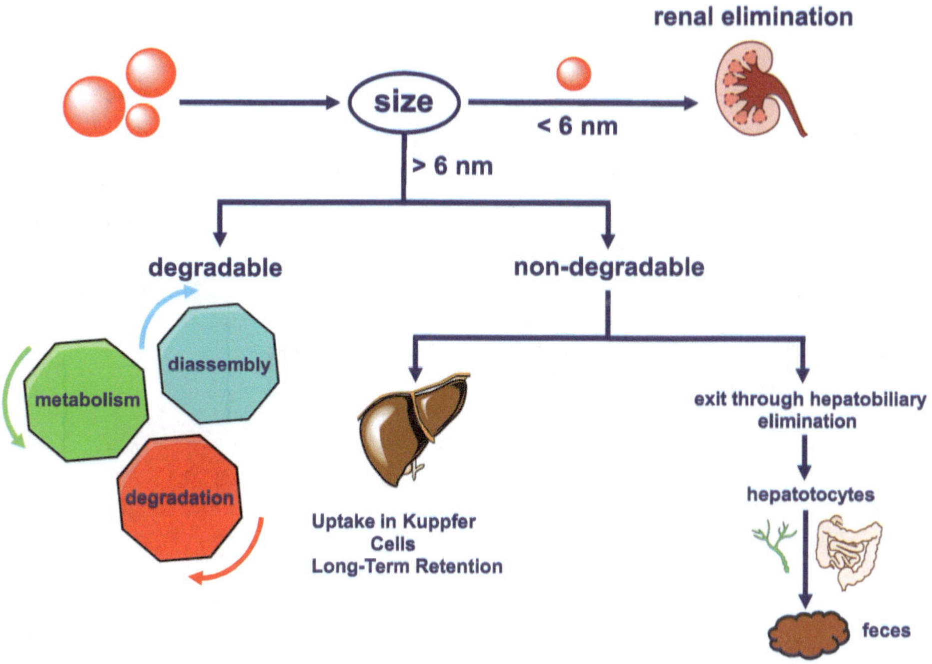

Figure 170: Role of size as a checkpoint at the entry of nanoparticles in the human body.

tance to large pH fluctuations and possible enzymatic degradation. The size of nanoparticles has a significant influence on their distribution and elimination. Nanoparticles smaller than 100 nm can be translocated into organs and other tissues, while nanoparticles smaller than 6 nm are excreted via the kidneys (see Section 5.2.1) (Figure 170) [816, 880]. Biodegradable nanoparticles are subject to diassembly or degradation and metabolism, while chemically stable nanoparticles are finally excreted via the liver and gastrointestinal tract in feces.

The internalization of nanoparticles by tumor cells may occur due to the enhanced permeability and retention effect (EPR). EPR is a phenomenon that significantly influences the distribution and accumulation of nanoparticles in tumor tissues. The EPR effect results from the unique pathophysiological properties of solid tumors and is attributed to the hyperpermeability of the rapidly growing vasculature in the absence of functioning lymphatic vessels [595, 603, 886]. The reason for this is the high oxygen and nutrient demand of cancer cells during tumor growth, which is met by the rapid and chaotic formation of new blood vessels. This neovascularization is characterized by structural and functional abnormalities of the blood vessels. The basement membrane is defective, and the endothelial tubes of immature vessels have fenestrated structures. Nanoparticles extravasate from the bloodstream into the tumor not only through gaps between the endothelial cells of the blood vessels, but also via transcellular pathways through vesiculo–vacuolar organelles. It is now known that gaps between endothelial

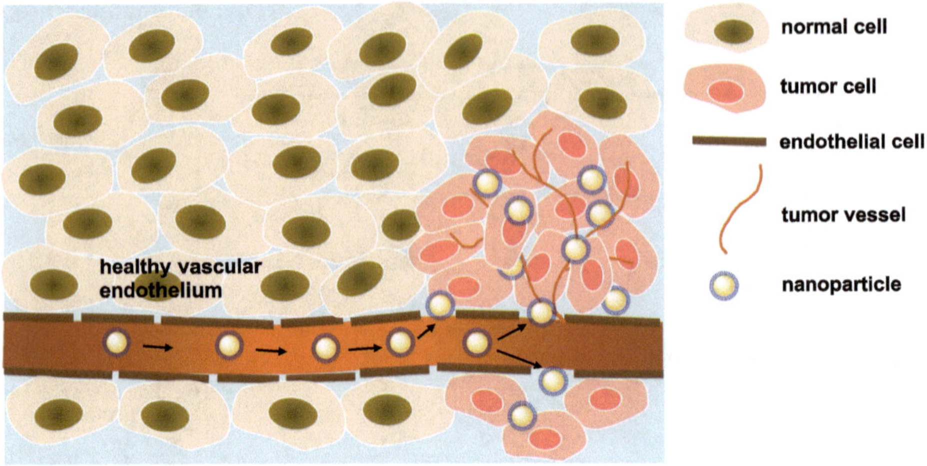

Figure 171: Schematic representation of the EPR effect.

cells in tumor vessels are rare, so that nanoparticle extravasation into the tumor occurs mainly via the transendothelial pathway. In addition, both the hyperpermeability of the blood vessels and the impaired lymphatic drainage contribute to increased interstitial fluid pressure, which in turn prevents convective removal of nanoparticles and thus ensures retention of nanoparticles in the tumor tissue (Figure 171). In addition, the EPR effect is promoted by the increased expression of inflammatory factors such as prostaglandins, bradykinin, nitric oxide, peroxynitrite, interleukin 1 beta, interleukin 2, interleukin 6, proteases, interferon gamma, VEGF, and HIF-1 alpha in tumor cells.

The circulation time of nanoparticles in the blood circuit increases with the size in the range of 10 to 100 nm. The EPR effect and therewith the uptake in the tumor tissue increases with rising circulation time. The pH, charge, and hydrophilicity of nanoparticles can influence their hydrodynamic diameter and thus their distribution in the organ system and their excretion. The filtration and elimination of nanoparticles by the vascular and lymphatic systems is also determined by the hydrodynamic diameter.

In general, positively charged nanoparticles exhibit higher cytotoxicity due to strong interactions with negatively charged biological surfaces, particularly intestinal mucus [816]. This hinders their absorption into the intestinal epithelium. Neutral nanoparticles, on the other hand, diffuse more efficiently through the intestinal mucosa. The uptake of nanoparticles into the intestinal epithelium depends on the interactions between the nanoparticles and the host cells and intestinal microbiota. Nanoparticles administered by inhalation can cross the alveolar–blood barrier and directly enter the systemic circulation [887]. Nanoparticles with a size of 20 nm are preferentially deposited in the alveolar region. Nanoparticles smaller than 55 nm penetrate the alveoli more efficiently than 200-nm nanoparticles. Nanoparticles injected intravenously enter the bloodstream directly. Once nanoparticles enter the systemic circulation, they can adsorb serum proteins, resulting in the formation of a protein corona. The formation of the protein corona interferes with the function of the nanomaterial and leads to opsonization, recognition by the immune system, and rapid clearance. However, surface-adsorbed serum proteins can reduce steric and electrostatic repulsive interactions between nanoparticles, leading to aggregation. Nanoparticle agglomerates can lead to vascular occlusion or thrombosis and ultimately embolism.

Nanotoxicology studies use in vitro, ex vivo, and in vivo models to assess the potential hazards associated with exposure to nanomaterials [881, 885]. In vitro studies employ cell culture systems to investigate cellular uptake, cytotoxicity, and the intracellular signaling pathways involved. Ex vivo studies utilize tissue slices or organotypic cultures to better mimic the physiological conditions of living organisms. In vivo experiments in animal models provide valuable insights into the systemic effects, biodistribution, and long-term toxicity profiles of nanomaterials. For risk assessment of nanoparticles, nanoparticles with a defined composition, narrow size distribution, uniform morphology, and surface chemistry should be produced as reproducibly as possible [888, 889]. The environmental and exposure conditions should also be as standardized and reproducible as possible. In vitro assays are used to gain an initial understanding of the complexity of the interaction of nanoparticles with 2D cell cultures. The problem is the lack of instrumentation, consistent protocols, and theoretical understanding, which limits the reproducibility of synthesis procedures, and the lack of standardized experiments with biomarkers, descriptive functional analysis, dosing, and statistics. In addition, cell cultures are sensitive to temperature fluctuations and changes in pH, nutrient, and waste concentrations [890].

Two-dimensional (2D) cell cultures are still the most commonly used in vitro tumor or tissue models for studying nanoparticle toxicity [888, 891, 892]. In adherent 2D cultures, cells grow as a monolayer in a culture flask or flat Petri dish [883]. These cells are usually obtained from expanded primary tumor cell clones. The advantage is very easy handling and straightforward production of 2D cell monolayers on different substrates. The adherent cells can be exposed to nanoparticles in the culture medium in a reproducible manner. For this purpose, preferably hydrophilic nanoparticles are dispersed in the culture medium by sonication. Cellular uptake efficiency and toxicity depend on nanoparticle size and size distribution, surface chemistry, and degree of agglomeration. However, 2D cell cultures are inadequate tissue and tumor models because they lack cell heterogeneity, 3D cell-to-cell communication in living tissue, and metabolic gradients. The oxygen, nutrient, and pH gradients, as well as the exchange of metabolites between neighboring cells in the 3D cell network of tumors, are responsible for tumor growth and metastasis. 3D in vitro cell cultures bridge the gap between 2D cell monolayers, and ex vivo culture of tumor slices and animal models. Multicellular tumor spheroids (MCTS) and organoids are generated by self-aggregation of cancer cells, which can be either primary tumor cells or, most commonly, established immortalized tumor cell lines [893]. MCTS and organoids have a 3D cellular architecture and mimic the pathophysiological gradients of avascular tumors and metastases. MCTS with sizes greater than 500 μm are characterized by an outer proliferating zone, an inner quiescent zone caused by limited distribution of oxygen, nutrients and metabolites, and a necrotic core that mimics the cellular heterogeneity of solid tumors in vivo. Organoids are 3D cell cultures that self-organize into functional organ-like subunits in vitro. They can be grown from primary cells or differentiated from stem cells. The morphological heterogeneity of organoids results from the intrinsic variability of primary or stem cells. The integration of MCTS, organoids or tumor tissue slices into a microfluidic platform [894, 895] allows the in vitro evaluation of the morphological heterogeneity of organoids.

Zebrafish are increasingly used for in vivo toxicity assessment of nanoparticles [896]. The genomes of zebrafish and humans show an overlap of about 70 %. The major developmental and physiological processes as well as the digestive, nervous, and cardiovascular systems are similar to those of humans. Since zebrafish embryogenesis takes place outside the womb and the embryos are transparent, the developmental phase can be easily observed. In addition, the zebrafish genome has been completely sequenced, making it possible to produce standardized microarrays for nanotoxicology studies. This allows nanotoxicity experiments to be performed quickly and on a large scale. In addition, zebrafish are suitable for testing nanomaterials via different exposure routes.

Small rodents such as mice, rats, and rabbits are still preferred for in vivo assessment of nanotoxicity [897]. They are easier to keep than larger animals such as pigs, which are genetically very similar to humans. Small rodents are also suitable for studying different routes of exposure, including inhalation, intravenous or intraperitoneal injection, ingestion, and intratracheal implantation of nanoparticles. The administration of nanoparticles by inhalation and intravenous injection is common. For example, suspensions of nanoparticles are injected directly into the bloodstream of healthy rats and blood samples are taken 48 h after administration. Information on the distribution of nanoparticles in vital organs can be obtained by counting monocytes, granulocytes, lymphocytes, platelets, and red blood cells in the blood samples. Hemoglobin concentration is also determined.

There are several standardized toxicological tests that can be used to measure the biological response of a chemical compound [898]. However, there is no standardization for assessing the cytotoxicity of nanoparticles [883]. Cytotoxicity testing of nanoparticles can be performed in vivo and in vitro [890]. Numerous in vitro cytotoxicity studies on nanoparticles have been published, employing different cell lines, incubation times, and colorimetric assays. However, these studies use a variety of nanoparticle species, a wide range of nanoparticle concentrations and exposure times. This makes the detection and unambiguous assessment of physiologically relevant cytotoxicity virtually impossible. The use of different cell lines and culture conditions also dramatically weakens the validity of in vitro cytotoxicity studies. This complicates the necessarily direct comparison between studies published to date.

Most toxicity tests for nanoparticles have been conducted in vitro, using cultures of mammalian cells from different organs and tissues of the body (e. g., spleen, lung, heart, kidney, and liver) [899]. Unfortunately, the information obtained from these studies on cellular responses to nanoparticles does not provide concrete information on their effects on human health. The most commonly used assessment parameters are LC50, LOEC, and NOEC [883, 885]. The LC50 is the concentration of nanomaterial that causes the death of 50 % of the population, while the LOEC is the lowest concentration that has a noticeable effect on the organism. On the other hand, the NOEC is the maximum concentration at which no effect on the organism is observed.

Established in vitro assays are staining assays that measure cell membrane integrity and mitochondrial activity of viable cells [900]. Apoptosis results in the dissolution of the cell membrane, while necrotic cells are characterized by perforated or ruptured cell membranes. The cytotoxicity of nanoparticles can be assessed by measuring the activity of cytoplasmic enzymes released by damaged cells. A stable cytoplasmic enzyme present in all cells is lactate dehydrogenase (LDH). LDH activity scales with the concentration of NADH produced during the conversion of lactate to pyruvate. In the colorimetric LDH assay, LDH activity is quantified using the yellow tetrazolium salt (INT) added to the cell culture medium. NADH stoichiometrically reduces INT to a red water-soluble formazan-class dye. The formazan concentration, determined quantitatively by measuring the absorbance at 492 nm, is directly proportional to the LDH concentration in the cell medium, which scales with the number of dead or damaged cells. Other viability assays include the neutral red and trypan blue tests. Neutral red, or toluene red, is a weak cationic dye that can diffuse through the plasma membrane and then accumulate in the lysosomes of the cell. A damaged cell membrane allows the leakage of neutral red, which can be used to distinguish between living and dead cells. Cytotoxicity can be quantified by neutral red uptake under different exposure conditions. Trypan blue, a diazo dye, is only permeable to cells with damaged membranes. It stains dead cells blue, while living cells remain colorless. The metabolic activity of viable cells can be determined using the colorimetric MTT and MTS assays. The MTT assay uses the yellow dye 3-(4,5-dimethylthiazol-2-yl)-2,5-diphenyltetrazolium bromide (MTT), which is reduced to purple, insoluble formazan crystals by mitochondrial succinate dehydrogenases in viable cells. Only active mitochondria contain these enzymes. This means that the reaction only takes place in living cells. This method is directly related to the number of viable cells. In the MTS (3-(4,5-dimethylthiazol-2-yl)-5-(3-carboxymethoxy-phenyl)-2-(4-sulfophenyl)-2H-tetrazolium) assay, the tetrazolium salt MTS enters viable cells. The assay reagent also contains the electron transfer reagent phenazine ethosulfate (PES). PES is reduced intracellularly by reaction equivalents such as NADH or NADPH, resulting in the reduction of the tetrazolium salt MTS to the color-intensive formazan outside the cell. The tetrazolium MTS enters viable cells and is converted by mitochondrial dehydrogenase enzymes to a colored, soluble formazan product. A colorimetric product is formed in the MTS assay, similar to the MTT assay. The concentration of formazan formed in the MTT and MTS assays is directly proportional to the number of live cells in the culture. The LIVE/DEAD viability test uses two compounds, calcein acetoxymethyl (calcein AM) and ethidium homodimer, to measure the number of damaged cells. Calcein AM, the acetoxymethyl ester of calcein, is converted to calcein, a green fluorescent dye, in the cytosol of living cells by intracellular esterases. The nuclei of damaged or dead cells are then stained with the ethidium homodimer. Ethidium homodimer is a membrane-permeable molecule that emits red fluorescence when this dye binds to nucleic acids. Upon optical excitation at 495 nm, calcein AM and ethidium homodimer exhibit different fluorescence signatures at 515 nm and 635 nm, respectively.

Cell death by apoptosis or necrosis is triggered by excessive oxidative stress [901, 902]. Oxidative stress is caused by overproduction of reactive oxygen species (ROS) and reactive nitrogen species (RNS) [803], which can be stimulated by various nanoparticle species [903]. Oxidative stress can be detected using a glutathione assay. Glutathione (GSH) is an important antioxidant that is oxidized to glutathione disulfide (GSSG) in the presence of ROS. A high GSH/GSSG ratio is required for maintaining homeostasis and protection against oxidative stress, and this ratio is regulated by the enzyme glutathione reductase. The Glutathione Assay measures glutathione concentration using the Ellman's reagent 5,5′-dithio-bis-2-nitrobenzoic acid (DTNB), which reacts with the sulfhydryl group of GSH to form a yellow product, 5-thio-2-nitrobenzoic acid (TNB). Glutathione reductase also recycles GSH from the GSH-TNB complex and produces more TNB. The concentration of TNB is directly proportional to the concentration of GSH in the sample. TNB concentration is determined by measuring the absorbance of TNB at 405 or 412 nm. High ROS concentrations also lead to lipid peroxidation in the cell membrane. Lipid peroxidation can be quantified using the widely used thiobarbituric acid (TBA) assay. In the colorimetric TBA assay, malondialdehyde (MDA), a toxic by-product of lipid peroxidation, reacts with 2-thiobarbituric acid when heated at acidic pH to form a fluorescent pink chromagen. The concentration of chromagen is determined by measuring the fluorescence intensity at 532 nm.

Nanoparticle-induced oxidative stress is induced at the cellular level after nanoparticle administration. Oxidative stress occurs when the increased production of ROS and RNS cannot be balanced by the activity of antioxidants [903]. The most reactive RNS and ROS are peroxynitrite ($ONOO^-$), nitric oxide (NO), hydroxyl radical ($^{\bullet}OH$), hydrogen peroxide (H_2O_2), and superoxide radical ($O_2^{\bullet -}$). ROS and RNS are mostly generated as by-products of neutrophil-mediated phagocytosis, enzymatic cytochrome P450 metabolism, and mitochondrial respiration. ROS damage nucleic acids, proteins, lipids, and mitochondrial structure. Nanoparticles with redox-active surfaces, such as SPIONs, MnO_2, TiO_2 and ZnO NPs in particular, stimulate the formation of ROS or stabilize free radicals and their intermediates, leading to cytotoxicity when nanoparticle-generated ROS cause DNA strand breaks leading to genetic mutations.

Nanotoxicological studies typically focus on the evaluation of one or more toxicological endpoints, e. g., cytotoxicity, pro-inflammation, and/or genotoxicity. Nanoparticle-induced cytotoxicity results from the increased generation of ROS, which damages proteins, the cell membrane, and DNA, making oxidative stress a crucial factor in genotoxicity. Oxidative stress arises due to an intracellular redox imbalance between ROS, RNS, and antioxidants (e. g., SOD), which is normally maintained in a state of homeostasis.

Excess ROS in this way has the potential to cause DNA damage, as $^{\bullet}OH$ can readily react with DNA and DNA precursors, leading to the formation of DNA lesions [884]. SiO_2 NPs catalyze the formation of ROS, which can be explained by the reactivity of oxygen defect centers [904]. Genetic damage caused by nanoparticles can occur via primary (direct or indirect) or secondary mechanisms. Primary indirect genotoxicity is caused by the redox activity of nanoparticles, which increases the generation of ROS in the cytosol. This ROS overproduction indirectly damages DNA by interfering with DNA replication, cell division and the function of replication or repair enzymes. Primary direct genotoxicity, on the other hand, is caused by the penetration of very small nanoparticles (<6 nm) into the cell nucleus, where they intercalate or accumulate, for example, between base pairs of DNA. Such interactions between nanoparticles and DNA can lead to DNA lesions and mutagenesis, with mutagenesis resulting from impaired self-repair. So far, semiconductor quantum dots and gold nanoclusters have been detected in cell nuclei. However, secondary genotoxicity caused by nanoparticles through activation of immune cells such as macrophages or neutrophils is significantly more harmful to health. These phagocytic cells act in vivo as the first immune defense mechanism against the penetrating nanoparticles and contribute to the elimination of the foreign material. Nanoparticles ingested by immune cells induce chronic pro-inflammatory responses, resulting in the release of various inflammatory mediators such as cytokines, chemokines, histamines, prostaglandins, ROS, and reactive nitrogen species (RNS). This secondary genotoxicity subsequently leads to oxidatively damaged DNA in surrounding tissues [888].

The pharmacokinetics and toxicokinetics of nanoparticles are determined by four processes: in vivo absorption, distribution, metabolism, and excretion (ADME). Absorption of nanoparticles can occur through various routes of administration and is associated with their uptake into the bloodstream. Distribution involves the transport of nanoparticles from the bloodstream to various tissues and organs, the lymphatic system, and their intracellular uptake. Metabolism of nanoparticles mainly concerns orally ingested nanoparticles that enter the gastrointestinal tract. Degradable nanoparticles such as iron oxide or SiO_2 NPs, which are not sufficiently surface stabilized, are digested. For example, the highly acidic gastric juice can dissolve SPIONs, resulting in the release of iron cations [905]. Chemically stable nanoparticles, especially AuNPs, enter the systemic circulation by direct translocation into the portal blood. Lipophilic nanoparticles escape the hepatic first-pass metabolism via the lymphatic route. Metabolic elimination of nanoparticles occurs via the renal, hepatobiliary, or fecal route. Non-biodegradable nanoparticles or their degradation products may be excreted in the bile after digestion by hepatocytes. Renal clearance is the most efficient and rapid route for the elimination of nanoparticles, whereas hepatobiliary excretion can take hours to months. The differences between renal and hepatobiliary clearance are shown in Figure 172. Nanoparticles also interact with the glomerular capillary wall and can pass through the glomerular capillaries and into the Bowman cave before being reabsorbed by the renal tubule. Some of the nanoparticles is returned to the bloodstream, while the majority is excreted in the urine.

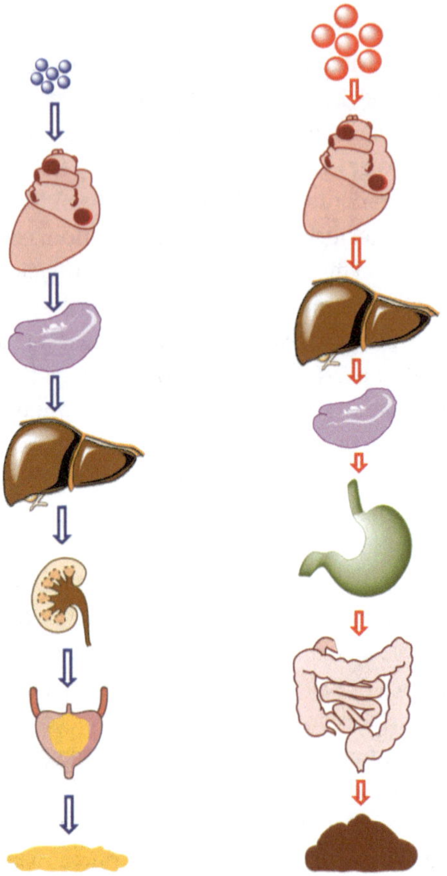

Figure 172: Routes of renal (left) and hepatobiliary excretion.

Predictive toxicology models such as structure–activity relationship (SAR) analysis and computational nanotoxicology approaches help predict the toxicity of nanomaterials based on their physicochemical properties and structure. These models contribute to the development of regulatory frameworks to assess the safety of nanomaterials and guide their responsible use in various applications, including medicine, electronics, cosmetics, and environmental remediation.

Physiology-based pharmacokinetic modeling (PBPK) of nanoparticle biokinetics allows the evaluation of absorption and biodistribution, xenobiotics, and toxicity variables in relation to external exposure, internal dose, and the likelihood of adverse health effects [816, 906]. The validity of a pharmacokinetic prediction model depends on the appropriate selection and formulation of the mathematical functions describing the kinetic processes involved. Mathematical models for simulating of drug absorption, distribution, and excretion are developed to characterize and predict drug concentrations in the body as a function of time. A powerful PBPK model for the use of nanoparticles in medicine will enable the calculation of the time course of nanoparticle accumulation in organisms and target tissues, as well as facilitate quantitative risk assessment. For this purpose, the parameters of the circulatory system, organ and tissue volumes, tissue distribution coefficients, and other anatomical and physiological parameters must be adapted to the biokinetic properties of nanoparticles.

While the exchange of small molecules between blood and tissue is often flow-limited, the exchange of nanoparticles between blood and tissue is typically diffusion-limited. Once in the bloodstream, nanoparticles are immediately taken up by phagocytic cells in organs such as the liver and spleen. PBPK models can be used to calculate quantitative data on the kinetic processes of ADME based on the physiological characteristics of animals and the physicochemical properties of nanoparticles. For example, a PBPK model developed for metallic nanoparticles was used to model the transit kinetics of nanoparticles in living organisms across biological barriers. Combined with pharmacodynamic models that correlate exposure in target tissues with pharmacological effects, PBPK modeling can predict efficacy and toxicity. The cytotoxicity of carbon nanoparticles, polymeric nanoparticles, nanocrystals, silver nanoparticles, liposomes, gold dendrimer composite nanoparticles, and other nanoparticles have all been evaluated using PBPK modeling.

The vascular circulation time of nanoparticles refers to the amount of time they remain in the bloodstream after being injected into the body. This can be influenced by a number of factors, including the size, shape, and surface properties of the nanoparticles, as well as the physiological environment in which they are delivered.

Gold nanoparticles and iron oxide nanoparticles are commonly used in medical applications due to their biocompatibility, stability, and ability to be functionalized with a variety of chemical groups. These properties make them suitable for a number of potential therapeutic applications such as drug delivery, imaging and cancer therapy. The circulation time of gold nanoparticles and iron oxide nanoparticles in blood vessels can vary depending on their specific characteristics and the physiological environment in which they are delivered. For example, smaller and more spherical nanoparticles tend to have longer circulation times in the blood vessels, while larger and more irregularly shaped nanoparticles may be more easily eliminated by the body. In addition, the surface properties of nanoparticles, such as their charge and chemical composition, can affect their interactions with proteins and biomolecules in the bloodstream, which can influence their circulation time. Understanding the factors that influence the circulation time of gold nanoparticles and iron oxide nanoparticles in blood vessels is important for the development of safe and effective nanoparticle-based therapies.

6 Nanoparticles for diagnostics and therapy of cancer

SPIONs and AuNPs and their nanoheterodimers have been and are being developed as contrast agents for diagnostic imaging [907–910]. Today, various complementary imaging modalities are used with excellent success to diagnose diseases such as cancer, cardiovascular syndromes, and neurological disorders. Contrast agents based on nanoparticles can be selectively enriched in diseased or vascular tissue, increasing the power of imaging. Nanoparticle-assisted imaging techniques include computed tomography (CT), magnetic resonance imaging (MRI), and photoacoustic imaging (PAI). In particular, Au–Fe_3O_4 NHDs are also suitable as contrast agents for multimodality imaging, including exactly these imaging techniques.

Compared to molecular contrast agents, Fe_3O_4 and Au nanostructures offer other therapeutic functions in cancer therapy and other medical treatments. They increase the X-ray dose in radiotherapy, SPIONs act as mediators in magnetic hyperthermia, and AuNPs serve as plasmonic heaters in photodynamic therapy. In addition, their surfaces can be functionalized not only with a variety of drugs, but also with tumor receptor-specific molecular ligands that allow selective targeting to the desired site (e. g., cancer tissue) for maximum contrast enhancement and for delivery of drugs, genes, or other therapeutics.

6.1 Tumor targeting

Nanoparticle-based methods for tumor targeting offer promising strategies for improving the precision and efficacy of cancer treatments. These tumor targeting methods aim to accumulate nanoparticles in tumor tissue as drug carriers, diagnostic tools, radiation dose enhancers, or plasmonic heaters, while minimizing side effects on healthy tissue. The main tumor-targeting methods are passive tumor targeting, active tumor targeting, and magnetic tumor targeting, each of which uses different mechanisms to achieve selective accumulation and retention of nanoparticles in the tumor microenvironment. Passive targeting exploits the EPR effect inherent in tumor physiology, active targeting uses specific ligand-receptor interactions to achieve selectivity, and magnetic targeting utilizes external magnetic fields to direct nanoparticles to the tumor. Each method has its own advantages and challenges, and ongoing research continues to refine these approaches to achieve better clinical outcomes.

6.1.1 Passive tumor targeting

Passive tumor targeting exploits unique anatomical and pathophysiological properties of tumor tissue that enable the EPR effect. This tumor targeting technique enables

https://doi.org/10.1515/9783110605495-006

the spontaneous accumulation of nanoparticulate diagnostics and therapeutics in tumor tissue by exploiting the EPR effect [911, 912]. The EPR effect was first described in 1986 by Maeda and Matsumura [594]: Leaky vasculature in tumor tissue allows permeation of macromolecular or nanoparticulate drug carriers, while inadequate lymphatic drainage ensures retention of nanoparticles in tumor tissue (Figure 173). The reason for the increased permeability of tumor vessels is the angiogenesis activated by metabolically active cancers, which leads to the formation of highly abnormal tumor vessels that often exhibit a chaotic and disorganized course and are structurally different from healthy vessels. Dvorak et al. [913] identified six different types of tumor vessels with characteristic architectures, which they categorized as; feeding artery, parent vessel, glomeruloid microvascular proliferation (GMP), vascular malformation, capillary, and draining vein. Parent vessels and GMPs have interendothelial gaps. The size of these gaps can range from a few nm to several hundred nm. In addition, mother vessels may have a fenestrated endothelium that allows extravasation of nanoparticles. Extravasation describes the transport of nanoparticles from the lumen of tumor blood vessels into the interstitium of the tumor. The extravasation of nanoparticles through interendothelial spaces is called paracellular transport. This passive transport mechanism requires a high concentration gradient between the blood and tumor interstitium. The nanoparticles are driven passively by diffusion through the endothelium into the tumor tissue. Thus, passive targeting that is based on a diffusion mechanism is therefore highly dependent on the size and shape of the nanoparticles. Spherical shape and sizes in the range of 20 to 200 nm not only benefit effective diffusive transport of extravasating nanoparticles, but also are favorable properties for long circulation times, both of which promote accumulation of nanoparticles in tumors [914].

Efficient accumulation of nanoparticles in the tumor requires the largest possible concentration gradient, which can be achieved with high nanoparticle concentrations. In addition, the efficiency of the passive tumor-targeting mechanism can be increased by allowing the administered nanoparticles to circulate in the blood for a long time. This ensures that the probability of paracellular nanoparticle transport into the tumor increases with both the concentration of, e. g., intravenously administered nanoparticles and their circulation time in the bloodstream.

Unfortunately, passive tumor targeting does not sufficiently selectively occur in tumor tissue. Nanoparticles can also, although to a lesser extent, extravasate from intact blood vessels into healthy tissue (Figure 173). Both active and passive tumor targeting exploit the EPR effect for nanoparticle transport into the tumor, whereas concentration-dependent passive paracellular transport across a discontinuous endothelium ensures nanoparticle accumulation in the tumor. The retention of nanoparticles in tumor tissue correlates with the inefficiency of lymphatic drainage. In healthy tissue, interstitial fluid balance is maintained by the lymphatic system, which drains excess fluid from the tissue. In solid tumors, however, the high density of proliferating cancer cells compresses the lymphatic vessels due to the high density of continuously proliferating cancer cells.

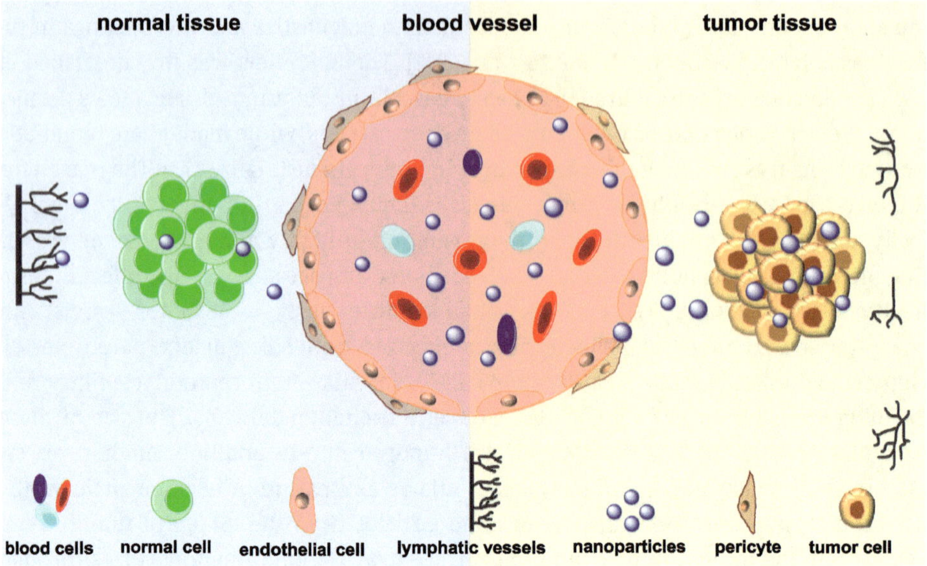

Figure 173: Schematics of the EPR effect in healthy or tumor tissue.

Compressed lymphatic vessels can collapse, resulting in inadequate lymphocyte infiltration. Collapsed lymphatic vessels are no longer able to drain fluid from the tumor tissue, which increases the pressure of the interstitial fluid in the tumor tissue. On the one hand, the increased interstitial pressure can promote the accumulation of nanoparticles in the tumor tissue; on the other hand, it makes it more difficult for the nanoparticles to diffuse into the tumor interstitium. The latter must then be compensated for by a concentration gradient between the blood and the tumor tissue that is as large as possible.

6.1.2 Active tumor targeting

In active tumor targeting, nanoparticles are functionalized with specific ligands that can recognize and bind to receptors overexpressed on tumor cells (described in detail in Section 4.2.2.3.2) [915]. Active tumor targeting utilizes ligand–receptor interactions where nanoparticles are conjugated with ligands such as antibodies, peptides, aptamers, or small molecules that specifically bind to receptors or antigens on the surface of tumor cells, enabling targeted delivery. After binding to the target receptor, the nanoparticles can be internalized by tumor cells through receptor-mediated endocytosis, enhancing intracellular delivery of therapeutics. Common targeting ligands include monoclonal antibodies that target specific tumor-associated antigens, short peptides that target integrins or other cell surface markers, and nucleic acid-based aptamers that exhibit high specificity and affinity for tumor markers. The efficiency of active targeting can be optimized by adjusting the density of targeting ligands on the surface of the nanoparticles and by using multivalency, where multiple ligands on a single nanoparticle enhance binding through multivalent interactions. The efficacy of tumor penetration of nanoparticles scales with the binding affinity of their ligands to tumor receptors.

6.1.3 Magnetic tumor targeting

Magnetic tumor targeting uses superparamagnetic nanoparticles, such as SPIONs and other ferrite nanoparticles that can be directed to tumor sites using external magnetic fields (Figure 174) [916–921]. In this method, external magnets create a magnetic field gradient that guides and retains the magnetic nanoparticles at the tumor site. To control magnetic tumor targeting, the magnetic properties of the nanoparticles, including their composition, size, and magnetic core structure, have to be tailored to the respective application. The strength and gradient of the applied magnetic field must be optimized to ensure effective targeting without damaging the surrounding tissue. Suitable magnetic nanoparticles are SPIONs that have highly reactive surfaces due to their Fe_3O_4 composition and, moreover, size-dependent magnetic properties due to their Fe_3O_4 composition [919, 920, 922]. Spherical SPIONs with sizes smaller than 30 nm consist of a single magnetic domain and are therefore superparamagnetic [916]. Larger Fe_3O_4 nanoparticles are ferrimagnetic since they consist of many magnetic domains separated by domain walls. The formation of the domain walls is energetically advantageous if the energy re-

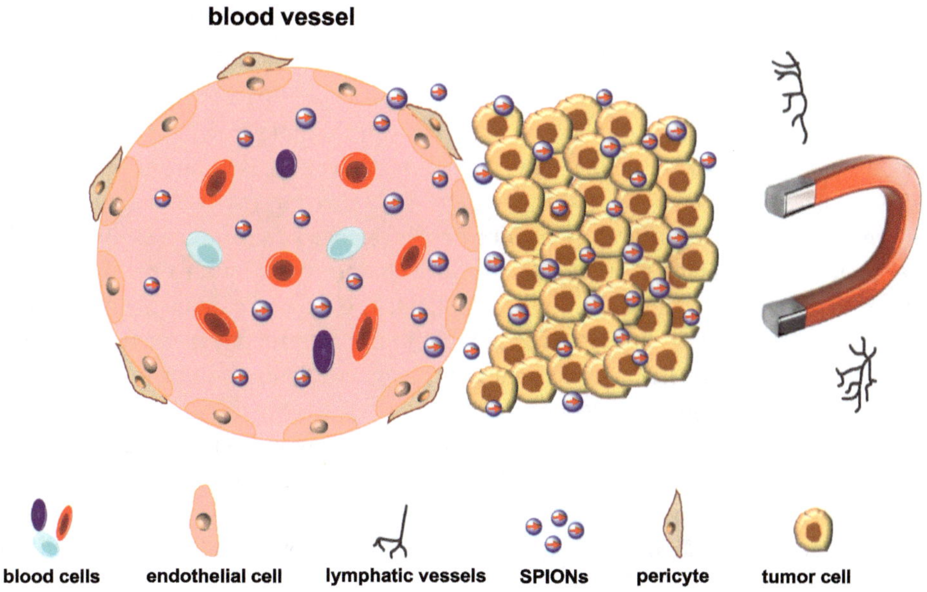

Figure 174: Schematics of the magnetic tumor targeting.

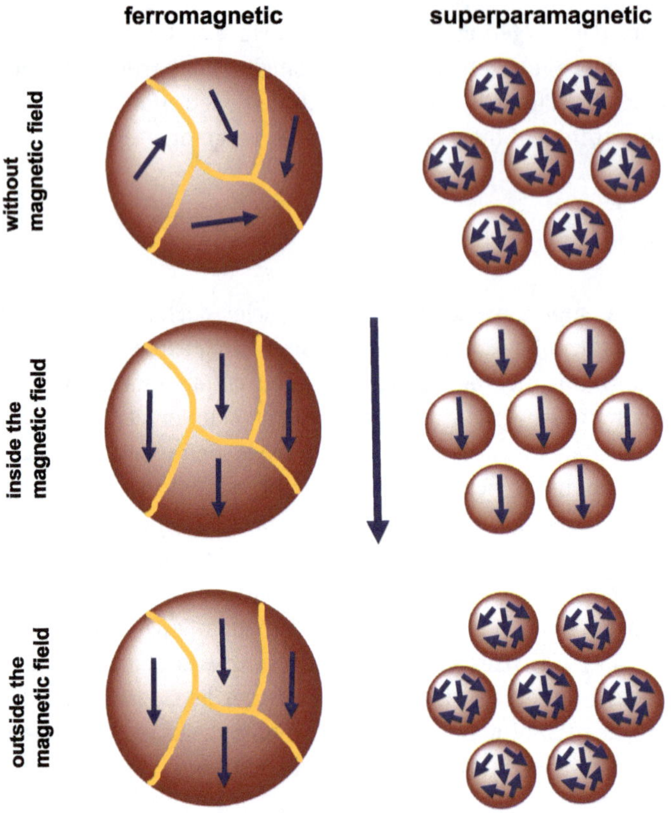

Figure 175: The influence of an external magnetic field on the net magnetization of ferromagnetic (left) and superparamagnetic (right) nanoparticles.

quired for the formation of the domain walls is lower than the growth of a single large domain with parallel ordered magnetic moments (Figure 175). Ferrimagnetic nanoparticles are like permanent magnets. They can be magnetized by external magnetic fields and remain magnetized after the magnetic field is removed. In contrast, SPIONs, which are smaller than a single magnetic domain, are characterized by zero net magnetization in the absence of an external magnetic field. The magnetic moments of the Fe^{2+} and Fe^{3+} cations, which are aligned in parallel in the external magnetic field, experience thermally induced fluctuations outside the magnetic field, resulting in an isotropic distribution of the magnetic field vectors.

Applications of SPIONs in biomedical fields require the use of suspensions of SPIONs in a physiological carrier fluid with concentrations in the range of 10^{21}–10^{23} SPIONs/m^3. The stability of such a magnetic colloid depends on the balance between attractive (van der Waals and dipole–dipole) and repulsive (steric and electrostatic) forces between the particles and the carrier liquid.

There are several advantages to using SPIONs for magnetic tumor targeting. Because the nanoparticles can be easily magnetized and demagnetized, they can be directed to specific locations in the body using an external magnetic field. One promising application of SPIONs to date is magnetic tumor targeting, in which they can be used to deliver therapeutic agents directly to cancerous tumors [319, 333, 631, 920, 923]. By attaching a therapeutic agent, such as a chemotherapy drug, to the surface of the SPIONs, it is possible to deliver the drug directly to the tumor cells. The SPIONs are magnetically directed to the tumor using an external magnetic field, and the drug is released directly into the tumor cells in response to environmental triggers or external stimuli. They can also be used to generate heat in the tumor microenvironment, increasing the sensitivity of tumor cells to chemotherapy or radiation [123, 192, 210, 924, 925]. One potential use of SPIONs is in magnetic resonance imaging (MRI), where they can be used to enhance the contrast of certain tissues or organs. When SPIONs are injected into the body and exposed to a magnetic field, they become magnetized and produce a signal that can be detected by the MRI scanner. This signal is used to create detailed images of the internal structures of the body.

6.2 Nanoparticle-based imaging techniques

6.2.1 CT contrast agents

Computed tomography is used for anatomical 3D imaging with high spatial resolution. Since soft tissues such as muscle, fat, and organ tissue have similarly low X-ray attenuation, X-ray contrast agents are required to differentiate between different tissue types [908]. The signal-to-noise ratio is increased while the radiation dose to the patient remains the same. X-ray contrast agents that accumulate preferentially in such diseased tissue are particularly important in oncology, i. e., for the detection of cancerous tissue and metastases. Suitable contrast media consist of or contain elements with a high atomic number, such as iodine ($Z = 53$) and gold ($Z = 79$). The most important properties of X-ray contrast agents include X-ray attenuation coefficient, colloidal stability in physiological media and during storage, vascular retention time, biodistribution, and biocompatibility. Nowadays, contrast agents are mostly highly water soluble, biocompatible, low molecular-weight iodinated aromatics [50]. Active targeting is achieved by introducing functional ligands into the aromatic rings. However, iodinated contrast agents have a relatively short blood retention time and are rapidly renally excreted. This results in a short imaging window that may require multiple injections.

Gold nanoparticles (AuNPs) can be equipped with surface functionalities that specifically bind to surface receptors on cancer cells or have other surface structures that promote cellular uptake into tumor cells [50]. In addition, they are nontoxic and exhibit high X-ray attenuation and longer retention time in the bloodstream compared to iodinated molecules. The X-ray attenuation of a contrast agent depends on Z, its density σ, and the energy spectrum of the X-ray source. X-ray imaging arises from the absorption or scattering of X-ray photons as a collimated X-ray beam passes through the sample. The intensity of X-rays (I_0) decreases linearly during their transmission through tissue according to Lambert–Beer's law

$$I = I_0 e^{-x\mu/\rho}, \tag{16}$$

where μ denotes the linear X-ray attenuation coefficient, ρ and x are the density and thickness of the tissue, respectively, and I is the attenuated X-ray intensity. Most CT scanners are calibrated with water. It is common practice in CT to express the attenuation of X-rays in Hounsfield units (HU). By definition, the HU value of water is set to 0, while air is assigned a value of −1,000 HU. Attenuation values for most soft tissues range from 30 to 100 HU. For a tissue with a known μ value, the corresponding HU value is calculated as follows:

$$\mathrm{HU} = 1000 \cdot \frac{\mu - \mu_{\text{water}}}{\mu_{\text{water}} - \mu_{\text{air}}}. \tag{17}$$

Dong et al. [239] synthesized PEGylated AuNPs in the size range of 4 nm to 153 nm and investigated the influence of AuNP size on CT contrast generation in CT phantom imaging, pharmacokinetics, and biodistribution in C57BL/6 mice. The experimental results showed that the size of the PEGylated AuNPs had no effect on the X-ray attenuation when the concentration of AuNPs introduced into the tissue model remained constant. All in vivo CT imaging experiments were performed using a MILabs U-CT scanner with a tube voltage of 50 kV, a tube current of 190 μA, a slice thickness of 100 μm, an in-plane resolution of 100 μm, an exposure time of 75 ms, and an analytical reconstruction kernel. Mice were intravenously injected with PEGylated AuNPs of various sizes (i. e., 4 nm, 15 nm, 50 nm, 79 nm, 100 nm, 152 nm) at a concentration of 500 mg/kg. Mice were scanned before injection and 5, 30, 60, and 120 min after injection. The CT images in Figure 176 were recorded 2 hours after injection. The strong contrast in the CT images demonstrates the X-ray attenuation effect, which scales with the accumulation of AuNPs in the various organs and blood vessels.

The blood vessels of mice injected with 4 nm and 15 nm AuNPs showed persistent strong CT contrast, demonstrating the long half-life of these AuNPs in the bloodstream. By comparison, the much weaker contrast in the liver and spleen of these mice indicates a significantly lower AuNP content. On the other hand, vascular contrast in the CT images of mice injected with 50-nm and larger AuNPs was minimal, indicating faster clearance of AuNPs from the bloodstream. Apparently, a large fraction of the AuNPs reached the liver and spleen of the mice, where the contrast increased over time. In short, the difference in pharmacokinetics is due to the size of the AuNPs: Large PEGylated AuNPs (>50 nm) were observed to leave the circulation earlier than the smaller ones, accumulating preferentially in the liver and spleen.

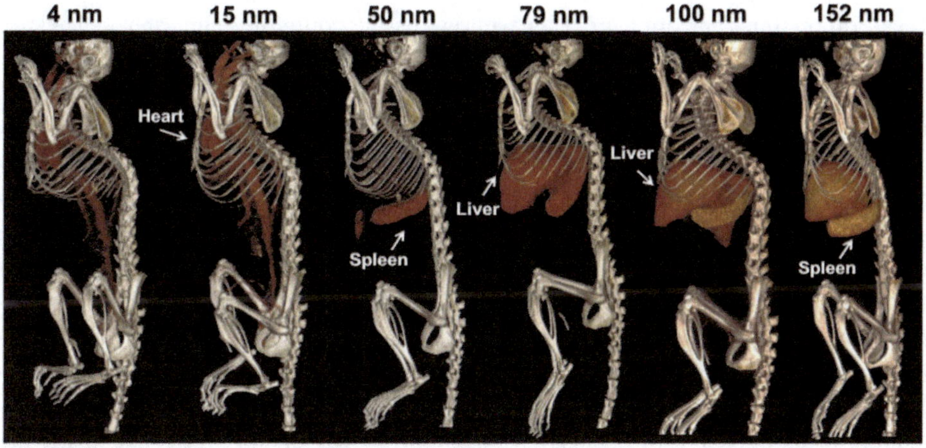

Figure 176: 3D volume CT images of mice taken 2 hours after injection of AuNPs; the bones of the animal are shown in a bone-like color; the contrast created by the injected AuNP is shown in red to orange-yellow; the CT images are displayed with a voxel size of 100 μm [239].

6.2.2 MRI contrast agents

Magnetic resonance imaging (MRI) is an indispensable imaging modality for clinical diagnosis due to its high depth sensitivity [57, 907]. As described in detail in Section 2.4, MRI is based on the alignment of the magnetic moments of unpaired nuclear spins in biological tissue along an external magnetic field and the reorientation of the nuclear spins by 90° and subsequently by 180° applying resonant radiofrequency (RF) pulses. The RF pulses emerge from an alternating high-frequency magnetic field that oscillates with the Larmor frequency of the nuclear spins orthogonally to the static external magnetic field. The 90° RF pulses flip the magnetization vector (∥ z), initially aligned along the magnetic field, into the x–y plane. The transverse magnetization decays by dephasing of the individual nuclear spins, which is characterized by the transverse (spin–spin) relaxation time T_2. The application of a second pulse, 180° RF pulse, provides the rephasing of the individual nuclear spins to reverse transverse magnetization, while the recovery time of the initial z alignment of the magnetization vector determines the longitudinal (spin–lattice) relaxation time T_1. Because of the high proton concentration in human tissue, the frequency of the RF pulses is mostly adjusted to the Larmor frequency of the nuclear spin precession of the proton. Different chemical environments of the nuclear spins of the protons in the tissue, as well as different proton densities in different tissue types (e. g., muscle tissue, adipose tissue, organ tissue, lymph, and blood) influence the MR signal strengths and thus create a contrast between adipose tissue, organ tissue, and bone. The values of T_1 and T_2 depend on the molecular composition of the tissue under study. For example, in T_1-weighted images, water and protein-rich tissues have longer

T_1 relaxation times and appear darker, whereas tissues containing cholesterol (like fatty tissue) have shorter T_1 times and appear brighter.

Contrast can be enhanced locally in the target tissue by using paramagnetic compounds such as gadolinium complexes and superparamagnetic iron oxide nanoparticles (SPIONs) [164]. When gadolinium complexes as positive contrast agents are introduced into tissues, they shorten the T_1 time constants in the tissue, resulting in a brightening of the surrounding area in the tissue. In the case of T_2-weighted MRI, negative contrast agents are used to make the surrounding tissue appear darker. These contrast agents are SPIONs and other superparamagnetic ferrite nanoparticles. Superparamagnetic nanoparticles interact with nuclear spins via their local magnetic fields, contributing to the loss of coherence and shortening T_2. These agents improve the contrast between different tissues, making it easier to detect abnormalities such as tumors, inflammation, or blood vessel problems. Paramagnetic Gd(III) chelates (e. g., gadopentetate dimeglumine, and gadobutrol) [926], which are paramagnetic, act as positive contrast agents because they interact with the proton nuclear spins by locally enhancing the magnetic field strength which shortens the recovery time of the longitudinal magnetization (T_1) of the tissue. Ultrasmall iron oxide nanoparticles (<5 nm) lose their solid-state properties and exhibit paramagnetic behavior [174, 927, 928] and, like gadolinium complexes, act as positive contrast agents by shortening the longitudinal relaxation time T_1. In contrast, iron oxide nanoparticles larger than 10 nm are superparamagnetic (SPIONs) and therefore have a very high magnetic susceptibility [177, 178, 929]. SPIONs are magnetized in the strong external magnetic field of the MRI scanner and thus increase the magnetic field strength in their environment. SPIONs thus create local magnetic field inhomogeneities that interfere with the external magnetic field. The transverse magnetization induced by the radiofrequency pulse undergoes phase loss in the local magnetic field of the SPIONs, which is caused by rapid dephasing of the proton spins in the surrounding tissue. The dephasing, accelerated by the influence of the local magnetic field, shortens the spin–spin relaxation time T_2, resulting in signal loss on T_2-weighted MR images. SPIONs thus act as negative MRI contrast agents as they cause the surrounding tissue to darken on the MR image.

SPION surfaces can be coated with biocompatible polymers (e. g., PEG) or biopolymers (e. g., dextran, chitosan) or functionalized with targeting ligands [178, 917, 923]. The surface modification strategy depends on the application. PEGylated or dextran-coated SPIONs are inert, do not agglomerate, and have long circulation times. They can only accumulate in cancer tissue due to the enhanced permeation and retention effect. On the other hand, SPIONs with targeting ligands bind specifically to cancer cell receptors, enabling more precise imaging of pathological conditions such as cancer and optimizing MRI-based cancer diagnostics.

A commonly used T_1 contrast agent in clinical MRI is gadolinium (III) diethylenetriamine pentaacetic acid (Gd-DTPA). Since there is still a probability, albeit lower than with previous Gd-based T_1 contrast agents, that Gd-DTPA may induce nephrogenic systemic fibrosis and lead to Gd deposition in the brain, there is an urgent need to develop

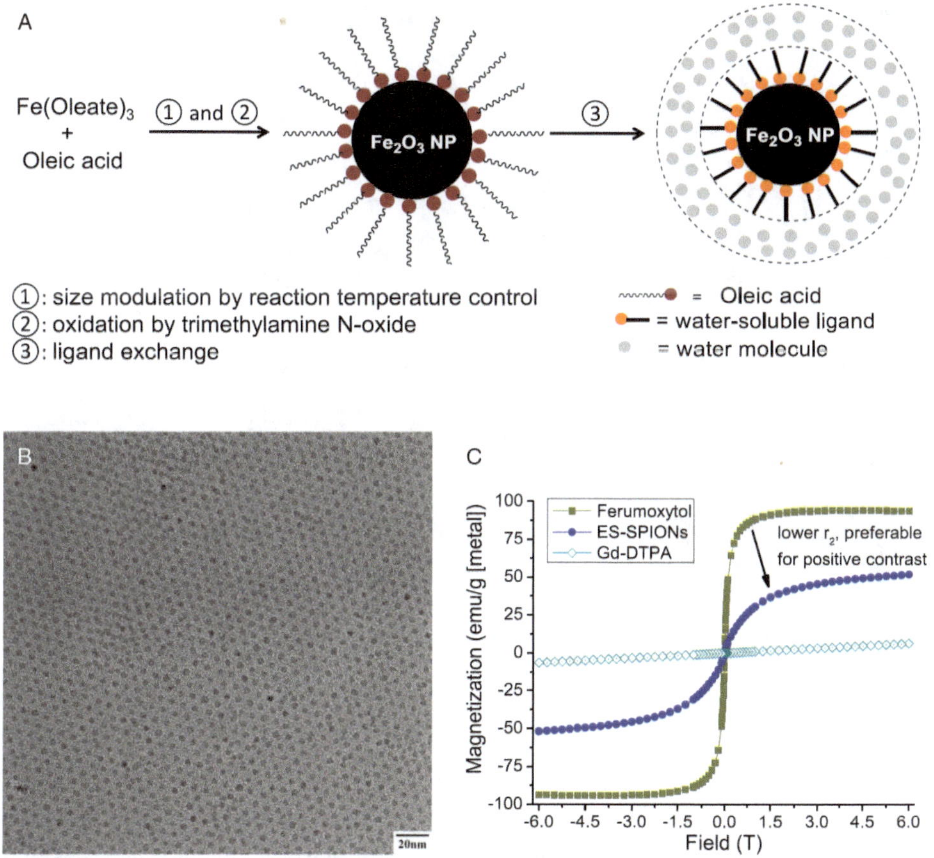

Figure 177: Schematic of the synthesis route of the of ZES-coated SPIONs (A), HR-TEM images of the 3-nm SPION cores (B), and magnetization curves of 3-nm ZES-SPIONs, ferumoxytol and Gd-DTPA (Magnevist) (C) [164].

similarly high performance but Gd-free T_1 contrast agents [164]. To achieve the paramagnetism required for T_1 imaging, ultrasmall iron oxide nanoparticles were synthesized by thermal decomposition of Fe(III) oleate. The hydrophobic oleic acid ligands were replaced by zwitterionic dopamine sulfonates (Figure 177(A)). The water-soluble biocompatible ZES-SPIONs (zwitterionic exceeding small SPIONs) obtained by this synthesis route were monodisperse and had core diameters of 3 nm, while the zwitterionic coating had a thickness of 1 nm (Figure 177(B)). The total diameter of the ZES-SPIONs is 5 nm, which is small enough to be excreted renally.

The preclinical potential of ZES-SPIONs for use as a positive contrast agent in magnetic resonance angiography (MRA) imaging was verified in a mouse model [164]. The magnetization curve of ZES-SPIONs compared to that of the FDA-approved contrast agents ferumoxytol (negative contrast agent) and Gd-DTPA (positive contrast agent),

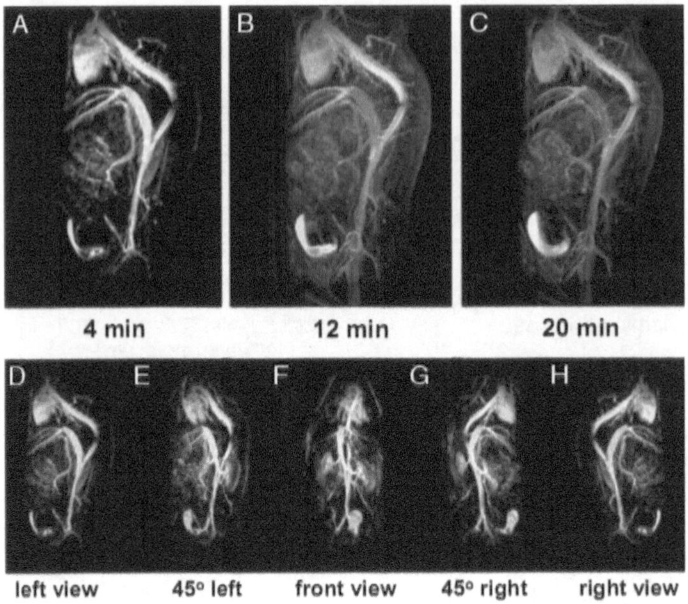

Figure 178: T_1-weighted 7 T magnetic resonance angiography (MRA) images of a mouse (A-C); the time points below each image indicate the time after injection of the ZES-SPIONs; MRA images from five different perspectives of the MRA (extracted from the 4 min. 3D scan) (D-H) [164].

shows neither clear superparamagnetic nor paramagnetic behavior (Figure 177(C)). Gd-DTPA, a paramagnetic Gd (III) chelate, is used as a positive contrast agent (T_1-weighted MRI), while ferumoxytol, which consists of 3–10-nm SPIONs, functions as a negative contrast agent (T_2-weighted MRI). ZES-SPIONs were administered to mice at a concentration of 0.2 mmol Fe/kg by tail vein injection, which is appropriate for preclinical studies in mouse models. The mice were imaged using a MRI scanner. The T_1-weighted magnetic resonance angiography (MRA) images of one mouse in Figure 178 were acquired at different times after injection of ZES-SPIONs. The strong T_1 contrast of the ZES-SPIONs and their long circulation time enabled MRA imaging with a spatial resolution of about 0.2 mm. Figures 178(A)–(C) shows a strong positive contrast of the heart and blood vessels, which fades with time, while the MR signal of the bladder increases, again indicating a renal excretory pathway. Figures 178(D)–(H) present different spatial perspectives of the MRA extracted from the 4-min 3D MRA scan (Figure 178(A)). In summary, these results show that ZES SPIONs are sufficiently small and paramagnetic to function as a T_1 contrast agent, are inert and small enough to be excreted via the kidneys.

6.2.3 Photoacoustic contrast agents

Photoacoustic imaging has found applications in a wide range of biomedical fields [536, 907, 930]. It has been used for brain imaging to enhance understanding of neural activity and map functional connectivity. In cancer research, it has proven valuable for tumor detection, characterization, and treatment monitoring. Cardiovascular applications include imaging atherosclerotic plaques and measuring blood flow. It has also been applied to ophthalmology, dermatology, and many other fields. The multiparameter imaging capability of photoacoustic techniques, combined with the use of nonionizing radiation, makes it a powerful and safe tool for both preclinical research and clinical diagnostics. As the technology continues to advance with improvements in laser technology, detection sensitivity, and image reconstruction algorithms, photoacoustic imaging will play an increasingly important role in biomedical research and clinical practice. The functionality of photoacoustic imaging is remarkably versatile. It enables high-resolution imaging deep within tissue, reaching depths of several cm with sub-mm resolution. It can provide anatomical, functional, molecular and metabolic information, often in real time. For example, it can quantify hemoglobin concentration and oxygen saturation in blood vessels, detect and monitor cancer using both endogenous and exogenous contrast agents, and even track brain activity through hemodynamic changes.

Photoacoustic imaging benefits from the high temporal and spatial resolution of ultrasound in combination with the good contrast and spectral tunability of optical imaging [931]. While photoacoustic imaging (PAI) achieves greater depths than purely optical

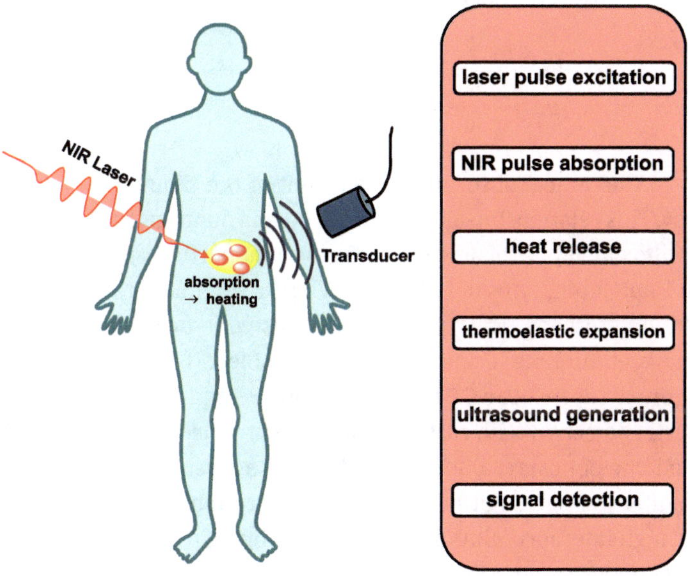

Figure 179: Schematic of photoacoustic imaging using AuNPs (red eggs) as contrast agent.

imaging methods, one limitation is that the penetration depth of NIR light is less than that of conventional ultrasound imaging due to optical scattering and absorption in tissue. Typically, PAI can image up to several centimeters deep, depending on the tissue type and the optical wavelength used [932]. The basic principle of this method is the photoacoustic effect, where pulsed laser light is absorbed by hemoglobin or melanin in the tissue. The absorbed energy is converted into heat, causing a rapid and localized increase in temperature. This temperature rise (ΔT) leads to thermoelastic expansion, which in turn causes an increase in pressure (p_0), generating broadband ultrasound waves. These ultrasound waves then propagate through the tissue and are detected by ultrasound transducers placed around or on the surface of the tissue (Figure 179).

For laser pulse durations shorter than the thermal relaxation time and stress relaxation time the initial pressure rise p_0 scales with the local temperature rise ΔT:

$$p_0 = \frac{\beta \Delta T}{\kappa}, \tag{18}$$

where β and κ denote the thermal expansion coefficient (K^{-1}) and the isothermal compressibility (Pa^{-1}), respectively. As emerging from light pulse absorption ΔT is directly proportional to the product $\eta_{th} A_e$:

$$\Delta T = \frac{\eta_{th} A_e}{\rho C_V}, \tag{19}$$

η_{th} is the fraction of absorbed light converted into heat, A_e is the specific optical energy (J/m^3), C_V is the specific heat capacity, and ρ is the specific density. Substituting ΔT in Eq. (19) by Eq. (18) results in

$$p_0 = \frac{\beta}{\kappa \rho C_V} \eta_{th} A_e. \tag{20}$$

Equation (20) relates p_0 to the optical absorption coefficient of the tissue and the deposited laser pulse energy. This relationship allows PAI to provide quantitative information about the concentration of light-absorbing molecules or nanoparticles in the tissue.

There are two main implementations of PAI: photoacoustic computed tomography (PACT) and photoacoustic microscopy (PAM) [907]. PACT uses wide-field illumination and detects acoustic signals at multiple locations around the object. The image is then reconstructed computationally, as in X-ray CT. PAM, on the other hand, uses focused optical excitation and acoustic detection with raster scanning to produce 3D images. PAM can be further categorized into optical resolution (OR-PAM) and acoustic resolution (AR-PAM), depending on whether the optical or acoustic focus determines the lateral resolution. OR-PAM achieves high lateral resolution down to 5 μm by tight optical focusing, while the resolution of AR-PAM is about 45 μm and the maximum penetration depth reaches 3 mm.

The imaging device typically consists of a pulsed laser source for excitation, ultrasonic transducers or arrays for detection, scanning mechanisms, data acquisition systems, and image reconstruction algorithms. The choice of components can be tailored to the specific application, enabling scalable imaging from microscopic to macroscopic scales. One of the key strengths of photoacoustic imaging is its ability to utilize a wide range of contrast agents, both endogenous and exogenous [910, 933]. Endogenous contrast agents include: hemoglobin, which allows imaging of blood vessels and oxygen saturation; melanin, useful for melanoma detection; and even DNA/RNA for label-free imaging of cell nuclei. Lipids and water can also provide contrast for tissue characterization. Exogenous contrast agents further expand the capabilities, including inorganic nanoparticles, organic dyes such as methylene blue, genetically encoded proteins, and even activatable and switchable agents for dynamic molecular imaging.

In particular, photoacoustic contrast agents are used to enhance the contrast of the target organ or tumor tissue [931]. In particular, AuNPs have emerged as excellent contrast agents for photoacoustic imaging due to their unique optical properties and versatility. These nanoparticles can exhibit absorption in the NIR window (700–1,350 nm), high molar extinction coefficient, excellent photostability, and high biocompatibility. The underlying process of photoacoustic signal generation involves the absorption of light by the AuNPs. The absorbed optical radiation is converted into plasmonic heating by excitation of the localized surface-plasmon resonance (LSPR) [75]. The absorbed energy creates nonequilibrium hot electrons and holes that rapidly thermalize through electron–phonon interactions, raising the temperature of the nanoparticle. The heat generated is then transferred to the surrounding medium, resulting in localized heating. This heating causes thermoelastic expansion in the surrounding tissue, generating ultrasound waves that can be detected and used to form high-resolution images. The LSPR effect of AuNPs enables tunable and exceptional optical absorption, making them ideal for generating strong photoacoustic signals. Their high photothermal conversion efficiency also contributes to their effectiveness as contrast agents. The ability to tune the optical properties of AuNPs through adjustments in shape, size, and surface modifications makes them highly versatile photoacoustic contrast agents that can be optimized for specific applications and target tissues. Several shapes have shown promise, including gold nanorods (AuNRs), nanostars (AuNSt), nanocages (AuNCs), nanoshells (AuNSh), and nanoprisms (AuNPrs). Each of these shapes offers unique advantages. For example, AuNRs allow tunable near-infrared (NIR) absorption based on their aspect ratio, with higher aspect ratios (>3.5) showing stronger photoacoustic imaging performance. AuNSt, with their star-shaped nanoconformation, provide tunable plasmon bands in the NIR. AuNCs, typically about 50 nm in size, allow tuning of their LSPR frequency by adjusting size or shell thickness. Surface coatings and modifications further enhance the functionality of AuNPs as contrast agents. PEGylation is often used to improve biocompatibility and stability. Silica coating can facilitate cellular uptake and increase photoacoustic intensity. Conjugation of targeting ligands, such as cyclic Arg-Gly-Asp (cRGD) peptide, can enhance specific uptake in tumors. The functionalization of AuNP surfaces with targeting ligands that specifically bind to the receptors of cancer cells, for example, is also easily achieved experimentally. Most of the AuNPs used for photoacoustic imaging are larger than 20 nm, in order to achieve a sufficiently long residence time in the target tissue and the highest possible signal strength.

6.3 Nanoparticle-based cancer therapy modalities

6.3.1 Nanoparticle-enhanced radiotherapy and chemotherapy

6.3.1.1 Radiosensitization

X-rays, primarily used in radiotherapy and CT imaging, are high-energy electromagnetic radiation that can penetrate and ionize organic tissue [54]. X-rays are attenuated by interaction with atoms of high atomic number ($Z > 20$). The probability of such interaction processes depends not only on the atomic number Z of the atoms contained in the materials, but also on the X-ray photon energy. X-rays with photon energies between 1 eV and 1 MeV are mainly attenuated during their passage by three different processes: the photoelectric effect (1 eV to 100 keV), the Compton effect (50 keV to 1 MeV), and also, although to a much lesser extent, Rayleigh scattering (Figure 180).

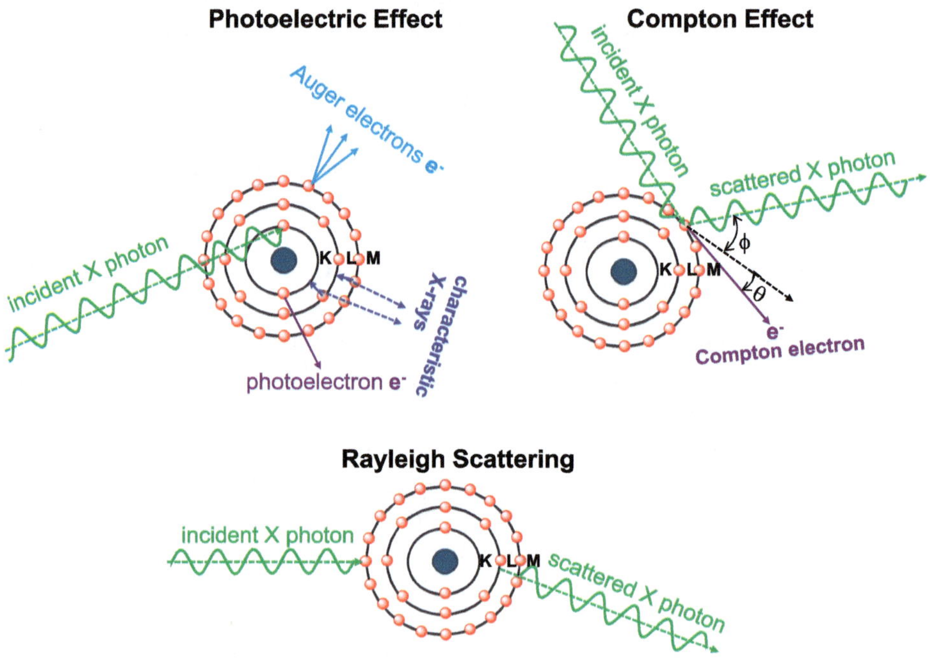

Figure 180: Schematic representation of X-ray–atom interaction processes: photoelectric effect, Compton effect, and Rayleigh scattering.

The effective cross-section is a measure of the probability of these interaction processes occurring, which increases with Z of the elements. The total mass attenuation coefficient (μ/ρ) corresponds to the sum of the individual attenuation coefficients. It represents the probability of interaction per unit mass of a particular material (in cm^2/g), which is

$$\frac{\mu}{\rho} = \frac{\tau}{\rho} + \frac{\sigma}{\rho} + \frac{\sigma_R}{\rho}, \tag{21}$$

where ρ is the density of the material and τ, σ, and σ_R are the cross-sections of the photoelectric, Compton, and Rayleigh scattering effects, respectively. The photoelectric effect occurs when a X-ray photon colliding inelastically with an electron in one of the inner orbitals (usually the K-shell) of an atom. The X-ray photon transfers its energy to the electron, which is ejected as a so-called (photoelectron) with the kinetic energy E_K. The photoelectron leaves a gap in the inner shell that is filled by an electron from the outer shell. The released transition energy causes the emission of characteristic X-rays or the ejection of Auger electrons. The probability of the photoelectric effect increases with the fourth to fifth power of the atomic number ($Z^{4\text{-}5}$) and and decreases with the cube of the photon energy. The Compton effect occurs when an X-ray photon collides with a free or weakly bound electron. The X-ray photon transfers some of its energy to the electron that is ejected from the atom as a Compton electron. The X-ray photon is scattered at an angle φ and loses some of its energy, increasing its wavelength (Compton shift). In the case of Rayleigh scattering, the incoming X-ray photon interacts coherently with all the electrons in the atom, causing them to oscillate in phase and emit secondary X-ray photons of the same energy but in a different direction. The probability of Rayleigh scattering increases with atomic number and is greater for lower energy X-ray photons. Rayleigh scattering means that an X-ray photon is scattered (elastically) by an entire atom without loss of energy, so that the direction of the photon changes. While Rayleigh-scattered photons are less significant for dose calculations in medical applications because they have no ionizing effect and contribute minimally to overall absorption of radiation. In conventional external beam radiotherapy with 100–250 keV photons, the Compton effect dominates in biological tissues, while the photoelectric effect prevails for interactions of X-rays with gold. The emitted Compton, photo-, and Auger electrons migrate into the surrounding tissue, which they can ionize.

X-rays interact with living tissue by damaging the plasma membrane, proteins, and DNA in the nucleus at the cellular level. The secondary radiation as arising from interaction with high Z elements includes the emission of Compton electrons, photoelectrons, Auger electrons and characteristic X-rays. This secondary radiation causes further cellular damage. The emitted Compton, photoelectrons, and Auger electrons ionize intra- and extracellular water molecules and oxygen in their immediate environment, leading to the formation of ROS such as $O_2^{\bullet-}$ and $^{\bullet}OH$ which initiate the chemical phase. While X-rays damage DNA through single- and double-strand breaks, ROS cause oxidative stress in cells. Hydroxyl radicals destroy lipids and DNA by oxidation. Intracellular

AuNPs raise this oxidative stress by causing the formation of ROS through the emission of secondary radiation, particularly the emission of Compton photoelectrons and Auger electrons. The probability of electron emission increases as the size of the AuNPs decreases due to increased surface curvature, leading to a higher concentration of reactive Au atoms at the surface. Au surface atoms are activated by radiation-induced ROS, further amplifying oxidative stress. AuNPs used as radiosensitizers increase the local radiation dose deposition.

Radiation therapy delivers ionizing radiation to tumor tissue with high precision. Ionizing radiation is either electromagnetic, such as γ rays or X-rays, or particulate radiation consisting of protons, electrons, or neutrons. The most commonly used radiation today is X-rays. For this reason, the interaction between X-rays and nanoparticles such as AuNPs, SiO_2 NPs, and SPIONs is the most intensively studied and characterized [934, 935]. X-ray radiation interacts either directly with DNA by causing single- or double-strand breaks, or indirectly by generating ROS through water radiolysis. However, radiation therapy is not selective for tumor cells. Therefore, the dose of radiation delivered is limited by the damage inflicted on surrounding healthy tissue. The radiation dose necessary to destroy cancer cells is fractionated into small, low doses that limit damage to the surrounding healthy tissue. However, intrinsic or acquired radioresistance reduces the sensitivity of tumor tissue to ionizing radiation. In addition, the poor vascularity of tumor tissues limits the diffusion of oxygen, which is important for the response of cells to ionizing radiation. These hypoxic or anoxic tumor regions are less sensitive to radiation damage and are one of the main reasons for failure of radiotherapy [936].

The function of radiosensitizing agents or radiosensitizers is to increase the susceptibility of tumors to radiation-induced damage. Radiosensitizers can be broadly classified into three categories: small molecules such as oxygen or nitric oxide; macromolecules such as peptides, microRNA or siRNAs; and noble metal or metal oxide nanoparticles [937]. The combination of radiosensitizers and radiation enhances tumor inactivation through various additive effects. The main strategies for the design of efficient radiosensitizers are listed next.

Increase in intracellular ROS

X-rays increase intracellular ROS generation by radiolysis of water or damage to mitochondria. Some radiosensitizers further enhance ROS generation by producing additional ROS upon irradiation. Single doses of 1–2 Gy in radiosensitizer-treated cancer cells produce elevated levels of ROS compared to untreated cells.

Inhibition of the cellular defense system

A cellular defense system counterbalances the generation of ROS and reduces the effects of oxidative damage. This defense system, which consists of low-molecular-weight antioxidants or ROS detoxifying enzymes, degrades the ROS produced by radiotherapy. Effective inhibition of the cellular defense system prevents ROS detoxification.

Production of cytotoxic substances

X-rays interact not only with intracellular water or biomolecules, but also with the radiosensitizer itself. Radiolysis of the radiosensitizer produces cytotoxic substances that cause cell damage or death.

Inhibition of DNA repair systems

DNA is damaged by ionizing radiation or intracellular ROS. This generates radicals within DNA (e. g., DNA bases), causes DNA single- or double-strand breaks, intra- or interstrand DNA-DNA cross-links, or cross-links with proteins. Several types of DNA repair mechanisms are capable of repairing DNA. Inhibition of repair mechanisms by radiosensitizers increases DNA damage.

6.3.1.2 Hypoxia and radiosensitizer oxygen

The aggressive growth of tumor cells combined with the overexpression of pro-angiogenic factors leads to the development of a disorganized vascular network in solid tumors [2]. This inadequate vasculature causes tumor heterogeneity in oxygen, pH, and glucose distribution. In particular, oxygen is rapidly metabolized by cells and the diffusion length is extremely shortened. This imbalance between oxygen supply and consumption in tumors is responsible for regions of mild (hypoxia) and severe (anoxia) oxygen deficiency [938–940]. Hypoxia in particular is strongly associated with X-radiation resistance but can also cause chemoresistance. Direct resistance effects are caused by the lack of oxygen, which prevents the maximal cytotoxicity of radiation and some drugs. Indirect effects are due to changes in cellular metabolism or an increase in genetic instability.

Multicellular tumor spheroids (MCTS) (Figure 181) are a good in vitro tumor model because they mimic the heterogeneity and radiosensitivity of a growing tumor. Expanding the size of MCTS limits the access of cells to nutrients and oxygen [941, 942] or hinders the penetration of drugs or nanoparticles [943]. Therefore, MCTS simulate certain in vivo characteristics of a tumor. The diffusion limit of oxygen and nutrients leads to the development of pH, metabolic, and proliferation gradients [944]. Large MCTS ($\geq$500 μm in diameter) consist of a highly proliferating surface surrounding a quiescent, hypoxic region and a central necrotic core (Figures 181(A), (C), and (D)) [945]. While both the proliferating surface and the quiescent region have high levels of viable cells, the formation of the necrotic core in the center of the MCTS leads to increased cell death over time (Figure 181(B)).

Well-oxygenated cells exhibit different radiosensitivity compared to hypoxic or anoxic cells [946]. This difference in radiosensitivity is measured by the oxygen en-

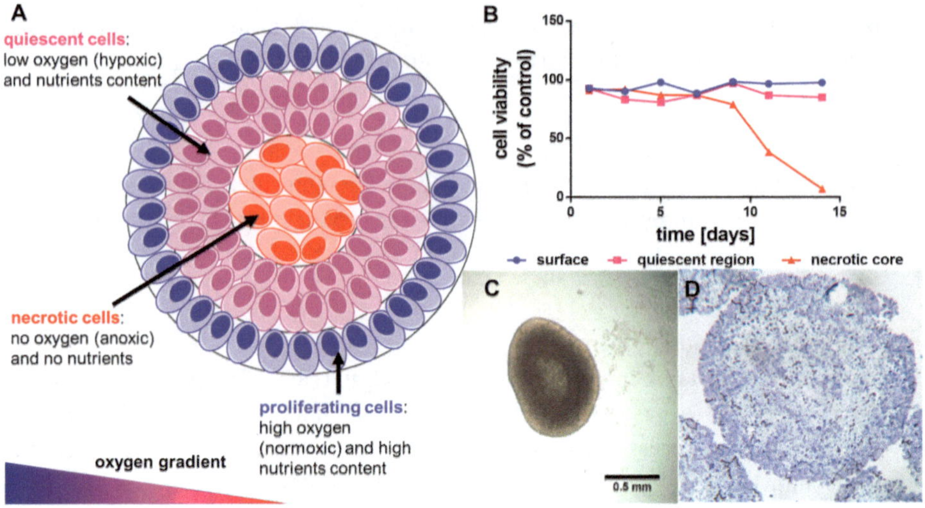

Figure 181: Schematic cross section of a MCTS showing the regions with different oxygen concentrations; B) Percentage of viable cells assigned to the normoxic, hypoxic and anoxic regions; C) MCTS grown for 11 days and expanding to a size of 500 nm, the lighter region in the center of the MCTS represents the necrotic core; D) Cross-section of a MCTS: the compact surface (normoxic) surrounds a region of loosely packed cells (quiescent).

hancement ratio (OER) (Eq. (22)). The OER is defined as the radiation dose of doses under hypoxic and normoxic conditions that produce the same lethal effect [947]. The OER of anoxic tumor regions is set to 1, while the OER for normoxic regions of a tumor ranges from 2.5 to 3.0.

$$\mathrm{OER} = \frac{\mathrm{Dose\ (hypoxia)}}{\mathrm{Dose\ (normoxic)}}. \tag{22}$$

Therefore, oxygen is considered to be a potent radiosensitizer. The oxygen-dependent enhancement effect is partly explained by the oxygen fixation hypothesis (Figure 182). During X-ray irradiation, the radiolysis of water produces hydroxyl radicals. The reaction of these free hydroxyl radicals with DNA results in the formation of DNA (base) radicals. In the presence of molecular oxygen, the reaction of DNA radicals with oxygen results in the formation of DNA peroxides. These DNA peroxides cannot be easily restored or repaired ("fixation" of the DNA damage). Under hypoxic or anoxic conditions, DNA radicals can be repaired by hydrogen donation from sulfhydryl-containing molecules (e. g., glutathione (GSH), cysteine) or other reducing molecules [947].

However, the oxygen fixation hypothesis has limitations. The direct interaction of ionizing radiation with DNA causes several types of DNA damage that do not result in the formation of DNA radicals. In addition, different levels of oxygen also affect DNA repair pathways. Radiosensitivity depends on the ability of hypoxic cells to adapt to this

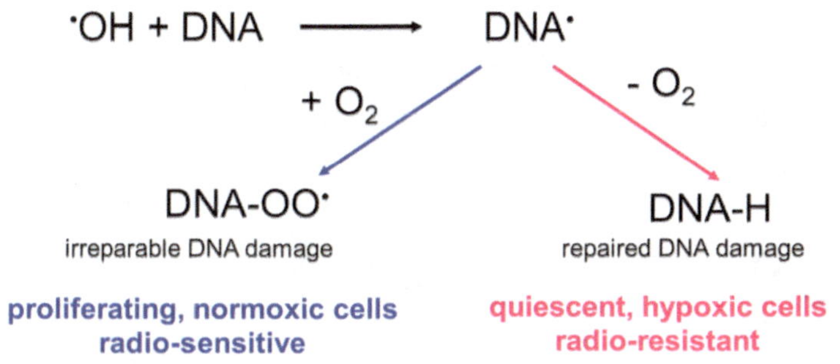

Figure 182: Reaction of DNA radicals under normoxic or hypoxic conditions according to the oxygen fixation hypothesis.

condition by inducing certain signaling pathways, such as the hypoxia-inducible factor pathway. Nevertheless, molecular oxygen is a potent radiosensitizer.

6.3.1.3 Fractionated radiation therapy

In fractionated radiation therapy (FRT), a single dose is divided into smaller doses and delivered to the patient over a period of time. FRT is designed to maximize the eradicating of tumor cells while reducing damage to normal tissue. This advantage of FRT is described by the **5 R's** of radiation therapy: **R**epair, **R**edistribution, **R**epopulation, **R**eoxygenation, and **R**adiation Sensitivity. In conventional FRT, daily doses of 2 Gy are administered 5 days per week, typically for 4 to 8 weeks. This allows normal tissue cells to regenerate through repair and repopulation, while the tumor tissue regains radiosensitivity through reoxygenation and redistribution. With hyperfractionated radiotherapy, the single dose is further reduced (<2 Gy), and the treatment is divided into multiple fractions per day (2–3 doses/day). This modified fractionation schedule takes into account the different radiosensitivities of tissues and tumors. In hypofractionated radiotherapy, a higher daily single dose (>2 Gy) is given to the patient. This reduces the total treatment time [948, 949].

The five Rs of radiation therapy

(1) *R*epair

As already described, X-rays can damage DNA directly or indirectly by generating ROS in the cytosol. ROS can enter the nucleus and oxidize DNA. Radiation-induced DNA damage (Figure 183) occurs in many types, from simple lesions such as base or sugar damage to complex lesions such as double-strand breaks [950]. These different types of DNA damage can lead to lethal, sublethal, or potentially lethal damage. Lethally damaged cells cannot be repaired and perish. A cell that is potentially lethally damaged can recover under certain circumstances. Sublethal DNA damage can be repaired if the cells have enough time to recover.

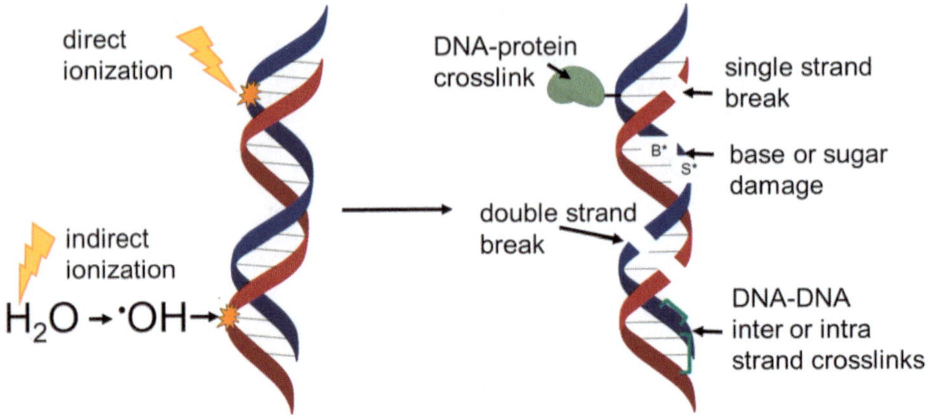

Figure 183: Multiple types of X-ray radiation-induced DNA damage.

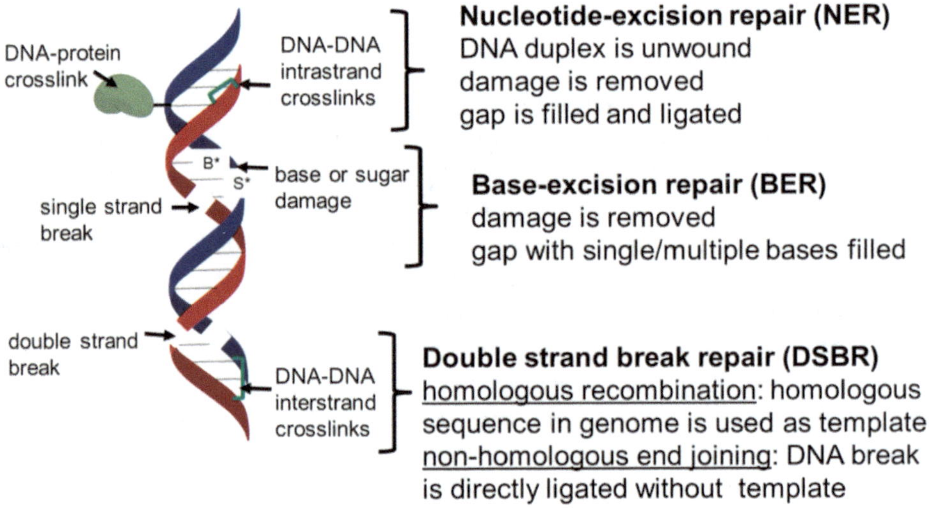

Figure 184: Fundamental process of DNA repair for the various types of radiation-induced damage.

Every cell has the ability to repair DNA damage. There are several repair pathways (Figure 184) that specifically rebuild DNA after irradiation to restore cell function and maintain genomic stability [951]. Fractionated radiation therapy gives the healthy tissue cells surrounding the tumor enough time to repair the X-radiation-induced damage. Cancer cells show many dysregulations of DNA repair pathways and have a lower DNA repair capacity compared to tissue cells [952]. In the given time, cancer cells cannot efficiently repair DNA damage and recover from radiation-induced damage.

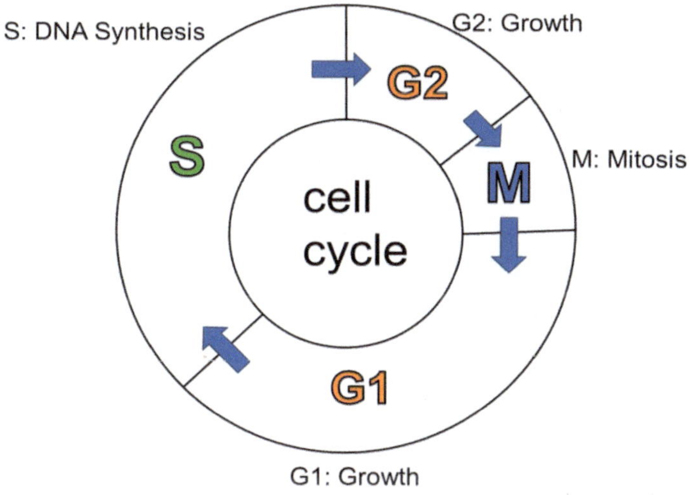

Figure 185: Cell cycle phases.

(2) *R*edistribution

Each cell undergoes cell divisions that follow the same four phases (Figure 185): 1) S phase: the cell copies its DNA; 2) G2 phase: the cell prepares to divide and condenses its DNA; 3) M phase: mitosis occurs and the cell divides; 4) G1 phase: the cell grows. Because each cell follows its own rhythm through these four phases, cells are in different phases of the cell cycle when radiation occurs. The cell cycle phases have different levels of radiosensitivity. The most radiosensitive phase is the G2/M phase, while the most resistant phase is the S/G0 phase [953]. The G0 phase is a resting or quiescent phase in which cells are not dividing or preparing to divide. In fractionated radiotherapy, cells in the radiosensitive G2/M phase die during irradiation. In the time between radiation doses, cells in the resistant phases continue their cell cycle into more radiosensitive phases. In addition, drugs can arrest cells in radiosensitive phases of the cell cycle.

(3) *R*epopulation

After exposure to X-rays, cancer and tissue cells increase their proliferation rate to replace the fatally damaged cells. The rate of repopulation varies depending on the specific characteristics of the tumor or tissue. Early responding tissues and tumors accelerate their repopulation. Surviving cells increase their doubling time and decrease their cell loss, which hinders the effectiveness of radiation therapy and induces radioresistance. Late-responding tissues and tumors start their repopulation after radiation treatment. Repopulation in slowly proliferating tumors is not an obstacle to radiation response [954].

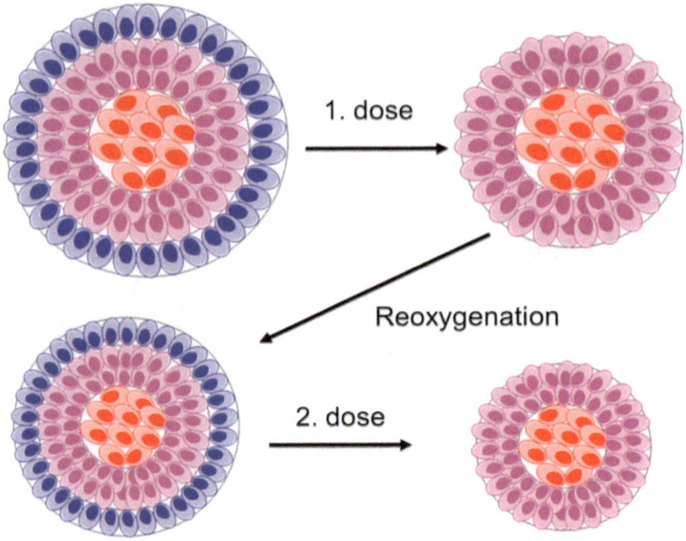

Figure 186: After each dose of radiation, the surface layer of the tumor is destroyed and the underlying cells are reoxygenated in the time between doses.

(4) *R*eoxygenation

Hypoxic regions of the tumor tissue show higher radioresistance compared to well-oxygenated tumor regions. After irradiation, the normoxic proliferating regions of the tumor are obliterated. In the time between the single radiation doses, the hypoxic regions gain access to the blood vessels [938]. These regions increase their oxygen and nutrient levels and begin to proliferate again. This reoxygenation (Figure 186) increases the radiosensitivity of the new surface layer, which is then damaged in the next radiation cycle. With each dose of radiation, the surface layer is destroyed, followed by shrinkage of the tumor and reoxygenation of the next layer of cancer cells, now facing the blood vessels.

(5) *R*adiosensitivity

Intrinsic radioresistance depends on the tumor- or patient-specific response to radiation therapy. Certain cancer cells are more sensitive to radiation than others. This radiosensitivity determines the intensity and amount of radiation needed to destroy a tumor. However, higher doses of radiation also increase toxicity and severe side effects in patients [951].

6.3.1.4 Acriflavine: a hypoxia-inducible factor 1 inhibitor

To adapt to hypoxic environments, cancer cells reprogram their metabolism, protein synthesis, and cell cycle progression through the synergy of transcription factors. This

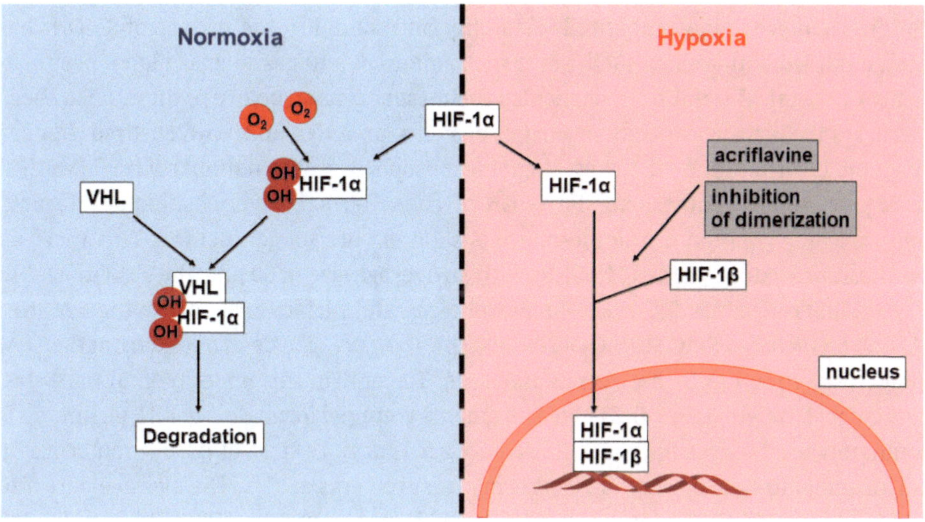

Figure 187: In normoxic cells, HIF-1α is continuously degraded; in response to hypoxia, HIF-1α combines with HIF-1β and initiates transcription of target genes; acriflavine inhibits the dimerization of HIF-1α and HIF-1β.

hypoxia-response program is controlled by oxygen-regulated transcription factors, among which hypoxia-inducible factor 1 (HIF-1) is of high importance. HIF-1 consists of an α and β subunit [955]. In normoxic cells, the HIF-1α subunit is constantly degraded via oxygen-dependent hydroxylation (Figure 187). The von Hippel–Lindau tumor suppressor protein (VHL) binds to the hydroxylated HIF-1α, leading to proteasomal degradation of HIF-1α [956]. Under hypoxic conditions, HIF-1α is stabilized and translocated to the nucleus. In the nucleus, the combination of HIF-1α with its β-subunit constitutes the transcriptionally active HIF-1 that binds to specific target genes [957]. The fluorescent drug acriflavine (ACF) inhibits the dimerization of HIF-1α and HIF-1β. The adaptation of cancer cells to hypoxia is hindered, which increases the chemo- and radiosensitivity of these cells [939, 958]. Unfortunately, the hydrophilic and cationic ACF exhibits rapid clearance and consequently poor tumor accumulation.

6.3.1.5 Drug delivery system: magnetic mesoporous silica nanoparticles

Nanoparticle delivery systems overcome the limitations of systemic drug delivery to targeted tumor tissues through cell-specific targeting, molecular transport to specific organelles, or other approaches. The physicochemical and biological properties of nanoparticles can be tailored to increase drug-loading capacity, prolong blood circulation time, and enhance target-specific drug delivery.

SPIONs are used as magnetic drug delivery systems that enhance tumor targeting in an inhomogeneous magnetic field. Despite their advantages, the application of SPIONs in drug delivery faces two major challenges: rapid agglomeration and surface oxidation. To improve biocompatibility, stability, and solubility, the surface of SPIONs has been modified with a silica shell. Silica shells offer advantages such as high biocompatibility or chemical stability. In particular, the mesoporous structure of these silica shells increases the loading capacity for drugs, proteins, or siRNA and, with customized pore sizes, the incorporation efficiency. Magnetic mesoporous silica nanoparticles (MMSNPs) were prepared by synthesizing the SPION core via thermal decomposition of $Fe(acac)_3$, followed by formation of the mesoporous shell via the sol-gel method. This two-step synthesis procedure yielded MMSNPs with an average size of 60 nm. The TEM images in (Figure 188) reflect the differences in morphology and surface structure of the MMSNPs (A) and SiO_2-coated SPIONs (SiO_2@SPIONs) (B). The pore size determined from the TEM images was found to be between 3 and 4 nm. To confirm this experimental pore size, Brunauer–Emmett–Teller (BET) measurements were performed. The BET technique is commonly used to calculate specific surface areas or pore sizes of porous materials by performing nitrogen adsorption isotherm measurements at 77 K. The average pore size obtained from the BET measurements was 46 Å or 4.6 nm. The MMSNPs were loaded with the drug ACF. In 3D calculations of the ACF molecule, the 3D size was 13.419 Å × 8.142 Å × 4.148 Å. Compared to the pore size of the MMSNPs, the size of the ACF molecule is much smaller than the diameter of the pore. Therefore, multiple ACF molecules could be accommodated in each pore, which should result in increased drug-loading capacity. The amount of drug loaded on the MMSNPs was measured by thermogravimetric analysis and fluorescence spectroscopy. The drug loading capacity (LC, Eq. (23)) and encapsulation efficiency (EC, Eq. (24)) were then calculated when 1 mg/mL MMSNP solution was functionalized with 0.5 mg/mL ACF solution.

$$\mathrm{LC}\,\% = \frac{\text{amount of entrapped drug}}{\text{nanoparticle weight}}, \tag{23}$$

$$\mathrm{EC}\,\% = \frac{\text{total drug added} - \text{free drug}}{\text{total drug}}. \tag{24}$$

The drug-loading capacity of MMSNPs in terms of LC% and EC% was compared with that of SiO_2-coated SPIONs (SiO_2@SPIONs). The nanoparticles were loaded with ACF by mixing 1 mg/ml MMSNP or SiO_2@SPION solution with 0.5 mg/ml ACF solution. As expected, ACF adsorbed at a significantly higher concentration on the MMSNP surface (0.19 mg ACF/mg MMSNP) than on the SiO_2@SPIONs (0.053 mg ACF/mg SiO_2@SPIONs). Consistently, the values of LC% (19 %) and EC% (38 %) of MMSNPs are higher than those of SiO_2@SPIONs with LC% = 5.3 % and EC% = 10.6 %.

MMSNPs exhibit a temporal evolution of drug (ACF) release that depends on the binding strength of the surface-adsorbed ACF. A significant amount of loaded ACF leaves the MMSNP surface pores right at the beginning, which is referred to as the initial burst of drug release (Figure 189) [959]. This phase of rapid drug release is followed by a very slow release of the remaining drug load. To control the drug release kinetics, the MMSNP surfaces were coated with a blocking agent or “gatekeeper” system [623]. These

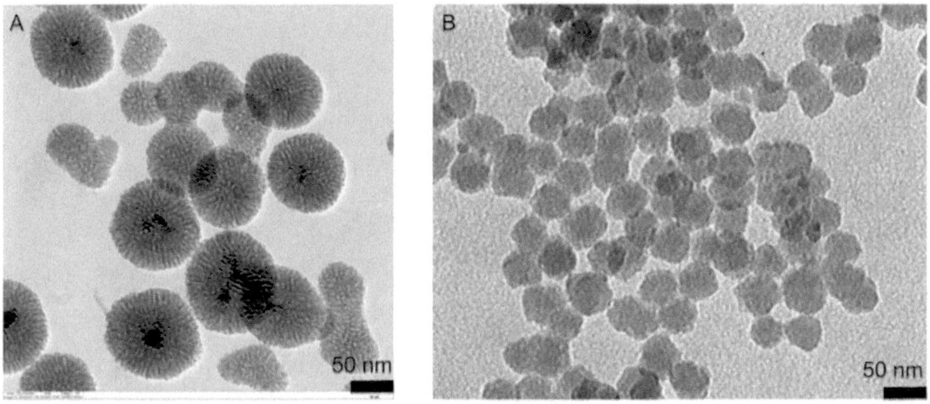

Figure 188: TEM images of the MMSNP (A) and SiO_2@SPIONs (B).

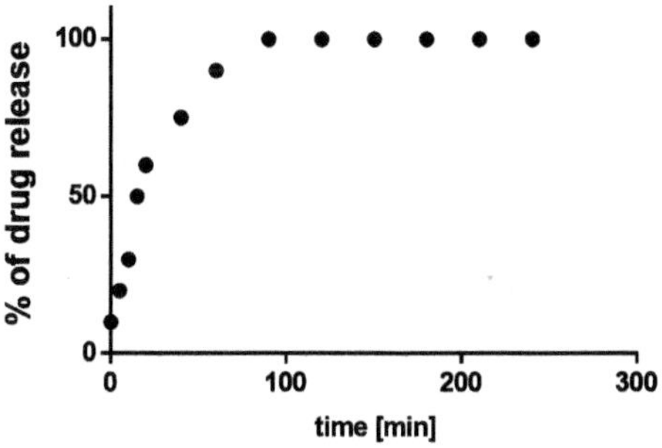

Figure 189: Drug (ACF) releasing profile of MMSNPs.

gatekeeper systems are sensitive to internal or external triggers that can stimulate drug release into the targeted cells/tissues in a controlled manner. Internal stimuli include pH-responsive, redox-responsive blockers, or blockers that are cleaved by certain enzymes [960]. External stimuli are also used to activate drug release [961]. External triggers such as light, alternating magnetic fields, ultrasound, or X-rays interact with the gatekeeper systems to open the drug-loaded pores.

Internal stimulus: pH responsiveness

pH-sensitive blockers respond to changes in the pH of the environment, particularly when the pH falls below a certain level. Drug release can be stimulated either during intracellular trafficking or by changes in the tumor microenvironment. Along the endosomal pathway, pH decreases, followed by endosomal escape of the loaded drug. Tumor hypoxia causes alterations in tumor metabolism, including the Warburg effect, which is associated with efflux of lactate into the tumor microenvironment. Therefore, pH-responsive surface coatings are used to control the drug release of MMSNPs.

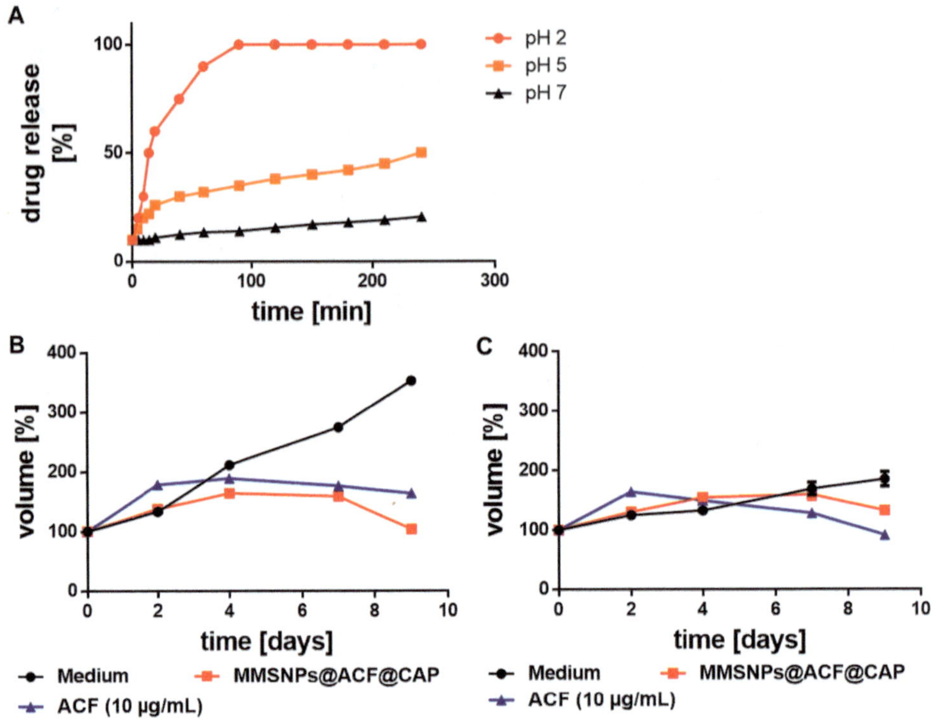

Figure 190: pH-depending release of ACF from MMSNPs@ACF@CAP (A), MCF-7 MCTS (B) or MCF-10A MCS (C), both incubated with MMSNPs@ACF@CAP or the drug ACF.

The surfaces of ACF-loaded MMSNPs (MMSNPs@ACF) were modified with (N-phenylaminoethyl)triethoxysilane (PhAMTES) and α-cyclodextrin (α-CD), which in combination form a pH-responsive capping agent (CAP) [962]. At neutral pH, α-CDs are tightly bound to PhAMTES through host–guest interactions, blocking the pores and preventing ACF release (Figure 190(A)). With decreasing pH, the amino group of PhAMTES is protonated and its binding to α-CD is weakened. The α-CDs leave the surface of the MMSNPs@ACF@CAP. The MMSNP surface pores are now open and ACF is released. The ACF release process was observed to start under mildly acidic conditions at pH 5 and to accelerate with decreasing pH (pH = 2; Figure 190(A)). These experimental results demonstrate the pH-responsiveness of the sequestrant.

Tumor spheroids grown from the breast cancer cell line MCF-7 (MCTS) and spheroids grown from the non-tumorigenic mammary epithelial cell line MCF-10A (MCS) were treated with MMSNPs@ACF@CAP or unbound drug. MCF-7-MCTSs were incubated with 50 µg/mL MMSNPs@ACF@CAP, which binds 10 µg/mL ACF, or with a solution of 10 µg/mL ACF. MCTS growth was measured every 2 days. MCTS in pure culture medium (Figure 190(B), black curve) grew exponentially. Exposure to MMSNPs@ACF@CAP or ACF solutions slowed their growth and eventually led to a decrease in MCTS size. After 9 days,

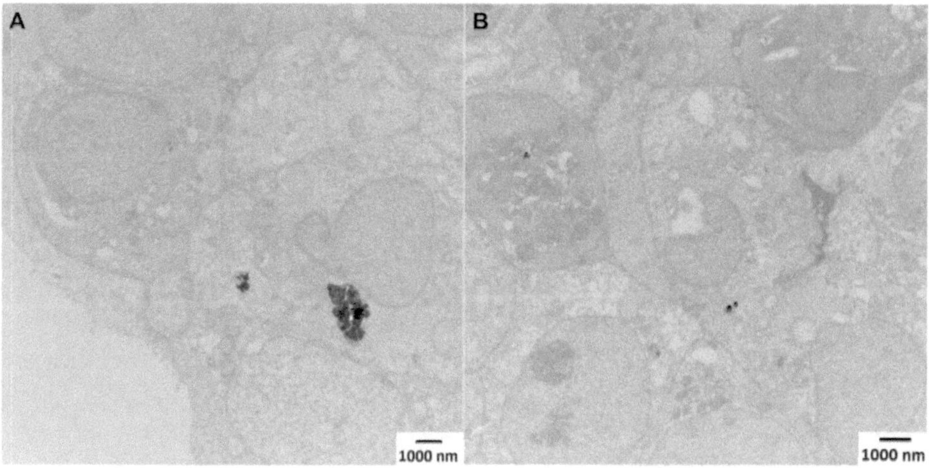

Figure 191: Cross-sections of MCTS treated with MMSNPs@CAP: agglomerates of MMSNPS@CAP in surface region (A) and deeper inside the MCTS (B).

the volume of the untreated MCTS increased by 253 %, while a reduced growth was observed for the treated MCTS (3 % MMSNPs@ACF@CAP, 64 % ACF). A significantly larger MCTS volume remained when the MCTS were treated with the free drug ACF.

These results were rationalized by the ability of the MMSNPs@ACF@CAP to penetrate into the inner zones of the MCTS and also reached the hypoxic regions where they efficiently released ACF due to the acidic environment. This rationalization was confirmed by TEM images of MCTS apparently containing MMSNPs@CAP (Figure 191). Large agglomerates of MMSNPs were found in the surface region of the MCTS (Figure 191(A)). In addition, smaller aggregates of MMSNPs were found deeper inside the MCTS (Figure 191(B)), indicating the accumulation of MMSNPs inside the MCTS.

On the other hand, MCS grown from non-tumorigenic MCF-10A cell line showed slower growth compared to MCTS (Figure 190(C), black curve). Interestingly, the growth curves of the MMSNPs@ACF@CAP treated MCS were significantly different from those of the MCTS. The volume of MCS exposed to ACF solution decreased from day 2 to day 9, indicating a toxic effect of the drug ACF on non-cancerous cells. In contrast, the growth curves of untreated MCS and MMSNPs@ACF@CAP loaded MCS were similar until day 7. On day 9, the volume of MMSNPs@ACF@CAP loaded MCS was smaller than that of untreated MCS. Apparently, MMSNPs@ACF@CAP also penetrated the MCS and released ACF into the hypoxic zones, which, similar to the MCTS, developed under hypoxia a glycolytic metabolism with a lowered pH.

External stimulus: X-rays

In the second example, SPIONs with mesoporous and non-mesoporous SiO_2 coatings (i. e., MMSNPs and MSNPs) were loaded with ACF, and X-ray radiation was applied as

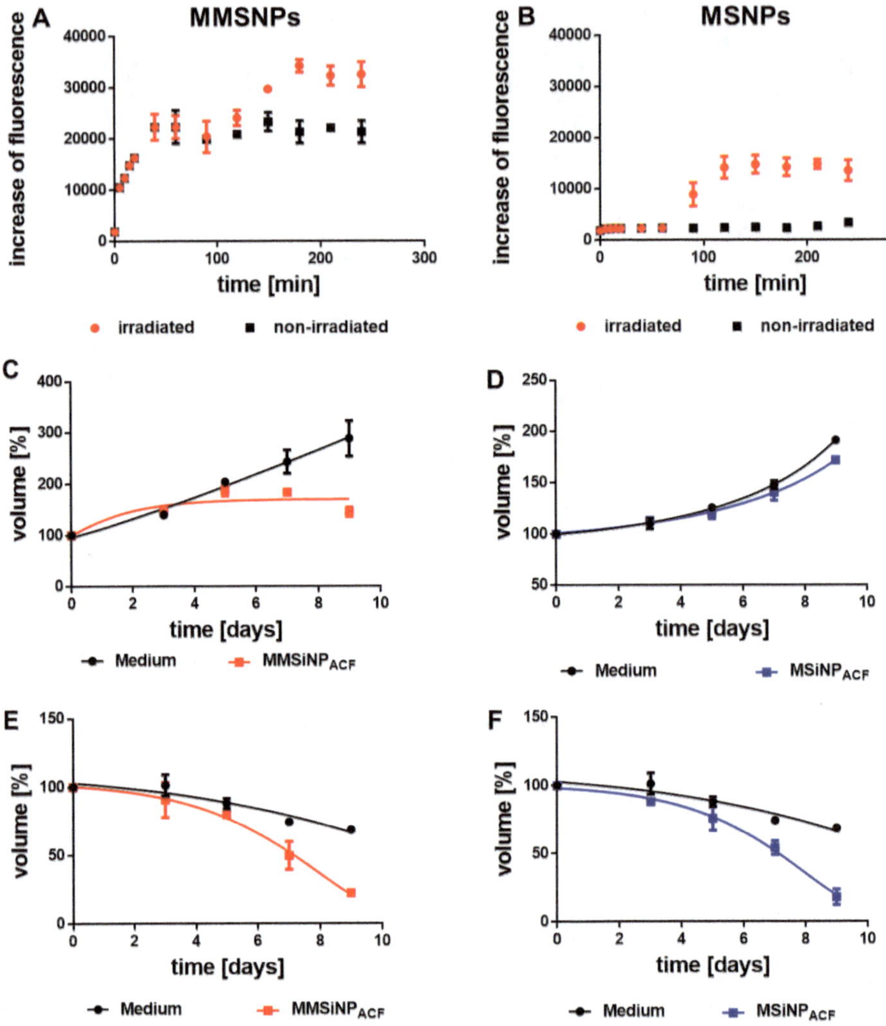

Figure 192: Drug-releasing profile of ACF-loaded MMSNPs (A) and MSNPs (B) without and with X-ray irradiation with a single dose of 1 Gy; growth curves of MCF-7 MCTS in culture medium (black curves) (C–F) and growth curves of MCF-7 MCTS incubated with ACF loaded MMSNPs (C) or MSNPs (D); growth curves of MCTS with ACF-loaded MMSNP(E) or MSNP (D) X-ray irradiated with two single doses of 5 Gy.

an external stimulus for drug release at a single dose of 1 Gy. MMSNPs and MSNPs were found to exhibit different drug loading capacities and consistently different drug release profiles. ACF-loaded MMSNPs without a capping agent showed an initial burst release when dispersed in water (Figure 193(A)). Positively charged ACF molecules formed strong electrostatic bonds with the negatively charged pore surface of MMSNPs. However, there are also ACF molecules that were unable to interact with the pore surface. These loosely bound ACF molecules were immediately released into the cell culture medium, while the more stable, electrostatically bound ACF molecules remained in the pores of the MMSNP surface. X-ray irradiation stimulated the release of the remaining ACF molecules (Figure 193(A), red curve). In the case of the non-mesoporous MSNPs, the ACF molecules were also electrostatically bound to the MSNP surface. When ACF-loaded MSNPs were dispersed in water, the ACF molecules did not desorb from the surface of the MSNPs (Figure 193(B), black curve). Upon exposure to X-rays, the ACF molecules were released from the surface of the MSNPs (Figure 193(B), red curve).

Both nanocarrier systems were tested in MCF-7 MCTS. In the first experiment, MCTS were incubated with ACF-loaded MMSNPs or MSNPs. A volume decrease was observed in MCTS treated with MMSNPs (Figure 192(C), red curve). The MMSNPs underwent drug release with and without X-ray stimulation (Figure 192(A)), causing the MCTS to shrink in volume. The growth curves of untreated and MSNPs-treated MCTS showed similar behavior, with a slight decrease in size of the MSNPs-loaded MCTS (Figure 192(D)). This result was explained by a small release of ACF molecules into the MCTS at longer incubation times. Different results were observed when the MSNP-treated MCTS were irradiated with two consecutive doses of 5 Gy X-ray radiation as anticancer treatment. This treatment caused a decrease in the volume of MCTS that were not incubated with either nanocarrier system (Figure 192(E)–(F), black curve). However, the cells in the hypoxic regions were observed to be resistant to X-ray radiation since only a small shrinkage of MCTS occurred. Exposure of the MMSNP- or MSNP-treated MCTS to two single X-ray doses apparently caused the release of the electrostatically bound ACF from the MMSNP- or MSNP-surface inside the MCTS. Further reduction in MCTS volume was observed. At day 9, 22 % of the initial volume remained in the MMSNP-loaded MCTS and 18 % of the initial volume remained in the MSNP-treated MCTS.

6.3.1.6 Nanoparticles as radiosensitizers

SPIONs

SPIONS with a Fe_3O_4 composition were synthesized by co-precipitation under an argon atmosphere to prevent oxidation and used as radiosensitizers [187]. Fe^{2+} and Fe^{3+} cations in the SPION surface catalyze the Fenton (Eq. (25)) and Haber–Weiss reactions (Eqs. (26)–(28)). As a byproduct of the oxidative phosphorylation, $O_2^{\bullet -}$ and H_2O_2 are generated. These ROS are converted to the highly toxic $^{\bullet}OH$ at the surfaces of the SPIONs according to Eq. (25) and Eqs. (26)–(28). Therefore, uncoated SPIONs show an increased intracellular ROS production and toxicity. SPIONs were coated with citrate to cover the surfaces and improve the biocompatibility.

$$Fe^{2+} + H_2O_2 \rightarrow Fe^{3+} + OH^- + {}^{\bullet}OH, \tag{25}$$

$$Fe^{3+} + O_2^{\bullet -} \rightarrow Fe^{2+} + O_2, \tag{26}$$

$$O_2^{\bullet -} + 2H^+ \rightarrow H_2O_2 + O_2, \tag{27}$$

$$Fe^{2+} + H_2O_2 \rightarrow Fe^{3+} + OH^- + {}^{\bullet}OH. \tag{28}$$

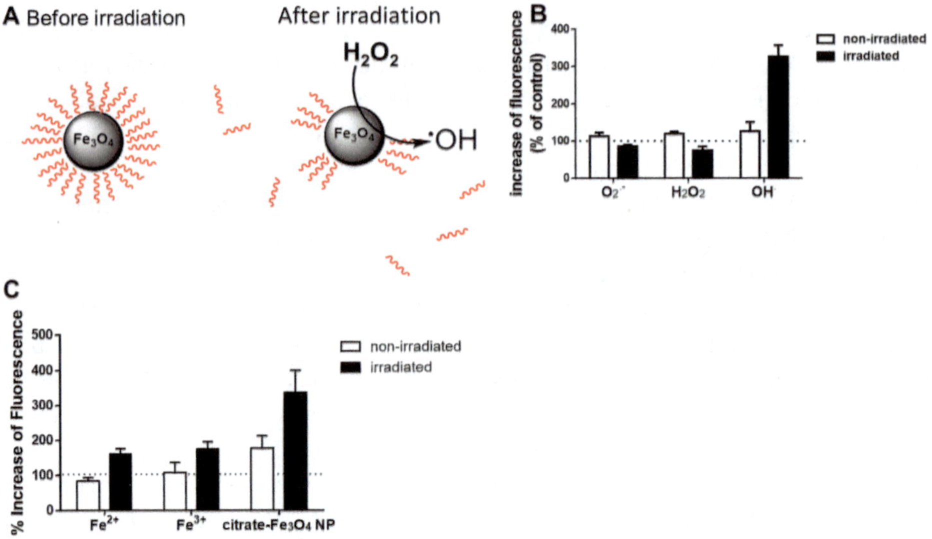

Figure 193: After X-ray irradiation, the SPION surfaces are activated (A), the surface Fe^{2+} and Fe^{3+} cations produce $^{\bullet}OH$ according to the Fenton and Haber–Weiss reactions (B), MCF-7 cells incubated with SPIONs produce more $^{\bullet}OH$ after X-ray irradiation compared to the treatment with ferric or ferrous chloride (C).

X-ray irradiation with a single dose of 1 Gy freed SPION surfaces from citrate ligands and thus provided the catalytic activity of the Fe^{2+} and Fe^{3+} cations (Figure 193). Surface-bound Fe^{2+} or Fe^{3+} ions can now catalyze the Fenton or Haber–Weiss reaction. According to the $O_2^{\bullet-}$-driven Haber–Weiss reaction, the intracellular $O_2^{\bullet-}$ concentration decreases after irradiation (Figure 193(B)). H_2O_2 is converted by the Fe^{2+} ions to $^{\bullet}OH$. After irradiation, the intracellular concentration of H_2O_2 concentration decreases, while the concentration of $^{\bullet}OH$ radicals increases dramatically (Figure 193(B)). This huge enhancement in $^{\bullet}OH$ generation made SPIONs potential candidates as radiosensitizers. To demonstrate the great potential of SPIONs, the $^{\bullet}OH$ production inside MCF-7 cells was compared to cells treated with Fe(II)chloride and Fe(III)chloride in the same concentrations. Due to their high surface-to-volume ratio and enhanced catalytic activity, SPIONs generate $^{\bullet}OH$ more efficiently than the corresponding iron salts (Figure 193(C)).

In addition to the breast cancer cells, non-tumorigenic mammary epithelial cells MCF-10A were treated with uncoated and citrate-coated SPIONs and irradiated with a single dose of 1 Gy (Figure 194). Interestingly, there is a huge difference in $^{\bullet}OH$ production between MCF-7 and MCF-10A cells when they incorporated citrate-coated SPIONs (Figure 194, red circle). To explain this difference, the intracellular uptake of the SPIONs as well as the redox homeostasis of the MCF-7 and MCF-10A cell lines were investigated. The citrate-coated SPIONs were internalized by endocytosis into both cell lines [188], but the amount of SPIONs inside the cells differed enormously (Figure 195). While the breast cancer MCF-7 cells internalized a huge amount of SPIONs, the non-tumorigenic

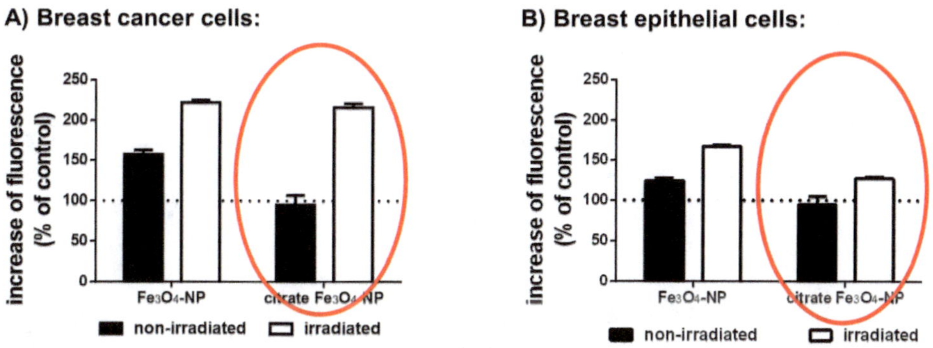

Figure 194: Breast cancer MCF-7 cells (A) and non-tumorigenic breast epithelial MCF-10A cells (B) show a huge difference (red circle) in their •OH production when treated with citrate-coated SPIONs.

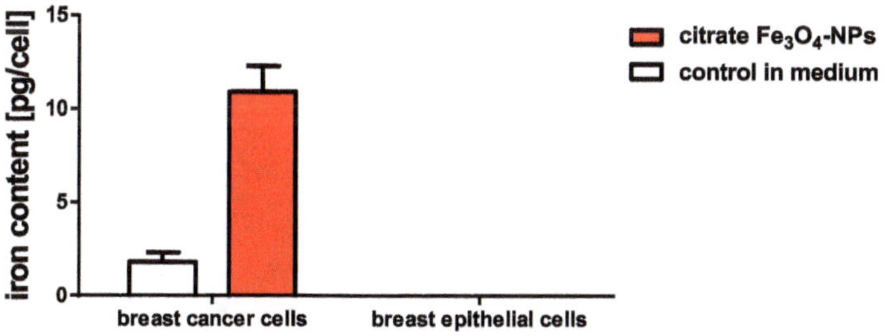

Figure 195: Intracellular SPION uptake capacity in MCF-7 and MCF-10A cells determined by the intracellular iron concentration.

MCF-10A cells internalized only a small amount of SPIONs within the same time. Many more SPIONs inside the cancer cells can catalyze the Fenton and Haber–Weiss reactions.

The formation of ROS and their degradation defines the redox environment of cells. $O_2^{\bullet -}$, which is mainly generated in mitochondria, is immediately converted to H_2O_2 via superoxide dismutase (SOD) (Figure 196(A)). H_2O_2 is detoxified mainly via the enzyme catalase or glutathione peroxidase (GPx), whereby water is produced. Noncancerous cells thus maintain their redox homeostasis. In contrast, cancer is associated with an oxidative environment and increased intracellular oxidative stress [963–965]. This oxidative environment is caused by elevated levels of intracellular ROS that arise either from enhanced ROS production or a disturbance of the ROS degradation system. In many types of cancer cells or tissues the levels of ROS-scavenging enzymes are significantly altered [966, 967]. For MCF-7 cells, the differences in enzyme superoxide dismutase (SOD) and GPx activity were examined in comparison to the non-tumorigenic MCF-7 cells. The SOD levels and activity are elevated inside MCF-7 cells (Figure 196(B)), while the GPx level and activity are drastically reduced (Figure 196(D)). This imbalance in ROS scavenging enzymes leads to increased H_2O_2 levels and a pro-oxidative state inside the MCF-7 cells (Figure 196(C)). $O_2^{\bullet -}$ emerging in the MCF-7 cells was rapidly converted to H_2O_2. Due

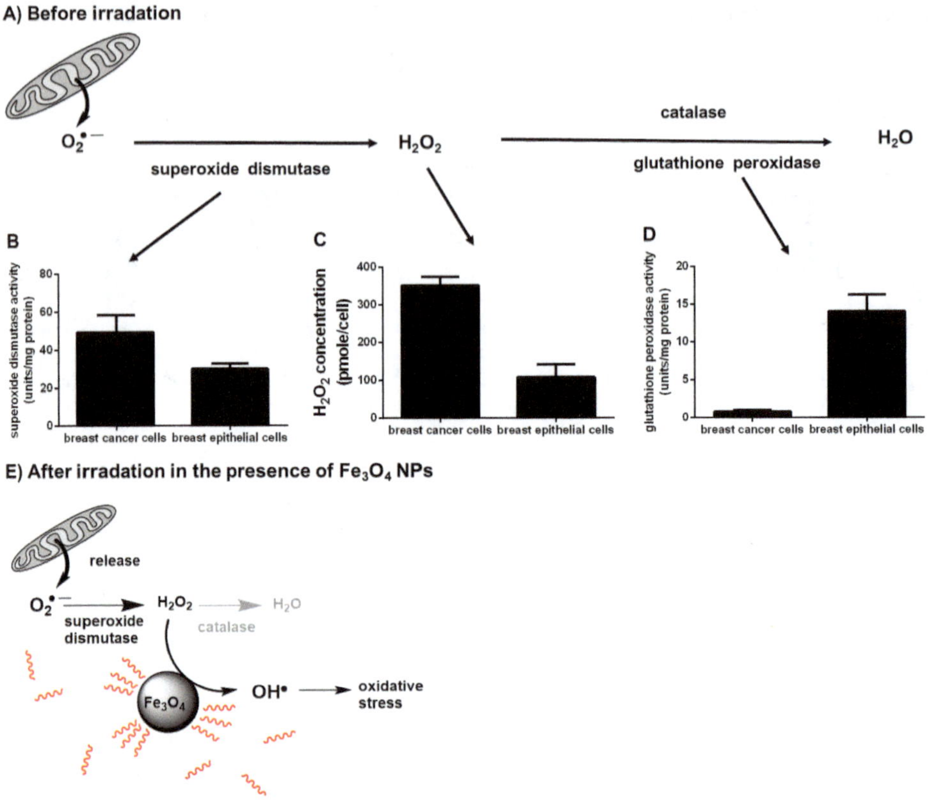

Figure 196: Enzymatic superoxide degradation to water (A), differences in SOD activity (A), intracellular hydrogen-peroxide concentration (B) and GPx activity (C) inside MCF-7 and MCF-10A cells, hydroxyl radical generation in the presence of SPIONs after X-ray irradiation (E).

to the low levels of cytoplasmic GPx, H_2O_2 is degraded very slowly, resulting in elevated intracellular H_2O_2 levels. This type of enzymatic alteration can also be found in other cancer cells, such as lung cancer cells A549 or colon cancer cells Caco-2. These high levels of H_2O_2 inside cancer cells have severe consequences, when SPIONs are present in the cells. After X-ray irradiation, the SPIONs, freed from surface structures and thus activated, catalyzed the Fenton reaction which reduced H_2O_2 to $^{\bullet}OH$ (Figure 196(B)). The H_2O_2 driven Fenton reaction is the favored reaction that occur at the SPION surfaces. Due to the low levels of $O_2^{\bullet -}$, the Haber–Weiss reaction was catalyzed to a lesser extent.

In short, the higher intracellular concentrations of SPIONs and hydrogen peroxide in cancer cells lead to a dramatic increase in the production of hydroxyl radicals. Low concentrations of SPIONs and hydrogen peroxide prevent the generation of hydroxyl radicals in non-tumorigenic cells.

The co-precipitation method performed under an argon atmosphere yielded magnetite SPIONs (Fe_3O_4). The same synthesis performed under ambient conditions pro-

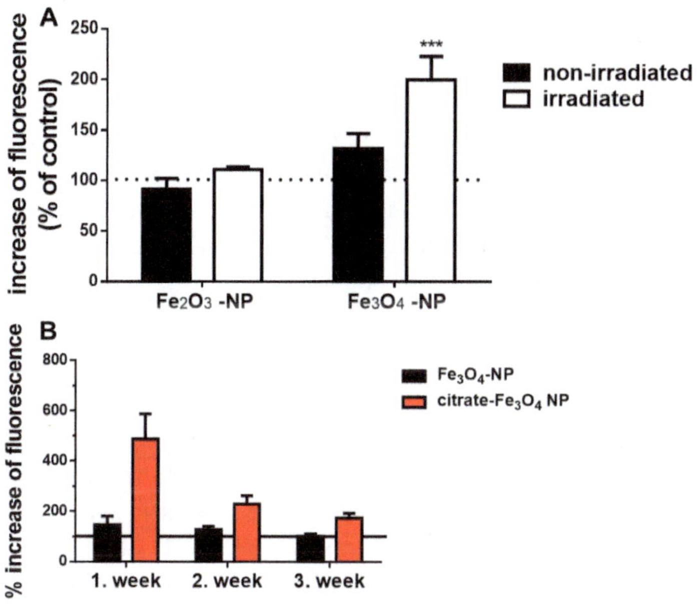

Figure 197: Difference in hydroxyl radical production of Fe_3O_4 NPs or Fe_2O_3 NPs (A); Fe_3O_4 NPs gradually oxidize to Fe_2O_3 NPs losing their efficiency for hydroxyl radical production depending on the storage time under ambient conditions (B).

duces mainly γ-maghemite SPIONs (Fe_2O_3). While magnetite SPIONs contain both Fe^{2+} and Fe^{3+} cations on the surface, maghemite SPIONs are composed of only Fe^{3+} cations. Fe^{3+} cations catalyze the superoxide-driven Haber–Weiss reaction, which is less efficient due to the low intracellular $O_2^{\bullet-}$ superoxide concentrations in MCF-7 cancer cells. Therefore, the magnetite SPIONs did not generate ROS after X-ray irradiation (Figure 197(A)). In contrast, the magnetite SPIONs produced high levels of hydroxyl radicals via the Fenton reaction catalyzed by Fe^{2+} cations (Figure 197(A)). Unfortunately, storage of the magnetite SPIONs under ambient conditions leads to oxidation of surface Fe^{2+} ions to Fe^{3+} ions. With increasing storage time the SPIONs lost their ability to generate hydroxyl radicals when exposed to X-rays (Figure 197(B)). Coating magnetite SPIONs with citric acid, for example, retarded but did not prevent surface oxidation (Figure 197(B)). After one week, citrate-coated SPIONs were still very efficient radiosensitizers. However, their ability to generate hydroxyl radicals decreased over time.

As promising as the results are, the application of SPIONs as radiosensitizers has its limitations. Although SPIONs are very efficient at producing ROS in response to X-rays, especially in cancer cells, their sensitivity to oxidation is problematic, so it is important to find ways to overcome this limitation. Alternative materials with similar X-radiation dose enhancing effects that are stable against oxidation should be developed.

Ferrite and metal oxide nanoparticles

Ferrite nanoparticles ($MeFe_2O_4$ NPs, Me = Co^{2+}, Zn^{2+}, Cu^{2+} or Mn^{2+}) and metal oxide nanoparticles (M_xO_y NPs) have been investigated for their radiosensitizing and radioprotective effects. These ferrite nanoparticles have in common that they catalyze the Fenton-like formation of hydroxyl radicals. Replacing the Fe^{2+} ions in magnetite SPIONs with other Me^{2+} cations, which are more stable against environmental oxidation, should overcome the limitations of magnetite SPIONs. The radiosensitization capabilities of $MeFe_2O_4$ NPs are subsequently discussed, along with their own advantages and limitations.

(a) Cobalt ferrite ($CoFe_2O_4$) NPs

$CoFe_2O_4$ NPs are superparamagnetic nanoparticles that are stable to oxidation. Co^{2+} can catalyze the formation of hydroxyl radicals in a Fenton-like reaction. What is problematic is their high toxicity mainly due to leakage of Co^{2+} ions into the cytoplasm. To prevent Co^{2+} release and increase therewith their biocompatibility, $CoFe_2O_4$ NPs were coated with self-assembling monolayers of 1-methyl-3-(dodecylphosphonic acid)imidazolium bromide (imidazolium PA) or (2-{2-(2-hydroxy-ethoxy)-ethoxy}-ethylphosphonic acid ($H(OC_2H_4)_3$ PA)[968].

$$Co^{2+} + H_2O_2 \rightarrow Co^{3+} + OH^- + {}^{\bullet}OH. \qquad (29)$$

MCF-7 and MCF-10A cells were treated with uncoated Imidazolium-PA and $H(OC_2H_4)_3$-PA $CoFe_2O_4$ NPs for 24 h to ensure their internalization. Afterwards both cell lines were irradiated with a 1 Gy X-ray dose. In MCF-7 cells the ROS production increased in the presence of all three $CoFe_2O_4$ NPs (Figure 198(A)). Uncoated $CoFe_2O_4$ NPs already produced ROS in nonirradiated cells. X-ray irradiation further enhanced the ROS production. In contrast, Imidazolium-PA and $H(OC_2H_4)_3$-PA coated $CoFe_2O_4$ NPs generated significantly lower ROS concentrations in non-irradiated cells. The ROS generation drastically increased after X-ray irradiation of MCF-7 cells with coated $CoFe_2O_4$ NPs (Figure 198(A)).

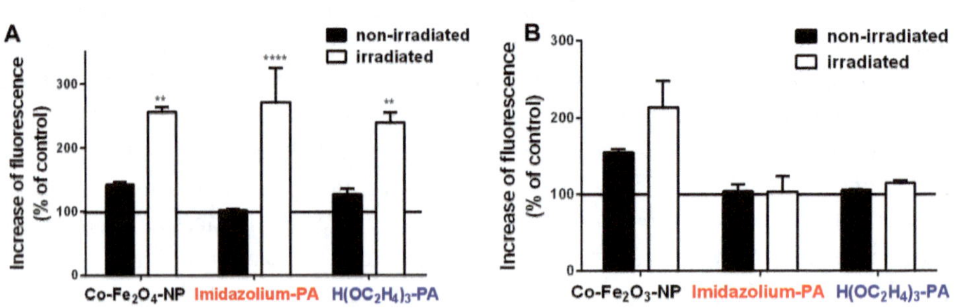

Figure 198: Hydroxyl radical production of uncoated and coated $CoFe_2O_4$ NPs in MCF-7 (A) and MCF-10A (B) cells.

Elevated ROS production was also observed in nonirradiated MCF-10A cells when containing uncoated $CoFe_2O_4$ NPs. X-ray irradiation of these MCF-10A cells led to further enhancement of ROS production. Different results were found for coated $CoFe_2O_4$ NPs. These $CoFe_2O_4$ NPs did not exhibit any change of ROS formation in MCF-10 cells either before or after X-ray irradiation (Figure 198(B)). The findings are similar to those of citrate-coated SPIONs (Figure 193). ROS generation was detected only in cancer cells, but not in non-tumorigenic cells. Thus, appropriately coated $CoFe_2O_4$ NPs are suitable for use as radiosensitizers.

To compare the radiosensitizing potential of both Fe_3O_4 and $CoFe_2O_4$ NPs, a clonogenic cell survival assay was conducted [969]. Cell surviving fractions (SF) were calculated for X-ray doses of 0–3 Gy. The SF values were then plotted against the radiation doses. The survival curves were fitted to the linear quadratic function: $\ln SF = -(\alpha D + \beta D^2)$. The different survival curves are shown in Figure 199. The black curve in each graph shows cell survival of cells without Fe_3O_4 and $CoFe_2O_4$ NPs. Curves below this black median curve indicate a radiosensitizing effect. Since the difference between cancer cells and healthy cells was observed only for the coated NPs, these were further considered. The survival curves of MCF-7 cells treated with Imidazolium-PA and $H(OC_2H_4)_3$-PA $CoFe_2O_4$ NPs lie far below the medium curve (Figure 199(A)). A similar

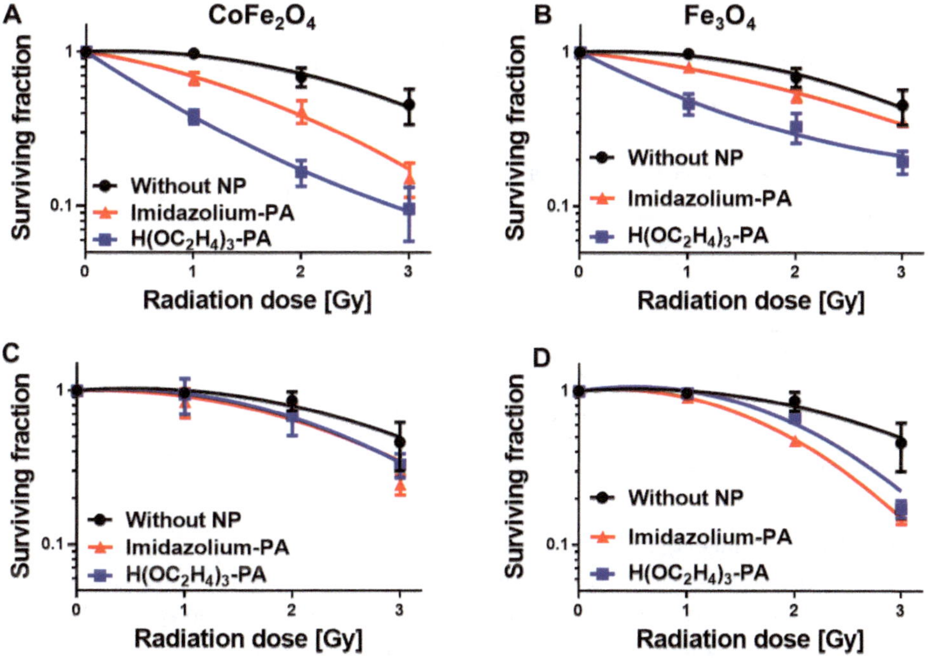

Figure 199: Survival curves of MCF-7 cells incubated with uncoated and coated $CoFe_2O_4$ NPs (A) or Fe_3O_4 NPs (B) and irradiated with X-ray doses of 0–3 Gy; survival curves of MCF-10A cells treated with uncoated or coated $CoFe_2$O4 NPs (C) or Fe_3O_4 NPs (D) and irradiated with X-ray doses of 0–3 Gy.

result was found for MCF-7 cells treated with Imidazolium-PA and $H(OC_2H_4)_3$-PA NPs (Figure 199(B)). In both cases, the $H(OC_2H_4)_3$-PA coated Fe_3O_4 NPs turned out to be more efficient than the Imidazolium-PA coated NPs. Unfortunately, the survival curves of MCF-10A cells treated with Imidazolium-PA and $H(OC_2H_4)_3$-PA $CoFe_2O_4$ NPs (Figure 199(C)) or Imidazolium-PA and $H(OC_2H_4)_3$-PA Fe_3O_4 NPs (Figure 199(D)) lie below the cell culture medium curve, but not as clear as for the cancer cells.

The Dose Modifying Factor (DMF) accounts for the radiosensitizing or radioprotective potential of nanoparticles. The DMF values are calculated from the survival curves. The radiation dose required to reduce the survival rate to 50 % was determined for nanoparticle-treated and untreated cells. The ratio of these doses then defines the DMF values (Eq. (30)). A DMF value below 1 indicates a radiosensitizing effect and a DMF value above 1 indicates a radioprotective effect.

$$\mathrm{DMF} = \frac{\text{dose to produce } a \text{ survival rate of 50 \% with NPs}}{\text{dose to produce } a \text{ survival rate of 50 \% in controls}}. \tag{30}$$

The determined DMF values are summarized in Table 15. All values are below 1 confirming the radiosensitizing potential of all nanoparticles under study. The lowest value was found for the $H(OC_2H_4)_3$-PA coated $CoFe_2O_4$ NPs in MCF-7 cells. In contrast, MCF-10A cells treated with the same nanoparticles have a much higher value, but still below 1. This difference in DMF values between MCF-7 and MCF-10A cells is very promising, as it suggests selective radiosensitization of cancer cells while sparing healthy cells. Therefore, $H(OC_2H_4)_3$-PA coated $CoFe_2O_4$ NPs could be excellent candidates for the application in radiation cancer therapy.

Table 15: DMF values calculated from the survival curves in Figure 199.

A) MCF-7	**Co-Fe_2O_4-NP**	**Fe_3O_4-NP**
Imidazolium-PA	0.52	0.59
$H(OC_2H_4)_3$-PA	0.23	0.32

B) MCF-10A	**Co-Fe_2O_4-NP**	**Fe_3O_4-NP**
Imidazolium-PA	0.65	0.67
$H(OC_2H_4)_3$-PA	0.75	0.75

(b) Manganese ($MnFe_2O_4$) ferrite NPs

Surface manganese ions can catalyze the Fenton reaction similar to Fe^{2+} (Eqs. (31)–(32)). $MnFe_2O_4$ NPs were synthesized by co-precipitation and were coated with self-assembling monolayers of Imidazolium-PA to increase water solubility.

$$Mn^n + O_2^{\bullet -} \rightarrow Mn^{n-1} + O_2, \tag{31}$$

$$Mn^{n-1} + H_2O_2 \rightarrow Mn^n + OH^- + {}^{\bullet}OH. \tag{32}$$

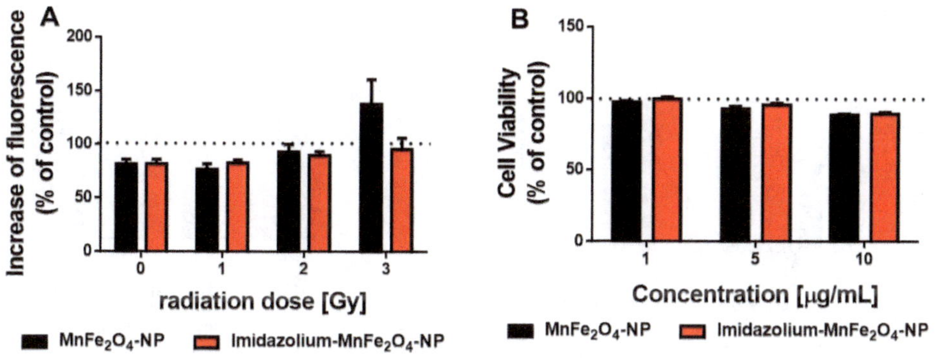

Figure 200: ROS generation in MCF-7 cells treated with 25 µg/mL uncoated and coated $MnFe_2O_4$ NPs and irradiated with X-ray doses of 0–3 Gy (A); cell viability of these NPs (B).

While the uncoated $MnFe_2O_4$ NPs were observed to enhance the ROS formation when exposed to a single X-ray dose of Gy, Imidazolium-PA coated $MnFe_2O_4$ NPs did not produce any ROS (Figure 200(A)). Despite their high biocompatibility (Figure 200(B)), the coated and uncoated $MnFe_2O_4$ NPs are not suitable for use as radiosensitizers.

(c) Zinc ferrite ($ZnFe_2O_4$) NPs

Zn^{2+} is not a typical Fenton catalyst. However, Zn^{2+} ions can generate ROS in a Fenton-like manner yielding hydroxyl radicals. Reaction equations (Eqs. (33)–(35)) show a possible mechanism for the ROS formation [813].

$$Zn^{2+} + H_2O_2 \rightarrow Zn^{II}(O_2H^-)^+ + H^+, \tag{33}$$

$$Zn^{II}(O_2H^-)^+ + H_2O_2 \rightarrow Zn^{II}(O_2H^-)(H_2O_2)^+, \tag{34}$$

$$Zn^{II}(O_2H^-)(H_2O_2)^+ \rightarrow Zn^{2+} + OH^- + {}^{\bullet}OH + HO_2^{\bullet}. \tag{35}$$

$ZnFe_2O_4$ NPs were synthesized by alkaline co-precipitation and coated with Imidazolium-PA. Uncoated $ZnFe_2O_4$ NPs generate ROS even in nonirradiated cells (Figure 201(A)). With increasing radiation doses, ROS production rises in MCF-7 cells treated with uncoated $ZnFe_2O_4$ NPs. In contrast, in MCF-7 cells incubated with Imidazolium-PA coated $ZnFe_2O_4$ NPs, ROS generation started to increase after exposure to a 2 Gy X-ray dose. Unfortunately, the uncoated $ZnFe_2O_4$ NPs are cytotoxic (Figure 201(B)). By contrast, the Imidazolium-Pa coated $ZnFe_2O_4$ NPs exhibited a higher cell viability. In comparison, coated $CoFe_2O_4$ NPs or Fe_3O_4 NPs produce ROS after irradiation with a single dose of 1 Gy, while in MCF-7 with coated $ZnFe_2O_4$ NPs no ROS generation was observed. $CoFe_2O_4$ NPs or Fe_3O_4 NPs are more suitable for X-ray radiation dose enhancement because they do generate ROS even at smaller X-ray doses.

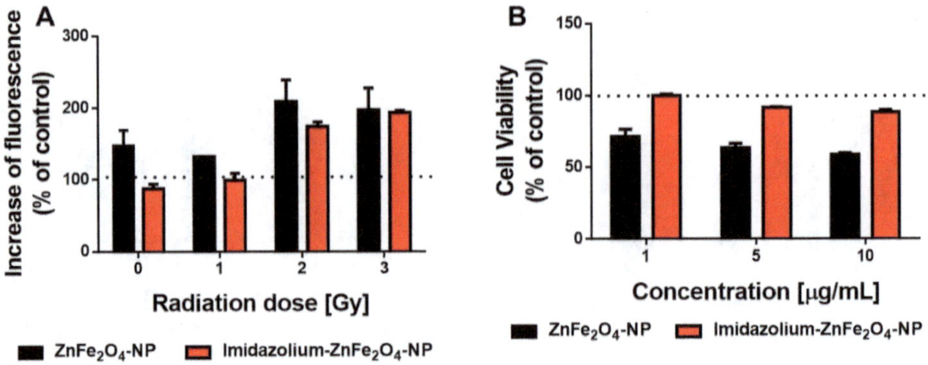

Figure 201: ROS generation in MCF-7 cells treated with 25 µg/mL uncoated and coated $ZnFe_2O_4$ NPs and irradiated with X-ray doses of 0–3 Gy (A), cell viability studies of these NPs (B).

(d) Copper ferrite ($CuFe_2O_4$), Copper(I)oxide (Cu_2O) and Copper(II)oxide (CuO) NPs

Fenton-like reactions can be catalyzed by Cu^{2+} (Eq. (36)) as well as by Cu^{+} (Eq. (39)) to yield hydroxyl radicals. As in the superoxide driven Haber–Weiss reaction, Cu^{2+} can be reduced to Cu^{+} by superoxide (Eq. (37)). Cu^{2+} could also be reduced by hydrogen peroxide producing superoxide (Eq. (37)) [814, 900]. Thus, Cu(II)-ferrite ($CuFe_2O_4$) NPs could be used as Fenton catalysts in radiation therapy. In comparison Cu(I)oxide (Cu_2O) NPs and Cu(II)oxide (CuO) NPs were also examined to determine whether the Fenton reaction based on Cu^{2+} or Cu^{+} is more efficient. All three types of copper oxide NPs were coated with PVP to increase water dispersibility and biocompatibility.

$$Cu^{2+} + H_2O_2 \rightarrow Cu^{3+} + OH^- + {}^{\bullet}OH, \quad (36)$$

$$Cu^{2+} + H_2O_2 \rightarrow Cu^{+} + 2H^{+} + O_2^{\bullet -}, \quad (37)$$

$$Cu^{2+} + O_2^{\bullet -} \rightarrow Cu^{+} + O_2, \quad (38)$$

$$Cu^{1+} + H_2O_2 \rightarrow Cu^{2+} + OH^- + {}^{\bullet}OH. \quad (39)$$

Twenty-five µg/mL of the various nanoparticle solutions were added to MCF-7 cells, followed by overnight incubation. After 24 h, the cells were irradiated with a single dose of 1 Gy. Nonirradiated MCF-7 treated with nanoparticles have already increased ROS levels compared to MCF-7 without nanoparticles (Figure 202(A)). After irradiation the ROS concentration inside the MCF-7 cells increased further (Figure 202(A), white bars). Compared to the Cu(II)-ferrite and Cu(II)oxide NPs, the Cu(I)oxide NPs led to the highest increase in ROS generation after irradiation. Therefore, the Fenton reaction catalyzed by Cu^{+} was more efficient compared to Cu^{2+}. Cu^{+} is easily oxidized to Cu^{2+} under ambient conditions, which is indicated by a color change of the Cu_2O NPs. Cu_2O NPs are not suitable for use as radiosensitizers. Another limitation arises from the huge intracellular ROS production in non-tumorigenic MCF-10A cells (Figure 202(B)). While only the Cu_2O NPs are toxic for the MCF-7 cells (Figure 203(C)), all three copper oxide NPs are toxic to

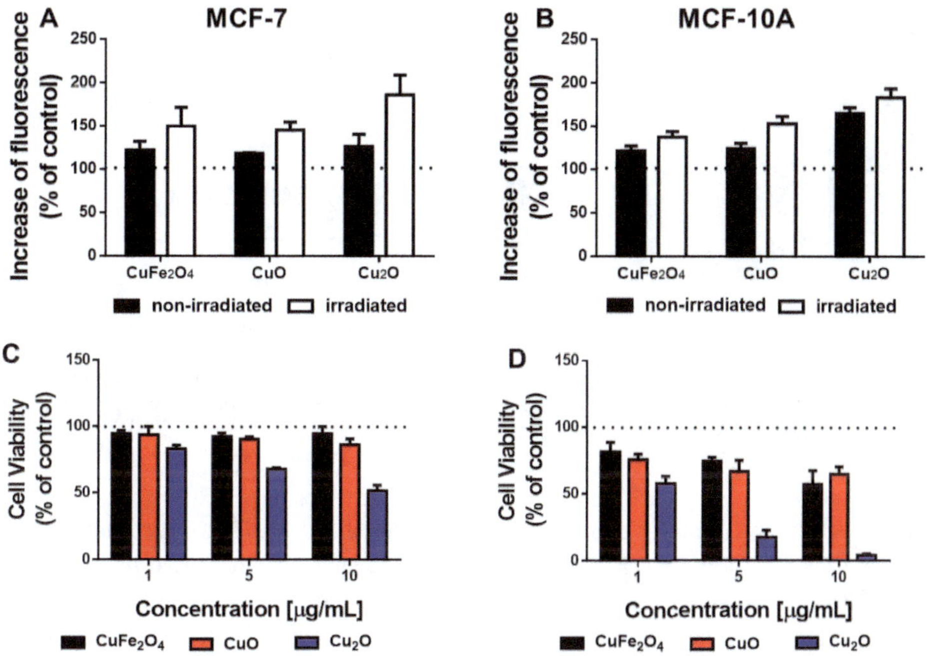

Figure 202: ROS generation in MCF-7 (A) and MCF-10A (B) cells treated with 25 μg/mL of the $CuFe_2O_4$, CuO, and Cu_2O NPs, biocompatibility assay for MCF-7 (C) and MCF-10A (D) cells incubated with various NP concentrations.

MCF-10A cells (Figure 203(D)). Because of this increased toxicity, copper oxide NPs were not further investigated for application in cancer radiation therapy.

Noble metal nanoparticles

As just described, high Z nanomaterials are known to have a radiosensitizing effect. This effect has also been observed in patients with metal implants [970]. In fact, gold is the preferred nanoparticulate material for radiosensitizers because it has a higher Z-number and is less toxic than iodine. Other noble metal nanoparticles such as platinum nanoparticles (PtNPs) [934, 971, 972] or palladium nanoparticles (PdNPs) [973] show similar radiosensitizing effects as AuNPs.

Gold nanoparticles (AuNPs)

AuNPs possess a high X-ray absorption coefficient [974]. When exposed to X-rays, AuNPs emit photo- and Auger electrons and secondary X-rays [975]. Charge transfer of electrons to surface-adsorbed oxygen results in the production of superoxide. Secondary X-rays induce secondary water radiolysis, which yields the same products as primary water radiolysis.

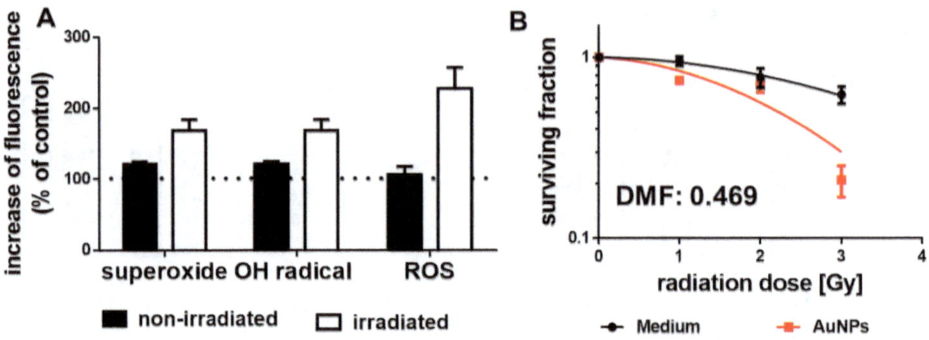

Figure 203: MCF-7 cells incubated with 25 µg/mL citrate-stabilized AuNPs and X-ray irradiated with 1 Gy: formation of superoxide, hydroxyl radicals, and ROS (A); survival curves of MCF-7 cells with and without AuNPs, X-ray irradiated with doses of 0–3 Gy (B).

MCF-7 cells were treated with 25 µg/mL citrate-stabilized AuNPs and irradiated with a single X-ray dose of 1 Gy. ROS generation in these cells was increased by 125 % (Figure 203(A)). The emission of electrons after irradiation led to the formation of superoxide, as evidenced by elevated intracellular superoxide levels (Figure 203(A)). As noted for SPIONs, the generation of highly toxic hydroxyl radicals is far more dangerous to cells than the generation of superoxide. The secondary water radiolysis produces a variety of ROS species that can lead to the formation of the hydroxyl radical. The levels of hydroxyl radicals were also increased in X-ray irradiated MCF-7 cells when containing AuNPs (Figure 204(A)). In a clonogenic cell survival assay, the survival fractions and survival curves of MCF-7 cells with and without AuNPs at multiple radiation doses were measured and determined (Figure 203(B)). The survival curve of AuNPs-loaded MCF-7 cells was observed to decrease faster than the survival curve of MCF-7 cells that was not incubated with AuNP solutions (control MCF-7 cells). The DMF value of 0.469 calculated from the survival curves was surprisingly lower than that of SPIONs. AuNPs are excellent radiosensitizers.

The influence of the shape of AuNPs on the radiosensitizing effect was investigated. Figure 204(A) shows the TEM images of CTAB-coated AuNRs with average lengths and widths of 38 nm and 10 nm, respectively, and thus an average high aspect ratio (HAR) of 3.8. The TEM image in Figure 204(B) shows CTAB-coated AuNRs with a length of 40 nm, a width of 15 nm, and an aspect ratio of 2.5 (low aspect ratio (LAR)).

Both types of AuNRs were functionalized with L-lysine (Lys). The cytotoxicity of the differently coated and sized AuNRs was investigated using human umbilical-vein endothelial cells (HUVEC) (Figure 204(C)). Within the first 24 h of incubation, AuNRs developed negligible cytotoxicity to HUVECs. After 72 h of incubation with CTAB-coated AuNRs, the viability of HUVECs decreased to values around 50 %. This is explained by the desorption of the cationic surfactant CTAB from the AuNP surface [976]. Free CTAB molecules are known to induce cell death by disrupting the cell membrane [977].

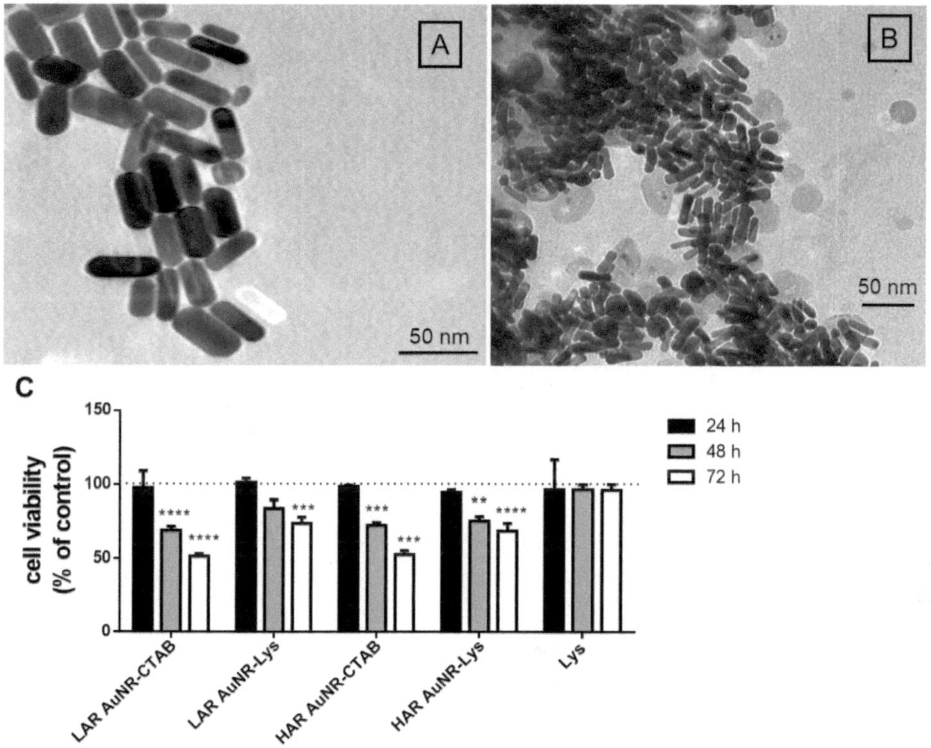

Figure 204: TEM images of low aspect ratio CTAB-coated AuNRs (A), high aspect ratio CTAB-coated AuNRs (B), cytotoxicity studies of CTAB- and L-lysine-coated AuNRs using the MTT assay, n = 6, *p > 0.05.

L-lysine-coated AuNRs were significantly less toxic to cells, with cell viability values between 70 and 80 %.

MCF-7 cells were incubated with the CTAB- or L-Lys-coated LAR and HAR AuNRs and X-ray irradiated with a single dose of 1 Gy. LAR and HAR AuNRs significantly differed in their ROS formation (Figure 205(A)). In comparison to CTAB-coated HAR AuNRs, the CTAB-coated LARs slightly increased the ROS level. The ROS formation appeared to be significantly more pronounced for LAR and HAR L-lysine-coated AuNRs as the relative ROS level was increased by 232 % and 142 %, respectively. Superoxide formation was already enhanced in nonirradiated MCF cells when having ingested CTAB-LAR and HAR AuNRs and Lys-HAR AuNRs (Figure 205(B)). CTAB- and L-lysine-coated AuNRs preferentially accumulated in various cell organelles. Incorporation of AuNRs into the mitochondrial membrane can induce superoxide production as a result of decoupling the electron-transport chain by transferring an electron to oxygen. By contrast, X-ray treatment of MCF-7 cells was found to elevate the superoxide level with values of about 120 % for LAR CTAB-coated AuNRs and even 210 % for LAR L-lysine-coated AuNRs. In particular, this large increase of superoxide concentration caused by L-lysine-coated AuNRs

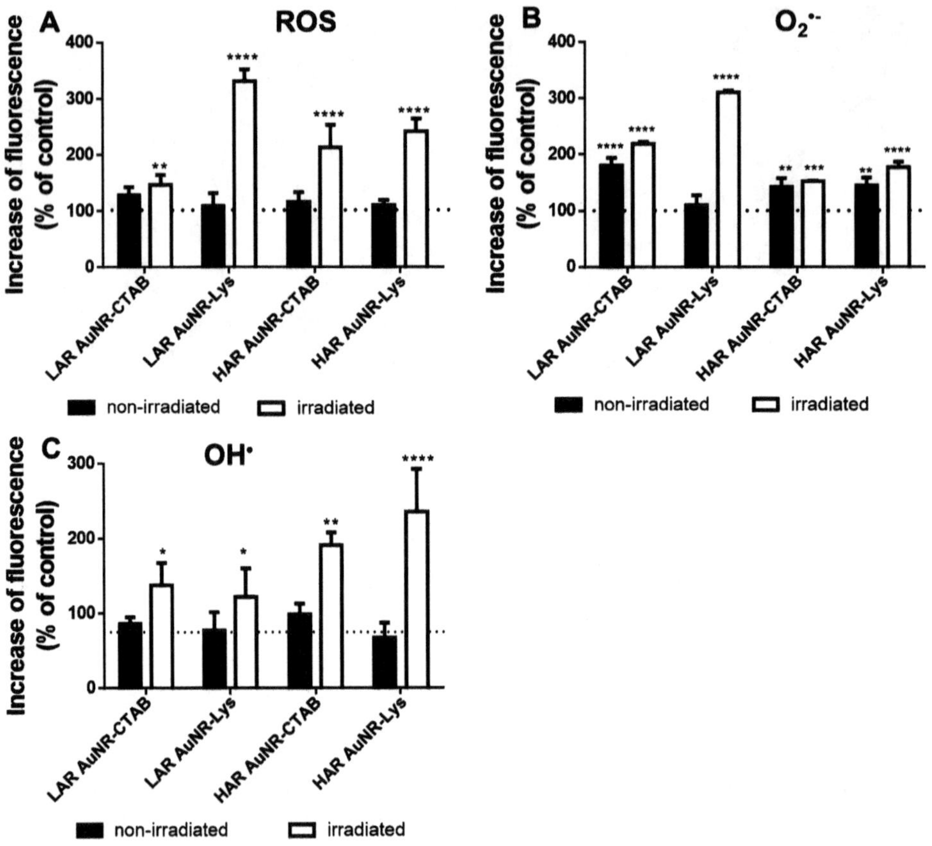

Figure 205: Determination of the relative ROS concentration: (A) superoxide concentration; (B) or hydroxyl radical concentration; (C) in MCF-7 cells incubated with differently coated and sized AuNRs and exposed to X-rays at a single dose of 1 Gy or left nonirradiated; all values are expressed as increase in relative fluorescence intensity (%), n = 4, *$p > 0.05$.

reached almost that of ROS concentration. HAR AuNRs with the different coating produced less superoxide after X-ray irradiation compared to the LAR AuNRs.

The hydroxyl radicals generated by secondary water radiolysis were measured for X-ray treated MCF-7 cells incubated with differently sized AuNRs (Figure 205(C)). AuNR-induced hydroxyl radical formation in these MCF-7 cells resulted in an increase in hydroxyl radical levels from 40 to 140 %, whereas intracellular AuNRs did not alter hydroxyl radical levels in nonirradiated MCF-7 cells. It is noteworthy that the largest hydroxyl level increase of 140 % was observed for HAR L-lysine-coated AuNRs in X-ray irradiated MCF-7 cells. This finding correlates well with the ROS formation in X-ray exposed MCF-7 incubated with this type of AuNRs. On the other hand, the relative ROS and superoxide levels in irradiated MCF-7 cells containing HAR L-lysine-coated AuNRs were

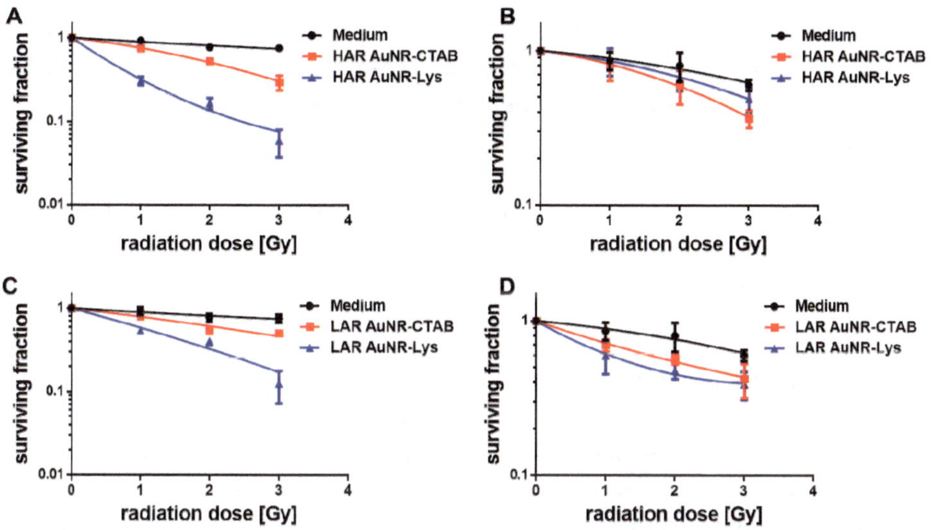

Figure 206: X-Radiation dose survival curves of MCF-7 cells incubated with low (A) and high aspect ratio AuNRs (C) and HUVECs incubated with low (B) and high aspect ratio AuNRs (D), n = 6, *p > 0.05.

found to increase to 232 % and 210 %, respectively, while the change in hydroxyl radical level was only about 30 %.

Exposure of cells to X-radiation causes DNA damage either through radical formation or direct strand breaks, both of which lead to cell death. This cell death can occur immediately or after one or more cell divisions. To demonstrate the cytotoxicity of ionizing radiation and the efficacy of radiosensitizers, clonogenic cell survival assays were performed for both cell lines, cancerous MCF-7 cells and healthy HUVECs. The percentage of surviving cells was measured following the application of 0–3 Gy of X-ray radiation. In Figure 206, the cell survival curves of HUVECs and MCF-7 cells cultured in cell medium are compared with these cells loaded with the LAR or HAR AuNRs. For HUVECs (Figures 206(B) and (C)), LAR AuNRs coated with CTAB or L-lysine showed a significant decrease in cell survival, while the survival curves of cells containing HAR AuNRs differed from the reference (medium) survival curve (cells without AuNRs) at higher radiation doses. Both types of AuNRs significantly decreased the surviving frac-

Table 16: DMF values calculated from the survival curves (Figure 206) at a survival rate of 50 %.

	MCF-7	HUVEC
HAR AuNR-CTAB	0.34	0.61
HAR AuNR-Lys	0.10	0.75
LAR AuNR-CTAB	0.45	0.61
LAR AuNR-Lys	0.21	0.41

tion of MCF-7 cells (Figures 206(A) and (D)), with the L-lysine-coated LAR and HAR AuNRs showing the lowest survival curves. This is consistent with the DMF values (Table 16). A remarkably low DMF value of 0.1 was calculated for the LAR L-Lys coated AuNR. The LAR AuNRs mainly generated superoxide due to x-ray irradiation, while the HAR AuNRs mainly generated hydroxyl radicals. Hydroxyl radicals cause the greatest cellular damage in oxidative stress, which is consistent with the progression of the survival curves. In summary, L-lysine-coated AuNRs achieved the highest capacity to sensitize tumor cells to a single clinical X-ray dose.

Platinum nanoparticles (PtNPs)

Caffeic acid stabilized PtNPs were shown to exhibit good biocompatibility in MCF-7cells and MCF-10A cells. Both cell lines were incubated with 25 µg/mL of the PtNPs and irradiated with a single X-ray radiation dose of 1 Gy. Noble metal nanoparticles such as PtNPs generate photo- and Auger electrons and secondary X-rays upon exposure to X-rays. In X-ray irradiated MCF-7 cells, when incubated with PtNPs, superoxide and hydroxyl radical formation was found to increase, confirming the radiosensitizing effect of PtNPs (Figure 207(A)). With a DMF value of 0.43, PtNPs turned out to act slightly more efficient as radiosensitizers than AuNPs (0.469) (Figure 207(C)). In MCF-10A cells, only the superoxide level was increased in X-ray treated cells. By comparison, the hydroxyl radical levels was found to lie below that of cells without PtNPs (control medium cells) (Figure 207(B)). The calculated DMF value of MCF-10A cells exceeded that of the MCF-7 cells

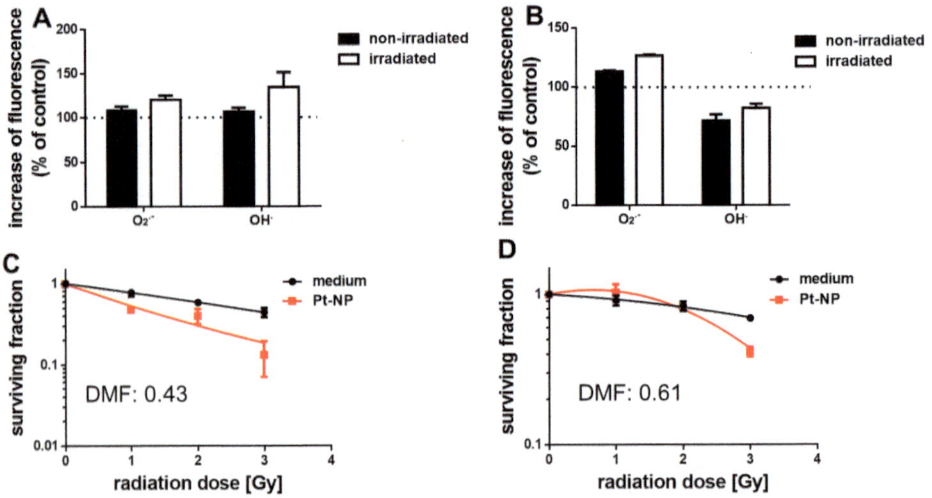

Figure 207: Determination of the superoxide and hydroxyl levels in MCF-7 (A) and MCF-10A cells (B) incubated with PtNPs; survival curves of MCF-7 (C) and MCF-10A cells (D) incubated with PtNPs and X-ray irradiated with single doses of 0–3 Gy compared to survival curves of cells in medium.

(Figure 207(D)). However, the DNF value smaller than 1 indicated a radiosensitizing effect in cancerous and also, unfortunately, in healthy cells.

Palladium nanoparticles (PdNPs)

Caffeic acid stabilized PdNPs are less biocompatible for MCF-7 and MCF-10A cells compared to PtNPs and AuNPs. To quantify the ability of PdNPs to induce ROS formation in X-ray exposed cancer and healthy cells, the change in superoxide and hydroxyl radical formation in MCF-7 and MCF-10A cells was measured. MCF-7 cells were incubated with PdNPs and irradiated with a single dose of 1 Gy. Apparently, the PdNPs acted as radiosensitizers by enhancing both superoxide and hydroxyl production (Figure 208(A)). X-ray exposed MCF-10A cells incubated with PdNPs also showed increased superoxide and hydroxyl radical formation (Figure 208(B)). In addition, PdNPs were shown to inhibit the activity of intracellular enzymes such as SOD and glutathione peroxidases (GPs). This inhibition of enzyme activity indirectly contributes to the increased hydroxyl and superoxide levels since these enzymes are responsible for the degradation of these ROS. The cell survival curves of the MCF-10A cells containing PdNPs showed a drastic drop to zero (Figure 208(D)). The very low DMF value of 0.30 indicates a strong radiosensitization effect of PdNPs in non-tumorigenic MCF-10A cells.

By contrast, the survival curves of MCF-7 cells without PdNPs and MCF-7 cells with PdNPs show a more similar, slowly decreasing course (Figure 208(C)). The DMF calculated from these survival curves is slightly less than one. Therefore, PdNPs are not suitable for use in cancer radiotherapy.

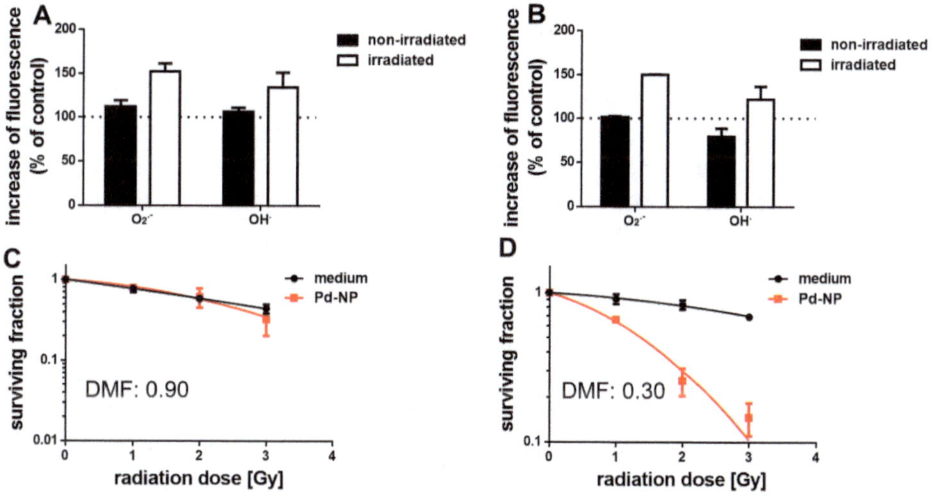

Figure 208: Superoxide and hydroxyl levels in MCF-7 (A) and MCF-10A cells treated with PdNPs; survival curves of MCF-7 (C) and MCF-10A cells (D) treated with PdNPs and irradiated with single X-radiation doses of 0–3 Gy compared to survival curves of cells in medium.

Noble metal-iron oxide nanoheterostructures

As described in detail in Section 3.3.1, the fusion of different types of nanomaterials into hybrid nanostructures can take on complex morphologies such as core-shell and onion, heterodimers and multimers, or dumbbell and snowman-shaped nanostructures [970, 978]. Dumbbell or snowman-shaped nanoheterodimers (NHDs) are advantageous because they are composed of two different spherical nanomaterials with complementary surface properties due to their elemental compositions and reactivities. Dumbbell- or snowman-shaped gold-iron oxide (Au–Fe_3O_4) NHDs not only exhibit superparamagnetism and surface-plasmon resonance, but also synergistically combine a Fenton catalyst with a noble metal-based radiosensitizer. In addition, the Au component in Au–Fe_3O_4 NHDs, when epitaxially grown on the Fe_3O_4 surface, was shown to maintain the Fe_3O_4 composition, especially the surface, in the NHDs [244]. This the preservation of the Fenton catalyst activity of the Fe_3O_4 component in the NHD.

Au–Fe_3O_4 NHDs functionalized with $NOBF_4$

Water-soluble $NOBF_4$-functionalized Au–Fe_3O_4 NHDs were evaluated for their application in radiotherapy. The Au–Fe_3O_4 NHDs were compared with AuNPs and Fe_3O_4 NPs also stabilized with a mixed oleic acid-$NOBF_4$ coating. MCF-7 cells were incubated with 25 µg/mL of the NHDs and NPs and X-ray irradiated with a single dose of 1 Gy. MCF-7 cells containing either Au–Fe_3O_4 NHDs, AuNPs, or Fe_3O_4 NPs were incubated with different fluorescent dyes that selectively react with the different ROS and RNS. In all assays, the dotted line represents the percentage of cells in the cell culture medium (control) at 0 or 1 Gy, respectively.

Superoxide can be generated either by single-electron transfer from AuNPs to surface-adsorbed oxygen due to activation by X-ray irradiation (Figure 209(A)). In X-ray irradiated MCF-7 cells loaded with AuNPs, the superoxide concentration was found to increase (Figure 209(B)). By contrast, Fe^{3+} cations in the X-ray activated Fe_3O_4 surface catalyze Haber–Weiss reaction and therepy provide superoxide degradation. A slight decrease in superoxide concentration was measured in MCF-7 cells due to the Haber–Weiss catalyzing activity of internalized Fe_3O_4 NPs (Figure 209(B)). Since MCF-7 cells possess elevated SOD activities, the intracellular superoxide levels are already very low, which limits superoxide scavenging by intracellular Fe_3O_4 NPs. The superoxide concentration in X-ray irradiated MCF-7 cells containing Au–Fe_3O_4 NHDs was observed to be smaller than that in nonirradiated MCF-7 cells. While intracellular AuNPs increased superoxide formation in X-ray irradiated MCF-7 cells, the Fe_3O_4 component in the Au–Fe_3O_4 NHD provided superoxide scavenging via the Haber–Weiss mechanism.

In addition, the interaction between X-radiation and AuNPs led to the formation of secondary X-rays that provoke secondary water radiolysis, producing hydroxyl radicals (Figure 210(A)). The level of hydroxyl radicals increased in X-ray irradiated MCF-7 cells when loaded with Fe_3O_4 NPs, AuNPs, or Au–Fe_3O_4 NHDs (Figure 210(B)). Surface Fe^{2+} cations catalyze the Fenton reaction that produces hydroxyl radicals (Figure 210(A)).

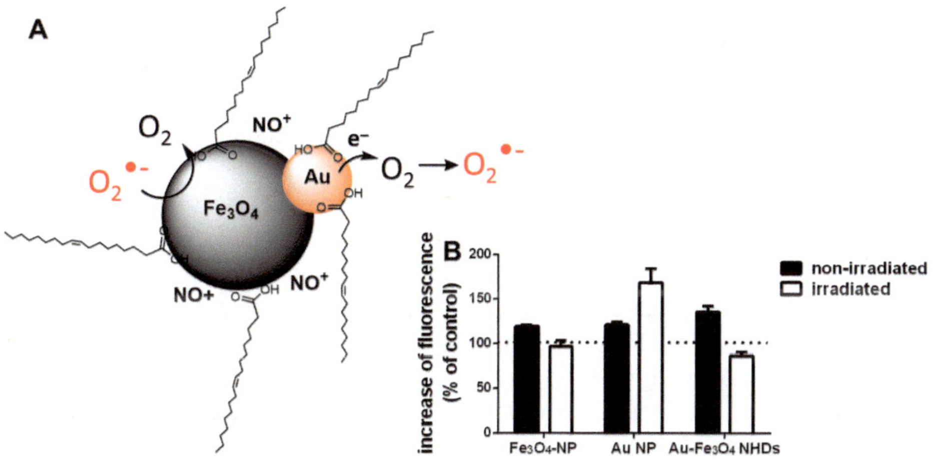

Figure 209: Superoxide produced by the AuNP component of the Au–Fe_3O_4 NHDs or degraded by Fe^{3+} cations in the Fe_3O_4 NP surface (A), the intracellular superoxide level (B) in MCF-7 cells containing Fe_3O_4 NPs, AuNPs, and Au–Fe_3O_4 NHDs.

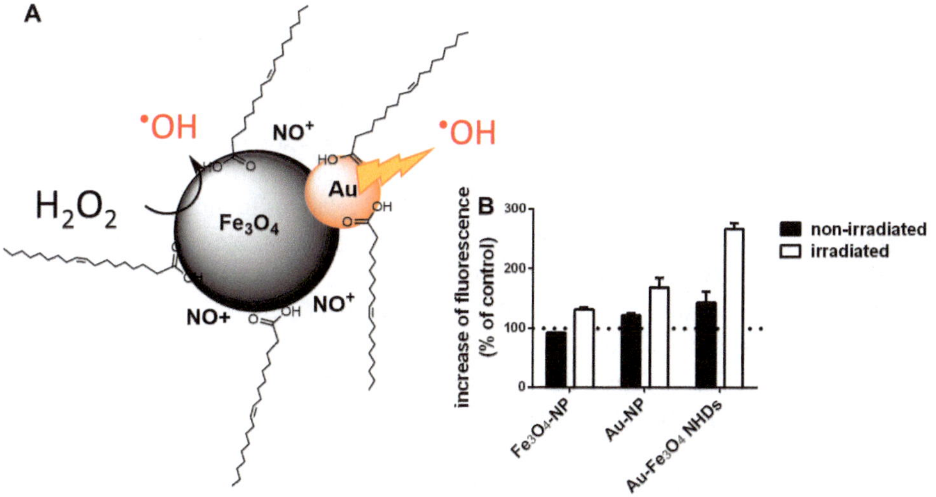

Figure 210: Hydroxyl radical can be produced by Fenton reaction or secondary water radiolysis (A), determination of the hydroxyl radical levels (B) in MCF-7 cells containing Fe_3O_4 NPs, AuNPs, and Au–Fe_3O_4 NHDs.

Only a slight increase in hydroxyl radical production was measured in MCF-7 cells containing Fe_3O_4 NPs (Figure 210(B)). Because Fe_3O_4 NPs were synthesized under ambient conditions and stored under ambient conditions, the surface Fe^{2+} cations gradually oxidized to Fe^{3+} cations, limiting the ability to catalyze the Fenton reaction. Compared with AuNPs and Fe_3O_4 NPs, the Au–Fe_3O_4 NHDs were shown to develop the highest increase in hydroxyl radical production in X-ray irradiated MCF-7 cells. Since both the AuNP and Fe_3O_4 NP component generate hydroxyl radicals in X-ray irradiated MCF-7 cells, a synergistic hydroxyl radical formation was observed. The fusion of AuNPs and Fe_3O_4 NPs in a Au–Fe_3O_4 NHD provided the maintenance of surface Fe^{2+} cations, even after several months of storage at ambient conditions.

Nitrosonium anions (NO^+) and tetrafluoroborate anions (BF_4^-) were presumably embedded in the oleic acid coating of the Au–Fe_3O_4 NHDs. Normally, NO^+ is rapidly hydrolyzed in aqueous solution, but was found intracellularly in nitrosyl complexes with Fe^{3+} centers [979]. This suggests that the oleic acid surface structures of Fe_3O_4 NPs and Au–Fe_3O_4 NHDs shield the NO^+ cations and prevent them from hydrolysis. NO radicals were formed by reduction of NO^+ by Fe^{2+} cations in the Fe_3O_4 surface (Figure 211(A)). Reduction of NO^+ to NO radicals also may occur on the AuNP surface by electron transfer. X-ray irradiation of MCF-7 cells containing Fe_3O_4 NPs resulted in a very small increase in NO concentration but significantly increased the NO level in MCF-7 cells that had internalized Au–Fe_3O_4 NHDs (Figure 211(B)). The Fe_3O_4 NPs were observed to oxidize during the experiments, as evidenced by the gradual color change from black to brown, thus gradually losing their reducing power. Instead, Au–Fe_3O_4 NHDs retained their Fe_3O_4 composition due to the Au component which can additionally emit photo- or Auger-electrons under X-ray irradiation. To confirm the NO formation in Au–Fe_3O_4 NHD loaded MCF-7 cells, MCF-7 cells were additionally incubated with the NO scavenger vitamin B12 [980] or the $ONOO^-$ scavenger uric acid [981, 982]. The NO radical production was reduced in the presence of vitamin B12, whereas the $ONOO^-$ scavenger uric acid did not alter the NO radical levels in MCF-7 cells (Figure 211(B)).

X-ray irradiation of Au–Fe_3O_4 NHD-loaded MCF-7 cells was found to enhanc the formation of intracellular NO, superoxide, and hydroxy radicals. When both superoxide and NO are generated in close proximity on the Au–Fe_3O_4 NHD surface, they can spontaneously react to the highly oxidative RONS peroxynitrite ($ONOO^-$) (Figure 212(A)) [945]. Peroxynitrite formation was demonstrated in X-ray irradiated MCF-7 cells with internalized Au–Fe_3O_4 NHDs (Figure 212(B)). As expected, the Au–Fe_3O_4 NHDs were more effective in peroxynitrite generation than intracellular Fe_3O_4 NPs or AuNPs. This demonstrates the spatially confined formation of both NO radicals and superoxide on the surface of Au–Fe_3O_4 NHDs and further underlines the ability of Au–Fe_3O_4 NHDs to provide simultaneous formation of ROS and RNS (ROS + RNS = RONS). In the presence of vitamin B12, a decrease in peroxynitrite level was observed, reflecting the NO scavenging activity of vitamin B12. The scavenging effect of uric acid on peroxynitrite was observed to be more pronounced since it reacts directly with peroxynitrite. This confirmed that peroxynitrite was the most dominant RONS generated in Au–Fe_3O_4 NHD-loaded MCF-7 cells after X-ray irradiation.

Increased peroxynitrite production, if not balanced by the cellular antioxidant defense system, should give rise to oxidative stress and cell damage. Peroxynitrite causes oxidation and nitration of biomolecules such as proteins, lipids, and DNA. The oxidative degradation of lipids, membrane lipid peroxidation (MLP), is induced when free RONS

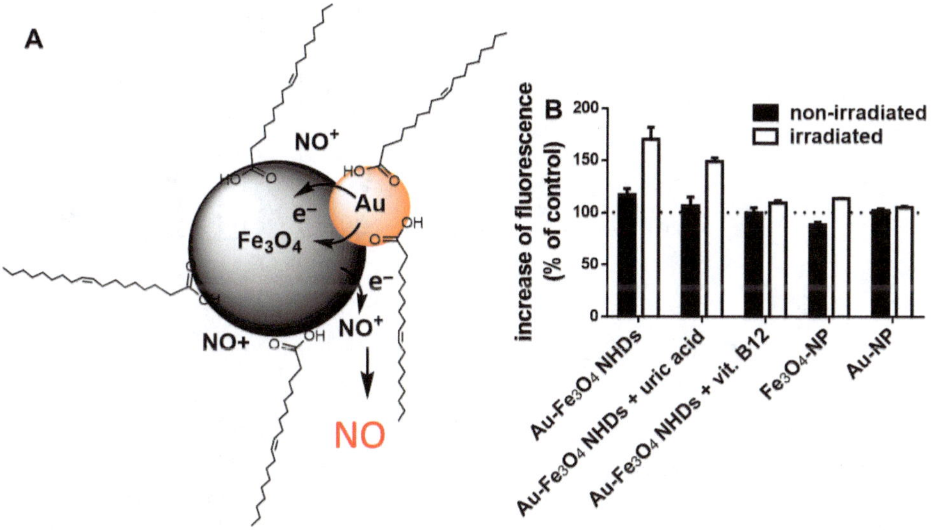

Figure 211: Electron transfer to surface-attached nitrosonium ions (NO^+) yield NO radicals(A); determination of the NO generation (B) in MCF-7 cells with the various NPs; additionally, the generation of NO was confirmed by adding the NO scavenger vitamin B12.

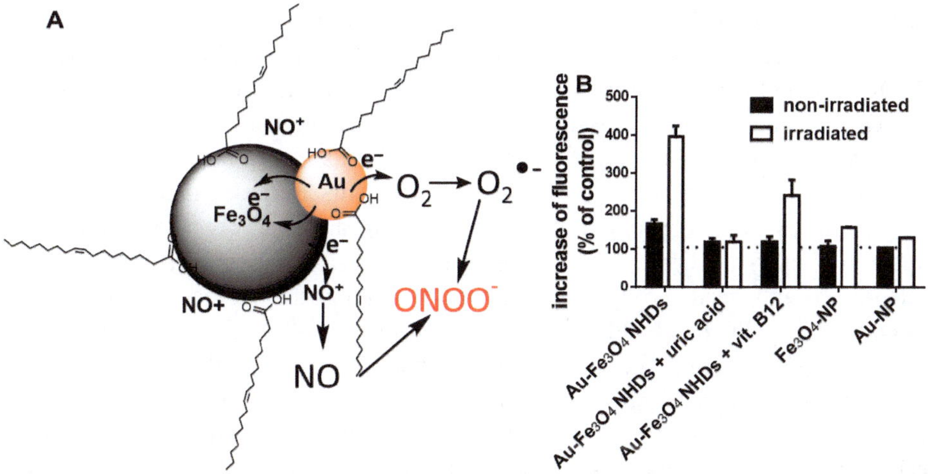

Figure 212: Simultaneous generation of superoxide and NO led to the formation or peroxynitrite ($ONOO^-$) (A); formation of $ONOO^-$ in MCF-7 cells containing Fe_3O_4 NPs, AuNPs, or Au–Fe_3O_4 NHDs, and, in addition, vitamin B12 or uric acid (B).

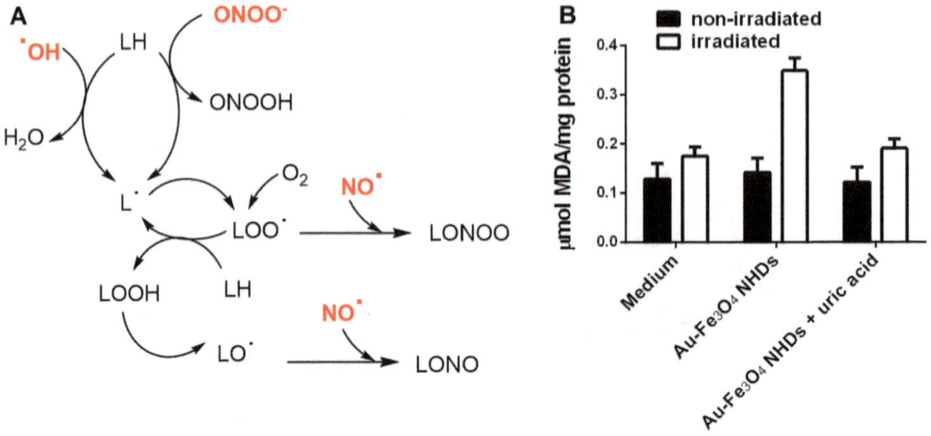

Figure 213: Membrane lipid peroxidation (MLP) (A), quantification of the MLP by measuring the production of MDA in MCF-7 cells with and without Au–Fe_3O_4 NHDs, before and after X-ray irradiation with a single 1 Gy dose (B).

such as hydroxyl radicals or peroxynitrite react with polyunsaturated fatty acids to form water and a fatty acid radical (Figure 213(A)). The oxidizing potential of peroxynitrite is sufficient to directly initiate MLP, which occurs by abstracting a hydrogen atom from the polyunsaturated fatty acid. If the concentration in lipids is high enough, this can lead to the initiation of a self-sustaining peroxidation reaction.

In addition to lipid oxidation, lipid nitration also may occur. The extent in MLP was quantified by measuring the amount of malondialdehyde (MDA), which is a product of MLP. Figure 213(B) shows the increase of MDA in X-ray irradiated MCF-7 cells with Au–Fe_3O_4 NHDs. Compared to the control MCF-7 cells, the concentration of MDA in Au–Fe_3O_4 NHD-loaded MCF-7 cells increased tremendously. This effect was diminished when uric acid was added to the Au–Fe_3O_4 NHD-loaded MCF-7 cells.

Peroxynitrite can induce DNA damage by causing deoxyribose oxidation and strand breaks, thereby inducing apoptosis (Figure 214(A)). It can also trigger the release of pro-apoptotic factors and cytochrome c-dependent apoptosis in the cytoplasm [983]. Consequently, the enhanced peroxynitrite formation should lead to an increase in cell death as was detected by the TUNEL assay. After exposure of MCF-7 cells to Au–Fe_3O_4 NHDs (Figure 214(B)), 44 % of the cells are TUNEL positive compared to only 13 % of the control cells. The TUNEL assay fluorescently labels DNA strand breaks. The increase in TUNEL-positive cells is attributed to cells with damaged DNA, which is likely to induce apoptotic cell death. The ratio of damaged cells is reduced to 24 % in the presence of uric acid, clearly demonstrating that DNA damage in Au–Fe_3O_4 NHD-treated MCF-7 cells is caused by peroxynitrite and X-rays together.

The effect of peroxynitrite generation on MLP and associated apoptotic cell death confirmed that cells were oxidatively damaged. The clonogenic cell survival assay (Figure 215(A)) was used to compare the efficacy of Au–Fe_3O_4 NHDs with that of Fe_3O_4 NPs

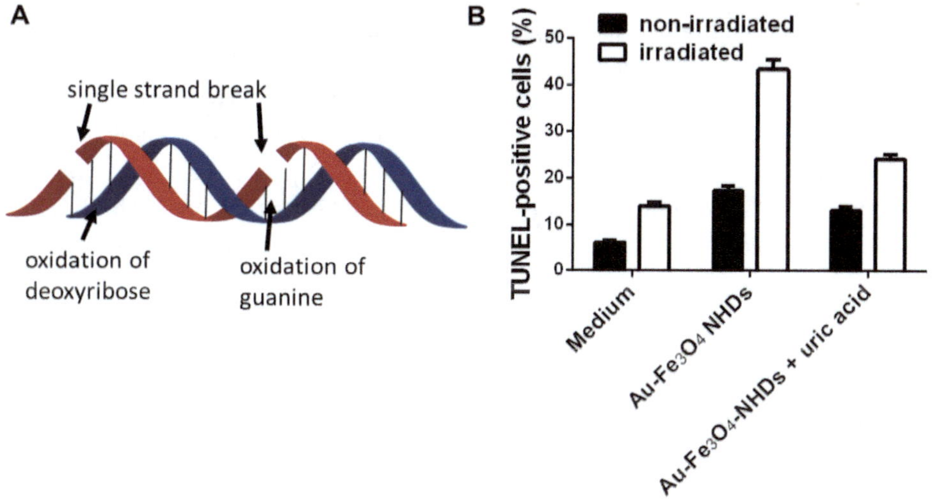

Figure 214: Peroxynitrite causes DNA damage such as DNA strand breaks (A); DNA strand breaks were fluorescently labeled with the TUNEL assay (B).

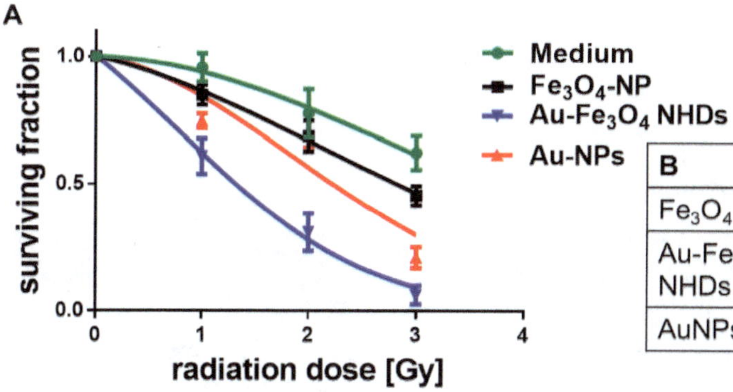

B	DMF
Fe_3O_4 NPs	0.783
Au-Fe_3O_4 NHDs	0.366
AuNPs	0.469

Figure 215: Survival curves of MCF-7 cells containing Fe_3O_4 NPs, AuNPs, or Au–Fe_3O_4 NHDs (A); the DMF values calculated from the survival curves at a survival rate of 50 % (B).

or AuNPs. The survival curves of the MCF-7 cells containing Fe_3O_4 NPs, AuNPs, or Au–Fe_3O_4 NHDs were lower than those of the control MCF-7 cells in medium. All calculated DMF values (Figure 215(B)) were smaller than 1, indicating that Fe_3O_4 NPs, AuNPs, and Au–Fe_3O_4 NHDs acted as radiosensitizers. Partially oxidized Fe_3O_4 NPs were limited in the generation of RONS and therefore exhibited the highest DMF value, which was still less than 1. AuNPs, when containing $NOBF_4$ in their surface layer of oleic acid, proved to be very effective in radiosensitization. Au–Fe_3O_4 NHDs exhibited the lowest DMF values, confirming the synergistic RONS generation by both Fe_3O_4 and Au components of the Au–Fe_3O_4 NHD. Au–Fe_3O_4 NHDs also benefitted from their functionalization with $NOBF_4$, which facilitated the generation of the highly reactive peroxynitrite.

The encouraging radiosensitizing potential of Au–Fe_3O_4 NHDs in 2D cell culture was verified in a 3D tumor model. Multicellular tumor spheroids (MCTS) were grown from MCF-7 cells and incubated for 5 days in cell culture medium, which was replaced every day. After 5 days, a subset of MCTS were incubated with a solution of Au–Fe_3O_4 NHDs at a concentration of 25 µg/mL. To ensure successful Au–Fe_3O_4 NHD internalization, the MCTS were incubated with the Au–Fe_3O_4 NHD solution for an additional 48 h. The MCTS were then irradiated with two 5 Gy doses of X-rays. The first X-radiation dose was ap-

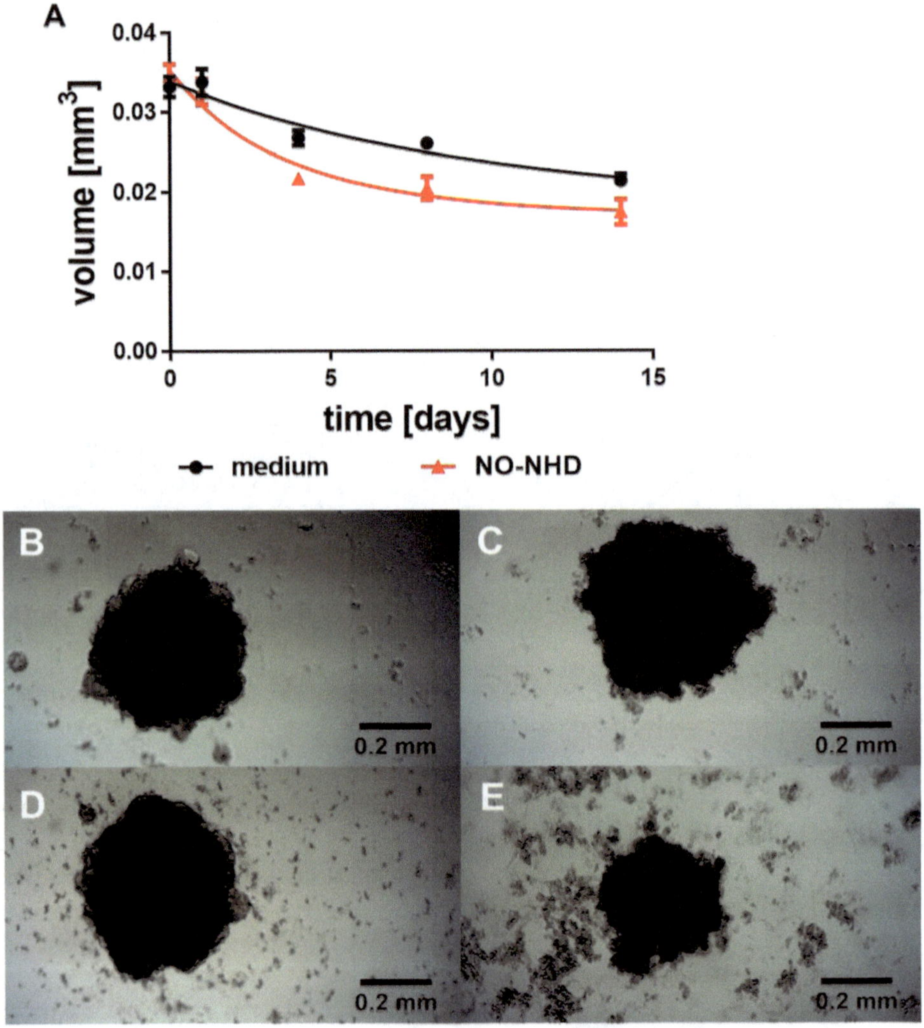

Figure 216: Growth curves of MCF-7 MCTS in medium or incubated with 25 µg/ml Au–Fe_3O_4 NHDs and irradiated with two X-radiation doses of 5 Gy (A), optical microscopy images of the MCTS at day 0 (B: medium; D: Au–Fe_3O_4 NHDs) and after 14 days (C: medium, E: Au–Fe_3O_4 NHDs).

plied at 0 (Figure 216(B) and (D)). After 14 days, the volume of Au–Fe_3O_4 NHD-treated MCTS (Figure 216(A) and (E)) was significantly reduced compared to the volume of untreated MCTS (Figure 216(A) and (C)). Apparently, the radiosensitizing effect of $NOBF_4$-functionalized Au–Fe_3O_4 NHDs was observed to impede the MCTS growth process, making Au–Fe_3O_4 NHDs ideal candidates for radiosensitizers to be tested in vivo.

Au–Fe_3O_4 NHDs functionalized with caffeic acid

Increased expression of mitochondrial manganese superoxide dismutase (MnSOD) protects healthy tissue cells from radiation-induced apoptosis by scavenging superoxide generation. This scavenging ability stabilizes the mitochondrial membrane while preventing cytochrome c release and inhibiting mitochondrial DNA strand breaks. These are critical factors that normally lead to apoptosis after X-ray exposure [984]. Antioxidants such as caffeic acid (CA) reduce superoxide formation (Figure 217(A)) [985]. Using the superoxide indicator MitoSOX, superoxide generation was measured in non-tumorigenic MCF-10A cells for 60 min (Figure 217(B)). Mitochondrial superoxide production decreased in the presence of CA, demonstrating the scavenging ability of CA. Au–Fe_3O_4 NHDs were functionalized with CA in a ligand-exchange procedure. The CA-functionalized Au–Fe_3O_4 NHDs (CA@Au–Fe_3O_4 NHDs) were water soluble with high biocompatibility [233]. MCF-10A cells were incubated with CA and CA@Au–Fe_3O_4 NHDs to evaluate their radioprotective potential. Prior to X-ray irradiation, a decrease in superoxide levels was observed in the presence of both CA and CA@Au–Fe_3O_4 NHDs (Fig-

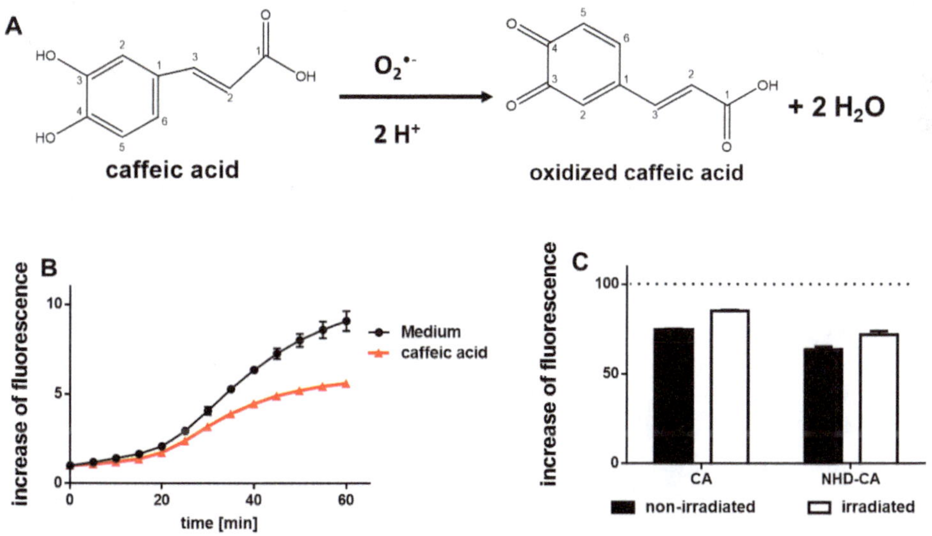

Figure 217: Superoxide scavenging by caffeic acid (CA) (A); incubation of MCF-10A cells with CA reduces superoxide production (B); MCF-10A cells incubated with CA or CA@Au–Fe_3O_4 NHDs (NHD-CA) resulted in decreased superoxide concentrations (C), before and after X-ray radiation.

ure 217(C)). After exposure, superoxide levels increased slightly in both cases, but were still below the dotted line representing control cells in medium. Thus, CA@Au–Fe_3O_4 NHDs were shown to act as radioprotectors in MCF-10A cells.

In addition to their radioprotective properties, CA possess antiproliferative, antimetastatic, and radiosensitizing capabilities [986, 987]. Therefore, CA and its water-soluble ester (caffeic acid phenethyl ester (CAPE)) were also used in cancer treatment. In CA@Au–Fe_3O_4 NHDs, when internalized by healthy and cancer cells, the anticancer drug CA may synergistically interact with the ROS production of the Au–Fe_3O_4 NHDs. Hence, CA@Au–Fe_3O_4 NHDs have an encouraging potential as radiosensitizers in breast cancer MCF-7 cells. The clonogenic cell survival assay was performed to evaluate the radiosensitizing and radioprotective efficacy of the CA-Au–Fe_3O_4 NHDs in MCF-7 and MCF-10 cells. The survival curves of MCF-7 cells treated with CA@Au–Fe_3O_4 NHDs are below the control survival curve of cells in medium, demonstrating the great radiosensitizing potential of CA@Au–Fe_3O_4 NHDs (Figure 218(A)). The DMF value of 0.49 confirms that the CA-NHDs effectively sensitize MCF-7 cells to X-radiation. By contrast, the survival curve of MCF-10A cells with CA-NHDs lies above the survival curve of control cells (Figure 218(B)). The DMF value of 1.09 confirms that the CA-NHDs protect the healthy MCF-10A cells from X-rays. The unique properties of CA@Au–Fe_3O_4 NHDs are to sensitize cancer cells to X-rays and to protect healthy tissues from X-rays.

MCF-7 MCTS were cultured and treated with a solution of CA-Au–Fe_3O_4 NHDs at a concentration of 25 µg/mL. As a first step, the uptake and distribution of CA-Au–Fe_3O_4 NHDs in the MCTS was investigated. The TEM image of the MCF-7 MCTS (Figure 219(B)) shows the CA-Au–Fe_3O_4 NHDs inside the MCTS as dark clusters. The surface region, peri-necrotic region, and necrotic core of the MCTS were subsequently separated (Figure 219(A)), and the iron concentration in these regions was determined. The iron concentration of the surface region initially increased during the cellular uptake of the CA-Au–Fe_3O_4 NHDs. After 3 days, the iron concentration of the other regions was observed to gradually grow at the expense of that of the surface region. The time-space evolution

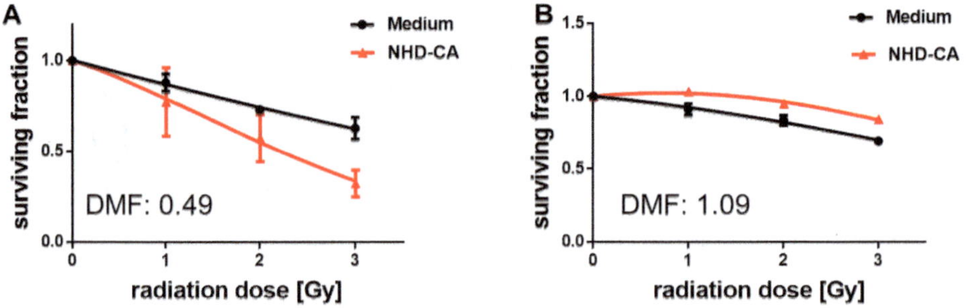

Figure 218: Survival curves of MCF-7 (A) and MCF-10A (B) cells either incubated with CA-Au–Fe_3O_4 NHDs or untreated and irradiated with X-radiation doses of 0–3 Gy.

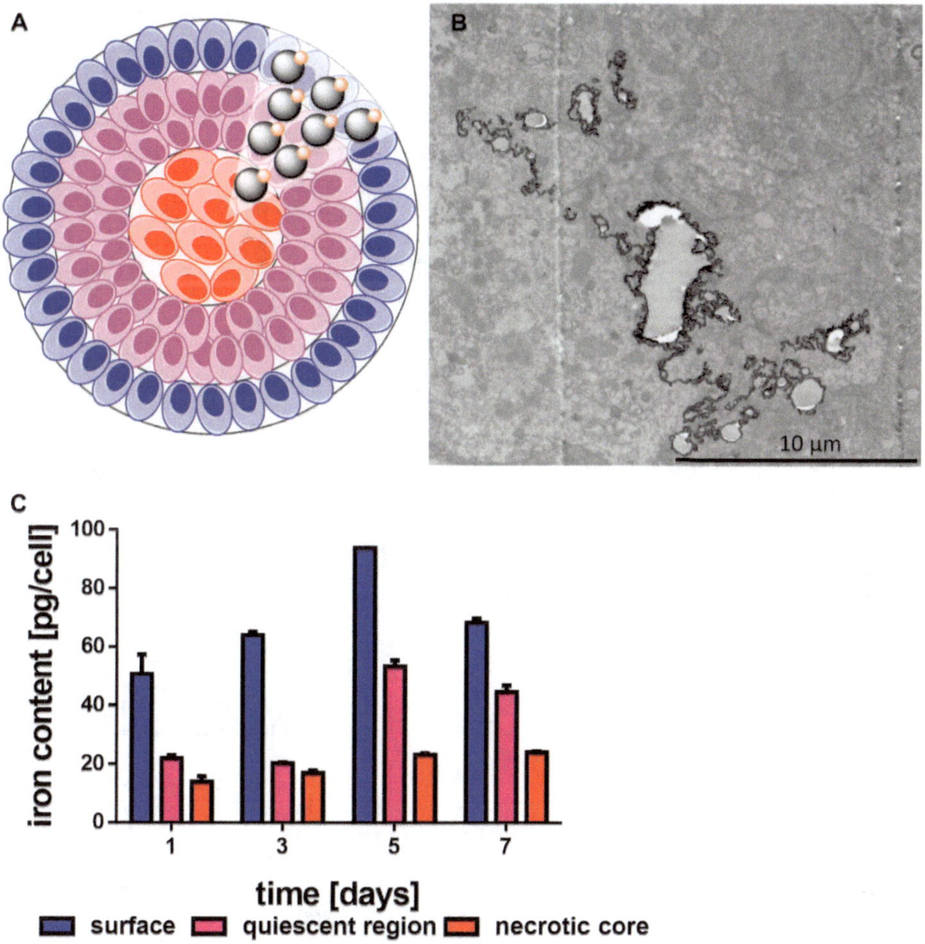

Figure 219: Schematic of the NHD-CA uptake (A); TEM image of NHD-CA internalization into MCTS (B); distribution of the NHD-CA (C) in the various regions of the MCTS.

of cellular uptake of the CA-Au–Fe_3O_4 NHDs demonstrates the time-dependent invasion of the CA-Au–Fe_3O_4 NHDs into the MCTS (Figure 219(C)).

The development of hypoxic regions within cancer tissue is typically accompanied by a shift in metabolic pathways toward increased glycolysis, a phenomenon known as the Warburg effect. Due to the lack of oxygen, cancer cells preferentially utilize glycolysis for energy production, resulting in elevated glucose consumption and increased secretion of lactate [988]. Therefore, the excretion of lactate into the surrounding medium is an indicator of the metabolic state of MCTS. CA is known for its ability to block monocarboxylate transporters. When lactate transport is inhibited, intracellular concentration of lactate rises, prompting a metabolic shift to oxidative phosphorylation—a reversal of the Warburg effect (Figure 220(A)). Incubation of the

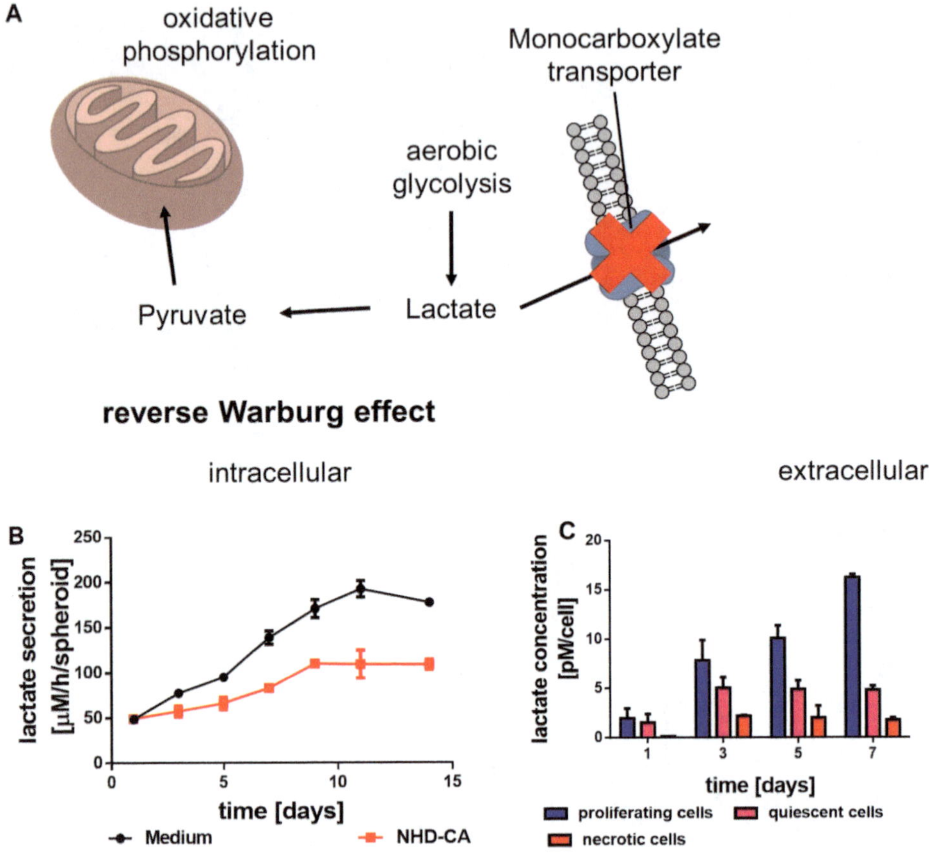

Figure 220: The reverse Warburg effect (A) induced by CA-Au–Fe_3O_4 NHDs (NHD-CA) which leads to a decreased lactate secretion (B) and increased intracellular lactate concentration (C).

MCTS with CA-Au–Fe_3O_4 NHDs resulted in an enormous decrease of the lactate concentration in the outer medium (Figure 220(B)) and an increase of the intracellular lactate concentration in the surface region of the MCTS (Figure 220(C)). Over time, the lactate concentration continued to rise even within the quiescent, hypoxic regions of the MCTS. This reverse Warburg effect reduces the radioresistance of glycolytic cancer cells by making mitochondria more susceptible to radiation-induced damage. Enhanced oxidative phosphorylation in these cells leads to increased ROS production upon X-ray exposure, thereby amplifying radiation-induced cellular damage.

6.3.2 Magnetic hyperthermia

Cancer cells are more sensitive to heat than healthy cells. When cancer cells are exposed to a temperature above 42 °C, they undergo the process of apoptosis or necrosis [989]. Healthy cells, on the other hand, can survive such temperatures without any effects. The higher heat sensitivity of cancer cells compared to healthy cells is due to the fact that cancer cells grow faster and more chaotically. Blood vessel formation is irregular. This makes blood flow around and into the tumor less efficient, which is detrimental to heat transport. The temperature in the tumor rises faster, triggering apoptosis or necrosis. The fundamentals of hyperthermia in cancer therapy have long been recognized as a valuable adjunct to conventional treatments. The basic principle is to raise the temperature of cancer cells to 40–43 °C and maintain it for 30 to 60 min. While hyperthermia alone is usually not sufficient to replace conventional treatments such as radiation and chemotherapy for most cancers, it has proven to be an effective adjunct therapy. Mild hyperthermia below 43 °C enhances the effectiveness of standard treatments in several ways. It increases the effectiveness of certain chemotherapy drugs, allowing for lower doses and less damage to healthy tissue. Heating the tumor causes blood vessels to dilate, improving blood flow and oxygen availability to the area. It can also increase the thermosensitivity of cancer cells, making them more susceptible to other treatments.

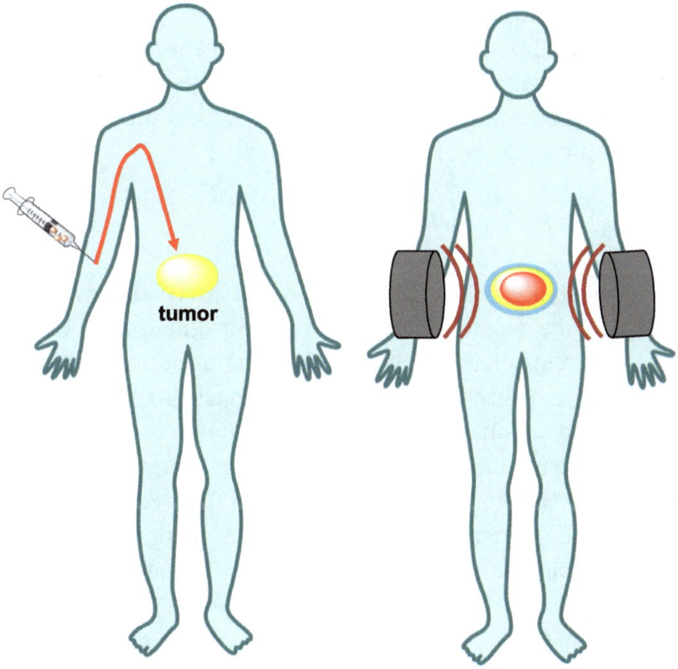

Figure 221: Illustration of magnetic hyperthermia therapy.

One modality of hyperthermia therapy, the magnetic hyperthermia, uses magnetic nanoparticles to convert magnetic energy into thermal energy [990, 991]. Magnetic hyperthermia represents a significant advancement in the field of hyperthermia therapy. This technique utilizes superparamagnetic iron oxide nanoparticles (SPIONs) as its cornerstone. The process begins with the targeting of SPIONs to the tumor tissue, either through direct injection (Figure 221), where SPION enrichment in the tumor tissue can be enhanced under the guidance of an external magnetic field. The biocompatible and often functionalized surfaces of these nanoparticles allow for specific binding to cancer cells. Once the SPIONs are localized within the tumor tissue, they are exposed to an alternating magnetic field. This exposure causes the SPIONs or their magnetic moments inside to oscillate, generating heat through a process known as magnetic induction heating. This localized heating raises the temperature of the tumor tissue to approximately 41–45 °C, producing several therapeutic effects, including direct killing of cancer cells, increased thermosensitivity of cancer cells to conventional treatments, and enhanced immune response against tumor cells. The key advantage of magnetic hyperthermia is that it heats only the tumor tissue while sparing the surrounding healthy tissue, thereby minimizing side effects.

To understand the efficacy of magnetic hyperthermia, it is essential to delve into the underlying physical principles [992]. The different heating mechanisms depend on the magnetic properties of the nanoparticles. Heat generation in this therapy relies on two primary mechanisms: Néel and Brownian relaxation. In the case of ferri- and ferromagnetic materials in multi-domain configuration, the heat generation is proportional to the hysteresis behavior under the influence of an alternating magnetic field. The cycle of alignment of the domains to the varying uniaxial magnetic field generates hysteresis losses. For each cycle, the heat generated, P, can be estimated from the area of the hysteresis loop using the following equation (Eq. (40))

$$P = \mu_0 \nu \oint H dM, \tag{40}$$

where μ_0 is the magnetic permeability, ν is the frequency of the alternating magnetic field strength, H is the strength of the magnetic field, and M is the magnetization. Heat is generated by Neel and Brownian relaxation [925, 991]. Néel relaxation occurs within the magnetic structure of the nanoparticle itself. When exposed to an alternating magnetic field, the magnetic moment of the nanoparticle aligns with the field. As the magnetic field alternates, the magnetic moment must realign, overcoming an energy barrier. This realignment process generates heat due to internal friction. Factors affecting Néel relaxation include particle size, material properties, and temperature. Néel relaxation tends to dominate in smaller particles and at higher frequencies of the alternating magnetic field. The rotation time of the magnetic moments is described by the Neel relaxation time in the following equation (Eq. (41))

$$\tau_N = \tau_0 \mathrm{e}^{KV/k_B T}, \tag{41}$$

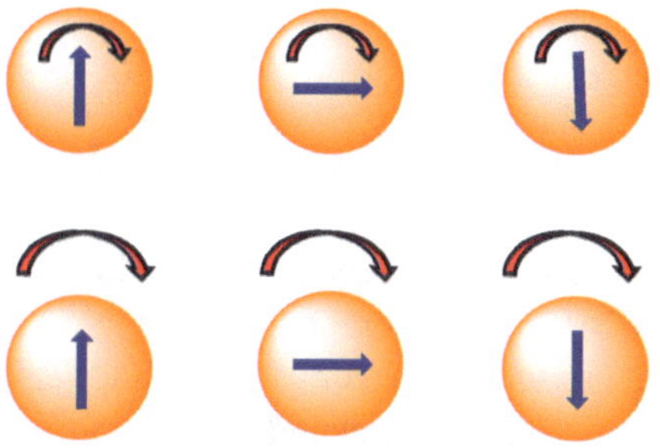

Figure 222: Schematic representation of Néel and Brownian relaxation: In Néel relaxation, the magnetic moment inside the nanoparticle rotates; in Brownian relaxation, the nanoparticle rotates as a whole.

where $\tau_0 \approx 10^{-9}$ s, K is the anisotropy constant of the material, V is the volume of the magnetic particle, k_B is the Boltzmann constant, and T is the temperature. Brownian relaxation, on the other hand, involves the physical rotation of the entire nanoparticle in its surrounding medium. When an alternating magnetic field is applied, the nanoparticle rotates to align its magnetic moment with the field (Figure 222). This rotation creates friction with the surrounding medium, which generates heat. Brownian relaxation is more pronounced in larger nanoparticles and at lower frequencies of the alternating magnetic field. It is also strongly influenced by the viscosity of the surrounding medium. The rotation is defined by the Brownian relaxation time τ_B with the following equation (Eq. (42)):

$$\tau_B = \frac{3\eta V_H}{k_B T}, \tag{42}$$

where η is the viscosity of the surrounding medium, V_H is the hydrodynamic volume of the particles, k_B the Boltzmann constant, and T the temperature.

The interplay between these two mechanisms determines the overall heating efficiency of the nanoparticles. Both relaxation processes are strongly dependent on the size and the magnetic anisotropy constant of the nanoparticle. In smaller nanoparticles, Néel relaxation contributes more to heat generation than Brownian relaxation because less energy is required to realign the magnetic moment. For larger nanoparticles, Brownian relaxation dominates. However, highly viscous fluids can hinder the rotational motion of the nanoparticles. Neel and Brownian relaxation occur independently (Eq. (43))

$$\tau_{\text{eff}} = \frac{\tau_B \tau_N}{\tau_B + \tau_N}. \tag{43}$$

In larger multi-domain particles, heat is generated due to the movement of domain walls. The heating power generated by nanoparticles in the AC magnetic field is quantified by the Specific Absorption Rate (SAR) (Eq. (44))

$$\text{SAR} = \frac{\Delta TC}{\Delta t m_{\text{Fe}}}, \tag{44}$$

where C is the heat capacity of the solvent of the nanoparticles, T is the temperature, is the time, and m_{Fe} is the iron concentration in the solution.

The core of magnetic hyperthermia therapy is the device that generates the alternating magnetic field. This complex system consists of several key components, including a magnetic field generator, cooling system, treatment bed or area, control unit, and temperature monitoring system. The effectiveness of the treatment depends on several key parameters, such as magnetic field strength, frequency, treatment duration, coil configuration, duty cycle, field geometry, treatment volume, target temperature, and safety limits.

Magnetic hyperthermia therapy has shown promise in various applications beyond cancer treatment. It can be used for drug delivery, where heat triggers the release of drugs from thermosensitive carriers. In theranostics, SPIONs act as both therapeutic agents and contrast agents for imaging. The technique also has potential in gene therapy, where heat can activate temperature-sensitive promoters for gene expression. Additionally, it shows promise in combating antibiotic-resistant bacterial infections.

The noninvasive nature of magnetic hyperthermia, its ability to target specific tissues, and the potential for repeated treatments make it an attractive option for cancer therapy. However, challenges remain in optimizing nanoparticle design, ensuring uniform heating, and managing potential side effects. As research and clinical trials continue, magnetic hyperthermia holds the promise of becoming a critical tool in the fight against cancer, potentially improving outcomes and quality of life for patients worldwide. This innovative approach represents a significant advancement in cancer treatment, offering a precise and effective method of treating tumors through targeted heating. Ongoing research into magnetic hyperthermia therapy continues to push the boundaries of medical science, paving the way for more effective and less invasive cancer treatments in the future.

Bibliography

[1] Vitor, T., et al., *PET/MRI: a novel hybrid imaging technique. Major clinical indications and preliminary experience in Brazil*. Einstein (Sao Paulo), 2017. **15**(1): p. 115–118.

[2] Hanahan, D. and R. A. Weinberg, *Hallmarks of cancer: the next generation*. Cell, 2011. **144**(5): p. 646–674.

[3] Neidle, S. and D. E. Thurston, *Chemical approaches to the discovery and development of cancer therapies*. Nat Rev Cancer, 2005. **5**(4): p. 285–296.

[4] Thakor, A. S. and S. S. Gambhir, *Nanooncology: the future of cancer diagnosis and therapy*. CA Cancer J Clin, 2013. **63**: p. 395–418.

[5] Riggio, C., et al., *Nano-oncology: clinical application for cancer therapy and future perspectives*. J Nanomater, 2011. **2011**: p. 1–10.

[6] El-Karim, S. S. A., M. I. El-Zahar, and M. M. Anwar, *Nanotechnology in cancer diagnosis and treatment*. J Pharm Pharmacol, 2015. **3**(7): p. 299–315.

[7] Jain, K. K., *Advances in the field of nanooncology*. BMC Med, 2010. **8**: p. 83.

[8] Jain, K. K., *Recent advances in nanooncology*. Technol Cancer Res Treat, 2008. **7**(1): p. 1–13.

[9] McCarthy, J. R., *The future of theranostic nanoagents*. Nanomedicine, 2009. **4**: p. 693–695.

[10] Cherry, S. R., *Multimodality in vivo imaging systems: twice the power or double the trouble?* Annu Rev Biomed Eng, 2006. **8**: p. 35–62.

[11] Bertram, J. S., *Biology of cancer*. Mol Aspects Med, 2001. **21**: p. 167–223.

[12] Nussinov, R., C. J. Tsai, and H. Jang, *A new view of activating mutations in cancer*. Cancer Res, 2022. **82**(22): p. 4114–4123.

[13] Knudson, A. G., *Two genetic hits (more or less) to cancer*. Nat Rev Cancer, 2001. **1**: p. 157–170.

[14] Vogelstein, B. and K. W. Kinzler, *Cancer genes and the pathways they control*. Nat Med, 2004. **10**(8): p. 789–799.

[15] Fearon, E. R., *A genetic model for colorectal tumorigenesis*. Cell, 1990. **61**: p. 759–767.

[16] Kontomanolis, E. N., et al., *Role of oncogenes and tumor-suppressor genes in carcinogenesis: a review*. Anticancer Res, 2020. **40**(11): p. 6009–6015.

[17] Esteller, M., *Epigenetic changes in cancer*. F1000 Biol Rep, 2011. **3**: p. 9.

[18] Hanahan, D., *Hallmarks of cancer: new dimensions*. Cancer Discov, 2022. **12**(1): p. 31–46.

[19] Hanahan, D. and R. A. Weinberg, *Hallmarks of cancer*, in *Holland–Frei cancer medicine, ninth edition*, C. M. C. Robert, C. Bast Jr., W. N. Hait, W. K. Hong, D. W. Kufe, M. Piccart-Gebhart, R. E. Pollock, R. R. Weichselbaum, H. Wang, and J. F. Holland, Editor. 2017, JohnWiley & Sons, Inc. p. 6–16.

[20] Nguyen, T. H., *Mechanisms of metastasis*. Clin Dermatol, 2004. **22**(3): p. 209–216.

[21] Suhail, Y., et al., *Systems biology of cancer metastasis*. Cell Syst, 2019. **9**(2): p. 109–127.

[22] Kalluri, R. and R. A. Weinberg, *The basics of epithelial-mesenchymal transition*. J Clin Invest, 2009. **119**(6): p. 1420–1428.

[23] Lai, X., et al., *Epithelial-mesenchymal transition and metabolic switching in cancer: lessons from somatic cell reprogramming*. Front Cell Dev Biol, 2020. **8**: p. 760.

[24] Lamouille, S., J. Xu, and R. Derynck, *Molecular mechanisms of epithelial-mesenchymal transition*. Nat Rev Mol Cell Biol, 2014. **15**(3): p. 178–196.

[25] Dongre, A. and R. A. Weinberg, *New insights into the mechanisms of epithelial-mesenchymal transition and implications for cancer*. Nat Rev Mol Cell Biol, 2019. **20**(2): p. 69–84.

[26] Zhang, J., X. J. Tian, and J. Xing, *Signal transduction pathways of eMT induced by TGF-beta, SHH, and WNT and their crosstalks*. J Clin Med, 2016. **5**(4): p. 41.

[27] Yilmaz, M. and G. Christofori, *EMT, the cytoskeleton, and cancer cell invasion*. Cancer Metastasis Rev, 2009. **28**(1–2): p. 15–33.

[28] Lambert, A. W., D. R. Pattabiraman, and R. A. Weinberg, *Emerging biological principles of metastasis*. Cell, 2017. **168**(4): p. 670–691.

https://doi.org/10.1515/9783110605495-007

[29] Fares, J., et al., *Molecular principles of metastasis: a hallmark of cancer revisited*. Signal Transduct Targeted Ther, 2020. **5**(1): p. 28.

[30] Strilic, B. and S. Offermanns, *Intravascular survival and extravasation of tumor cells*. Cancer Cell, 2017. **32**(3): p. 282–293.

[31] DeBerardinis, R. J. and N. S. Chandel, *Fundamentals of cancer metabolism*. Sci Adv, 2016. **2**: p. e1600200.

[32] Pavlova, N. N. and C. B. Thompson, *The emerging hallmarks of cancer metabolism*. Cell Metab, 2016. **23**(1): p. 27–47.

[33] Nguyen, T.-L. and R. V. Durán, *Glutamine metabolism in cancer therapy*. Cancer Drug Resist, 2018. **1**: 126–138.

[34] Finley, L. W. S., *What is cancer metabolism?* Cell, 2023. **186**(8): p. 1670–1688.

[35] de Villiers, D., et al., *The role of reactive oxygen species in adipogenic differentiation*. Adv Exp Med Biol, 2018. **1083**: p. 125–144.

[36] Martinez-Outschoorn, U. E., et al., *Cancer metabolism: a therapeutic perspective*. Nat Rev Clin Oncol, 2017. **14**(1): p. 11–31.

[37] Liberti, M. V. and J. W. Locasale, *The Warburg effect: how does it benefit cancer cells?* Trends Biochem Sci, 2016. **41**(3): p. 211–218.

[38] Semenza, G. L., *Pharmacologic targeting of hypoxia-inducible factors*. Annu Rev Pharmacol Toxicol, 2019. **59**: p. 379–403.

[39] Yu, L., et al., *The glycolytic switch in tumors: how many players are involved?* J Cancer, 2017. **8**(17): p. 3430–3440.

[40] Conte, M. and V. Cauda, *Multimodal therapies against pancreatic ductal adenocarcinoma: a review on synergistic approaches toward ultimate nanomedicine treatments*. Adv Ther, 2022. **5**(11): p. 2200079.

[41] Goetz, L. H. and N. J. Schork, *Personalized medicine: motivation, challenges, and progress*. Fertil Steril, 2018. **109**(6): p. 952–963.

[42] Paul, S., et al., *Cancer therapy with antibodies*. Nat Rev Cancer, 2024. **24**(6): p. 399–426.

[43] Hussain, S., et al., *Modern diagnostic imaging technique applications and risk factors in the medical field: a review*. BioMed Res Int, 2022. **2022**: p. 5164970.

[44] Castro-Giner, F., et al., *Cancer diagnosis using a liquid biopsy: challenges and expectations*. Diagnostics (Basel), 2018. **8**(2): p. 31.

[45] Shariat, S. F. and C. G. Roehrborn, *Using biopsy to detect prostate cancer*. Rev Urol, 2008. **10**: p. 262–280.

[46] Katsanis, S. H. and N. Katsanis, *Molecular genetic testing and the future of clinical genomics*. Nat Rev Genet, 2013. **14**(6): p. 415–426.

[47] Fan, W., et al., *Nanotechnology for multimodal synergistic cancer therapy*. Chem Rev, 2017. **117**(22): p. 13566–13638.

[48] Seibert, J. A., *X-ray imaging physics for nuclear medicine technologists. Part 1: basic principles of X-ray production*. J Nucl Med Technol, 2004. **32**: p. 139–147.

[49] Seibert, J. A. and J. M. Boone, *X-ray imaging physics for nuclear medicine technologists. Part 2: X-ray interactions and image formation*. J Nucl Med Technol, 2005. **33**: p. 3–18.

[50] Lusic, H. and M. W. Grinstaff, *X-ray-computed tomography contrast agents*. Chem Rev, 2013. **113**(3): p. 1641–1666.

[51] Wang, G., H. Yu, and B. De Man, *An outlook on X-ray CT research and development*. Med Phys, 2008. **35**(3): p. 1051–1064.

[52] Kalender, W. A., *X-ray computed tomography*. Phys Med Biol, 2006. **51**(13): p. R29–R43.

[53] Spahn, M., *X-ray detectors in medical imaging*. Nucl Instrum Methods Phys Res, Sect A, Accel Spectrom Detect Assoc Equip, 2013. **731**: p. 57–63.

[54] Ou, X., et al., *Recent development in X-ray imaging technology: future and challenges*. Research (Wash D C), 2021. **2021**: p. 9892152.

[55] Moser, E., et al., *Magnetic resonance imaging methodology*. Eur J Nucl Med Mol Imaging, 2009. **36**(Suppl 1): p. S30–S41.

[56] Grover, V. P., et al., *Magnetic resonance imaging: principles and techniques: lessons for clinicians*. J Clin Exp Hepatol, 2015. **5**(3): p. 246–255.

[57] Bruno, F., et al., *Advanced magnetic resonance imaging (MRI) techniques: technical principles and applications in nanomedicine*. Cancers (Basel), 2022. **14**(7): p. 1626.

[58] Nutt, R., *The history of positron emission tomography*. Mol Imaging Biol, 2002. **4**: p. 11–26.

[59] Townsend, D. W., *Positron emission tomography/computed tomography*. Semin Nucl Med, 2008. **38**(3): p. 152–166.

[60] Jones, T. and D. Townsend, *History and future technical innovation in positron emission tomography*. J Med Imag (Bellingham), 2017. **4**(1): p. 011013.

[61] Gallamini, A., C. Zwarthoed, and A. Borra, *Positron emission tomography (PET) in Oncology*. Cancers (Basel), 2014. **6**(4): p. 1821–1889.

[62] Choi, J. Y., et al., *Improved detection of second primary cancer using integrated [18F] fluorodeoxyglucose positron emission tomography and computed tomography for initial tumor staging*. J Clin Oncol, 2005. **23**(30): p. 7654–7659.

[63] Jensen, J. A., *Medical ultrasound imaging*. Prog Biophys Mol Biol, 2007. **93**(1–3): p. 153–165.

[64] Christensen-Jeffries, K., et al., *Super-resolution ultrasound imaging*. Ultrasound Med Biol, 2020. **46**(4): p. 865–891.

[65] Ng, A. and J. Swanevelder, *Resolution in ultrasound imaging*. Contin Educ Anaesth Crit Care Pain, 2011. **11**(5): p. 186–192.

[66] Carovac, A., F. Smajlovic, and D. Junuzovic, *Application of ultrasound in medicine*. Acta Inform Med, 2011. **19**(3): p. 168–171.

[67] Alamzadeh, Z., et al., *Gold nanoparticles promote a multimodal synergistic cancer therapy strategy by co-delivery of thermo-chemo-radio therapy*. Eur J Pharm Sci, 2020. **145**: p. 105235.

[68] Bansal, S. A., et al., *Role of gold nanoparticles in advanced biomedical applications*. Nanoscale Adv, 2020. **2**(9): p. 3764–3787.

[69] Elahi, N., M. Kamali, and M. H. Baghersad, *Recent biomedical applications of gold nanoparticles: a review*. Talanta, 2018. **184**: p. 537–556.

[70] Mikhailova, E. O., *Gold nanoparticles: biosynthesis and potential of biomedical application*. J Funct Biomater, 2021. **12**(4): p. 70.

[71] Gehrke, H., et al., *Platinum nanoparticles and their cellular uptake and DNA platination at non-cytotoxic concentrations*. Arch Toxicol, 2011. **85**(7): p. 799–812.

[72] Yaqoob, S. B., et al., *Gold, silver, and palladium nanoparticles: a chemical tool for biomedical applications*. Front Chem, 2020. **8**: p. 376.

[73] Huang, Z., et al., *Silver ion-enhanced particle-specific cytotoxicity of silver nanoparticles and effect on the production of extracellular secretions of Phanerochaete chrysosporium*. Chemosphere, 2018. **196**: p. 575–584.

[74] Giljohann, D. A., et al., *Gold nanoparticles for biology and medicine*. Angew Chem, Int Ed Engl, 2010. **49**(19): p. 3280–3294.

[75] Amendola, V., et al., *Surface plasmon resonance in gold nanoparticles: a review*. J Phys Condens Matter, 2017. **29**(20): p. 203002.

[76] Chen, H., et al., *Shape- and size-dependent refractive index sensitivity of gold nanoparticles*. Langmuir, 2008. **24**: p. 5233–5237.

[77] Chen, H., et al., *Gold nanorods and their plasmonic properties*. Chem Soc Rev, 2013. **42**(7): p. 2679–2724.

[78] Zheng, J., et al., *Gold nanorods: the most versatile plasmonic nanoparticles*. Chem Rev, 2021. **121**(21): p. 13342–13453.

[79] Herizchi, R., et al., *Current methods for synthesis of gold nanoparticles*. Artif Cells Nanomed Biotechnol, 2016. **44**(2): p. 596–602.

[80] Liang, H.-P., et al., *Gold hollow nanospheres: tunable surface plasmon resonance controlled by interior-cavity sizes*. J Phys Chem B, 2005. **109**: p. 7795–7800.

[81] Xia, Y., et al., *Shape-controlled synthesis of metal nanocrystals: simple chemistry meets complex physics?* Angew Chem, Int Ed Engl, 2009. **48**(1): p. 60–103.

[82] Daruich De Souza, C., B. Ribeiro Nogueira, and M. E. C. M. Rostelato, *Review of the methodologies used in the synthesis gold nanoparticles by chemical reduction*. J Alloys Compd, 2019. **798**: p. 714–740.

[83] Samadi, A., et al., *Platinum nanoparticles: a non-toxic, effective and thermally stable alternative plasmonic material for cancer therapy and bioengineering*. Nanoscale, 2018. **10**(19): p. 9097–9107.

[84] Liu, Y., et al., *Platinum nanoparticles: efficient and stable catechol oxidase mimetics*. ACS Appl Mater Interfaces, 2015. **7**(35): p. 19709–19717.

[85] Hullo, M., et al., *Radiation enhancer effect of platinum nanoparticles in breast cancer cell lines: in vitro and in silico analyses*. Int J Mol Sci, 2021. **22**(9): p. 4436.

[86] Cai, K., et al., *Aqueous synthesis of porous platinum nanotubes at room temperature and their intrinsic peroxidase-like activity*. Chem Commun (Camb), 2013. **49**(54): p. 6024–6026.

[87] Jeyaraj, M., et al., *A comprehensive review on the synthesis, characterization, and biomedical application of platinum nanoparticles*. Nanomaterials (Basel), 2019. **9**(12): p. 1719.

[88] Jain, P. K., *Calculated absorption and scattering properties of gold nanoparticles of different size, shape, and composition: applications in biological imaging and biomedicine*. J Phys Chem B, 2006. **110**(14): p. 7238–7248.

[89] Liu, S., et al., *Synthesis of monodisperse Au, Ag, and Au–Ag alloy nanoparticles with tunable size and surface plasmon resonance frequency*. Chem Mater, 2011. **23**(18): p. 4098–4101.

[90] Link, S. and M. A. El-Sayed, *Shape and size dependence of radiative, non-radiative and photothermal properties of gold nanocrystals*. Int Rev Phys Chem, 2000. **19**: p. 409–453.

[91] Mie, G., *Beiträge zur Optik trüber Medien, speziell kolloidaler Metalllösungen*. Ann Phys, 1908. **25**: p. 377–445.

[92] Draine, B. T. and P. J. Flatau, *Discrete-dipole approximation for scattering calculations*. J Opt Soc Am A, 1994. **11**: p. 1491–1499.

[93] Yang, W. H., G. C. Schatz, and R. P. Van Duyne, *Discrete dipole approximation for calculating extinction and Raman intensities for small particles with arbitrary shapes*. J Chem Phys, 1995. **103**(3): p. 869–875.

[94] Zhao, J., et al., *Methods for describing the electromagnetic properties of silver and gold nanoparticles*. Acc Chem Res, 2008. **41**: p. 1710–1720.

[95] Wiley, B. J., et al., *Maneuvering the surface plasmon resonance of silver nanostructures through shape-controlled synthesis*. J Phys Chem B, 2006. **110**: p. 15666–15675.

[96] Niu, J., T. Zhu, and Z. Liu, *One-step seed-mediated growth of 30–150 nm quasispherical gold nanoparticles with 2-mercaptosuccinic acid as a new reducing agent*. Nanotechnology, 2007. **18**(32): p. 325607.

[97] Guan, Y., et al., *One-pot synthesis of size-tunable hollow gold nanoshells via APTES-in-water suspension*. Colloids Surf A, Physicochem Eng Asp, 2016. **502**: p. 6–12.

[98] Prevo, B. G., et al., *Scalable routes to gold nanoshells with tunable sizes and response to near-infrared pulsed-laser irradiation*. Small, 2008. **4**(8): p. 1183–1195.

[99] Hirsch, L. R., et al., *Nanoshell-mediated near-infrared thermal therapy of tumors under magnetic resonance guidance*. Proc Natl Acad Sci USA, 2003. **100**: p. 13549–13554.

[100] Zhao, J., M. Wallace, and M. P. Melancon, *Cancer theranostics with gold nanoshells*. Nanomedicine (Lond), 2014. **9**(13): p. 2041–2057.

[101] Myroshnychenko, V., et al., *Modelling the optical response of gold nanoparticles*. Chem Soc Rev, 2008. **37**(9): p. 1792–1805.

[102] Jabeen, F., et al., *Au-nanomaterials as a superior choice for near-infrared photothermal therapy*. Molecules, 2014. **19**(12): p. 20580–20593.

[103] Bucharskaya, A., et al., *Towards effective photothermal/photodynamic treatment using plasmonic gold nanoparticles*. Int J Mol Sci, 2016. **17**(8): p. 1295.

[104] Riley, R. S. and E. S. Day, *Gold nanoparticle-mediated photothermal therapy: applications and opportunities for multimodal cancer treatment*. Wiley Interdiscip Rev Nanomed Nanobiotechnol, 2017. **9**(4): p. e1449.

[105] Vines, J. B., et al., *Gold nanoparticles for photothermal cancer therapy*. Front Chem, 2019. **7**: p. 167.

[106] Jauffred, L., et al., *Plasmonic heating of nanostructures*. Chem Rev, 2019. **119**(13): p. 8087–8130.

[107] Bailly, A. L., et al., *In vivo evaluation of safety, biodistribution and pharmacokinetics of laser-synthesized gold nanoparticles*. Sci Rep, 2019. **9**(1): p. 12890.

[108] Zhao, P., N. Li, and D. Astruc, *State of the art in gold nanoparticle synthesis*. Coord Chem Rev, 2013. **257**(3–4): p. 638–665.

[109] Grzelczak, M., et al., *Shape control in gold nanoparticle synthesis*. Chem Soc Rev, 2008. **37**(9): p. 1783–1791.

[110] Hühn, J., et al., *Selected standard protocols for the synthesis, phase transfer, and characterization of inorganic colloidal nanoparticles*. Chem Mater, 2016. **29**(1): p. 399–461.

[111] Tan, B. J. Y., et al., *Fabrication of size-tunable gold nanoparticles array with nanosphere lithography, reactive ion etching, and thermal annealing*. J Phys Chem B, 2005. **109**: p. 11100–11109.

[112] Ziefuß, A. R., et al., *Laser fragmentation of colloidal gold nanoparticles with high-intensity nanosecond pulses is driven by a single-step fragmentation mechanism with a defined educt particle-size threshold*. J Phys Chem C, 2018. **122**(38): p. 22125–22136.

[113] Ziefuss, A. R., et al., *In situ structural kinetics of picosecond laser-induced heating and fragmentation of colloidal gold spheres*. Phys Chem Chem Phys, 2020. **22**(9): p. 4993–5001.

[114] Al-Kattan, A., et al., *Recent advances in laser-ablative synthesis of bare Au and Si nanoparticles and assessment of their prospects for tissue engineering applications*. Int J Mol Sci, 2018. **19**(6): p. 1563.

[115] Amendola, V. and M. Meneghetti, *Laser ablation synthesis in solution and size manipulation of noble metal nanoparticles*. Phys Chem Chem Phys, 2009. **11**(20): p. 3805–3821.

[116] Gonzalez-Rubio, G., A. Guerrero-Martinez, and L. M. Liz-Marzan, *Reshaping, fragmentation, and assembly of gold nanoparticles assisted by pulse lasers*. Acc Chem Res, 2016. **49**(4): p. 678–686.

[117] Kabashin, A. V. and M. Meunier, *Synthesis of colloidal nanoparticles during femtosecond laser ablation of gold in water*. J Appl Phys, 2003. **94**(12): p, 7941–7943.

[118] Mat Isa, S. Z., R. Zainon, and M. Tamal, *State of the art in gold nanoparticle synthesisation via pulsed laser ablation in liquid and its characterisation for molecular imaging: a review*. Materials (Basel), 2022. **15**(3): p. 875.

[119] Levy, A., et al., *Surface chemistry of gold nanoparticles produced by laser ablation in pure and saline water*. Langmuir, 2021. **37**(19): p. 5783–5794.

[120] Maximova, K., et al., *Size-controllable synthesis of bare gold nanoparticles by femtosecond laser fragmentation in water*. Nanotechnology, 2015. **26**(6): p. 065601.

[121] Priyadarshini, M., et al., *Optical nonlinearities in chemically synthesized and femtosecond laser fabricated gold nanoparticle colloidal solutions*. Opt Laser Technol, 2021. **139**: p. 107008.

[122] Sylvestre, J. P., et al., *Femtosecond laser ablation of gold in water: influence of the laser-produced plasma on the nanoparticle size distribution*. Appl Phys A, 2005. **80**(4): p. 753–758.

[123] Beola, L., et al., *Critical parameters to improve pancreatic cancer treatment using magnetic hyperthermia: field conditions, immune response, and particle biodistribution*. ACS Appl Mater Interfaces, 2021. **13**(11): p. 12982–12996.

[124] Huang, X., S. Neretina, and M. A. El-Sayed, *Gold nanorods: from synthesis and properties to biological and biomedical applications*. Adv Mater, 2009. **21**(48): p. 4880–4910.

[125] Murphy, C. J., et al., *The many faces of gold nanorods*. J Phys Chem Lett, 2010. **1**(19): p. 2867–2875.

[126] Murphy, C. J., et al., *Gold nanorod crystal growth: from seed-mediated synthesis to nanoscale sculpting*. Curr Opin Colloid Interface Sci, 2011. **16**(2): p. 128–134.

[127] Su, G., C. Yang, and J. J. Zhu, *Fabrication of gold nanorods with tunable longitudinal surface plasmon resonance peaks by reductive dopamine*. Langmuir, 2015. **31**(2): p. 817–823.

[128] Alkilany, A. M., et al., *Cellular uptake and cytotoxicity of gold nanorods: molecular origin of cytotoxicity and surface effects*. Small, 2009. **5**(6): p. 701–708.

[129] Correard, F., et al., *Gold nanoparticles prepared by laser ablation in aqueous biocompatible solutions: assessment of safety and biological identity for nanomedicine applications*. Int J Nanomed, 2014. **9**: p. 5415–5430.

[130] De Anda Villa, M., et al., *Assessing the surface oxidation state of free-standing gold nanoparticles produced by laser ablation*. Langmuir, 2019. **35**(36): p. 11859–11871.

[131] Merk, V., et al., *In situ non-DLVO stabilization of surfactant-free, plasmonic gold nanoparticles: effect of Hofmeister's anions*. Langmuir, 2014. **30**(15): p. 4213–4222.

[132] Sajti, C. L., et al., *Delay time and concentration effects during bioconjugation of nanosecond laser-generated nanoparticles in a liquid flow*. J Phys Chem C, 2011. **115**(12): p. 5094–5101.

[133] Kanitz, A., et al., *Review on experimental and theoretical investigations of the early stage, femtoseconds to microseconds processes during laser ablation in liquid-phase for the synthesis of colloidal nanoparticles*. Plasma Sources Sci Technol, 2019. **28**(10): p. 103001.

[134] Turkevich, J., P. C. Stevenson, and J. Hillier, *A study of the nucleation and growth processes in the synthesis of colloidal gold*. Discuss Faraday Soc, 1951. **11**: p. 55–75.

[135] Turkevich, J., P. C. Stevenson, and J. Hillier, *The formation of colloidal gold*. J Phys Chem, 1953. **57**: p. 670–673.

[136] Wuithschick, M., et al., *Turkevich in new robes: key questions answered for the most common gold nanoparticle synthesis*. ACS Nano, 2015. **9**: p. 7052–7071.

[137] Tyagi, H., et al., *A facile pH controlled citrate-based reduction method for gold nanoparticle synthesis at room temperature*. Nanoscale Res Lett, 2016. **11**(1): p. 362.

[138] Brust, M., *Synthesis of thiol-derivatised gold nanoparticles in a two-phase liquid-liquid system*. J Chem Soc, Chem Commun, 1994. **1994**: p. 801–802.

[139] Goulet, P. J. G., *New insights into brust-schiffrin metal nanoparticle synthesis*. J Am Chem Soc, 2010. **132**: p. 9582–9584.

[140] Perala, S. R. and S. Kumar, *On the mechanism of metal nanoparticle synthesis in the Brust–Schiffrin method*. Langmuir, 2013. **29**(31): p. 9863–9873.

[141] Zheng, J., et al., *Different sized luminescent gold nanoparticles*. Nanoscale, 2012. **4**(14): p. 4073–4083.

[142] Chen, L. Y., et al., *Fluorescent gold nanoclusters: recent advances in sensing and imaging*. Anal Chem, 2015. **87**(1): p. 216–229.

[143] Jana, N. R., G. Latha, and C. J. Murphy, *Wet chemical synthesis of high aspect ratio cylindrical gold nanorods*. J Phys Chem B, 2001. **105**: p. 4065–4067.

[144] Sau, T. K. and C. J. Murphy, *Self-assembly patterns formed upon solvent evaporation of aqueous cetyltrimethylammonium bromide-coated gold nanoparticles of various shapes*. Langmuir, 2005. **21**: p. 2923–2929.

[145] Nikoobakht, B. and M. A. El-Sayed, *Preparation and growth mechanism of gold nanorods (NRs) using seed-mediated growth method*. Chem Mater, 2003. **15**: p. 1957–1962.

[146] Murphy, C. J., et al., *Anisotropic metal nanoparticles: synthesis, assembly, and optical applications*. J Phys Chem B, 2005. **109**: p. 13857–13870.

[147] Burrows, N. D., et al., *Understanding the seed-mediated growth of gold nanorods through a fractional factorial design of experiments*. Langmuir, 2017. **33**(8): p. 1891–1907.

[148] Ali, M. R., B. Snyder, and M. A. El-Sayed, *Synthesis and optical properties of small Au nanorods using a seedless growth technique*. Langmuir, 2012. **28**(25): p. 9807–9815.

[149] Liu, K., et al., *Biocompatible gold nanorods: one-step surface functionalization, highly colloidal stability, and low cytotoxicity*. Langmuir, 2015. **31**(17): p. 4973–4980.

[150] Santhosh, P. B., et al., *Phospholipid stabilized gold nanorods: towards improved colloidal stability and biocompatibility*. Phys Chem Chem Phys, 2017. **19**(28): p. 18494–18504.

[151] Jia, Y. P., et al., *Effects of cetyltrimethylammonium bromide on the toxicity of gold nanorods both in vitro and in vivo: molecular origin of cytotoxicity and inflammation*. Small Methods, 2020. **4**(3): p. 1900799.

[152] Sauerbeck, C., et al., *Shedding light on the growth of gold nanoshells*. ACS Nano, 2014. **8**: p. 3088–3096.

[153] Wang, Y. C., et al., *Synthetic methodologies to gold nanoshells: an overview*. Molecules, 2018. **23**(11): p. 2851.

[154] Pham, T., et al., *Preparation and characterization of gold nanoshells coated with self-assembled monolayers*. Langmuir, 2002. **18**: p. 4915–4920.

[155] Kah, J. C. Y., et al., *Synthesis of gold nanoshells based on the deposition-precipitation process*. Gold Bull, 2008. **41**: p. 23–36.

[156] Wu, W., Q. He, and C. Jiang, *Magnetic iron oxide nanoparticles: synthesis and surface functionalization strategies*. Nanoscale Res Lett, 2008. **3**(11): p. 397–415.

[157] Wu, W., et al., *Recent progress on magnetic iron oxide nanoparticles: synthesis, surface functional strategies and biomedical applications*. Sci Technol Adv Mater, 2015. **16**(2): p. 023501.

[158] Guo, H. and A. S. Barnard, *Naturally occurring iron oxide nanoparticles: morphology, surface chemistry and environmental stability*. J Mater Chem A, 2013. **1**(1): p. 27–42.

[159] Zheng, B., et al., *Magnetic particle imaging tracks the long-term fate of in vivo neural cell implants with high image contrast*. Sci Rep, 2015. **5**: p. 14055.

[160] Yu, E. Y., et al., *Magnetic particle imaging: a novel in vivo imaging platform for cancer detection*. Nano Lett, 2017. **17**(3): p. 1648–1654.

[161] Xi, L., et al., *Molecular photoacoustic tomography of breast cancer using receptor targeted magnetic iron oxide nanoparticles as contrast agents*. J Biophotonics, 2014. **7**(6): p. 401–409.

[162] Weinstein, J. S., et al., *Superparamagnetic iron oxide nanoparticles: diagnostic magnetic resonance imaging and potential therapeutic applications in neurooncology and central nervous system inflammatory pathologies, a review*. J Cereb Blood Flow Metab, 2010. **30**(1): p. 15–35.

[163] Wei, H., et al., *Single-nanometer iron oxide nanoparticles as tissue-permeable MRI contrast agents*. Proc Natl Acad Sci USA, 2021. **118**(42): p. e2102340118.

[164] Wei, H., et al., *Exceedingly small iron oxide nanoparticles as positive MRI contrast agents*. Proc Natl Acad Sci USA, 2017. **114**(9): p. 2325–2330.

[165] van de Loosdrecht, M. M., et al., *A novel characterization technique for superparamagnetic iron oxide nanoparticles: the superparamagnetic quantifier, compared with magnetic particle spectroscopy*. Rev Sci Instrum, 2019. **90**(2): p. 024101.

[166] Tromsdorf, U. I., et al., *a highly effective, nontoxic T_1 MR contrast agent based on ultrasmall pegylated iron oxide nanoparticles*. Nano Lett, 2009. **9**: p. 4434–4440.

[167] Thorek, D. L., et al., *Superparamagnetic iron oxide nanoparticle probes for molecular imaging*. Ann Biomed Eng, 2006. **34**(1): p. 23–38.

[168] Thawani, J. P., et al., *Photoacoustic-guided surgery with indocyanine green-coated superparamagnetic iron oxide nanoparticle clusters*. Small, 2017. **13**(37).

[169] Tassa, C., S. Y. Shaw, and R. Weissleder, *Dextran-coated iron oxide nanoparticles: a versatile platform for targeted molecular imaging, molecular diagnostics, and therapy*. Acc Chem Res, 2011. **44**: p. 842–852.

[170] Poller, W. C., et al., *Magnetic particle spectroscopy reveals dynamic changes in the magnetic behavior of very small superparamagnetic iron oxide nanoparticles during cellular uptake and enables determination of cell-labeling efficacy*. J Biomed Nanotechnol, 2016. **12**(2): p. 337–346.

[171] Makela, A. V., et al., *Magnetic particle imaging of magnetotactic bacteria as living contrast agents is improved by altering magnetosome arrangement*. Nano Lett, 2022. **22**(12): p. 4630–4639.

[172] Kucheryavy, P., et al., *Superparamagnetic iron oxide nanoparticles with variable size and an iron oxidation state as prospective imaging agents*. Langmuir, 2013. **29**(2): p. 710–716.

[173] Javed, Y., et al., *MRI based on iron oxide nanoparticles contrast agents: effect of oxidation state and architecture*. J Nanopart Res, 2017. **19**(11): p. 366.

[174] Iqbal, M. Z., et al., *Silica-coated super-paramagnetic iron oxide nanoparticles (SPIONPs): a new type contrast agent of T_1 magnetic resonance imaging (MRI)*. J Mater Chem B, 2015. **3**(26): p. 5172–5181.

[175] Hola, K., et al., *Tailored functionalization of iron oxide nanoparticles for MRI, drug delivery, magnetic separation and immobilization of biosubstances*. Biotechnol Adv, 2015. **33**(6 Pt 2): p. 1162–1176.

[176] Harvell-Smith, S., L. D. Tung, and N. T. K. Thanh, *Magnetic particle imaging: tracer development and the biomedical applications of a radiation-free, sensitive, and quantitative imaging modality*. Nanoscale, 2022. **14**(10): p. 3658–3697.

[177] Estelrich, J., M. J. Sanchez-Martin, and M. A. Busquets, *Nanoparticles in magnetic resonance imaging: from simple to dual contrast agents*. Int J Nanomed, 2015. **10**: p. 1727–1741.

[178] Deng, L. H., et al., *Size and PEG length-controlled pegylated monocrystalline superparamagnetic iron oxide nanocomposite for MRI contrast agent*. Int J Nanomed, 2021. **16**: p. 201–211.

[179] Bell, G., et al., *Functionalised iron oxide nanoparticles for multimodal optoacoustic and magnetic resonance imaging*. J Mater Chem B, 2019. **7**(13): p. 2212–2219.

[180] Bauer, L. M., et al., *Magnetic particle imaging tracers: state-of-the-art and future directions*. J Phys Chem Lett, 2015. **6**(13): p. 2509–2517.

[181] Zhi, D., et al., *Targeting strategies for superparamagnetic iron oxide nanoparticles in cancer therapy*. Acta Biomater, 2020. **102**: p. 13–34.

[182] Wydra, R. J., et al., *The role of ROS generation from magnetic nanoparticles in an alternating magnetic field on cytotoxicity*. Acta Biomater, 2015. **25**: p. 284–290.

[183] Wahajuddin and S. Arora, *Superparamagnetic iron oxide nanoparticles: magnetic nanoplatforms as drug carriers*. Int J Nanomed, 2012. **7**: p. 3445–3471.

[184] Piehler, S., et al., *Iron oxide nanoparticles as carriers for dox and magnetic hyperthermia after intratumoral application into breast cancer in mice: impact and future perspectives*. Nanomaterials (Basel), 2020. **10**(6): p. 1016.

[185] Murase, K., et al., *Simulation and experimental studies on magnetic hyperthermia with use of superparamagnetic iron oxide nanoparticles*. Radiol Phys Technol, 2011. **4**(2): p. 194–202.

[186] Mahmoudi, M., et al., *Superparamagnetic iron oxide nanoparticles (SPIONs): development, surface modification and applications in chemotherapy*. Adv Drug Deliv Rev, 2011. **63**(1–2): p. 24–46.

[187] Klein, S., et al., *Superparamagnetic iron oxide nanoparticles as radiosensitizer via enhanced reactive oxygen species formation*. Biochem Biophys Res Commun, 2012. **425**(2): p. 393–397.

[188] Klein, S., et al., *Superparamagnetic iron oxide nanoparticles as novel X-ray enhancer for low-dose radiation therapy*. J Phys Chem B, 2014. **118**(23): p. 6159–6166.

[189] Frtus, A., et al., *Analyzing the mechanisms of iron oxide nanoparticles interactions with cells: a road from failure to success in clinical applications*. J Control Release, 2020. **328**: p. 59–77.

[190] Fortin, J.-P., et al., *Size-sorted anionic iron oxide nanomagnets as colloidal mediators for magnetic hyperthermia*. J Am Chem Soc, 2007. **129**: p. 2628–2635.

[191] Estelrich, J. and M. A. Busquets, *Iron oxide nanoparticles in photothermal therapy*. Molecules, 2018. **23**(7): p. 1567.

[192] Espinosa, A., et al., *Duality of iron oxide nanoparticles in cancer therapy: amplification of heating efficiency by magnetic hyperthermia and photothermal bimodal treatment*. ACS Nano, 2016. **10**(2): p. 2436–2446.

[193] Dilnawaz, F., et al., *Dual drug loaded superparamagnetic iron oxide nanoparticles for targeted cancer therapy*. Biomaterials, 2010. **31**(13): p. 3694–3706.

[194] Ashikbayeva, Z., et al., *Application of nanoparticles and nanomaterials in thermal ablation therapy of cancer*. Nanomaterials (Basel), 2019. **9**(9): p. 1195.

[195] Zandieh, M. and J. Liu, *Nanozyme catalytic turnover and self-limited reactions*. ACS Nano, 2021. **15**(10): p. 15645–15655.

[196] Gao, L., et al., *Intrinsic peroxidase-like activity of ferromagnetic nanoparticles*. Nat Nanotechnol, 2007. **2**(9): p. 577–583.

[197] Chen, Q., et al., *Tumor-specific expansion of oxidative stress by glutathione depletion and use of a fenton nanoagent for enhanced chemodynamic therapy*. ACS Appl Mater Interfaces, 2019. **11**(34): p. 30551–30565.

[198] Lu, A. H., E. L. Salabas, and F. Schuth, *Magnetic nanoparticles: synthesis, protection, functionalization, and application*. Angew Chem, Int Ed Engl, 2007. **46**(8): p. 1222–1244.

[199] Peng, S., et al., *Synthesis and stabilization of monodisperse Fe nanoparticles*. J Am Chem Soc, 2006. **128**: p. 10676–10677.

[200] Ho, C.-H., et al., *Shape-Controlled growth and shape-dependent cation site occupancy of monodisperse Fe_3O_4 nanoparticles*. Chem Mater, 2011. **23**(7): p. 1753–1760.

[201] Muro-Cruces, J., et al., *Precise Size control of the growth of Fe_3O_4 nanocubes over a wide size range using a rationally designed one-pot synthesis*. ACS Nano, 2019. **13**(7): p. 7716–7728.

[202] Caruntu, D., et al., *Synthesis of variable-sized nanocrystals of Fe_3O_4 with high surface reactivity*. Chem Mater, 2004. **16**: p. 5527–5534.

[203] Bronstein, L. M., et al., *Influence of iron oleate complex structure on iron oxide nanoparticle formation*. Chem Mater, 2007. **19**: p. 3624–3632.

[204] Kampferbeck, M., et al., *Little adjustments significantly simplify the gram-scale synthesis of high-quality iron oxide nanocubes*. Langmuir, 2021. **37**(32): p. 9851–9857.

[205] Massart, R., *Preparation of aqueous magnetic liquids in alkaline and acidic media*. IEEE Trans Magn, 1981. **17**: p. 1247–1248.

[206] Williams, D. N., et al., *Surface modification of magnetic nanoparticles using gum arabic*. J Nanopart Res, 2006. **8**(5): p. 749–753.

[207] Ho, D., X. Sun, and S. Sun, *Monodisperse magnetic nanoparticles for theranostic applications*. Acc Chem Res, 2011. **44**: p. 875–882.

[208] Kovalenko, M. V., et al., *Fatty acid salts as stabilizers in size- and shape-controlled nanocrystal synthesis: the case of inverse spinel iron oxide*. J Am Chem Soc, 2007. **129**: p. 6352–6353.

[209] Zhu, H., et al., *Monodisperse M(x)Fe(3-x)O4 (M = Fe, Cu, Co, Mn) nanoparticles and their electrocatalysis for oxygen reduction reaction*. Nano Lett, 2013. **13**(6): p. 2947–2951.

[210] Islam, K., et al., *Manganese ferrite nanoparticles ($MnFe_2O_4$): size dependence for hyperthermia and negative/positive contrast enhancement in MRI*. Nanomaterials (Basel), 2020. **10**(11): p. 2297.

[211] Sun, S., et al., *Monodisperse MFe_2O_4 (M = Fe, Co, Mn) nanoparticles*. J Am Chem Soc, 2004. **126**: p. 273–279.

[212] Hemery, G., et al., *Tuning sizes, morphologies, and magnetic properties of monocore versus multicore iron oxide nanoparticles through the controlled addition of water in the polyol synthesis*. Inorg Chem, 2017. **56**(14): p. 8232–8243.

[213] Hachani, R., et al., *Polyol synthesis, functionalisation, and biocompatibility studies of superparamagnetic iron oxide nanoparticles as potential MRI contrast agents*. Nanoscale, 2016. **8**(6): p. 3278–3287.

[214] Dong, H., Y. C. Chen, and C. Feldmann, *Polyol synthesis of nanoparticles: status and options regarding metals, oxides, chalcogenides, and non-metal elements*. Green Chem, 2015. **17**(8): p. 4107–4132.

[215] Cheng, C., F. Xu, and H. Gu, *Facile synthesis and morphology evolution of magnetic iron oxide nanoparticles in different polyol processes*. New J Chem, 2011. **35**(5): p. 1072–1079.

[216] Cheah, P., et al., *The key role of reaction temperature on a polyol synthesis of water-dispersible iron oxide nanoparticles*. J Magn Magn Mater, 2021. **540**: p. 168481.

[217] Raeisi Shahraki, R., et al., *Structural characterization and magnetic properties of superparamagnetic zinc ferrite nanoparticles synthesized by the coprecipitation method*. J Magn Magn Mater, 2012. **324**(22): p. 3762–3765.

[218] Samrot, A. V., et al., *A review on synthesis, characterization and potential biological applications of superparamagnetic iron oxide nanoparticles*. Curr Res Green Sustain Chem, 2021. **4**: p. 100042.

[219] Gupta, A. K. and M. Gupta, *Synthesis and surface engineering of iron oxide nanoparticles for biomedical applications*. Biomaterials, 2005. **26**(18): p. 3995–4021.

[220] Zaloga, J., et al., *Development of a lauric acid/albumin hybrid iron oxide nanoparticle system with improved biocompatibility*. Int J Nanomed, 2014. **9**: p. 4847–4866.

[221] Xu, Z., et al., *Oleylamine as both reducing agent and stabilizer in a facile synthesis of magnetite nanoparticles*. Chem Mater, 2009. **21**: p. 1778–1780.

[222] Unni, M., et al., *Thermal Decomposition synthesis of iron oxide nanoparticles with diminished magnetic dead layer by controlled addition of oxygen*. ACS Nano, 2017. **11**(2): p. 2284–2303.

[223] Kim, D., et al., *Synthesis of uniform ferrimagnetic magnetite nanocubes*. J Am Chem Soc, 2009. **131**: p. 454–455.

[224] Nobile, C. and P. D. Cozzoli, *Synthetic approaches to colloidal nanocrystal heterostructures based on metal and metal-oxide materials*. Nanomaterials (Basel), 2022. **12**(10): p. 1729.

[225] Cozzoli, P. D., T. Pellegrino, and L. Manna, *Synthesis, properties and perspectives of hybrid nanocrystal structures*. Chem Soc Rev, 2006. **35**(11): p. 1195–1208.

[226] Carbone, L. and P. D. Cozzoli, *Colloidal heterostructured nanocrystals: Synthesis and growth mechanisms*. Nano Today, 2010. **5**(5): p. 449–493.

[227] Pineider, F., et al., *Spin-polarization transfer in colloidal magnetic-plasmonic Au/iron oxide hetero-nanocrystals*. ACS Nano, 2013. **7**: p. 857–866.

[228] Ali Dheyab, M., et al., *Recent advances in synthesis, medical applications and challenges for gold-coated iron oxide: comprehensive study*. Nanomaterials (Basel), 2021. **11**(8): p. 2147.

[229] Shi, W., *A general approach to binary and ternary hybrid nanocrystals*. Nano Lett, 2006. **4**: p. 875–881.

[230] Lee, Y., et al., *Synthetic tuning of the catalytic properties of Au–Fe_3O_4 nanoparticles*. Angew Chem, Int Ed Engl, 2010. **49**(7): p. 1271–1274.

[231] Tancredi, P., et al., *Exploring the synthesis conditions to control the morphology of gold-iron oxide heterostructures*. Nano Res, 2019. **12**(8): p. 1781–1788.

[232] Shevchenko, E. V., et al., *Gold/iron oxide core/hollow-shell nanoparticles*. Adv Mater, 2008. **20**(22): p. 4323–4329.

[233] Klein, S., et al., *Bifunctional Au-Fe_3O_4 nanoheterodimers acting as x-ray protector in healthy cells and as X-ray enhancer in tumor cells*. ACS Appl Mater Interfaces, 2019. **11**(43): p. 39613–39623.

[234] Klein, S., et al., *Pt–Fe_3O_4, Pd–Fe_3O_4, and Au-Fe_3O_4 nanoheterodimers and their efficacy as radiosensitizers in cancer therapy*. ACS Appl Bio Mater, 2021. **4**(11): p. 7879–7892.

[235] Das, R., et al., *Boosted hyperthermia therapy by combined ac magnetic and photothermal exposures in Ag/Fe_3O_4 nanoflowers*. ACS Appl Mater Interfaces, 2016. **8**(38): p. 25162–25169.

[236] Umut, E., et al., *Magnetic, optical and relaxometric properties of organically coated gold–magnetite (Au–Fe_3O_4) hybrid nanoparticles for potential use in biomedical applications*. J Magn Magn Mater, 2012. **324**(15): p. 2373–2379.

[237] Muzzi, B., et al., *Star-shaped magnetic-plasmonic Au@Fe_3O_4 nano-heterostructures for photothermal therapy*. ACS Appl Mater Interfaces, 2022. **14**(25): p. 29087–29098.

[238] Yu, H., et al., *Dumbbell-like bifunctional Au–Fe_3O_4 nanoparticles*. Nano Lett, 2005. **5**: p. 379–382.

[239] Dong, Y. C., et al., *Effect of gold nanoparticle size on their properties as contrast agents for computed tomography*. Sci Rep, 2019. **9**(1): p. 14912.

[240] Yang, Y., et al., *Multifunctional gold nanoparticles in cancer diagnosis and treatment*. Int J Nanomed, 2022. **17**: p. 2041–2067.

[241] Singh, P., et al., *Gold nanoparticles in diagnostics and therapeutics for human cancer*. Int J Mol Sci, 2018. **19**(7): p. 1979.

[242] Lucky, S. S., K. C. Soo, and Y. Zhang, *Nanoparticles in photodynamic therapy*. Chem Rev, 2015. **115**(4): p. 1990–2042.

[243] Mackey, M. A., et al., *The most effective gold nanorod size for plasmonic photothermal therapy: theory and in vitro experiments*. J Phys Chem B, 2014. **118**(5): p. 1319–1326.

[244] Klein, S., et al., *$NOBF_4$-functionalized Au–Fe_3O_4 nanoheterodimers for radiation therapy: synergy effect due to simultaneous reactive oxygen and nitrogen species formation*. ACS Appl Mater Interfaces, 2018. **10**(20): p. 17071–17080.

[245] Fantechi, E., et al., *Seeded growth synthesis of Au–Fe_3O_4 heterostructured nanocrystals: rational design and mechanistic insights*. Chem Mater, 2017. **29**(9): p. 4022–4035.

[246] Klein, S., et al., *Understanding the role of surface charge in cellular uptake and X-ray-induced ROS enhancing of Au–Fe_3O_4 nanoheterodimers*. ACS Appl Bio Mater, 2018. **1**(6): p. 2002–2011.

[247] Klein, S., et al., *Caffeic acid, quercetin and 5-fluorocytidine-functionalized Au–Fe_3O_4 nanoheterodimers for x-ray-triggered drug delivery in breast tumor spheroids*. Nanomaterials (Basel), 2021. **11**(5): p. 1167.

[248] Asefa, T. and Z. Tao, *Biocompatibility of mesoporous silica nanoparticles*. Chem Res Toxicol, 2012. **25**(11): p. 2265–2284.
[249] Croissant, J. G., Y. Fatieiev, and N. M. Khashab, *Degradability and clearance of silicon, organosilica, silsesquioxane, silica mixed oxide, and mesoporous silica nanoparticles*. Adv Mater, 2017. **29**(9): p. 1604634.
[250] Chang, B., et al., *Surface functionalization of magnetic mesoporous silica nanoparticles for controlled drug release*. J Mater Chem, 2010. **20**(44): p. 9941–9947.
[251] Wu, S. H., Y. Hung, and C. Y. Mou, *Mesoporous silica nanoparticles as nanocarriers*. Chem Commun (Camb), 2011. **47**(36): p. 9972–9985.
[252] Lu, J., et al., *Biocompatibility, biodistribution, and drug-delivery efficiency of mesoporous silica nanoparticles for cancer therapy in animals*. Small, 2010. **6**(16): p. 1794–1805.
[253] Igaz, N., et al., *Functionalized mesoporous silica nanoparticles for drug-delivery to multidrug-resistant cancer cells*. Int J Nanomed, 2022. **17**: p. 3079–3096.
[254] Kim, J., et al., *Magnetic fluorescent delivery vehicle using uniform mesoporous silica spheres embedded with monodisperse magnetic and semiconductor nanocrystals*. J Am Chem Soc, 2006. **128**: p. 688–689.
[255] Knežević, N. Ž., et al., *Magnetic mesoporous silica-based core/shell nanoparticles for biomedical applications*. RSC Adv, 2013. **3**(25): p. 9584–9593.
[256] Knežević, N. Ž., I. I. Slowing, and V. S. Y. Lin, *tuning the release of anticancer drugs from magnetic iron oxide/mesoporous silica core/shell nanoparticles*. ChemPlusChem, 2012. **77**(1): p. 48–55.
[257] Darfarin, G., et al., *The effect of SiO_2/Au core-shell nanoparticles on breast cancer cell's radiotherapy*. Artif Cells Nanomed Biotechnol, 2018. **46**(sup2): p. 836–846.
[258] Niculescu, V.-C., *Mesoporous silica nanoparticles for bio-applications*. Front Mater, 2020. **7**: p. 36.
[259] Vallet-Regi, M., *Our contributions to applications of mesoporous silica nanoparticles*. Acta Biomater, 2022. **137**: p. 44–52.
[260] Xu, G., et al., *Fluorescent mesoporous silica nanomaterials for targeted tumor chemotherapy*. Mater Lett, 2021. **304**: p. 130520.
[261] Yamamoto, E. and K. Kuroda, *Colloidal mesoporous silica nanoparticles*. Bull Chem Soc Jpn, 2016. **89**(5): p. 501–539.
[262] Lin, Y.-S., et al., *Well-ordered mesoporous silica nanoparticles as cell markers*. Chem Mater, 2005. **17**: p. 4570–4573.
[263] Wan, Y. and D. Zhao, *On the controllable soft-templating approach to mesoporous silicates*. Chem Rev, 2007. **107**: p. 2821–2860.
[264] Neville, F., et al., *Preparation and characterization of colloidal silica particles under mild conditions*. J Chem Educ, 2012. **89**(7): p. 940–942.
[265] Stöber, W., A. Fink, and E. Bohn, *Controlled growth of monodisperse silica spheres in the micron size range*. J Colloid Interface Sci, 1968. **26**(1): p. 62–69.
[266] Han, Y., et al., *Unraveling the growth mechanism of silica particles in the stober method: in situ seeded growth model*. Langmuir, 2017. **33**(23): p. 5879–5890.
[267] Qiao, Z.-A., et al., *Synthesis of mesoporous silica nanoparticles via controlled hydrolysis and condensation of silicon alkoxide*. Chem Mater, 2009. **21**(16): p. 3823–3829.
[268] Olivieri, F., et al., *Mesoporous silica nanoparticles as carriers of active agents for smart anticorrosive organic coatings: a critical review*. Nanoscale, 2021. **13**(20): p. 9091–9111.
[269] Sun, R., et al., *Recent Advance on mesoporous silica nanoparticles-based controlled release system: intelligent switches open up new horizon*. Nanomaterials (Basel), 2015. **5**(4): p. 2019–2053.
[270] Song, Y., et al., *Mesoporous silica nanoparticles for stimuli-responsive controlled drug delivery: advances, challenges, and outlook*. Int J Nanomed, 2017. **12**: p. 87–110.
[271] Sreenivasan, V. K., A. V. Zvyagin, and E. M. Goldys, *Luminescent nanoparticles and their applications in the life sciences*. J Phys Condens Matter, 2013. **25**(19): p. 194101.
[272] Montalti, M., A. Cantelli, and G. Battistelli, *Nanodiamonds and silicon quantum dots: ultrastable and biocompatible luminescent nanoprobes for long-term bioimaging*. Chem Soc Rev, 2015. **44**(14): p. 4853–4921.

[273] Sapsford, K. E., et al., *Biosensing with luminescent semiconductor quantum dots*. Sensors, 2006. **6**: p. 925–953.

[274] Chen, N., et al., *The cytotoxicity of cadmium-based quantum dots*. Biomaterials, 2012. **33**(5): p. 1238–1244.

[275] Rosario, F., et al., *In vitro hepatotoxic and neurotoxic effects of titanium and cerium dioxide nanoparticles, arsenic and mercury co-exposure*. Int J Mol Sci, 2022. **23**(5): p. 2737.

[276] Zhou, M., et al., *The critical number of gold atoms for a metallic state nanocluster: resolving a decades-long question*. ACS Nano, 2021. **15**(9): p. 13980–13992.

[277] Klein, S., et al., *Oxidized silicon nanoparticles for radiosensitization of cancer and tissue cells*. Biochem Biophys Res Commun, 2013. **434**(2): p. 217–222.

[278] Romero, J. J., et al., *Photoluminescent 1–2 nm sized silicon nanoparticles: a surface-dependent system*. Chem Mater, 2013. **25**(17): p. 3488–3498.

[279] Li, Q. and R. Jin, *Photoluminescence from colloidal silicon nanoparticles: significant effect of surface*. Nanotechnol Rev, 2017. **6**(6): p. 601–612.

[280] Wilcoxon, J. P., G. A. Samara, and P. N. Provencio, *Optical and electronic properties of Si nanoclusters synthesized in inverse micelles*. Phys Rev B, 1999. **60**: p. 2704–2714.

[281] Tilley, R. D., et al., *Micro-emulsion synthesis of monodisperse surface stabilized silicon nanocrystals*. Chem Commun (Camb), 2005. **2005**(14): p. 1833–1835.

[282] Warner, J. H., et al., *Water-soluble photoluminescent silicon quantum dots*. Angew Chem, Int Ed Engl, 2005. **44**(29): p. 4550–4554.

[283] Wilcoxon, J. P. and G. A. Samara, *Tailorable, visible light emission from silicon nanocrystals*. Appl Phys Lett, 1999. **74**(21): p. 3164–3166.

[284] Chinnathambi, S., et al., *Silicon quantum dots for biological applications*. Adv Healthc Mater, 2014. **3**(1): p. 10–29.

[285] He, X. and N. Ma, *An overview of recent advances in quantum dots for biomedical applications*. Colloids Surf B, Biointerfaces, 2014. **124**: p. 118–131.

[286] Medintz, I. L., H. Mattoussi, and A. R. Clapp, *Potential clinical applications of quantum dots*. Int J Nanomed, 2008. **3**: p. 151–167.

[287] Wegner, K. D. and N. Hildebrandt, *Quantum dots: bright and versatile in vitro and in vivo fluorescence imaging biosensors*. Chem Soc Rev, 2015. **44**(14): p. 4792–4834.

[288] Batu Ghosh, B. and N. Shirahata, *Colloidal silicon quantum dots: synthesis and luminescence tuning from the near-UV to the near-IR range*. Sci Technol Adv Mater, 2014. **15: p. 014207**.

[289] Doğan, İ. and M. C. M. van de Sanden, *Gas-phase plasma synthesis of free-standing silicon nanoparticles for future energy applications*. Plasma Process Polym, 2016. **13**(1): p. 19–53.

[290] Cimpean, C., et al., *Ultrafast exciton relaxation dynamics in silicon quantum dots*. Laser Photonics Rev, 2009. **3**(1–2): p. 138–145.

[291] Groenewegen, V., et al., *Excited-state relaxation dynamics of 3-vinylthiophene-terminated silicon quantum dots*. J Phys Chem C, 2010. **114**: p. 11693–11698.

[292] Sommer, A., et al., *Ultrafast excitation energy transfer in vinylpyridine terminated silicon quantum dots*. J Phys Chem C, 2011. **115**(46): p. 22781–22788.

[293] Klein, S., *Einsatz von funktionalisierten Siliciumquantenpunkten als Transfektionsmittel für ABCB1-siRNA in Caco-2-Zellen*, in *Dept. Chemistry and Pharmacy*. 2009, Friedrich-Alexander University Erlangen.

[294] Buriak, J. M., *Organometallic chemistry on silicon and germanium surfaces*. Chem Rev, 2002. **102**: p. 1271–1308.

[295] Dasog, M., et al., *Chemical insight into the origin ofred and blue photoluminescence arising from freestanding silicon nanocrystals*. ACS Nano, 2013. **7**: p. 2676–2685.

[296] Jurbergs, D., et al., *Silicon nanocrystals with ensemble quantum yields exceeding 60 %*. Appl Phys Lett, 2006. **88**(23): p. 233116.

[297] Mangolini, L., E. Thimsen, and U. Kortshagen, *High-yield plasma synthesis of luminescent silicon nanocrystals*. Nano Lett, 2005. **5**: p. 655–659.

[298] Hwang, J., et al., *Simple preparation of fluorescent silicon nanoparticles from used si wafers*. Ind Eng Chem Res, 2015. **54**(22): p. 5982–5989.

[299] Ledoux, G., et al., *Photoluminescence properties of silicon nanocrystals as a function of their size*. Phys Rev B, 2000. **62**: p. 15942–15951.

[300] Knipping, J., et al., *Synthesis of high purity silicon nanoparticles in a low pressure microwave reactor*. J Nanosci Nanotechnol, 2004. **4**(8): p. 1039–1044.

[301] Kunze, F., et al., *Synthesis of silicon nanoparticles in a pilot-plant-scale microwave plasma reactor: impact of flow rates and precursor concentration on the nanoparticle size and aggregation*. Powder Technol, 2019. **342**: p. 880–886.

[302] Gupta, A. and H. Wiggers, *Freestanding silicon quantum dots: origin of red and blue luminescence*. Nanotechnology, 2011. **22**(5): p. 055707.

[303] Atkins, T. M., et al., *Femtosecond ligand/core dynamics of microwave-assisted synthesized silicon quantum dots in aqueous solution*. J Am Chem Soc, 2011. **133**(51): p. 20664–20667.

[304] Holmes, J. D., et al., *Highly luminescent silicon nanocrystals with discrete optical transitions*. J Am Chem Soc, 2001. **123**: p. 3743–3748.

[305] McVey, B. F. and R. D. Tilley, *Solution synthesis, optical properties, and bioimaging applications of silicon nanocrystals*. Acc Chem Res, 2014. **47**(10): p. 3045–3051.

[306] Neiner, D., H. W. Chiu, and S. M. Kauzlarich, *Low-temperature solution route to macroscopic amounts of hydrogen terminated silicon nanoparticles*. J Am Chem Soc, 2006. **128**: p. 11016–11017.

[307] Zhang, X., et al., *A new solution route to hydrogen terminated silicon nanoparticles: synthesis, functionalization, and water stability*. Nanotechnology, 2007. **18**(9): p. 095601.

[308] Li, X., et al., *Process for preparing macroscopic quantities of brightly photoluminescent silicon nanoparticles with emission spanning the visible spectrum*. Langmuir, 2003. **19**: p. 8490–8496.

[309] Hua, F., M. T. Swihart, and E. Ruckenstein, *Efficient surface grafting of luminescent silicon quantum dots by photoinitiated hydrosilylation*. Langmuir, 2005. **21**: p. 6054–6062.

[310] Giesen, B., et al., *Formation of Si-nanoparticles in a microwave reactor: comparison between experiments and modelling*. J Nanopart Res, 2005. **7**(1): p. 29–41.

[311] Klein, S., et al., *Functionalized silicon quantum dots tailored for targeted siRNA delivery*. Biochem Biophys Res Commun, 2009. **387**(1): p. 164–168.

[312] Mangolini, L., et al., *Plasma synthesis and liquid-phase surface passivation of brightly luminescent Si nanocrystals*. J Lumin, 2006. **121**(2): p. 327–334.

[313] Veinot, J. G., *Synthesis, surface functionalization, and properties of freestanding silicon nanocrystals*. Chem Commun (Camb), 2006. **2006**(40): p. 4160–4168.

[314] Yang, C.-S., *Synthesis of alkyl-terminated silicon nanoclusters by a solution route*. J Am Chem Soc, 1999. **121**: p. 5191–5195.

[315] Conde, J., et al., *Revisiting 30 years of biofunctionalization and surface chemistry of inorganic nanoparticles for nanomedicine*. Front Chem, 2014. **2**: p. 48.

[316] Subbiah, R., M. Veerapandian, and K. S. Yun, *Nanoparticles: functionalization and multifunctional applications in biomedical sciences*. Curr Med Chem, 2010. **17**: p. 4559–4577.

[317] Wang, C., et al., *Composition of intracellular protein corona around nanoparticles during internalization*. ACS Nano, 2021. **15**(2): p. 3108–3122.

[318] Sperling, R. A. and W. J. Parak, *Surface modification, functionalization and bioconjugation of colloidal inorganic nanoparticles*. Philos Trans A, Math Phys Eng Sci, 2010. **368**(1915): p. 1333–1383.

[319] Amstad, E., M. Textor, and E. Reimhult, *Stabilization and functionalization of iron oxide nanoparticles for biomedical applications*. Nanoscale, 2011. **3**(7): p. 2819–2843.

[320] Thuenemann, A. F., et al., *Maghemite nanoparticles protectively coated with poly(ethylene imine) and poly(ethylene oxide)-block-poly(glutamic acid)*. Langmuir, 2006. **22**: p. 2351–2357.

[321] Chastellain, M., A. Petri, and H. Hofmann, *Particle size investigations of a multistep synthesis of PVA coated superparamagnetic nanoparticles*. J Colloid Interface Sci, 2004. **278**(2): p. 353–360.

[322] Hauser, A. K., et al., *The effects of synthesis method on the physical and chemical properties of dextran coated iron oxide nanoparticles*. Mater Chem Phys, 2015. **160**: p. 177–186.

[323] Lee, I. C., et al., *Apoptotic cell death in rat epididymis following epichlorohydrin treatment*. Hum Exp Toxicol, 2013. **32**(6): p. 640–646.

[324] Karakoti, A. S., et al., *PEGylated inorganic nanoparticles*. Angew Chem, Int Ed Engl, 2011. **50**(9): p. 1980–1994.

[325] Wang, W. and H. Mattoussi, *Engineering the bio-nano interface using a multifunctional coordinating polymer coating*. Acc Chem Res, 2020. **53**(6): p. 1124–1138.

[326] Wang, J., M. Sui, and W. Fan, *Nanoparticles for tumor targeted therapies and their pharmacokinetics*. Curr Drug Metab, 2010. **11**(2): p. 129–141.

[327] Amoozgar, Z. and Y. Yeo, *Recent advances in stealth coating of nanoparticle drug delivery systems*. Wiley Interdiscip Rev Nanomed Nanobiotechnol, 2012. **4**(2): p. 219–233.

[328] Gref, R., et al., *'Stealth' corona-core nanoparticles surface modified by polyethylene glycol (PEG): influences of the corona (PEG chain length and surface density) and of the core composition on phagocytic uptake and plasma protein adsorption*. Colloids Surf B, 2000. **18**: p. 301–313.

[329] Hadjidemetriou, M., et al., *The human in vivo biomolecule corona onto pegylated liposomes: a proof-of-concept clinical study*. Adv Mater, 2019. **31**(4): p. e1803335.

[330] Dubaj, T., et al., *Pharmacokinetics of PEGylated gold nanoparticles: in vitro-in vivo correlation*. Nanomaterials (Basel), 2022. **12**(3): 511.

[331] Verhoef, J. J. F. and T. J. Anchordoquy, *Questioning the use of PEGylation for drug delivery*. Drug Deliv Transl Res, 2013. **3**(6): p. 499–503.

[332] Prakash, A., et al., *Bilayers as phase transfer agents for nanocrystals prepared in nonpolar solvents*. ACS Nano, 2009. **3**: p. 2139–2146.

[333] Kim, J. E., J. Y. Shin, and M. H. Cho, *Magnetic nanoparticles: an update of application for drug delivery and possible toxic effects*. Arch Toxicol, 2012. **86**(5): p. 685–700.

[334] Zhang, Y., et al., *Silica-based nanocapsules: synthesis, structure control and biomedical applications*. Chem Soc Rev, 2015. **44**: p. 315–335.

[335] Jankiewicz, B. J., et al., *Silica-metal core-shell nanostructures*. Adv Colloid Interface Sci, 2012. **170**(1–2): p. 28–47.

[336] Liu, S. and M. Y. Han, *Silica-coated metal nanoparticles*. Chem Asian J, 2010. **5**(1): p. 36–45.

[337] Guerrero-Martinez, A., J. Perez-Juste, and L. M. Liz-Marzan, *Recent progress on silica coating of nanoparticles and related nanomaterials*. Adv Mater, 2010. **22**(11): p. 1182–1195.

[338] Graf, C., et al., *A general method to coat colloidal particles with silica*. Langmuir, 2003. **19**: p. 6693–6700.

[339] Kobayashi, Y., et al., *Silica coating of silver nanoparticles using a modified Stober method*. J Colloid Interface Sci, 2005. **283**(2): p. 392–396.

[340] Jana, N. R., C. Earhart, and J. Y. Ying, *Synthesis of water-soluble and functionalized nanoparticles by silica coating*. Chem Mater, 2007. **19**: p. 5074–5082.

[341] Liu, S. H. and M. Y. Han, *Synthesis, functionalization, and bioconjugation of monodisperse, silica-coated gold nanoparticles: robust bioprobes*. Adv Funct Mater, 2005. **15**(6): p. 961–967.

[342] Zhang, M., B. L. Cushing, and C. J. O'Connor, *Synthesis and characterization of monodisperse ultra-thin silica-coated magnetic nanoparticles*. Nanotechnology, 2008. **19**(8): p. 085601.

[343] Ding, H. L., et al., *Fe_3O_4@SiO2 core/shell nanoparticles: the silica coating regulations with a single core for different core sizes and shell thicknesses*. Chem Mater, 2012. **24**(23): p. 4572–4580.

[344] Ahmad, T., et al., *Particle size dependence of relaxivity for silica-coated iron oxide nanoparticles*. Curr Appl Phys, 2012. **12**(3): p. 969–974.

[345] Kralj, S., et al., *Producing ultra-thin silica coatings on iron-oxide nanoparticles to improve their surface reactivity*. J Magn Magn Mater, 2010. **322**(13): p. 1847–1853.

[346] Mine, E., et al., *Direct coating of gold nanoparticles with silica by a seeded polymerization technique*. J Colloid Interface Sci, 2003. **264**(2): p. 385–390.

[347] Reczynska, K., et al., *Superparamagnetic iron oxide nanoparticles modified with silica layers as potential agents for lung cancer treatment*. Nanomaterials (Basel), 2020. **10**(6): p. 1076.

[348] Yi, D. K., et al., *Nanoparticle architectures templated by SiO_2/Fe_2O_3 nanocomposites*. Chem Mater, 2006. **18**: p. 614–619.

[349] Ahire, J. H., et al., *Synthesis of carbohydrate capped silicon nanoparticles and their reduced cytotoxicity, in vivo toxicity, and cellular uptake*. Adv Healthc Mater, 2015. **4**(12): p. 1877–1886.

[350] Petit, C., P. Lixon, and M.-P. Pileni, *In situ synthesis of silver nanocluster in AOT reverse micelles*. J Phys Chem, 1993. **97**: p. 12974–12983.

[351] Han, Y., et al., *Reverse microemulsion-mediated synthesis of silica-coated gold and silver nanoparticles*. Langmuir, 2008. **24**: p. 5842–5848.

[352] Erathodiyil, N. and J. Y. Ying, *Functionalization of inorganic nanoparticles for bioimaging applications*. Acc Chem Res, 2011. **44**: p. 925–935.

[353] Dragoman, R. M., et al., *Surface-engineered cationic nanocrystals stable in biological buffers and high ionic strength solutions*. Chem Mater, 2017. **29**(21): p. 9416–9428.

[354] Dong, A., et al., *A generalized ligand-exchange strategy enabling sequential surface functionalization of colloidal nanocrystals*. J Am Chem Soc, 2011. **133**(4): p. 998–1006.

[355] Palma, S., et al., *Effects of phase transfer ligands on monodisperse iron oxide magnetic nanoparticles*. J Colloid Interface Sci, 2015. **437**: p. 147–155.

[356] Woehrle, G. H., L. O. Brown, and J. E. Hutchison, *Thiol-functionalized*, 1.5-*nm gold nanoparticles through ligand exchange reactions: scope and mechanism of ligand exchange*. J Am Chem Soc, 2005. **127**: p. 2172–2183.

[357] Li, W., et al., *Surface engineering superparamagnetic nanoparticles for aqueous applications: design and characterization of tailored organic bilayers*. Environ Sci Nano, 2016. **3**(1): p. 85–93.

[358] Verma, A. and F. Stellacci, *Effect of surface properties on nanoparticle-cell interactions*. Small, 2010. **6**(1): p. 12–21.

[359] Verma, A., et al., *Surface-structure-regulated cell-membrane penetration by monolayer-protected nanoparticles*. Nat Mater, 2008. **7**(7): p. 588–595.

[360] Shen, A. Q., et al., *Fiber coating with surfactant solutions*. Phys Fluids, 2002. **14**: p. 4055–4068.

[361] Chen, Z. P., et al., *Preparation and characterization of water-soluble monodisperse magnetic iron oxide nanoparticles via surface double-exchange with DMSA*. Colloids Surf A, Physicochem Eng Asp, 2008. **316**(1–3): p. 210–216.

[362] Davis, K., et al., *Quantitative measurement of ligand exchange with small-molecule ligands on iron oxide nanoparticles via radioanalytical techniques*. Langmuir, 2016. **32**(51): p. 13716–13727.

[363] Jishkariani, D., et al., *Preparation and self-assembly of dendronized Janus Fe_3O_4–Pt and Fe_3O_4–Au heterodimers*. ACS Nano, 2017. **11**(8): p. 7958–7966.

[364] Klein, S., et al., *$NOBF_4$-functionalized Au–Fe_3O_4 nanoheterodimers for radiation therapy: synergy effect due to simultaneous reactive oxygen and nitrogen species formation*. ACS Appl Mater Interfaces, 2018. **10**(20): p. 17071–17080.

[365] Korpany, K. V., et al., *Iron oxide surface chemistry: effect of chemical structure on binding in benzoic acid and catechol derivatives*. Langmuir, 2017. **33**(12): p. 3000–3013.

[366] Schechtel, E., et al., *Mixed ligand shell formation upon catechol ligand adsorption on hydrophobic TiO_2 nanoparticles*. Langmuir, 2019. **35**(38): p. 12518–12531.

[367] Hiramatsu, H. and F. E. Osterloh, *A simple large-scale synthesis of nearly monodisperse gold and silver nanoparticles with adjustable sizes and with exchangeable surfactants*. Chem Mater, 2004. **16**: p. 2509–2511.

[368] Yang, Y., L. A. Serrano, and S. Guldin, *A versatile AuNP synthetic platform for decoupled control of size and surface composition*. Langmuir, 2018. **34**(23): p. 6820–6826.

[369] Lundqvist, M., et al., *The evolution of the protein corona around nanoparticles: a test study*. ACS Nano, 2011. **5**: p. 7503–7509.

[370] Fleischer, C. C. and C. K. Payne, *Nanoparticle-cell interactions: molecular structure of the protein corona and cellular outcomes*. Acc Chem Res, 2014. **47**(8): p. 2651–2659.

[371] Galmarini, S., et al., *Beyond unpredictability: the importance of reproducibility in understanding the protein corona of nanoparticles*. Bioconjug Chem, 2018. **29**(10): p. 3385–3393.

[372] Ke, P. C., et al., *A decade of the protein corona*. ACS Nano, 2017. **11**(12): p. 11773–11776.

[373] Ren, J., et al., *Chemical and biophysical signatures of the protein corona in nanomedicine*. J Am Chem Soc, 2022. **144**(21): p. 9184–9205.

[374] Piella, J., N. G. Bastus, and V. Puntes, *Size-dependent protein-nanoparticle interactions in citrate-stabilized gold nanoparticles: the emergence of the protein corona*. Bioconjug Chem, 2017. **28**(1): p. 88–97.

[375] Mohammad-Beigi, H., et al., *Mapping and identification of soft corona proteins at nanoparticles and their impact on cellular association*. Nat Commun, 2020. **11**(1): p. 4535.

[376] Galdino, F. E., et al., *Inside the protein corona: from binding parameters to unstained hard and soft coronas visualization*. Nano Lett, 2021. **21**(19): p. 8250–8257.

[377] Hoang, K. N. L., K. E. Wheeler, and C. J. Murphy, *Isolation methods influence the protein corona composition on gold-coated iron oxide nanoparticles*. Anal Chem, 2022. **94**(11): p. 4737–4746.

[378] Pederzoli, F., et al., *Protein corona and nanoparticles: how can we investigate on?* Wiley Interdiscip Rev Nanomed Nanobiotechnol, 2017. **9**(6).

[379] del Pino, P., et al., *Protein corona formation around nanoparticles – from the past to the future*. Mater Horiz, 2014. **1**(3): p. 301–313.

[380] Latreille, P. L., et al., *Scratching the surface of the protein corona: challenging measurements and controversies*. ACS Nano, 2022. **16**(2): p. 1689–1707.

[381] Gupta, M. N. and I. Roy, *How corona formation impacts nanomaterials as drug carriers*. Mol Pharm, 2020. **17**(3): p. 725–737.

[382] Ashby, J., S. Pan, and W. Zhong, *Size and surface functionalization of iron oxide nanoparticles influence the composition and dynamic nature of their protein corona*. ACS Appl Mater Interfaces, 2014. **6**(17): p. 15412–15419.

[383] Madathiparambil Visalakshan, R., et al., *The Influence of Nanoparticle Shape on Protein Corona Formation*. Small, 2020. **16**(25): p. e2000285.

[384] Marichal, L., et al., *Protein-nanoparticle interactions: what are the protein-corona thickness and organization?* Langmuir, 2019. **35**(33): p. 10831–10837.

[385] Marichal, L., et al., *Protein corona composition of silica nanoparticles in complex media: nanoparticle size does not matter*. Nanomaterials (Basel), 2020. **10**(2): p. 240.

[386] Spinozzi, F., et al., *Structural and thermodynamic properties of nanoparticle-protein complexes: a combined SAXS and SANS study*. Langmuir, 2017. **33**(9): p. 2248–2256.

[387] Cedervall, T., et al., *Understanding the nanoparticle–protein corona using methods to quantify exchange rates and affinities of proteins for nanoparticles*. Proc Natl Acad Sci USA, 2007. **104**: p. 2050–2055.

[388] Rocker, C., et al., *A quantitative fluorescence study of protein monolayer formation on colloidal nanoparticles*. Nat Nanotechnol, 2009. **4**(9): p. 577–580.

[389] Lai, Z. W., et al., *Emerging techniques in proteomics for probing nano bio interactions*. ACS Nano, 2012. **6**: p. 10438–10448.

[390] Anderson, N. L. and N. G. Anderson, *The human plasma proteome: history, character, and diagnostic prospects*. Mol Cell Proteomics, 2002. **1**(11): p. 845–867.

[391] Garcia-Alvarez, R., et al., *In vivo formation of protein corona on gold nanoparticles. The effect of their size and shape*. Nanoscale, 2018. **10**(3): p. 1256–1264.

[392] Nice, E. C., et al., *Use of multidimensional separation protocols for the purification of trace components in complex biological samples for proteomics analysis*. J Chromatogr A, 2007. **1168**(1–2): p. 190–210; discussion 189.

[393] Zhu, Z. J., et al., *The interplay of monolayer structure and serum protein interactions on the cellular uptake of gold nanoparticles*. Small, 2012. **8**(17): p. 2659–2663.

[394] Blume, J. E., et al., *Rapid, deep and precise profiling of the plasma proteome with multi-nanoparticle protein corona*. Nat Commun, 2020. **11**(1): p. 3662.
[395] Wang, J., et al., *Soft interactions at nanoparticles alter protein function and conformation in a size dependent manner*. Nano Lett, 2011. **11**(11): p. 4985–4991.
[396] Park, S. J., *Protein-nanoparticle interaction: corona formation and conformational changes in proteins on nanoparticles*. Int J Nanomed, 2020. **15**: p. 5783–5802.
[397] Zhang, Y., et al., *An analysis of the binding function and structural organization of the protein corona*. J Am Chem Soc, 2020. **142**(19): p. 8827–8836.
[398] Li, Y. and J. S. Lee, *Insights into characterization methods and biomedical applications of nanoparticle-protein corona*. Materials (Basel), 2020. **13**(14): p. 3093.
[399] Erickson, H. P., *Size and shape of protein molecules at the nanometer level determined by sedimentation, gel filtration, and electron microscopy*. Biol Proced Online, 2009. **11**: p. 32–51.
[400] Roach, P., D. Farrar, and C. C. Perry, *Surface tailoring for controlled protein adsorption: effect of topography at the nanometer scale and chemistry*. J Am Chem Soc, 2006. **128**: p. 3939–3945.
[401] Gagner, J. E., et al., *Effect of gold nanoparticle morphology on adsorbed protein structure and function*. Biomaterials, 2011. **32**(29): p. 7241–7252.
[402] Nguyen, V. H. and B. J. Lee, *Protein corona: a new approach for nanomedicine design*. Int J Nanomed, 2017. **12**: p. 3137–3151.
[403] Kokkinopoulou, M., et al., *Visualization of the protein corona: towards a biomolecular understanding of nanoparticle-cell-interactions*. Nanoscale, 2017. **9**(25): p. 8858–8870.
[404] Nel, A. E., et al., *Understanding biophysicochemical interactions at the nano-bio interface*. Nat Mater, 2009. **8**(7): p. 543–557.
[405] Gref, R., et al., *'Stealth' corona-core nanoparticles surface modified by polyethylene glycol (PEG): influences of the corona (PEG chain length and surface density) and of the core composition on phagocytic uptake and plasma protein adsorption*. Colloids Surf B, Biointerfaces, 2000. **18**: p. 301–313.
[406] He, Q., et al., *The effect of PEGylation of mesoporous silica nanoparticles on nonspecific binding of serum proteins and cellular responses*. Biomaterials, 2010. **31**(6): p. 1085–1092.
[407] Brandenberger, C., et al., *Quantitative evaluation of cellular uptake and trafficking of plain and polyethylene glycol-coated gold nanoparticles*. Small, 2010. **6**(15): p. 1669–1678.
[408] Moyano, D. F., et al., *Fabrication of corona-free nanoparticles with tunable hydrophobicity*. ACS Nano, 2014. **7**: p. 6748–6755.
[409] Wei, H., et al., *Compact zwitterion-coated iron oxide nanoparticles for biological applications*. Nano Lett, 2012. **12**(1): p. 22–25.
[410] Erik Holmlin, R., et al., *Zwitterionic SAMs that resist nonspecific adsorption of protein from aqueous buffer*. Langmuir, 2001. **17**: p. 2841–2850.
[411] Liu, X., et al., *Minimizing nonspecific phagocytic uptake of biocompatible gold nanoparticles with mixed charged zwitterionic surface modification*. J Mater Chem, 2012. **22**(5): p. 1916–1927.
[412] Nuckowski, L., et al., *Application of magnetic nanoparticles coated with crosslinked zwitterionic poly(ionic liquid)s for the extraction of oligonucleotides*. Materials (Basel), 2021. **14**(12): p. 3146.
[413] Agasti, S. S., et al., *Photoregulated release of caged anticancer drugs from gold nanoparticles*. J Am Chem Soc, 2009. **131**: p. 5728–5729.
[414] Behzadi, S., et al., *Cellular uptake of nanoparticles: journey inside the cell*. Chem Soc Rev, 2017. **46**(14): p. 4218–4244.
[415] Chithrani, B. D. and W. C. W. Chan, *Elucidating the mechanism of cellular uptake and removal of protein-coated gold nanoparticles of different sizes and shapes*. Nano Lett, 2007. **6**: p. 1542–1550.
[416] Ho, Y. T., R. D. Kamm, and J. C. Y. Kah, *Influence of protein corona and caveolae-mediated endocytosis on nanoparticle uptake and transcytosis*. Nanoscale, 2018. **10**(26): p. 12386–12397.
[417] Huefner, A., et al., *Gold nanoparticles explore cells: cellular uptake and their use as intracellular probes*. Methods, 2014. **68**(2): p. 354–363.

[418] Cheng, X., et al., *Protein Corona influences cellular uptake of gold nanoparticles by phagocytic and nonphagocytic cells in a size-dependent manner*. ACS Appl Mater Interfaces, 2015. **7**(37): p. 20568–20575.

[419] Ding, L., et al., *Size, shape, and protein corona determine cellular uptake and removal mechanisms of gold nanoparticles*. Small, 2018. **14**(42): p. e1801451.

[420] Ding, H. M. and Y. Q. Ma, *Theoretical and computational investigations of nanoparticle-biomembrane interactions in cellular delivery*. Small, 2015. **11**(9–10): p. 1055–1071.

[421] Bohmer, N. and A. Jordan, *Caveolin-1 and CDC42 mediated endocytosis of silica-coated iron oxide nanoparticles in HeLa cells*. Beilstein J Nanotechnol, 2015. **6**: p. 167–176.

[422] Engelberg, S., et al., *Cancer cell-selective, clathrin-mediated endocytosis of aptamer decorated nanoparticles*. Oncotarget, 2018. **9**: p. 20993–21006.

[423] Hao, X., et al., *Caveolae-mediated endocytosis of biocompatible gold nanoparticles in living Hela cells*. J Phys Condens Matter, 2012. **24**(16): p. 164207.

[424] Kirkham, M. and R. G. Parton, *Clathrin-independent endocytosis: new insights into caveolae and non-caveolar lipid raft carriers*. Biochim Biophys Acta, 2005. **1745**(3): p. 273–286.

[425] Edidin, M., *Lipids on the frontier: a century of cell-membrane bilayers*. Nat Rev Mol Cell Biol, 2003. **4**: p. 414–418.

[426] Escriba, P. V., et al., *Membranes: a meeting point for lipids, proteins and therapies*. J Cell Mol Med, 2008. **12**(3): p. 829–875.

[427] Oh, N. and J. H. Park, *Endocytosis and exocytosis of nanoparticles in mammalian cells*. Int J Nanomed, 2014. **9**(Suppl 1): p. 51–63.

[428] Zhang, S., H. Gao, and G. Bao, *Physical principles of nanoparticle cellular endocytosis*. ACS Nano, 2015. **9**: p. 8655–8671.

[429] Nakamura, H. and S. Watano, *Direct permeation of nanoparticles across cell membrane: a review*. KONA Powder Part J, 2018. **35**: p. 49–65.

[430] Candeloro, P., et al., *Nanoparticle microinjection and Raman spectroscopy as tools for nanotoxicology studies*. Analyst, 2011. **136**(21): p. 4402–4408.

[431] Jen, C.-P., et al., *A nonviral transfection approach in vitro: the design of a gold nanoparticle vector joint with microelectromechanical systems*. Langmuir, 2004. **20**: p. 1369–1374.

[432] Conner, S. D. and S. L. Schmid, *Regulated portals of entry into the cell*. Nature, 2003. **422**: p. 37–44.

[433] Foroozandeh, P. and A. A. Aziz, *Insight into cellular uptake and intracellular trafficking of nanoparticles*. Nanoscale Res Lett, 2018. **13**(1): p. 339.

[434] Rosales, C. and E. Uribe-Querol, *Phagocytosis: a fundamental process in immunity*. BioMed Res Int, 2017. **2017**: p. 9042851.

[435] Uribe-Querol, E. and C. Rosales, *Phagocytosis: our current understanding of a universal biological process*. Front Immunol, 2020. **11**: p. 1066.

[436] Freeman, S. A. and S. Grinstein, *Phagocytosis: receptors, signal integration, and the cytoskeleton*. Immunol Rev, 2014. **262**: p. 193–215.

[437] Gordon, S., *Phagocytosis: an immunobiologic process*. Immunity, 2016. **44**(3): p. 463–475.

[438] Flannagan, R. S., V. Jaumouille, and S. Grinstein, *The cell biology of phagocytosis*. Annu Rev Pathol, 2012. **7**: p. 61–98.

[439] Ostrowski, P. P., S. Grinstein, and S. A. Freeman, *Diffusion barriers, mechanical forces, and the biophysics of phagocytosis*. Dev Cell, 2016. **38**(2): p. 135–146.

[440] Gutierrez, M. G., *Functional role(s) of phagosomal Rab GTPases*. Small GTPases, 2013. **4**(3): p. 148–158.

[441] Vieira, O. V., R. J. Botelho, and S. Grinstein, *Phagosome maturation: aging gracefully*. Biochem J, 2002. **366**: p. 689–704.

[442] Kinchen, J. M. and K. S. Ravichandran, *Phagosome maturation: going through the acid test*. Nat Rev Mol Cell Biol, 2008. **9**(10): p. 781–795.

[443] Marshansky, V. and M. Futai, *The V-type H+-ATPase in vesicular trafficking: targeting, regulation and function*. Curr Opin Cell Biol, 2008. **20**(4): p. 415–426.

[444] Babior, B. M., *NADPH oxidase*. Curr Opin Immunol, 2004. **16**(1): p. 42–47.

[445] Minakami, R. and H. Sumimotoa, *Phagocytosis-coupled activation of the superoxide-producing phagocyte oxidase, a member of the NADPH oxidase (nox) family*. Int J Hematol, 2006. **84**(3): p. 193–198.

[446] Metz, S., et al., *Capacity of human monocytes to phagocytose approved iron oxide MR contrast agents in vitro*. Eur Radiol, 2004. **14**(10): p. 1851–1858.

[447] Thorek, D. L. and A. Tsourkas, *Size, charge and concentration dependent uptake of iron oxide particles by non-phagocytic cells*. Biomaterials, 2008. **29**(26): p. 3583–3590.

[448] Shapiro, E. M., et al., *Antibody-mediated cell labeling of peripheral T cells with micron-sized iron oxide particles (MPIOs) allows single cell detection by MRI*. Contrast Media Mol Imag, 2007. **2**(3): p. 147–153.

[449] Shapiro, E. M., S. Skrtic, and A. P. Koretsky, *Sizing it up: cellular MRI using micron-sized iron oxide particles*. Magn Reson Med, 2005. **53**(2): p. 329–338.

[450] Clemens, D. L., et al., *Targeted intracellular delivery of antituberculosis drugs to Mycobacterium tuberculosis-infected macrophages via functionalized mesoporous silica nanoparticles*. Antimicrob Agents Chemother, 2012. **56**(5): p. 2535–2545.

[451] Franca, A., et al., *Macrophage scavenger receptor A mediates the uptake of gold colloids by macrophages in vitro*. Nanomedicine (Lond), 2011. **6**(7): p. 1175–1188.

[452] Tsai, C. Y., et al., *Size-dependent attenuation of TLR9 signaling by gold nanoparticles in macrophages*. J Immunol, 2012. **188**(1): p. 68–76.

[453] Sadauskas, E., et al., *Kupffer cells are central in the removal of nanoparticles from the organism*. Part Fibre Toxicol, 2007. **4**: p. 10.

[454] Sadauskas, E., et al., *Biodistribution of gold nanoparticles in mouse lung following intratracheal instillation*. Chem Cent J, 2009. **3**: p. 16.

[455] Arnida, et al., *Geometry and surface characteristics of gold nanoparticles influence their biodistribution and uptake by macrophages*. Eur J Pharm Biopharm, 2011. **77**(3): p. 417–423.

[456] Hutter, E., et al., *Microglial response to gold nanoparticles*. ACS Nano, 2010. **4**: p. 2595–2606.

[457] Champion, J. A. and S. Mitragotri, *Role of target geometry in phagocytosis*. Proc Natl Acad Sci USA, 2006. **103**: p. 4930–4934.

[458] Huang, X., et al., *The shape effect of mesoporous silica nanoparticles on biodistribution, clearance, and biocompatibility in vivo*. ACS Nano, 2011. **5**: p. 5390–5399.

[459] Park, J. H., et al., *Magnetic iron oxide nanoworms for tumor targeting and imaging*. Adv Mater, 2008. **20**(9): p. 1630–1635.

[460] Gustafson, H. H., et al., *Nanoparticle uptake: the phagocyte problem*. Nano Today, 2015. **10**: p. 487–510.

[461] Qie, Y., et al., *Surface modification of nanoparticles enables selective evasion of phagocytic clearance by distinct macrophage phenotypes*. Sci Rep, 2016. **6**: p. 26269.

[462] Swanson, J. A. and S. Yoshida, *Macropinosomes as units of signal transduction*. Philos Trans R Soc Lond B, Biol Sci, 2019. **374**(1765): p. 20180157.

[463] Kumari, S. and S. Mayor, *ARF1 is directly involved in dynamin-independent endocytosis*. Nat Cell Biol, 2008. **10**(1): p. 30–41.

[464] Choo, P., T. Liu, and T. W. Odom, *Nanoparticle shape determines dynamics of targeting nanoconstructs on cell membranes*. J Am Chem Soc, 2021. **143**(12): p. 4550–4555.

[465] Cho, E. C., et al., *Understanding the role of surface charges in cellular adsorption versus internalization by selectively removing gold nanoparticles on the cell surface with a I2/KI Etchant*. Nano Lett, 2009. **9**: p. 1080–1084.

[466] Rejman, J., et al., *Size-dependent internalization of particles via the pathways of clathrinand caveolae-mediated endocytosis*. Biochem J, 2004. **377**: p. 159–169.

[467] Chompoosor, A., et al., *The role of surface functionality on acute cytotoxicity, ROS generation and DNA damage by cationic gold nanoparticles*. Small, 2010. **6**(20): p. 2246–2249.

[468] Kerr, M. C. and R. D. Teasdale, *Defining macropinocytosis*. Traffic, 2009. **10**(4): p. 364–371.

[469] Amyere, M., et al., *Origin, originality, functions, subversions and molecular signalling of macropinocytosis*. Int J Med Microbiol, 2002. **291**: p. 487–494.

[470] McMahon, H. T. and E. Boucrot, *Molecular mechanism and physiological functions of clathrin-mediated endocytosis*. Nat Rev Mol Cell Biol, 2011. **12**(8): p. 517–533.

[471] Kaksonen, M. and A. Roux, *Mechanisms of clathrin-mediated endocytosis*. Nat Rev Mol Cell Biol, 2018. **19**(5): p. 313–326.

[472] Seto, E. S., H. J. Bellen, and T. E. Lloyd, *When cell biology meets development: endocytic regulation of signaling pathways*. Genes Dev, 2002. **16**(11): p. 1314–1336.

[473] Di Fiore, P. P. and P. De Camilli, *Endocytosis and signaling: an inseparable partnership*. Cell, 2001. **106**: p. 1–4.

[474] Mettlen, M., et al., *Regulation of clathrin-mediated endocytosis*. Annu Rev Biochem, 2018. **87**: p. 871–896.

[475] Ng, C. T., et al., *Clathrin-mediated endocytosis of gold nanoparticles in vitro*. Anat Rec (Hoboken), 2015. **298**(2): p. 418–427.

[476] Harush-Frenkel, O., et al., *Targeting of nanoparticles to the clathrin-mediated endocytic pathway*. Biochem Biophys Res Commun, 2007. **353**(1): p. 26–32.

[477] Parton, R. G. and K. Simons, *The multiple faces of caveolae*. Nat Rev Mol Cell Biol, 2007. **8**(3): p. 185–194.

[478] Pelkmans, L. and A. Helenius, *Endocytosis via caveolae*. Traffic, 2002. **3**(5): p. 311–320.

[479] Wang, Z., et al., *Delivery of nanoparticle: complexed drugs across the vascular endothelial barrier via caveolae*. IUBMB Life, 2011. **63**(8): p. 659–667.

[480] Parton, R. G. and A. A. Richards, *Lipid rafts and caveolae as portals for endocytosis: new insights and common mechanisms*. Traffic, 2003. **4**(11): p. 724–738.

[481] Anderson, H. A., Y. Chen, and L. C. Norkin, *Bound simian virus 40 translocates to caveolinenriched membrane domains, and its entry is inhibited by drugs that selectively disrupt caveolae*. Mol Biol Cell, 1996. **7**: p. 1825–1834.

[482] Kiss, A. L. and E. Botos, *Endocytosis via caveolae: alternative pathway with distinct cellular compartments to avoid lysosomal degradation?* J Cell Mol Med, 2009. **13**(7): p. 1228–1237.

[483] Sens, P. and M. S. Turner, *Budded membrane microdomains as tension regulators*. Phys Rev E, Stat Nonlinear Soft Matter Phys, 2006. **73**(3 Pt 1): p. 031918.

[484] Thomas, C. M. and E. J. Smart, *Caveolae structure and function*. J Cell Mol Med, 2008. **12**(3): p. 796–809.

[485] Nabi, I. R. and P. U. Le, *Caveolae/raft-dependent endocytosis*. J Cell Biol, 2003. **161**(4): p. 673–677.

[486] Damke, H., et al., *Clathrin-independent pinocytosis is induced in cells overexpressing a temperature-sensitive mutant of dynamin*. J Cell Biol, 1995. **131**: p. 69–80.

[487] Fra, A. M., et al., *Detergent-insoluble glycolipid microdomains in lymphocytes in the absence of caveolae*. J Biol Chem, 1994. **269**(49): p. 30745–30748.

[488] Sabharanjak, S., et al., *GPI-anchored proteins are delivered to recycling endosomes via a distinct cdc42-regulated, clathrin-independent pinocytic pathway*. Dev Cell, 2002. **2**: p. 411–423.

[489] Johannes, L., et al., *Building endocytic pits without clathrin*. Nat Rev Mol Cell Biol, 2015. **16**(5): p. 311–321.

[490] Lundmark, R., et al., *The GTPase-activating protein GRAF1 regulates the CLIC/GEEC endocytic pathway*. Curr Biol, 2008. **18**(22): p. 1802–1808.

[491] Naslavsky, N., R. Weigert, and J. G. Donaldson, *Convergence of non-clathrin- and clathrin-derived endosomes involves Arf6 inactivation and changes in phosphoinositides*. Mol Biol Cell, 2003. **14**(2): p. 417–431.

[492] Radhakrishna, H. and J. G. Donaldson, *ADP-ribosylation factor 6 regulates a novel plasma membrane recycling pathway*. J Cell Biol, 1997. **139**: p. 49–61.

[493] Lamaze, C., et al., *Interleukin 2 receptors and detergent-resistant membrane domains define a clathrin-independent endocytic pathway*. Mol Cell, 2001. **7**: p. 661–671.

[494] Sauvonnet, N., A. Dujeancourt, and A. Dautry-Varsat, *Cortactin and dynamin are required for the clathrin-independent endocytosis of gammac cytokine receptor*. J Cell Biol, 2005. **168**(1): p. 155–163.

[495] Ding, L., et al., *Intracellular fate of nanoparticles with polydopamine surface engineering and a novel strategy for exocytosis-inhibiting, lysosome impairment-based cancer therapy*. Nano Lett, 2017. **17**(11): p. 6790–6801.
[496] Ha, K. D., S. M. Bidlingmaier, and B. Liu, *Macropinocytosis exploitation by cancers and cancer therapeutics*. Front Physiol, 2016. **7**: p. 381.
[497] Lin, X. P., J. D. Mintern, and P. A. Gleeson, *Macropinocytosis in different cell types: similarities and differences*. Membranes (Basel), 2020. **10**(8): p. 177.
[498] Kamphorst, J. J., et al., *Human pancreatic cancer tumors are nutrient poor and tumor cells actively scavenge extracellular protein*. Cancer Res, 2015. **75**(3): p. 544–553.
[499] Vettore, L., R. L. Westbrook, and D. A. Tennant, *New aspects of amino acid metabolism in cancer*. Br J Cancer, 2020. **122**(2): p. 150–156.
[500] Song, S., et al., *The dual role of macropinocytosis in cancers: promoting growth and inducing methuosis to participate in anticancer therapies as targets*. Front Oncol, 2020. **10**: p. 570108.
[501] Chen, J.-J., et al., *H-Ras regulation of TRAIL death receptor mediated apoptosis*. Oncotarget, 2014. **5**: p. 5125.
[502] Reif, R., et al., *Activated ErbB3 translocates to the nucleus via clathrin-independent endocytosis, which is associated with proliferating cells*. J Biol Chem, 2016. **291**(8): p. 3837–3847.
[503] Wang, H., Y. You, and X. Zhu, *The role of exosomes in the progression and therapeutic resistance of hematological malignancies*. Front Oncol, 2022. **12**: p. 887518.
[504] Whiteside, T. L., *Tumor-derived exosomes and their role in cancer progression*. Adv Clin Chem, 2016. **74**: p. 103–141.
[505] Javeed, N., et al., *Pancreatic cancer-derived exosomes cause paraneoplastic beta-cell dysfunction*. Clin Cancer Res, 2015. **21**(7): p. 1722–1733.
[506] Zhang, Y., et al., *Synthesis and application of AS1411-functionalized gold nanoparticles for targeted therapy of gastric cancer*. ACS Omega, 2020. **5**(48): p. 31227–31233.
[507] Tong, X., et al., *Progress in cancer drug delivery based on AS1411 oriented nanomaterials*. J Nanobiotechnol, 2022. **20**(1): p. 57.
[508] Ghosh, S., et al., *Arginine-terminated, chemically designed nanoparticle for direct cell translocation*. ACS Appl Bio Mater, 2019. **2**(1): p. 339–348.
[509] Panja, P. and N. R. Jana, *Arginine-terminated nanoparticles of <10 nm size for direct membrane penetration and protein delivery for straight access to cytosol and nucleus*. J Phys Chem Lett, 2020. **11**(6): p. 2363–2368.
[510] Copolovici, D. M., et al., *Cell-penetrating peptides: design, synthesis, and applications*. ACS Nano, 2014. **8**: p. 1972–1994.
[511] Pal, N. K. and C. Kryschi, *A facile UV-light mediated synthesis of l-histidine stabilized silver nanocluster for efficient photodegradation of methylene blue*. J Mol Catal A, Chem, 2015. **404–405**: p. 27–35.
[512] Wu, M., et al., *Size-dependent cellular uptake and localization profiles of silver nanoparticles*. Int J Nanomed, 2019. **14**: p. 4247–4259.
[513] Arvizo, R. R., et al., *Effect of nanoparticle surface charge at the plasma membrane and beyond*. Nano Lett, 2010. **10**(7): p. 2543–2548.
[514] Jiang, X., et al., *Fast intracellular dissolution and persistent cellular uptake of silver nanoparticles in CHO-K1 cells: implication for cytotoxicity*. Nanotoxicology, 2015. **9**(2): p. 181–189.
[515] Lund, T., et al., *The influence of ligand organization on the rate of uptake of gold nanoparticles by colorectal cancer cells*. Biomaterials, 2011. **32**(36): p. 9776–9784.
[516] Van Lehn, R. C., et al., *Effect of particle diameter and surface composition on the spontaneous fusion of monolayer-protected gold nanoparticles with lipid bilayers*. Nano Lett, 2013. **13**(9): p. 4060–4067.
[517] Klein, S., et al., *APTES-terminated ultrasmall and iron-doped silicon nanoparticles as X-ray dose enhancer for radiation therapy*. Biochem Biophys Res Commun, 2018. **498**(4): p. 855–861.
[518] Mu, Q., et al., *Mechanism of cellular uptake of genotoxic silica nanoparticles*. Part Fibre Toxicol, 2012. **9**: p. 29.

[519] Chou, L. Y., K. Ming, and W. C. Chan, *Strategies for the intracellular delivery of nanoparticles*. Chem Soc Rev, 2011. **40**(1): p. 233–245.

[520] Muthaiyan Shanmugam, M. and H. Manoj, *Microinjection for single-cell analysis and therapy*, in *Handbook of single cell technologies*. 2021. p. 1–27.

[521] Derfus, A. M., W. C. W. Chan, and S. N. Bhatia, *Intracellular delivery of quantum dots for live cell labeling and organelle tracking*. Adv Mater, 2004. **16**(12): p. 961–966.

[522] Medintz, I. L., et al., *Intracellular delivery of quantum dot-protein cargos mediated by cell penetrating peptides*. Bioconjug Chem, 2008. **19**: p. 1785–1795.

[523] Kim, K. and W. G. Lee, *Electroporation for nanomedicine: a review*. J Mater Chem B, 2017. **5**(15): p. 2726–2738.

[524] Wei, Z., et al., *A laminar flow electroporation system for efficient DNA and siRNA delivery*. Anal Chem, 2011. **83**(15): p. 5881–5887.

[525] Olofsson, J., et al., *Single-cell electroporation*. Curr Opin Biotechnol, 2003. **14**(1): p. 29–34.

[526] Huang, S., et al., *Gold nanoparticles electroporation enhanced polyplex delivery to mammalian cells*. Electrophoresis, 2014. **35**(12–13): p. 1837–1845.

[527] Phonesouk, E., et al., *Increasing uptake of silica nanoparticles with electroporation: from cellular characterization to potential applications*. Materials (Basel), 2019. **12**(1): p. 179.

[528] Rezaee, Z., et al., *Gold nanoparticles and electroporation impose both separate and synergistic radiosensitizing effects in HT-29 tumor cells: an in vitro study*. Int J Nanomed, 2017. **12**: p. 1431–1439.

[529] Zu, Y., et al., *Gold nanoparticles enhanced electroporation for mammalian cell transfection*. J Biomed Nanotechnol, 2014. **10**: p. 982–992.

[530] Piao, J. G., et al., *Bioreducible cross-linked nanoshell enhances gene transfection of polycation/DNA polyplex in vivo*. Biomacromolecules, 2014. **15**(8): p. 2907–2913.

[531] Geers, B., et al., *Crucial factors and emerging concepts in ultrasound-triggered drug delivery*. J Control Release, 2012. **164**(3): p. 248–255.

[532] Snipstad, S., et al., *Sonoporation using nanoparticle-loaded microbubbles increases cellular uptake of nanoparticles compared to co-incubation of nanoparticles and microbubbles*. Pharmaceutics, 2021. **13**(5): p. 640.

[533] Piraner, D. I., et al., *Going deeper: biomolecular tools for acoustic and magnetic imaging and control of cellular function*. Biochemistry, 2017. **56**(39): p. 5202–5209.

[534] Blomley, M. J. K., et al., *Microbubble contrast agents: a new era in ultrasound*. BMJ, 2001. **322**: p. 1222–1225.

[535] Lee, H., et al., *Microbubbles used for contrast enhanced ultrasound and theragnosis: a review of principles to applications*. Biomed Eng Lett, 2017. **7**(2): p. 59–69.

[536] Hosseinaee, Z., et al., *Towards non-contact photoacoustic imaging [review]*. Photoacoustics, 2020. **20**: p. 100207.

[537] Lentacker, I., et al., *Understanding ultrasound induced sonoporation: definitions and underlying mechanisms*. Adv Drug Deliv Rev, 2014. **72**: p. 49–64.

[538] Zhou, Q. L., et al., *Ultrasound-mediated local drug and gene delivery using nanocarriers*. BioMed Res Int, 2014. **2014**: p. 963891.

[539] Yang, F., et al., *Controlled release of Fe_3O_4 nanoparticles in encapsulated microbubbles to tumor cells via sonoporation and associated cellular bioeffects*. Small, 2011. **7**(7): p. 902–910.

[540] Wang, Y. H., et al., *Synergistic delivery of gold nanorods using multifunctional microbubbles for enhanced plasmonic photothermal therapy*. Sci Rep, 2014. **4**: p. 5685.

[541] Du, X., et al., *Monitoring optoporated process on mammalian cells by real-time measurement of membrane resealing time*. J Biomed Opt, 2023. **28**(6): p. 065006.

[542] Yao, C., et al., *Cancer cell-specific protein delivery by optoporation with laser-irradiated gold nanorods*. J Biophotonics, 2020. **13**(7): p. e202000017.

[543] Bergeron, E., et al., *Cell-specific optoporation with near-infrared ultrafast laser and functionalized gold nanoparticles*. Nanoscale, 2015. **7**: p. 17836–17847.

[544] Baumgart, J., et al., *Off-resonance plasmonic enhanced femtosecond laser optoporation and transfection of cancer cells*. Biomaterials, 2012. **33**(7): p. 2345–2350.
[545] Patskovsky, S., M. Qi, and M. Meunier, *Single point single-cell nanoparticle mediated pulsed laser optoporation*. Analyst, 2020. **145**(2): p. 523–529.
[546] Oh, E., et al., *Cellular uptake and fate of PEGylated gold nanoparticles is dependent on both cell-penetration peptides and particle size*. ACS Nano, 2011. **5**: p. 6434–6448.
[547] Vogel, A., et al., *Mechanisms of femtosecond laser nanosurgery of cells and tissues*. Appl Phys B, 2005. **81**(8): p. 1015–1047.
[548] Zhu, M., et al., *Physicochemical properties determine nanomaterial cellular uptake, transport, and fate*. Acc Chem Res, 2013. **46**: p. 622–631.
[549] Wang, X., et al., *Influence of cell size on cellular uptake of gold nanoparticles*. Biomater Sci, 2016. **4**(6): p. 970–978.
[550] Meng, H., et al., *Engineered design of mesoporous silica nanoparticles to deliver doxorubicin and P-glycoprotein siRNA to overcome drug resistance in a cancer cell line*. ACS Nano, 2010. **4**: p. 4539–4550.
[551] Qiu, Y., et al., *Surface chemistry and aspect ratio mediated cellular uptake of Au nanorods*. Biomaterials, 2010. **31**(30): p. 7606–7619.
[552] Xia, T., et al., *polyethyleneimine coating enhances the cellular uptake of mesoporous silica nanoparticles and allows safe delivery of siRNA and DNA constructs*. ACS Nano, 2009. **3**: p. 3273–3286.
[553] Tree-Udom, T., et al., *Shape effect on particle-lipid bilayer membrane association, cellular uptake, and cytotoxicity*. ACS Appl Mater Interfaces, 2015. **7**(43): p. 23993–24000.
[554] *Delivering the promise of RNA therapeutics*. Nat Med, 2019. **25**(9): p. 1321.
[555] Contini, C., et al., *Nanoparticle–membrane interactions*. J Exp Nanosci, 2017. **13**(1): p. 62–81.
[556] Chen, K. L. and G. D. Bothun, *Nanoparticles meet cell membranes: probing nonspecific interactions using model membranes*. Environ Sci Technol, 2014. **48**(2): p. 873–880.
[557] Rascol, E., J. M. Devoisselle, and J. Chopineau, *The relevance of membrane models to understand nanoparticles-cell membrane interactions*. Nanoscale, 2016. **8**(9): p. 4780–4798.
[558] Jaskiewicz, K., et al., *Incorporation of nanoparticles into polymersomes: size and concentration effects*. ACS Nano, 2012. **6**: p. 7254–7262.
[559] Montis, C., et al., *Interaction of nanoparticles with lipid membranes: a multiscale perspective*. Nanoscale, 2014. **6**(12): p. 6452–6457.
[560] Michel, R., et al., *Internalization of silica nanoparticles into fluid liposomes: formation of interesting hybrid colloids*. Angew Chem, Int Ed Engl, 2014. **53**(46): p. 12441–12445.
[561] Roiter, Y., et al., *Interaction of nanoparticles with lipid membrane*. Nano Lett, 2008. **8**: p. 941–944.
[562] Rasch, M. R., et al., *Hydrophobic gold nanoparticle self-assembly with phosphatidylcholine lipid: membrane-loaded and Janus vesicles*. Nano Lett, 2010. **10**(9): p. 3733–3739.
[563] Paasonen, L., et al., *Gold nanoparticles enable selective light-induced contents release from liposomes*. J Control Release, 2007. **122**(1): p. 86–93.
[564] Yavlovich, A., et al., *Light-sensitive lipid-based nanoparticles for drug delivery: design principles and future considerations for biological applications*. Mol Membr Biol, 2010. **27**(7): p. 364–381.
[565] Amstad, E., I. Bilecka, M. Textor, and E. Reimhult, *Ultrastable iron oxide nanoparticle colloidal suspensions using dispersants with catechol-derived anchor groups*. Nano Lett, 2009. **9**: p. 4042–4048.
[566] Na, H. B., G. Palui, J. T. Rosenberg, X. Ji, S. C. Grant, and H. Mattoussi, *Multidentate catechol-based polyethylene glycol oligomers provide enhanced stability and biocompatibility to iron oxide nanoparticles*. ACS Nano, 2012. **6**: p. 389–399.
[567] Namiki, Y., et al., *A novel magnetic crystal-lipid nanostructure for magnetically guided in vivo gene delivery*. Nat Nanotechnol, 2009. **4**(9): p. 598–606.
[568] Lee, H. Y., et al., *Integration of gold nanoparticles into bilayer structures via adaptive surface chemistry*. J Am Chem Soc, 2013. **135**(16): p. 5950–5953.
[569] Lin, J., et al., *Penetration of lipid membranes by gold nanoparticles: insights into cellular uptake, cytotoxicity, and their relationship*. ACS Nano, 2010. **4**: p. 5421–5429.

[570] Lerouеil, P. R., et al., *Wide varieties of cationic nanoparticles induce defects in supported lipid bilayers*. Nano Lett, 2008. **8**: p. 420–424.
[571] Wang, B., et al., *Nanoparticle-induced surface reconstruction of phospholipid membranes*. Proc Natl Acad Sci USA, 2008. **105**: p. 18171–18175.
[572] Stiegler, L. M. S., et al., *Smart shell-by-shell nanoparticles with tunable perylene fluorescence in the organic interlayer*. Chemistry, 2021. **27**(5): p. 1655–1669.
[573] Gao, H., W. Shi, and L. B. Freund, *Mechanics of receptor-mediated endocytosis*. Proc Natl Acad Sci USA, 2005. **102**: p. 9469–9474.
[574] Yue, J., et al., *Gold nanoparticle size and shape effects on cellular uptake and intracellular distribution of sirna nanoconstructs*. Bioconjug Chem, 2017. **28**(6): p. 1791–1800.
[575] Nambara, K., et al., *Reverse size dependences of the cellular uptake of triangular and spherical gold nanoparticles*. Langmuir, 2016. **32**(47): p. 12559–12567.
[576] Albanese, A. and W. C. W. Chan, *Effect of gold nanoparticle aggregation on cell uptake and toxicity*. ACS Nano, 2011. **5**: p. 5478–5489.
[577] Xie, X., et al., *The effect of shape on cellular uptake of gold nanoparticles in the forms of stars, rods, and triangles*. Sci Rep, 2017. **7**(1): p. 3827.
[578] Huang, J., et al., *Effects of nanoparticle size on cellular uptake and liver MRI with polyvinylpyrrolidone-coated iron oxide nanoparticles*. ACS Nano, 2010. **4**: p. 7151–7160.
[579] Valdiglesias, V., et al., *Are iron oxide nanoparticles safe? Current knowledge and future perspectives*. J Trace Elem Med Biol, 2016. **38**: p. 53–63.
[580] Feng, Q., et al., *Uptake, distribution, clearance, and toxicity of iron oxide nanoparticles with different sizes and coatings*. Sci Rep, 2018. **8**(1): p. 2082.
[581] Larsen, E. K. U., et al., *Size-dependent accumulation of pegylated silane-coated magnetic iron oxide nanoparticles in murine tumors*. ACS Nano, 2009. **3**: p. 1947–1951.
[582] Ge, Y., et al., *Effect of surface charge and agglomerate degree of magnetic iron oxide nanoparticles on KB cellular uptake in vitro*. Colloids Surf B, Biointerfaces, 2009. **73**(2): p. 294–301.
[583] Kurzhals, S., et al., *Controlled aggregation and cell uptake of thermoresponsive polyoxazoline-grafted superparamagnetic iron oxide nanoparticles*. Nanoscale, 2017. **9**(8): p. 2793–2805.
[584] Li, X., et al., *Glycopolymer-coated iron oxide nanoparticles: shape-controlled synthesis and cellular uptake*. J Mater Chem B, 2014. **2**(34): p. 5569–5575.
[585] Lu, F., et al., *Size effect on cell uptake in well-suspended, uniform mesoporous silica nanoparticles*. Small, 2009. **5**(12): p. 1408–1413.
[586] Liu, H. M., et al., *Mesoporous silica nanoparticles improve magnetic labeling efficiency in human stem cells*. Small, 2008. **4**(5): p. 619–626.
[587] Hao, N., et al., *The shape effect of PEGylated mesoporous silica nanoparticles on cellular uptake pathway in Hela cells*. Microporous Mesoporous Mater, 2012. **162**: p. 14–23.
[588] Jiang, W., et al., *Nanoparticle-mediated cellular response is size-dependent*. Nat Nanotechnol, 2008. **3**(3): p. 145–150.
[589] Saikia, J., et al., *Differential protein adsorption and cellular uptake of silica nanoparticles based on size and porosity*. ACS Appl Mater Interfaces, 2016. **8**(50): p. 34820–34832.
[590] Bjornmalm, M., et al., *Bridging bio-nano science and cancer nanomedicine*. ACS Nano, 2017. **11**(10): p. 9594–9613.
[591] Maeda, H., *The 35th anniversary of the discovery of EPR effect: a new wave of nanomedicines for tumor-targeted drug delivery-personal remarks and future prospects*. J Pers Med, 2021. **11**(3): p. 229.
[592] Maeda, H., et al., *Vascular permeability enhancement in solid tumor: various factors, mechanisms involved and its implications*. Int Immunopharmacol, 2003. **3**(3): p. 319–328.
[593] Maeda, H., T. Sawa, and T. Konno, *Mechanism of tumor-targeted delivery of macromolecular drugs, including the EPR effect in solid tumor and clinical overview of the prototype polymeric drug SMANCS*. J Control Release, 2001. **74**: p. 47–61.

[594] Matsumura, Y. and H. Maeda, *A new concept for macromolecular therapeutics in cancer chemotherapy: mechanism of tumoritropic accumulation of proteins and the antitumor agent smancs*. Cancer Res, 1986. **46**: p. 6387–6392.

[595] Tahmasbi Rad, A., et al., *Combinational effects of active targeting, shape, and enhanced permeability and retention for cancer theranostic nanocarriers*. ACS Appl Mater Interfaces, 2019. **11**(11): p. 10505–10519.

[596] Li, S. D. and L. Huang, *Pharmacokinetics and biodistribution of nanoparticles*. Mol Pharm, 2008. **5**: p. 496–504.

[597] Akhtar, M. J., et al., *Targeted anticancer therapy: overexpressed receptors and nanotechnology*. Clin Chim Acta, 2014. **436**: p. 78–92.

[598] Kim, C. S., et al., *Triggered nanoparticles as therapeutics*. Nano Today, 2013. **8**(4): p. 439–447.

[599] Jia, R., et al., *Advances in multiple stimuli-responsive drug-delivery systems for cancer therapy*. Int J Nanomed, 2021. **16**: p. 1525–1551.

[600] Salvati, A., et al., *Transferrin-functionalized nanoparticles lose their targeting capabilities when a biomolecule corona adsorbs on the surface*. Nat Nanotechnol, 2013. **8**(2): p. 137–143.

[601] Feliu, N., et al., *In vivo degeneration and the fate of inorganic nanoparticles*. Chem Soc Rev, 2016. **45**(9): p. 2440–2457.

[602] Wu, J., *The enhanced permeability and retention (EPR) effect: the significance of the concept and methods to enhance its application*. J Pers Med, 2021. **11**(8): p. 771.

[603] Xu, L., et al., *Quantitative comparison of gold nanoparticle delivery via the enhanced permeation and retention (EPR) effect and mesenchymal stem cell (MSC)-based targeting*. ACS Nano, 2023. **17**(3): p. 2039–2052.

[604] Fang, J., H. Nakamura, and H. Maeda, *The EPR effect: unique features of tumor blood vessels for drug delivery, factors involved, and limitations and augmentation of the effect*. Adv Drug Deliv Rev, 2011. **63**(3): p. 136–151.

[605] Du, B., M. Yu, and J. Zheng, *Transport and interactions of nanoparticles in the kidneys*. Nat Rev Mater, 2018. **3**(10): p. 358–374.

[606] Zhu, G. H., et al., *Innate immune invisible ultrasmall gold nanoparticles-framework for synthesis and evaluation*. ACS Appl Mater Interfaces, 2021. **13**(20): p. 23410–23422.

[607] Li, X., et al., *Surface chemistry governs the sub-organ transfer, clearance and toxicity of functional gold nanoparticles in the liver and kidney*. J Nanobiotechnol, 2020. **18**(1): p. 45.

[608] Park, Y. C., et al., *Effect of PEG molecular weight on stability, $T(2)$ contrast, cytotoxicity, and cellular uptake of superparamagnetic iron oxide nanoparticles (SPIONs)*. Colloids Surf B, Biointerfaces, 2014. **119**: p. 106–114.

[609] Walkey, C. D., et al., *Protein corona fingerprinting predicts the cellular interaction of gold and silver nanoparticles*. ACS Nano, 2014. **8**: p. 2439–2455.

[610] Holmlin, R. E., et al., *Zwitterionic SAMs that resist nonspecific adsorption of protein from aqueous buffer*. Langmuir, 2001. **17**: p. 2841–2850.

[611] Mizuhara, T., et al., *Acylsulfonamide-functionalized zwitterionic gold nanoparticles for enhanced cellular uptake at tumor pH*. Angew Chem, Int Ed Engl, 2015. **54**(22): p. 6567–6570.

[612] Mori, D., et al., *Extracellular acidity in tumor tissue upregulates programmed cell death protein 1 expression on tumor cells via proton-sensing G protein-coupled receptors*. Int J Cancer, 2021. **149**(12): p. 2116–2124.

[613] Gustafson, H. H., et al., *Nanoparticle uptake: the phagocyte problem*. Nano Today, 2015. **10**(4): p. 487–510.

[614] Zhai, Y., et al., *Preparation and application of cell membrane-camouflaged nanoparticles for cancer therapy*. Theranostics, 2017. **7**(10): p. 2575–2592.

[615] Fang, R. H., et al., *Cancer cell membrane-coated nanoparticles for anticancer vaccination and drug delivery*. Nano Lett, 2014. **14**(4): p. 2181–2188.

[616] Fang, R. H., et al., *Cell membrane coating nanotechnology*. Adv Mater, 2018. **30**(23): p. e1706759.

[617] Ai, X., et al., *Emerging approaches to functionalizing cell membrane-coated nanoparticles*. Biochemistry, 2021. **60**(13): p. 941–955.

[618] Sosale, N. G., et al., *Macrophage engulfment of a cell or nanoparticle is regulated by unavoidable opsonization, a species-specific 'Marker of Self' CD47, and target physical properties*. Curr Opin Immunol, 2015. **35**: p. 107–112.

[619] Willingham, S. B., et al., *The CD47-signal regulatory protein alpha (SIRPa) interaction is a therapeutic target for human solid tumors*. Proc Natl Acad Sci USA, 2012. **109**(17): p. 6662–6667.

[620] Burger, P., et al., *CD47 functions as a molecular switch for erythrocyte phagocytosis*. Blood, 2012. **119**(23): p. 5512–5521.

[621] Jalil, A. R., M. P. Tobin, and D. E. Discher, *Suppressing or enhancing macrophage engulfment through the use of cd47 and related peptides*. Bioconjug Chem, 2022. **33**(11): p. 1989–1995.

[622] Rodriguez, P. L., et al., *Minimal "self" peptides that inhibit phagocytic clearance and enhance delivery of nanoparticles*. Science, 2013. **339**(6122): p. 971–975.

[623] Watermann, A. and J. Brieger, *Mesoporous silica nanoparticles as drug delivery vehicles in cancer*. Nanomaterials (Basel), 2017. **7**(7): p. 189.

[624] Muhamad, N., T. Plengsuriyakarn, and K. Na-Bangchang, *Application of active targeting nanoparticle delivery system for chemotherapeutic drugs and traditional/herbal medicines in cancer therapy: a systematic review*. Int J Nanomed, 2018. **13**: p. 3921–3935.

[625] Barui, S. and V. Cauda, *Multimodal decorations of mesoporous silica nanoparticles for improved cancer therapy*. Pharmaceutics, 2020. **12**(6): p. 527.

[626] Narmani, A., et al., *Folic acid functionalized nanoparticles as pharmaceutical carriers in drug delivery systems*. Drug Dev Res, 2019. **80**(4): p. 404–424.

[627] Wu, P. H., et al., *Targeting integrins with RGD-conjugated gold nanoparticles in radiotherapy decreases the invasive activity of breast cancer cells*. Int J Nanomed, 2017. **12**: p. 5069–5085.

[628] Shen, Y., et al., *Transferrin receptor 1 in cancer: a new sight for cancer therapy*. Am J Cancer Res, 2018. **8**(6): p. 916–931.

[629] El Hallal, R., N. Lyu, and Y. Wang, *Effect of cetuximab-conjugated gold nanoparticles on the cytotoxicity and phenotypic evolution of colorectal cancer cells*. Molecules, 2021. **26**(3): p. 567.

[630] Mohapatra, S., et al., *Synthesis of highly stable folic acid conjugated magnetite nanoparticles for targeting cancer cells*. Nanotechnology, 2007. **18**(38): p. 385102.

[631] Angelopoulou, A., et al., *Folic acid-functionalized, condensed magnetic nanoparticles for targeted delivery of doxorubicin to tumor cancer cells overexpressing the folate receptor*. ACS Omega, 2019. **4**(26): p. 22214–22227.

[632] Nasiri, R., et al., *In vitro evaluation of actively targetable superparamagnetic nanoparticles to the folate receptor positive cancer cells*. Mater Sci Eng C, Mater Biol Appl, 2016. **69**: p. 1147–1158.

[633] Zhang, Z., et al., *Conjugating folic acid to gold nanoparticles through glutathione for targeting and detecting cancer cells*. Bioorg Med Chem, 2010. **18**: p. 5528–5534.

[634] Daniels, T. R., et al., *The transferrin receptor and the targeted delivery of therapeutic agents against cancer*. Biochim Biophys Acta, 2012. **1820**(3): p. 291–317.

[635] Luo, M., et al., *Systematic evaluation of transferrin-modified porous silicon nanoparticles for targeted delivery of doxorubicin to glioblastoma*. ACS Appl Mater Interfaces, 2019. **11**(37): p. 33637–33649.

[636] Kciuk, M., et al., *Doxorubicin – an agent with multiple mechanisms of anticancer activity*. Cells, 2023. **12**(4): p. 659.

[637] Wu, P. H., et al., *Targeting integrins in cancer nanomedicine: applications in cancer diagnosis and therapy*. Cancers (Basel), 2019. **11**(11): p. 1783.

[638] Cox, D., M. Brennan, and N. Moran, *Integrins as therapeutic targets: lessons and opportunities*. Nat Rev Drug Discov, 2010. **9**(10): p. 804–820.

[639] Takada, Y., X. Ye, and S. Simon, *The integrins*. Genome Biol, 2007. **8**(5): p. 215.

[640] Pang, X., et al., *Targeting integrin pathways: mechanisms and advances in therapy*. Signal Transduct Targeted Ther, 2023. **8**(1): p. 1.

[641] Li, M., et al., *Integrins as attractive targets for cancer therapeutics*. Acta Pharmacol Sin B, 2021. **11**(9): p. 2726–2737.

[642] Desgrosellier, J. S. and D. A. Cheresh, *Integrins in cancer: biological implications and therapeutic opportunities*. Nat Rev Cancer, 2010. **10**(1): p. 9–22.

[643] Bianconi, D., M. Unseld, and G. W. Prager, *Integrins in the spotlight of cancer*. Int J Mol Sci, 2016. **17**(12): p. 2037.

[644] Zhang, C., et al., *Molecular magnetic resonance imaging of activated hepatic stellate cells with ultrasmall superparamagnetic iron oxide targeting integrin alphavbeta(3) for staging liver fibrosis in rat model*. Int J Nanomed, 2016. **11**: p. 1097–1108.

[645] Marques, A. C., et al., *Functionalizing nanoparticles with cancer-targeting antibodies: A comparison of strategies*. J Control Release, 2020. **320**: p. 180–200.

[646] Johnston, M. C. and C. J. Scott, *Antibody conjugated nanoparticles as a novel form of antibody drug conjugate chemotherapy*. Drug Discov Today Technol, 2018. **30**: p. 63–69.

[647] Kukkar, D., et al., *Recent advances in nanoscale materials for antibody-based cancer theranostics*. Biosens Bioelectron, 2020. **173**: p. 112787.

[648] Popovtzer, A., et al., *Actively targeted gold nanoparticles as novel radiosensitizer agents: an in vivo head and neck cancer model*. Nanoscale, 2016. **8**(5): p. 2678–2685.

[649] Zhang, X., et al., *Cetuximab-modified silica nanoparticle loaded with ICG for tumor-targeted combinational therapy of breast cancer*. Drug Deliv, 2019. **26**(1): p. 129–136.

[650] Fu, Z., et al., *Antibody drug conjugate: the "biological missile" for targeted cancer therapy*. Signal Transduct Targeted Ther, 2022. **7**(1): p. 93.

[651] Ruiz, G., et al., *pH impacts the orientation of antibody adsorbed onto gold nanoparticles*. Bioconjug Chem, 2019. **30**(4): p. 1182–1191.

[652] Herranz, F., et al., *Superparamagnetic nanoparticles for atherosclerosis imaging*. Nanomaterials (Basel), 2014. **4**(2): p. 408–438.

[653] Kumari, M., A. Acharya, and P. T. Krishnamurthy, *Antibody-conjugated nanoparticles for target-specific drug delivery of chemotherapeutics*. Beilstein J Nanotechnol, 2023. **14**: p. 912–926.

[654] Winter, M. C. and B. W. Hancock, *Ten years of rituximab in NHL*. Expert Opin Drug Saf, 2009. **8**(2): p. 223–235.

[655] Bou-Assaly, W. and S. Mukherji, *Cetuximab (erbitux)*. AJNR Am J Neuroradiol, 2010. **31**(4): p. 626–627.

[656] Merchant, S. J., et al., *Effectiveness of trastuzumab in routine clinical practice: a population-based study of patients with HER-2-positive oesophageal, gastroesophageal and gastric cancer*. Clin Oncol (R Coll Radiol), 2021. **33**(3): p. 202–207.

[657] Marcus, R. and A. Hagenbeek, *The therapeutic use of rituximab in non-Hodgkin's lymphoma*. Eur J Haematol, 2006. **78**(s67): p. 5–14.

[658] Goddard, Z. R., et al., *Active targeting of gold nanoparticles as cancer therapeutics*. Chem Soc Rev, 2020. **49**(23): p. 8774–8789.

[659] Bhattacharyya, S., et al., *Nanoconjugation modulates the trafficking and mechanism of antibody induced receptor endocytosis*. Proc Natl Acad Sci USA, 2010. **107**(33): p. 14541–14546.

[660] Goldsmith, S. J., *Targeted radionuclide therapy: a historical and personal review*. Semin Nucl Med, 2020. **50**(1): p. 87–97.

[661] Dziawer, L., et al., *Trastuzumab-modified gold nanoparticles labeled with ^{211}At as a prospective tool for local treatment of her2-positive breast cancer*. Nanomaterials (Basel), 2019. **9**(4): p. 632.

[662] Song, L., et al., *Apoptosis-promoting effect of rituximab-conjugated magnetic nanoprobes on malignant lymphoma cells with CD20 overexpression*. Int J Nanomed, 2019. **14**: p. 921–936.

[663] Morita, Y., et al., *Aptamer therapeutics in cancer: current and future*. Cancers (Basel), 2018. **10**(3): p. 80.

[664] Zhou, J. and J. Rossi, *Aptamers as targeted therapeutics: current potential and challenges*. Nat Rev Drug Discov, 2017. **16**(3): p. 181–202.

[665] Jo, H. and C. Ban, *Aptamer-nanoparticle complexes as powerful diagnostic and therapeutic tools*. Exp Mol Med, 2016. **48**(5): p. e230.

[666] Medley, C. D., et al., *Aptamer-conjugated nanoparticles for cancer cell detection*. Anal Chem, 2011. **83**(3): p. 727–734.
[667] Gilles, M. E., et al., *Nucleolin targeting impairs the progression of pancreatic cancer and promotes the normalization of tumor vasculature*. Cancer Res, 2016. **76**(24): p. 7181–7193.
[668] Reyes-Reyes, E. M. and S. K. Akiyama, *Cell-surface nucleolin is a signal transducing P-selectin binding protein for human colon carcinoma cells*. Exp Cell Res, 2008. **314**(11–12): p. 2212–2223.
[669] Lopes-Nunes, J., et al., *Aptamer-functionalized gold nanoparticles for drug delivery to gynecological carcinoma cells*. Cancers (Basel), 2021. **13**(16): p. 4038.
[670] Varakumar, P., et al., *Acridine as an anti-tumour agent: a critical review*. Molecules, 2022. **28**(1): p. 193.
[671] Schon, M. P., et al., *Death receptor-independent apoptosis in malignant melanoma induced by the small-molecule immune response modifier imiquimod*. J Invest Dermatol, 2004. **122**(5): p. 1266–1276.
[672] Donahue, N. D., H. Acar, and S. Wilhelm, *Concepts of nanoparticle cellular uptake, intracellular trafficking, and kinetics in nanomedicine*. Adv Drug Deliv Rev, 2019. **143**: p. 68–96.
[673] van der Zwaag, D., et al., *Super resolution imaging of nanoparticles cellular uptake and trafficking*. ACS Appl Mater Interfaces, 2016. **8**(10): p. 6391–6399.
[674] Vtyurina, N., C. Aberg, and A. Salvati, *Imaging of nanoparticle uptake and kinetics of intracellular trafficking in individual cells*. Nanoscale, 2021. **13**(23): p. 10436–10446.
[675] Yameen, B., et al., *Insight into nanoparticle cellular uptake and intracellular targeting*. J Control Release, 2014. **190**: p. 485–499.
[676] Elkin, S. R., A. M. Lakoduk, and S. L. Schmid, *Endocytic pathways and endosomal trafficking: a primer*. Wien Med Wochenschr, 2016. **166**(7–8): p. 196–204.
[677] Gruenberg, J., *The endocytic pathway: a mosaic of domains*. Nat Rev Mol Cell Biol, 2001. **2**: p. 721–730.
[678] Hu, Y. B., et al., *The endosomal-lysosomal system: from acidification and cargo sorting to neurodegeneration*. Transl Neurodegener, 2015. **4**: p. 18.
[679] Huotari, J. and A. Helenius, *Endosome maturation*. EMBO J, 2011. **30**(17): p. 3481–3500.
[680] He, Y., et al., *Research progress on the mechanism of nanoparticles crossing the intestinal epithelial cell membrane*. Pharmaceutics, 2023. **15**(7): p. 1816.
[681] Gu, F., C. M. Crump, and G. Thomas, *Trans-Golgi network sorting*. Cell Mol Life Sci, 2001. **58**: p. 1067–1084.
[682] Pelkmans, L., *Secrets of caveolae- and lipid raft-mediated endocytosis revealed by mammalian viruses*. Biochim Biophys Acta, 2005. **1746**(3): p. 295–304.
[683] Iversen, T.-G., T. Skotland, and K. Sandvig, *Endocytosis and intracellular transport of nanoparticles: Present knowledge and need for future studies*. Nano Today, 2011. **6**(2): p. 176–185.
[684] Lim, J. P. and P. A. Gleeson, *Macropinocytosis: an endocytic pathway for internalising large gulps*. Immunol Cell Biol, 2011. **89**(8): p. 836–843.
[685] Maekawa, M., R. Natsume, and M. Arita, *Functional significance of ion channels during macropinosome resolution in immune cells*. Front Physiol, 2022. **13**: p. 1037758.
[686] Buckley, C. M. and J. S. King, *Drinking problems: mechanisms of macropinosome formation and maturation*. FEBS J, 2017. **284**(22): p. 3778–3790.
[687] Scott, C. C. and J. Gruenberg, *Ion flux and the function of endosomes and lysosomes: pH is just the start: the flux of ions across endosomal membranes influences endosome function not only through regulation of the luminal pH*. BioEssays, 2011. **33**(2): p. 103–110.
[688] Steinberg, B. E., et al., *A cation counterflux supports lysosomal acidification*. J Cell Biol, 2010. **189**(7): p. 1171–1186.
[689] Portilla, Y., et al., *The surface coating of iron oxide nanoparticles drives their intracellular trafficking and degradation in endolysosomes differently depending on the cell type*. Biomaterials, 2022. **281**: p. 121365.
[690] Portilla, Y., et al., *Iron oxide nanoparticle coatings dictate cell outcomes despite the influence of protein coronas*. ACS Appl Mater Interfaces, 2021. **13**(7): p. 7924–7944.
[691] Jin, R., et al., *Iron oxide nanoparticles promote macrophage autophagy and inflammatory response through activation of toll-like Receptor-4 signaling*. Biomaterials, 2019. **203**: p. 23–30.

[692] Andrian, T., et al., *A super-resolution and transmission electron microscopy correlative approach to study intracellular trafficking of nanoparticles*. Nanoscale, 2023. **15**(35): p. 14615–14627.

[693] Liu, M., et al., *Real-time visualization of clustering and intracellular transport of gold nanoparticles by correlative imaging*. Nat Commun, 2017. **8**: p. 15646.

[694] Aberg, C., *Kinetics of nanoparticle uptake into and distribution in human cells*. Nanoscale Adv, 2021. **3**(8): p. 2196–2212.

[695] Garcia Romeu, H., S. Deville, and A. Salvati, *Time- and space-resolved flow-cytometry of cell organelles to quantify nanoparticle uptake and intracellular trafficking by cells*. Small, 2021. **17**(34): p. e2100887.

[696] Varkouhi, A. K., et al., *Endosomal escape pathways for delivery of biologicals*. J Control Release, 2011. **151**(3): p. 220–228.

[697] Qiu, C., et al., *Advanced Strategies for Overcoming Endosomal/Lysosomal Barrier in Nanodrug Delivery*. Research (Wash D C), 2023. **6**: p. 0148.

[698] Reilly, M. J., J. D. Larsen, and M. O. Sullivan, *Polyplexes traffic through caveolae to the Golgi and endoplasmic reticulum en route to the nucleus*. Mol Pharm, 2012. **9**(5): p. 1280–1290.

[699] Wang, G., et al., *KDEL peptide gold nanoconstructs: promising nanoplatforms for drug delivery*. Nanomedicine, 2013. **9**(3): p. 366–374.

[700] Selby, L. I., et al., *Nanoescapology: progress toward understanding the endosomal escape of polymeric nanoparticles*. Wiley Interdiscip Rev Nanomed Nanobiotechnol, 2017. **9**(5): p. e1452.

[701] Smith, S. A., et al., *The endosomal escape of nanoparticles: toward more efficient cellular delivery*. Bioconjug Chem, 2019. **30**(2): p. 263–272.

[702] Funhoff, A. M., et al., *Endosomal escape of polymeric gene delivery complexes is not always enhanced by polymers buffering at low pH*. Biomacromolecules, 2004. **5**: p. 32–39.

[703] Yue, Y., et al., *Revisit complexation between DNA and polyethylenimine – effect of length of free polycationic chains on gene transfection*. J Control Release, 2011. **152**(1): p. 143–151.

[704] Kongkatigumjorn, N., et al., *Probing endosomal escape using pHlexi nanoparticles*. Macromol Biosci, 2017. **17**(4): p. 1600248.

[705] Massignani, M., et al., *Controlling cellular uptake by surface chemistry, size, and surface topology at the nanoscale*. Small, 2009. **5**(21): p. 2424–2432.

[706] Massignani, M., et al., *Enhanced fluorescence imaging of live cells by effective cytosolic delivery of probes*. PLoS ONE, 2010. **5**(5): p. e10459.

[707] Li, J., Y. Yang, and L. Huang, *Calcium phosphate nanoparticles with an asymmetric lipid bilayer coating for siRNA delivery to the tumor*. J Control Release, 2012. **158**(1): p. 108–114.

[708] Qiu, C., et al., *Systemic delivery of siRNA by hyaluronan-functionalized calcium phosphate nanoparticles for tumor-targeted therapy*. Nanoscale, 2016. **8**(26): p. 13033–13044.

[709] Hu, Y., et al., *Cytosolic delivery of membrane-impermeable molecules in dendritic cells using pH-responsive core–shell nanoparticles*. Nano Lett, 2007. **7**: p. 3056–3064.

[710] Griset, A. P., et al., *Expansile nanoparticles: synthesis, characterization, and in vivo efficacy of an acid-responsive polymeric drug delivery system*. J Am Chem Soc, 2009. **131**: p. 2469–2471.

[711] Chiang, W. H., et al., *Dual-layered nanogel-coated hollow lipid/polypeptide conjugate assemblies for potential pH-triggered intracellular drug release*. PLoS ONE, 2014. **9**(3): p. e92268.

[712] Song, H. P., et al., *Gene transfer using self-assembled ternary complexes of cationic magnetic nanoparticles, plasmid DNA and cell-penetrating Tat peptide*. Biomaterials, 2010. **31**(4): p. 769–778.

[713] Akinc, A., et al., *Exploring polyethylenimine-mediated DNA transfection and the proton sponge hypothesis*. J Gene Med, 2005. **7**(5): p. 657–663.

[714] Moreira, C., et al., *Improving chitosan-mediated gene transfer by the introduction of intracellular buffering moieties into the chitosan backbone*. Acta Biomater, 2009. **5**(8): p. 2995–3006.

[715] Li, H. J., et al., *Stimuli-responsive clustered nanoparticles for improved tumor penetration and therapeutic efficacy*. Proc Natl Acad Sci USA, 2016. **113**(15): p. 4164–4169.

[716] Ahmad, A., J. M. Khan, and S. Haque, *Strategies in the design of endosomolytic agents for facilitating endosomal escape in nanoparticles*. Biochimie, 2019. **160**: p. 61–75.

[717] Huan, Y., et al., *Antimicrobial peptides: classification, design, application and research progress in multiple fields*. Front Microbiol, 2020. **11**: p. 582779.

[718] Jadhav, S. G. and S. F. Dowdy, *Overcoming delivery barriers with LNPsRestriction boosts piezoelectricity*. Nat Mater, 2021. **20**(5): p. 575–577.

[719] Kulkarni, J. A., P. R. Cullis, and R. van der Meel, *Lipid nanoparticles enabling gene therapies: from concepts to clinical utility*. Nucleic Acid Ther, 2018. **28**(3): p. 146–157.

[720] Plaza-Ga, I., et al., *pH-triggered endosomal escape of pore-forming Listeriolysin O toxin-coated gold nanoparticles*. J Nanobiotechnol, 2019. **17**(1): p. 108.

[721] Kwon, E. J., J. M. Bergen, and S. H. Pun, *Application of an HIV gp41-derived peptide for enhanced intracellular trafficking of synthetic gene and siRNA delivery vehicles*. Bioconjug Chem, 2008. **19**: p. 920–927.

[722] Lopez-Barbosa, N., et al., *Multifunctional magnetite nanoparticles to enable delivery of siRNA for the potential treatment of Alzheimer's*. Drug Deliv, 2020. **27**(1): p. 864–875.

[723] Blokpoel Ferreras, L. A., et al., *Rapidly transducing and spatially localized magnetofection using peptide-mediated non-viral gene delivery based on iron oxide nanoparticles*. ACS Appl Nano Mater, 2021. **4**(1): p. 167–181.

[724] Sadhukha, T., T. S. Wiedmann, and J. Panyam, *Inhalable magnetic nanoparticles for targeted hyperthermia in lung cancer therapy*. Biomaterials, 2013. **34**(21): p. 5163–5171.

[725] Lundberg, P., et al., *Delivery of short interfering RNA using endosomolytic cell-penetrating peptides*. FASEB J, 2007. **21**(11): p. 2664–2671.

[726] Idris, A. H., et al., *Nanostructured lipid carrier co-loaded with docetaxel and magnetic nanoparticles: physicochemical characterization and in vitro evaluation*. Pharmaceutics, 2023. **15**(5): p. 1319.

[727] Rayamajhi, S., et al., *pH-responsive cationic liposome for endosomal escape mediated drug delivery*. Colloids Surf B, Biointerfaces, 2020. **188**: p. 110804.

[728] Balazs, D. A. and W. Godbey, *Liposomes for use in gene delivery*. J Drug Deliv, 2011. **2011**: p. 326497.

[729] Gao, L. Y., et al., *Core-shell type lipid/rPAA-Chol polymer hybrid nanoparticles for in vivo siRNA delivery*. Biomaterials, 2014. **35**(6): p. 2066–2078.

[730] Shirsat, S. D., et al., *Endosomal escape in magnetic nanostructures: recent advances and future perspectives*. Mater Today Adv, 2024. **22**: p. 100484.

[731] Pucelik, B., A. Sułek, and J. M. Dąbrowski, *Bacteriochlorins and their metal complexes as NIR-absorbing photosensitizers: properties, mechanisms, and applications*. Coord Chem Rev, 2020. **416**: p. 213340.

[732] Wu, K., et al., *Urchin-like magnetic microspheres for cancer therapy through synergistic effect of mechanical force, photothermal and photodynamic effects*. J Nanobiotechnol, 2022. **20**(1): p. 224.

[733] Zou, R., et al., *^{68}Ga-labeled magnetic-nir persistent luminescent hybrid mesoporous nanoparticles for multimodal imaging-guided chemotherapy and photodynamic therapy*. ACS Appl Mater Interfaces, 2021. **13**(8): p. 9667–9680.

[734] Cabral, H., et al., *A photo-activated targeting chemotherapy using glutathione sensitive camptothecin-loaded polymeric micelles*. Pharm Res, 2009. **26**(1): p. 82–92.

[735] Wei, P., et al., *Ultrasound-responsive polymersomes capable of endosomal escape for efficient cancer therapy*. J Control Release, 2020. **322**: p. 81–94.

[736] Batchelor, D. V. B., et al., *The influence of nanobubble size and stability on ultrasound enhanced drug delivery*. Langmuir, 2022. **38**(45): p. 13943–13954.

[737] Alamoudi, K., et al., *Thermoresponsive pegylated bubble liposome nanovectors for efficient siRNA delivery via endosomal escape*. Nanomedicine (Lond), 2017. **12**(12): p. 1421–1433.

[738] Lin, Q., et al., *Brain tumor-targeted delivery and therapy by focused ultrasound introduced doxorubicin-loaded cationic liposomes*. Cancer Chemother Pharmacol, 2016. **77**(2): p. 269–280.

[739] Veeren, A., et al., *Liposome-tethered gold nanoparticles triggered by pulsed NIR light for rapid liposome contents release and endosome escape*. Pharmaceutics, 2022. **14**(4): p. 701.

[740] Fraire, J. C., et al., *Vapor nanobubble is the more reliable photothermal mechanism for inducing endosomal escape of siRNA without disturbing cell homeostasis*. J Control Release, 2020. **319**: p. 262–275.

[741] Lu, Y. J., et al., *Thermosensitive cationic magnetic liposomes for thermoresponsive delivery of CPT-11 and SLP2 shRNA in glioblastoma treatment.* Pharmaceutics, 2023. **15**(4): p. 1169.
[742] Domenech, M., et al., *Lysosomal membrane permeabilization by targeted magnetic nanoparticles in alternating magnetic fields.* ACS Nano, 2013. **7**: p. 5091–5101.
[743] Wang, F., et al., *Doxorubicin-tethered responsive gold nanoparticles facilitate intracellular drug delivery for overcoming multidrug resistance in cancer cells.* ACS Nano, 2011. **5**: p. 3679–3692.
[744] Sangtani, A., et al., *Intracellularly actuated quantum dot-peptide-doxorubicin nanobioconjugates for controlled drug delivery via the endocytic pathway.* Bioconjug Chem, 2018. **29**(1): p. 136–148.
[745] Acar, H., et al., *Cathepsin-mediated cleavage of peptides from peptide amphiphiles leads to enhanced intracellular peptide accumulation.* Bioconjug Chem, 2017. **28**(9): p. 2316–2326.
[746] Prasetyanto, E. A., et al., *Breakable hybrid organosilica nanocapsules for protein delivery.* Angew Chem, Int Ed Engl, 2016. **55**(10): p. 3323–3327.
[747] Yang, Y. S., et al., *Targeting small molecule drugs to T cells with antibody-directed cell-penetrating gold nanoparticles.* Biomater Sci, 2018. **7**(1): p. 113–124.
[748] Lu, J., et al., *Mesoporous silica nanoparticles as a delivery system for hydrophobic anticancer drugs.* Small, 2007. **3**(8): p. 1341–1346.
[749] Chikte, S., N. Panchal, and G. Warnes, *Use of LysoTracker dyes: a flow cytometric study of autophagy.* Cytometry, Part A, 2014. **85**(2): p. 169–178.
[750] Madani, F., A. Peralvarez-Marin, and A. Graslund, *Liposome model systems to study the endosomal escape of cell-penetrating peptides: transport across phospholipid membranes induced by a proton gradient.* J Drug Deliv, 2011. **2011**: p. 897592.
[751] El-Sayed, A., et al., *Octaarginine- and octalysine-modified nanoparticles have different modes of endosomal escape.* J Biol Chem, 2008. **283**(34): p. 23450–23461.
[752] Evans, B. C., et al., *Ex vivo red blood cell hemolysis assay for the evaluation of pH-responsive endosomolytic agents for cytosolic delivery of biomacromolecular drugs.* J Vis Exp, 2013(73): p. e50166.
[753] Wilson, J. T., et al., *pH-responsive nanoparticle vaccines for dual-delivery of antigens and immunostimulatory oligonucleotides.* ACS Nano, 2013. **7**(5): p. 3912–3925.
[754] McManus, M. T. and P. A. Sharp, *Gene silencing in mammals by small interfering RNAs.* Nat Rev Genet, 2002. **3**(10): p. 737–747.
[755] Wang, Z., et al., *RNA interference and cancer therapy.* Pharm Res, 2011. **28**(12): p. 2983–2995.
[756] Zhang, J., et al., *A comprehensive review of small interfering RNAs (siRNAs): mechanism, therapeutic targets, and delivery strategies for cancer therapy.* Int J Nanomed, 2023. **18**: p. 7605–7635.
[757] Liu, J., et al., *Exocytosis of nanoparticles: a comprehensive review.* Nanomaterials (Basel), 2023. **13**(15): p. 2215.
[758] Sakhtianchi, R., et al., *Exocytosis of nanoparticles from cells: role in cellular retention and toxicity.* Adv Colloid Interface Sci, 2013. **201–202**: p. 18–29.
[759] Bartczak, D., et al., *Exocytosis of peptide functionalized gold nanoparticles in endothelial cells.* Nanoscale, 2012. **4**(15): p. 4470–4472.
[760] Feng, Q., et al., *The dynamic interactions between nanoparticles and macrophages impact their fate in brain tumors.* Small, 2021. **17**(49): p. e2103600.
[761] Serda, R. E., et al., *Logic-embedded vectors for intracellular partitioning, endosomal escape, and exocytosis of nanoparticles.* Small, 2010. **6**(23): p. 2691–2700.
[762] Frohlich, E., *Cellular elimination of nanoparticles.* Environ Toxicol Pharmacol, 2016. **46**: p. 90–94.
[763] Irene Canton, I. and G. Battaglia, *Endocytosis at the nanoscale.* Chem Soc Rev, 2012. **41**: p. 2718–2739.
[764] Sipos, A., et al., *Alveolar epithelial cell processing of nanoparticles activates autophagy and lysosomal exocytosis.* Am J Physiol, Lung Cell Mol Physiol, 2018. **315**(2): p. L286–L300.
[765] Yanes, R. E., et al., *Involvement of lysosomal exocytosis in the excretion of mesoporous silica nanoparticles and enhancement of the drug delivery effect by exocytosis inhibition.* Small, 2013. **9**(5): p. 697–704.
[766] Al-Hajaj, N. A., et al., *Short ligands affect modes of QD uptake and elimination in human cells.* ACS Nano, 2011. **5**: p. 4909–4918.

[767] Cui, Y., et al., *The impact of receptor recycling on the exocytosis of αvβ3 integrin targeted gold nanoparticles*. Oncotarget, 2017. **8**: p. 38618–38630.

[768] Zhang, J., et al., *Effects of the surface charge of polyamidoamine dendrimers on cellular exocytosis and the exocytosis mechanism in multidrug-resistant breast cancer cells*. J Nanobiotechnol, 2021. **19**(1): p. 135.

[769] Zheng, Y., et al., *Tailored elasticity combined with biomimetic surface promotes nanoparticle transcytosis to overcome mucosal epithelial barrier*. Biomaterials, 2020. **262**: p. 120323.

[770] Du, B., et al., *Glomerular barrier behaves as an atomically precise bandpass filter in a sub-nanometre regime*. Nat Nanotechnol, 2017. **12**(11): p. 1096–1102.

[771] Bae, Y. M., et al., *Endocytosis, intracellular transport, and exocytosis of lanthanide-doped upconverting nanoparticles in single living cells*. Biomaterials, 2012. **33**(35): p. 9080–9086.

[772] Ekkapongpisit, M., et al., *Labeling and exocytosis of secretory compartments in RBL mastocytes by polystyrene and mesoporous silica nanoparticles*. Int J Nanomed, 2012. **7**: p. 1829–1840.

[773] Escrevente, C., et al., *Rab11 is required for lysosome exocytosis through the interaction with Rab3a, Sec15 and GRAB*. J Cell Sci, 2021. **134**(11): p. jcs246694.

[774] Liu, Y. Y., et al., *Fate of CdSe/ZnS quantum dots in cells: endocytosis, translocation and exocytosis*. Colloids Surf B, Biointerfaces, 2021. **208**: p. 112140.

[775] Liu, Y. Y., et al., *On the cellular uptake and exocytosis of carbon dots horizontal line significant cell type dependence and effects of cell division*. ACS Appl Bio Mater, 2022.

[776] Pi, Q. M., et al., *Degradation or excretion of quantum dots in mouse embryonic stem cells*. BMC Biotechnol, 2010. **10**: p. 36.

[777] Chu, Z., et al., *Cellular uptake, evolution, and excretion of silica nanoparticles in human cells*. Nanoscale, 2011. **3**(8): p. 3291–3299.

[778] Slowing, II, et al., *Exocytosis of mesoporous silica nanoparticles from mammalian cells: from asymmetric cell-to-cell transfer to protein harvesting*. Small, 2011. **7**(11): p. 1526–1532.

[779] Wang, L., et al., *Selective targeting of gold nanorods at the mitochondria of cancer cells: implications for cancer therapy*. Nano Lett, 2011. **11**(2): p. 772–780.

[780] Ohta, S., S. Inasawa, and Y. Yamaguchi, *Real time observation and kinetic modeling of the cellular uptake and removal of silicon quantum dots*. Biomaterials, 2012. **33**(18): p. 4639–4645.

[781] Hu, L., et al., *Influences of size of silica particles on the cellular endocytosis, exocytosis and cell activity of HepG2 cells*. J Nanosci Lett, 2011. **1**: p. 1–16.

[782] Panyam, J. and V. Labhasetwar, *Biodegradable nanoparticles for drug and gene delivery to cells and tissue*. Adv Drug Deliv Rev, 2003. **55**(3): p. 329–347.

[783] Zhuang, J., et al., *The influence of nanoparticle shape on bilateral exocytosis from Caco-2 cells*. Chin Chem Lett, 2018. **29**(12): p. 1815–1818.

[784] Oh, N., et al., *Macrophage-mediated exocytosis of elongated nanoparticles improves hepatic excretion and cancer phototherapy*. ACS Appl Mater Interfaces, 2018. **10**(34): p. 28450–28457.

[785] Kim, C. S., et al., *The role of surface functionality in nanoparticle exocytosis*. Adv Healthc Mater, 2014. **3**(8): p. 1200–1202.

[786] Ho, L. W. C., et al., *Effect of surface modification with hydrocarbyl groups on the exocytosis of nanoparticles*. Biochemistry, 2021. **60**(13): p. 1019–1030.

[787] Oh, N. and J. H. Park, *Endocytosis and exocytosis of nanoparticles in mammalian cells*. Int J Nanomed, 2014. **9**(Suppl 1): p. 51–63.

[788] Manshian, B. B., et al., *The role of intracellular trafficking of CdSe/ZnS QDs on their consequent toxicity profile*. J Nanobiotechnol, 2017. **15**(1): p. 45.

[789] Stayton, I., et al., *Study of uptake and loss of silica nanoparticles in living human lung epithelial cells at single cell level*. Anal Bioanal Chem, 2009. **394**(6): p. 1595–1608.

[790] Wang, Z., et al., *CuO nanoparticle interaction with human epithelial cells: cellular uptake, location, export, and genotoxicity*. Chem Res Toxicol, 2012. **25**(7): p. 1512–1521.

[791] Flannery, A. R., C. Czibener, and N. W. Andrews, *Palmitoylation-dependent association with CD63 targets the Ca2+ sensor synaptotagmin VII to lysosomes*. J Cell Biol, 2010. **191**(3): p. 599–613.

[792] Nguyen, T. D. K., et al., *Visualization of partial exocytotic content release and chemical transport into nanovesicles in cells*. ACS Nano, 2022. **16**(3): p. 4831–4842.

[793] Ren, L., et al., *Zinc regulates chemical-transmitter storage in nanometer vesicles and exocytosis dynamics as measured by amperometry*. Angew Chem, Int Ed Engl, 2017. **56**(18): p. 4970–4975.

[794] Redhai, S. and M. Boutros, *The role of organelles in intestinal function, physiology, and disease*. Trends Cell Biol, 2021. **31**(6): p. 485–499.

[795] Kuznetsov, A. V. and R. Margreiter, *Heterogeneity of mitochondria and mitochondrial function within cells as another level of mitochondrial complexity*. Int J Mol Sci, 2009. **10**(4): p. 1911–1929.

[796] Protasoni, M. and M. Zeviani, *Mitochondrial structure and bioenergetics in normal and disease conditions*. Int J Mol Sci, 2021. **22**(2): p. 586.

[797] Porporato, P. E., et al., *Mitochondrial metabolism and cancer*. Cell Res, 2018. **28**(3): p. 265–280.

[798] Turrens, J. F., *Mitochondrial formation of reactive oxygen species*. J Physiol, 2003. **552**(Pt 2): p. 335–344.

[799] Dan Dunn, J., et al., *Reactive oxygen species and mitochondria: a nexus of cellular homeostasis*. Redox Biol, 2015. **6**: p. 472–485.

[800] Lushchak, V. I. and O. Lushchak, *Interplay between reactive oxygen and nitrogen species in living organisms*. Chem-Biol Interact, 2021. **349**: p. 109680.

[801] Kannan, K. and S. K. Jain, *Oxidative stress and apoptosis*. Pathophysiology, 2000. **7**: p. 153–163.

[802] Schieber, M. and N. S. Chandel, *ROS function in redox signaling and oxidative stress*. Curr Biol, 2014. **24**(10): p. R453–R462.

[803] Sies, H., C. Berndt, and D. P. Jones, *Oxidative stress*. Annu Rev Biochem, 2017. **86**: p. 715–748.

[804] Bai, X., et al., *Small-molecule-based fluorescent sensors for selective detection of reactive oxygen species in biological systems*. Annu Rev Biochem, 2019. **88**: p. 605–633.

[805] Soh, N., *Recent advances in fluorescent probes for the detection of reactive oxygen species*. Anal Bioanal Chem, 2006. **386**(3): p. 532–543.

[806] Chang, M. C. Y., et al., *A selective, cell-permeable optical probe for hydrogen peroxide in living cells*. J Am Chem Soc, 2004. **126**: p. 15392–15393.

[807] Tirichen, H., et al., *Mitochondrial reactive oxygen species and their contribution in chronic kidney disease progression through oxidative stress*. Front Physiol, 2021. **12**: p. 627837.

[808] Palma, F. R., et al., *ROS production by mitochondria: function or dysfunction?* Oncogene, 2024. **43**(5): p. 295–303.

[809] Hernansanz-Agustin, P. and J. A. Enriquez, *Generation of reactive oxygen species by mitochondria*. Antioxidants (Basel), 2021. **10**(3): p. 415.

[810] Sindhu, R. K., et al., *Exploring multifunctional antioxidants as potential agents for management of neurological disorders*. Environ Sci Pollut Res Int, 2022. **29**(17): p. 24458–24477.

[811] Jiang, X. and X. Wang, *Cytochrome C-mediated apoptosis*. Annu Rev Biochem, 2004. **73**: p. 87–106.

[812] Ranji-Burachaloo, H., et al., *Cancer treatment through nanoparticle-facilitated fenton reaction*. ACS Nano, 2018. **12**(12): p. 11819–11837.

[813] Meyerstein, D., *What are the oxidizing intermediates in the fenton and fenton-like reactions? A perspective*. Antioxidants (Basel), 2022. **11**(7): p. 1368.

[814] Pham, A. N., et al., *Fenton-like copper redox chemistry revisited: hydrogen peroxide and superoxide mediation of copper-catalyzed oxidant production*. J Catal, 2013. **301**: p. 54–64.

[815] Valdés-Solís, T., et al., *Manganese ferrite nanoparticles synthesized through a nanocasting route as a highly active Fenton catalyst*. Catal Commun, 2007. **8**(12): p. 2037–2042.

[816] Haripriyaa, M. and K. Suthindhiran, *Pharmacokinetics of nanoparticles: current knowledge, future directions and its implications in drug delivery*. Future J Pharm Sci, 2023. **9**(1): p. 113.

[817] van den Anker, J., et al., *Developmental changes in pharmacokinetics and pharmacodynamics*. J Clin Pharmacol, 2018. **58**(Suppl 10): p. S10–S25.

[818] De Matteis, V., *Exposure to inorganic nanoparticles: routes of entry, immune response, biodistribution and in vitro/in vivo toxicity evaluation*. Toxics, 2017. **5**(4): p. 29.

[819] Yildirimer, L., et al., *Toxicology and clinical potential of nanoparticles*. Nano Today, 2011. **6**(6): p. 585–607.

[820] Bakand, S. and A. Hayes, *Toxicological considerations, toxicity assessment, and risk management of inhaled nanoparticles*. Int J Mol Sci, 2016. **17**(6): p. 929.

[821] Lin, Z., N. A. Monteiro-Riviere, and J. E. Riviere, *Pharmacokinetics of metallic nanoparticles*. Wiley Interdiscip Rev Nanomed Nanobiotechnol, 2015. **7**(2): p. 189–217.

[822] MacParland, S. A., et al., *Phenotype determines nanoparticle uptake by human macrophages from liver and blood*. ACS Nano, 2017. **11**(3): p. 2428–2443.

[823] Wang, L., et al., *Exploring and analyzing the systemic delivery barriers for nanoparticles*. Adv Funct Mater, 2023. **34**(8): p. 2308446.

[824] Mitchell, M. J., et al., *Engineering precision nanoparticles for drug delivery*. Nat Rev Drug Discov, 2021. **20**(2): p. 101–124.

[825] Nowak-Jary, J. and B. Machnicka, *Pharmacokinetics of magnetic iron oxide nanoparticles for medical applications*. J Nanobiotechnol, 2022. **20**(1): p. 305.

[826] Ngo, W., et al., *Why nanoparticles prefer liver macrophage cell uptake in vivo*. Adv Drug Deliv Rev, 2022. **185**: p. 114238.

[827] Tsoi, K. M., et al., *Mechanism of hard-nanomaterial clearance by the liver*. Nat Mater, 2016. **15**(11): p. 1212–1221.

[828] Hume, D. A., *The mononuclear phagocyte system*. Curr Opin Immunol, 2006. **18**(1): p. 49–53.

[829] Arami, H., et al., *In vivo delivery, pharmacokinetics, biodistribution and toxicity of iron oxide nanoparticles*. Chem Soc Rev, 2015. **44**(23): p. 8576–8607.

[830] Poon, W., et al., *Elimination pathways of nanoparticles*. ACS Nano, 2019. **13**(5): p. 5785–5798.

[831] Dixon, L. J., et al., *Kupffer cells in the liver*. Compr Physiol, 2013. **3**(2): p. 785–797.

[832] Mitten, E. K. and G. Baffy, *Mechanobiology in the development and progression of non-alcoholic fatty liver disease: an updated review*. Metab Target Organ Damage, 2023. **3**(1): p. 2.

[833] De Jong, W. H., et al., *Particle size-dependent organ distribution of gold nanoparticles after intravenous administration*. Biomaterials, 2008. **29**(12): p. 1912–1919.

[834] Fischer, H. C., et al., *Pharmacokinetics of nanoscale quantum dots: in vivo distribution, sequestration, and clearance in the rat*. Adv Funct Mater, 2006. **16**(10): p. 1299–1305.

[835] Xie, G., J. Sun, and G. Zhong, *Tissular localization and excretion of intravenously administered silica nanoparticles of different sizes*. J Nanopart Res, 2012. **14**(1): p. 671.

[836] He, X., et al., *In vivo study of biodistribution and urinary excretion of surface-modified silica nanoparticles*. Anal Chem, 2008. **80**: p. 9597–9603.

[837] Malfatti, M. A., et al., *Determining the pharmacokinetics and long-term biodistribution of SiO2 nanoparticles in vivo using accelerator mass spectrometry*. Nano Lett, 2012. **12**(11): p. 5532–5538.

[838] Choi, H. S., et al., *Renal clearance of quantum dots*. Nat Biotechnol, 2007. **25**(10): p. 1165–1170.

[839] Hauser, P. V., et al., *Nanotechnology, nanomedicine, and the kidney*. Appl Sci, 2021. **11**(16): p. 7187.

[840] Huang, Y., M. Yu, and J. Zheng, *Proximal tubules eliminate endocytosed gold nanoparticles through an organelle-extrusion-mediated self-renewal mechanism*. Nat Nanotechnol, 2023. **18**(6): p. 637–646.

[841] Mosleh-Shirazi, S., et al., *Renal clearable nanoparticles: an expanding horizon for improving biomedical imaging and cancer therapy*. Mater Today Commun, 2021. **26**: p. 102064.

[842] Naumenko, V., et al., *Intravital microscopy reveals a novel mechanism of nanoparticles excretion in kidney*. J Control Release, 2019. **307**: p. 368–378.

[843] Ruggiero, A., et al., *Paradoxical glomerular filtration of carbon nanotubes*. Proc Natl Acad Sci USA, 2010. **107**(27): p. 12369–12374.

[844] Jasim, D. A., et al., *The effects of extensive glomerular filtration of thin graphene oxide sheets on kidney physiology*. ACS Nano, 2016. **10**(12): p. 10753–10767.

[845] Sábio, R. M., et al., *Exploiting mesoporous silica nanoparticles as versatile drug carriers for several routes of administration*. Microporous Mesoporous Mater, 2021. **312**: p. 110774.

[846] McLennan, D. N., C. J. Porter, and S. A. Charman, *Subcutaneous drug delivery and the role of the lymphatics*. Drug Discov Today Technol, 2005. **2**(1): p. 89–96.
[847] Breslin, J. W., et al., *Lymphatic vessel network structure and physiology*. Compr Physiol, 2018. **9**(1): p. 207–299.
[848] Oghabian, M. A., et al., *Detection sensitivity of lymph nodes of various sizes using USPIO nanoparticles in magnetic resonance imaging*. Nanomedicine, 2010. **6**(3): p. 496–499.
[849] Johnson, L., S. E. Pinder, and M. Douek, *Deposition of superparamagnetic iron-oxide nanoparticles in axillary sentinel lymph nodes following subcutaneous injection*. Histopathology, 2013. **62**(3): p. 481–486.
[850] Leclerc, L., et al., *Testicular biodistribution of silica-gold nanoparticles after intramuscular injection in mice*. Biomed Microdevices, 2015. **17**(4): p. 66.
[851] Fu, C., et al., *The absorption, distribution, excretion and toxicity of mesoporous silica nanoparticles in mice following different exposure routes*. Biomaterials, 2013. **34**(10): p. 2565–2575.
[852] Morais, R. P., et al., *Skin interaction, permeation, and toxicity of silica nanoparticles: Challenges and recent therapeutic and cosmetic advances*. Int J Pharm, 2022. **614**: p. 121439.
[853] Larese Filon, F., et al., *Nanoparticles skin absorption: new aspects for a safety profile evaluation*. Regul Toxicol Pharmacol, 2015. **72**(2): p. 310–322.
[854] Singpanna, K., et al., *Gold nanoparticles and their applications in transdermal drug delivery: a review*. J Drug Deliv Sci Technol, 2023. **90**: p. 105174.
[855] Baroli, B., et al., *Penetration of metallic nanoparticles in human full-thickness skin*. J Invest Dermatol, 2007. **127**(7): p. 1701–1712.
[856] Zhao, Z., et al., *Noninvasive transdermal delivery of mesoporous silica nanoparticles using deep eutectic solvent*. J Control Release, 2022. **343**: p. 43–56.
[857] Palmer, B. C. and L. A. DeLouise, *Nanoparticle-enabled transdermal drug delivery systems for enhanced dose control and tissue targeting*. Molecules, 2016. **21**(12): p. 1719.
[858] Hsiao, P. F., et al., *Transdermal delivery of poly(ethylene glycol)-co-oleylamine modified gold nanoparticles: effect of size and shape*. Mater Chem Phys, 2019. **224**: p. 22–28.
[859] Lee, S. E., et al., *Penetration pathways induced by low-frequency sonophoresis with physical and chemical enhancers: iron oxide nanoparticles versus lanthanum nitrates*. J Invest Dermatol, 2010. **130**(4): p. 1063–1072.
[860] Paliwal, S., G. K. Menon, and S. Mitragotri, *Low-frequency sonophoresis: ultrastructural basis for stratum corneum permeability assessed using quantum dots*. J Invest Dermatol, 2006. **126**(5): p. 1095–1101.
[861] Yhee, J. Y., J. Im, and R. S. Nho, *Advanced therapeutic strategies for chronic lung disease using nanoparticle-based drug delivery*. J Clin Med, 2016. **5**(9): p. 82.
[862] Yang, W., J. I. Peters, and R. O. Williams, 3rd, *Inhaled nanoparticles – a current review*. Int J Pharm, 2008. **356**(1–2): p. 239–247.
[863] De Berardis, B., et al., *Exposure to airborne gold nanoparticles: a review of current toxicological data on the respiratory tract*. J Nanopart Res, 2021. **22**(8): p. 235.
[864] Frohlich, E. E. and E. Frohlich, *Cytotoxicity of nanoparticles contained in food on intestinal cells and the gut microbiota*. Int J Mol Sci, 2016. **17**(4): p. 509.
[865] Ruaro, B., et al., *The history and mystery of alveolar epithelial type II cells: focus on their physiologic and pathologic role in lung*. Int J Mol Sci, 2021. **22**(5): p. 2566.
[866] Oberdorster, G., E. Oberdorster, and J. Oberdorster, *Nanotoxicology: an emerging discipline evolving from studies of ultrafine particles*. Environ Health Perspect, 2005. **113**(7): p. 823–839.
[867] Miller, M. R., et al., *Inhaled nanoparticles accumulate at sites of vascular disease*. ACS Nano, 2017. **11**(5): p. 4542–4552.
[868] Kreyling, W. G., et al., *Age-dependent rat lung deposition patterns of inhaled 20 nanometer gold nanoparticles and their quantitative biokinetics in adult rats*. ACS Nano, 2018. **12**(8): p. 7771–7790.
[869] Tsuda, A., et al., *Age-dependent translocation of gold nanoparticles across the air-blood barrier*. ACS Nano, 2019. **13**(9): p. 10095–10102.

[870] Elder, A., S. Vidyasagar, and L. De Louise, *Physicochemical factors that affect metal and metal oxide nanoparticle passage across epithelial barriers*. WIREs Nanomed Nanobiotechnol, 2009. **1**(4): p. 434–450.

[871] McClements, D. J., H. Xiao, and P. Demokritou, *Physicochemical and colloidal aspects of food matrix effects on gastrointestinal fate of ingested inorganic nanoparticles*. Adv Colloid Interface Sci, 2017. **246**: p. 165–180.

[872] Ensign, L. M., R. Cone, and J. Hanes, *Oral drug delivery with polymeric nanoparticles: the gastrointestinal mucus barriers*. Adv Drug Deliv Rev, 2012. **64**(6): p. 557–570.

[873] Vitulo, M., et al., *Interactions between nanoparticles and intestine*. Int J Mol Sci, 2022. **23**(8): p. 4339.

[874] Cao, X., et al., *Evaluation of the cytotoxic and cellular proteome impacts of food-grade TiO_2 (E171) using simulated gastrointestinal digestions and a tri-culture small intestinal epithelial model*. NanoImpact, 2020. **17**.

[875] EFSA Panel on Food Additives and Flavourings (FAF), et al., *Safety assessment of titanium dioxide (E171) as a food additive*. EFSA J, 2021. **19**(5): p. e06585.

[876] Askri, D., et al., *Nanoparticles in foods? A multiscale physiopathological investigation of iron oxide nanoparticle effects on rats after an acute oral exposure: trace element biodistribution and cognitive capacities*. Food Chem Toxicol, 2019. **127**: p. 173–181.

[877] Hillyer, J. F. and R. M. Albrecht, *Gastrointestinal persorption and tissue distribution of differently sized colloidal gold nanoparticles*. J Pharm Sci, 2001. **90**(12): p. 1927–1936.

[878] Kaplan, G. G. and J. W. Windsor, *The four epidemiological stages in the global evolution of inflammatory bowel disease*. Nat Rev Gastroenterol Hepatol, 2021. **18**(1): p. 56–66.

[879] Zhu, S., et al., *Orally administered gold nanoparticles protect against colitis by attenuating Toll-like receptor 4- and reactive oxygen/nitrogen species-mediated inflammatory responses but could induce gut dysbiosis in mice*. J Nanobiotechnol, 2018. **16**(1): p. 86.

[880] Gubala, V., et al., *Engineered nanomaterials and human health: part 2. Applications and nanotoxicology (IUPAC Technical Report)*. Pure Appl Chem, 2018. **90**(8): p. 1325–1356.

[881] Krug, H. F. and P. Wick, *Nanotoxicology: an interdisciplinary challenge*. Angew Chem, Int Ed Engl, 2011. **50**(6): p. 1260–1278.

[882] Gunsolus, I. L. and C. L. Haynes, *Analytical aspects of nanotoxicology*. Anal Chem, 2016. **88**(1): p. 451–479.

[883] Zielinska, A., et al., *Nanotoxicology and nanosafety: safety-by-design and testing at a glance*. Int J Environ Res Public Health, 2020. **17**(13): p. 4657.

[884] Singh, A. V., et al., *Review of emerging concepts in nanotoxicology: opportunities and challenges for safer nanomaterial design*. Toxicol Mech Methods, 2019. **29**(5): p. 378–387.

[885] Domingues, C., et al., *Where is nano today and where is it headed? A review of nanomedicine and the dilemma of nanotoxicology*. ACS Nano, 2022. **16**(7): p. 9994–10041.

[886] Wong, A. D., et al., *Quantitative analysis of the enhanced permeation and retention (EPR) effect*. PLoS ONE, 2015. **10**(5): p. e0123461.

[887] Rothen-Rutishauser, B., et al., *In vitro models of the human epithelial airway barrier to study the toxic potential of particulate matter*. Expert Opin Drug Metab Toxicol, 2008. **4**(8): p. 1075–1089.

[888] Evans, S. J., et al., *Critical review of the current and future challenges associated with advanced in vitro systems towards the study of nanoparticle (secondary) genotoxicity*. Mutagenesis, 2017. **32**(1): p. 233–241.

[889] Kozics, K., et al., *Pharmacokinetics, biodistribution, and biosafety of pegylated gold nanoparticles in vivo*. Nanomaterials (Basel), 2021. **11**(7): p. 1702.

[890] Lewinski, N., V. Colvin, and R. Drezek, *Cytotoxicity of nanoparticles*. Small, 2008. **4**(1): p. 26–49.

[891] Duval, K., et al., *Modeling physiological events in 2D vs. 3D cell culture*. Physiology (Bethesda), 2017. **32**(4): p. 266–277.

[892] Kapalczynska, M., et al., *2D and 3D cell cultures – a comparison of different types of cancer cell cultures*. Arch Med Sci, 2018. **14**(4): p. 910–919.

[893] Goncalves, I. M., et al., *Organ-on-a-chip platforms for drug screening and delivery in tumor cells: a systematic review*. Cancers (Basel), 2022. **14**(4): p. 935.

[894] Kang, S., S. E. Park, and D. D. Huh, *Organ-on-a-chip technology for nanoparticle research*. Nano Converg, 2021. **8**(1): p. 20.

[895] Sontheimer-Phelps, A., B. A. Hassell, and D. E. Ingber, *Modelling cancer in microfluidic human organs-on-chips*. Nat Rev Cancer, 2019. **19**(2): p. 65–81.

[896] Haque, E. and A. C. Ward, *Zebrafish as a model to evaluate nanoparticle toxicity*. Nanomaterials (Basel), 2018. **8**(7): p. 561.

[897] Chrishtop, V. V., et al., *Nanosafety vs. nanotoxicology: adequate animal models for testing in vivo toxicity of nanoparticles*. Toxicology, 2021. **462**: p. 152952.

[898] Krewski, D., et al., *Toxicity testing in the 21st century: progress in the past decade and future perspectives*. Arch Toxicol, 2020. **94**(1): p. 1–58.

[899] Khalili Fard, J., S. Jafari, and M. A. Eghbal, *A review of molecular mechanisms involved in toxicity of nanoparticles*. Adv Pharm Bull, 2015. **5**(4): p. 447–454.

[900] Kumar, V., N. Sharma, and S. S. Maitra, *In vitro and in vivo toxicity assessment of nanoparticles*. Int Nano Lett, 2017. **7**(4): p. 243–256.

[901] Kannan, K. and S. K. Jain, *Oxidative stress and apoptosis*. Pathophysiology, 2000. **7**: p. 153–163.

[902] Hanus, J., et al., *Induction of necrotic cell death by oxidative stress in retinal pigment epithelial cells*. Cell Death Dis, 2013. **4**(12): p. e965.

[903] Sanati, M., et al., *Recent trends in the application of nanoparticles in cancer therapy: The involvement of oxidative stress*. J Control Release, 2022. **348**: p. 287–304.

[904] Huang, X., et al., *The promotion of human malignant melanoma growth by mesoporous silica nanoparticles through decreased reactive oxygen species*. Biomaterials, 2010. **31**: p. 6142–6153.

[905] Voss, L., et al., *Intestinal and hepatic effects of iron oxide nanoparticles*. Arch Toxicol, 2021. **95**(3): p. 895–905.

[906] Shao, K., et al., *Nanoparticle-based immunotherapy for cancer*. ACS Nano, 2015. **9**: p. 16–30.

[907] Attia, A. B. E., et al., *A review of clinical photoacoustic imaging: current and future trends*. Photoacoustics, 2019. **16**: p. 100144.

[908] Cole, L. E., et al., *Gold nanoparticles as contrast agents in X-ray imaging and computed tomography*. Nanomedicine (Lond), 2015. **10**: p. 321–341.

[909] Draack, S., M. Schilling, and T. Viereck, *Magnetic particle imaging of particle dynamics in complex matrix systems*. Phys Sci Rev, 2023. **8**(2): p. 213–237.

[910] Fu, Q., et al., *Photoacoustic imaging: contrast agents and their biomedical applications*. Adv Mater, 2019. **31**(6): p. e1805875.

[911] Attia, M. F., et al., *An overview of active and passive targeting strategies to improve the nanocarriers efficiency to tumour sites*. J Pharm Pharmacol, 2019. **71**(8): p. 1185–1198.

[912] Narum, S. M., et al., *Passive targeting in nanomedicine: fundamental concepts, body interactions, and clinical potential*, in *Nanoparticles for biomedical applications*. 2020. p. 37–53.

[913] Dvorak, H. F., *How tumors make bad blood vessels and stroma*. Am J Clin Pathol, 2003. **162**: p. 1747–1756.

[914] Subhan, M. A., et al., *Recent advances in tumor targeting via EPR effect for cancer treatment*. J Pers Med, 2021. **11**(6): p. 571.

[915] Danhier, F., O. Feron, and V. Preat, *To exploit the tumor microenvironment: passive and active tumor targeting of nanocarriers for anti-cancer drug delivery*. J Control Release, 2010. **148**(2): p. 135–146.

[916] Goya, G. F., V. Grazú, and M. R. Ibarra, *Magnetic nanoparticles for cancer therapy*. Curr Nanosci, 2008. **4**: p. 1–16.

[917] Schleich, N., et al., *Comparison of active, passive and magnetic targeting to tumors of multifunctional paclitaxel/SPIO-loaded nanoparticles for tumor imaging and therapy*. J Control Release, 2014. **194**: p. 82–91.

[918] Feng, L., et al., *Magnetic targeting, tumor microenvironment-responsive intelligent nanocatalysts for enhanced tumor ablation*. ACS Nano, 2018. **12**(11): p. 11000–11012.

[919] Belyanina, I., et al., *Targeted magnetic nanotheranostics of cancer*. Molecules, 2017. **22**(6): p. 975.

[920] Mody, V. V., et al., *Magnetic nanoparticle drug delivery systems for targeting tumor*. Appl Nanosci, 2013. **4**(4): p. 385–392.

[921] Prijic, S. and G. Sersa, *Magnetic nanoparticles as targeted delivery systems in oncology*. Radiol Oncol, 2011. **45**(1): p. 1–16.

[922] Wu, K., et al., *Magnetic nanoparticles in nanomedicine: a review of recent advances*. Nanotechnology, 2019. **30**(50): p. 502003.

[923] Dulinska-Litewka, J., et al., *Superparamagnetic iron oxide nanoparticles – current and prospective medical applications*. Materials (Basel), 2019. **12**(4): p. 617.

[924] Arriortua, O. K., et al., *RGD-functionalized Fe_3O_4 nanoparticles for magnetic hyperthermia*. Colloids Surf B, Biointerfaces, 2018. **165**: p. 315–324.

[925] Lian, Y., et al., *Recent advances on the magnetic nanoparticle-based nanocomposites for magnetic induction hyperthermia of tumor: a short review*. Adv Compos Hybrid Mater, 2021. **4**(4): p. 925–937.

[926] Peter Caravan, J. J. E., T. J. McMurry, and R. B. Lauffer, *Gadolinium(III) chelates as MRI contrast agents: structure, dynamics, and applications*. Chem Rev, 1999. **99**: p. 2293–2352.

[927] Shen, Z., A. Wu, and X. Chen, *Iron oxide nanoparticle based contrast agents for magnetic resonance imaging*. Mol Pharm, 2017. **14**(5): p. 1352–1364.

[928] Kim, B. H., et al., *Large-scale synthesis of uniform and extremely small-sized iron oxide nanoparticles for high-resolution T_1 magnetic resonance imaging contrast agents*. J Am Chem Soc, 2011. **133**(32): p. 12624–12631.

[929] Fatima, H. and K.-S. Kim, *Iron-based magnetic nanoparticles for magnetic resonance imaging*. Adv Powder Technol, 2018. **29**(11): p. 2678–2685.

[930] Hu, S. and L. V. Wang, *Photoacoustic imaging and characterization of the microvasculature*. J Biomed Opt, 2010. **15**(1): p. 011101.

[931] Zhou, Y., J. Yao, and L. V. Wang, *Tutorial on photoacoustic tomography*. J Biomed Opt, 2016. **21**(6): p. 61007.

[932] Sun, A., et al., *Evaluation of visible NIR-I and NIR-II light penetration for photoacoustic imaging in rat organs*. Opt Express, 2020. **28**(6): p. 9002–9013.

[933] Han, S., et al., *Contrast agents for photoacoustic imaging: a review focusing on the wavelength range*. Biosensors (Basel), 2022. **12**(8): p. 594.

[934] Hossain, M. and M. Su, *Nanoparticle location and material dependent dose enhancement in X-ray radiation therapy*. J Phys Chem C, Nanomater Interfaces, 2012. **116**(43): p. 23047–23052.

[935] Babaei, M. and M. Ganjalikhani, *The potential effectiveness of nanoparticles as radio sensitizers for radiotherapy*. Bioimpacts, 2014. **4**(1): p. 15–20.

[936] Wang, H., et al., *Hypoxic radioresistance: can ROS be the key to overcome it?* Cancers (Basel), 2019. **11**(1): p. 112.

[937] Gong, L., et al., *Application of radiosensitizers in cancer radiotherapy*. Int J Nanomed, 2021. **16**: p. 1083–1102.

[938] Hong, B. J., et al., *Tumor hypoxia and reoxygenation: the yin and yang for radiotherapy*. Radiat Oncol J, 2016. **34**(4): p. 239–249.

[939] Meng, L., et al., *Tumor oxygenation and hypoxia inducible factor-1 functional inhibition via a reactive oxygen species responsive nanoplatform for enhancing radiation therapy and abscopal effects*. ACS Nano, 2018. **12**(8): p. 8308–8322.

[940] Vaupel, P. and A. Mayer, *Hypoxia and anemia: effects on tumor biology and treatment resistance*. Transfus Clin Biol, 2005. **12**(1): p. 5–10.

[941] Zanoni, M., et al., *3D tumor spheroid models for in vitro therapeutic screening: a systematic approach to enhance the biological relevance of data obtained*. Sci Rep, 2016. **6**: p. 19103.

[942] Weiswald, L. B., D. Bellet, and V. Dangles-Marie, *Spherical cancer models in tumor biology*. Neoplasia, 2015. **17**(1): p. 1–15.

[943] Mikhail, A. S., S. Eetezadi, and C. Allen, *Multicellular tumor spheroids for evaluation of cytotoxicity and tumor growth inhibitory effects of nanomedicines in vitro: a comparison of docetaxel-loaded block copolymer micelles and Taxotere(R)*. PLoS ONE, 2013. **8**(4): p. e62630.

[944] Ho, W. Y., et al., *Development of multicellular tumor spheroid (MCTS) culture from breast cancer cell and a high throughput screening method using the MTT assay*. PLoS ONE, 2012. **7**(9): p. e44640.

[945] Al-Ramadan, A., et al., *Analysis of radiation effects in two irradiated tumor spheroid models*. Oncol Lett, 2018. **15**(3): p. 3008–3016.

[946] Oronsky, B. T., S. J. Knox, and J. Scicinski, *Six degrees of separation: the oxygen effect in the development of radiosensitizers*. Transl Oncol, 2011. **4**(4): p. 189–198.

[947] Liu, C., Q. Lin, and Z. Yun, *Cellular and molecular mechanisms underlying oxygen-dependent radiosensitivity*. Radiat Res, 2015. **183**(5): p. 487–496.

[948] Kepka, L. and J. Socha, *Dose and fractionation schedules in radiotherapy for non-small cell lung cancer*. Transl Lung Cancer Res, 2021. **10**(4): p. 1969–1982.

[949] Marcu, L. G., *Altered fractionation in radiotherapy: from radiobiological rationale to therapeutic gain*. Cancer Treat Rev, 2010. **36**(8): p. 606–614.

[950] Kavanagh, J. N., et al., *DNA double strand break repair: a radiation perspective*. Antioxid Redox Signal, 2013. **18**(18): p. 2458–2472.

[951] Bilynsky, C., N. Millot, and A. L. Papa, *Radiation nanosensitizers in cancer therapy-From preclinical discoveries to the outcomes of early clinical trials*. Bioeng Transl Med, 2022. **7**(1): p. e10256.

[952] Li, L. Y., et al., *DNA repair pathways in cancer therapy and resistance*. Front Pharmacol, 2020. **11**: p. 629266.

[953] Pawlik, T. M. and K. Keyomarsi, *Role of cell cycle in mediating sensitivity to radiotherapy*. Int J Radiat Oncol Biol Phys, 2004. **59**(4): p. 928–942.

[954] Ng, W. L., et al., *Molecular mechanisms involved in tumor repopulation after radiotherapy*. Transl Cancer Res, 2013. **2**(5): p. 442–448.

[955] Weijer, R., et al., *Inhibition of hypoxia inducible factor 1 and topoisomerase with acriflavine sensitizes perihilar cholangiocarcinomas to photodynamic therapy*. Oncotarget, 2016. **7**(3): p. 3341–3356.

[956] Broekgaarden, M., et al., *Inhibition of hypoxia-inducible factor 1 with acriflavine sensitizes hypoxic tumor cells to photodynamic therapy with zinc phthalocyanine-encapsulating cationic liposomes*. Nano Res, 2016. **9**(6): p. 1639–1662.

[957] Marti-Diaz, R., et al., *Acriflavine, a potent inhibitor of HIF-1alpha, disturbs glucose metabolism and suppresses ATF_4-protective pathways in melanoma under non-hypoxic conditions*. Cancers (Basel), 2020. **13**(1): p. 102.

[958] Lim, M. J., et al., *Acriflavine enhances radiosensitivity of colon cancer cells through endoplasmic reticulum stress-mediated apoptosis*. Int J Biochem Cell Biol, 2012. **44**(8): p. 1214–1222.

[959] Vallet-Regi, M., F. Balas, and D. Arcos, *Mesoporous materials for drug delivery*. Angew Chem, Int Ed Engl, 2007. **46**(40): p. 7548–7558.

[960] Rastegari, E., et al., *An update on mesoporous silica nanoparticle applications in nanomedicine*. Pharmaceutics, 2021. **13**(7): p. 1067.

[961] Eckenberger, E., et al., *Acriflavine-functionalized silica@manganese ferrite nanostructures for synergistic radiation and hypoxia therapies*. ACS Appl Bio Mater, 2023. **6**(8): p. 3089–3102.

[962] Li, Z., et al., *Measurement of uptake and release capacities of mesoporous silica nanoparticles enabled by nanovalve gates*. J Phys Chem C, Nanomater Interfaces, 2011. **115**(40): p. 19496–19506.

[963] Gibellini, L., et al., *Interfering with ROS metabolism in cancer cells: the potential role of quercetin*. Cancers (Basel), 2010. **2**(2): p. 1288–1311.

[964] Schumacker, P. T., *Reactive oxygen species in cancer cells: live by the sword, die by the sword*. Cancer Cell, 2006. **10**(3): p. 175–176.

[965] Benhar, M., D. Engelberg, and A. Levitzki, *ROS, stress-activated kinases and stress signaling in cancer.* EMBO Rep, 2002. **3**: p. 420–425.

[966] Trachootham, D., J. Alexandre, and P. Huang, *Targeting cancer cells by ROS-mediated mechanisms: a radical therapeutic approach?* Nat Rev Drug Discov, 2009. **8**(7): p. 579–591.

[967] Nishikawa, M., *Reactive oxygen species in tumor metastasis.* Cancer Lett, 2008. **266**(1): p. 53–59.

[968] Klein, S., et al., *Enhanced in vitro biocompatibility and water dispersibility of magnetite and cobalt ferrite nanoparticles employed as ROS formation enhancer in radiation cancer therapy.* Small, 2018. **14**(21): p. e1704111.

[969] Franken, N. A., et al., *Clonogenic assay of cells in vitro.* Nat Protoc, 2006. **1**(5): p. 2315–2319.

[970] Choi, J., et al., *Radiosensitizing high-Z metal nanoparticles for enhanced radiotherapy of glioblastoma multiforme.* J Nanobiotechnol, 2020. **18**(1): p. 122.

[971] Li, S., et al., *Platinum nanoparticles: an exquisite tool to overcome radioresistance.* Cancer Nanotechnol, 2017. **8**(1): p. 4.

[972] Pedone, D., et al., *Platinum nanoparticles in nanobiomedicine.* Chem Soc Rev, 2017. **46**(16): p. 4951–4975.

[973] Jiang, Y. W., et al., *Palladium nanosheets as safe radiosensitizers for radiotherapy.* Langmuir, 2020. **36**(39): p. 11637–11644.

[974] Penninckx, S., et al., *Gold nanoparticles as a potent radiosensitizer: a transdisciplinary approach from physics to patient.* Cancers (Basel), 2020. **12**(8): p. 2021.

[975] Misawa, M. and J. Takahashi, *Generation of reactive oxygen species induced by gold nanoparticles under X-ray and UV irradiations.* Nanomedicine, 2011. **7**(5): p. 604–614.

[976] Soenen, S. J., et al., *Cytotoxic effects of gold nanoparticles: a multiparametric study.* ACS Nano, 2012. **6**(7): p. 5767–5783.

[977] Orban, J. I. and J. A. Peterson, *Modification of the phosphoketolase assay for rapid identification of bifidobacteria.* J Microbiol Methods, 2000. **40**: p. 221–224.

[978] Hoskins, C., et al., *Hybrid gold-iron oxide nanoparticles as a multifunctional platform for biomedical application.* J Nanobiotechnol, 2012. **10**: p. 27.

[979] Hughes, M. N., *Relationships between nitric oxide, nitroxyl ion, nitrosonium cation and peroxynitrite.* Biochim Biophys Acta, 1999. **1411**: p. 263–272.

[980] Szabo, C., H. Ischiropoulos, and R. Radi, *Peroxynitrite: biochemistry, pathophysiology and development of therapeutics.* Nat Rev Drug Discov, 2007. **6**(8): p. 662–680.

[981] Cooper, D. C., et al., *Uric acid, a natural scavenger of peroxynitrite, in experimental allergic encephalomyelitis and multiple sclerosis.* Proc Natl Acad Sci, 1998. **95**: p. 675–680.

[982] Kean, R. B., et al., *The peroxynitrite scavenger uric acid prevents inflammatory cell invasion into the central nervous system in experimental allergic encephalomyelitis through maintenance of blood-central nervous system barrier integrity.* J Immunol, 2000. **165**(11): p. 6511–6518.

[983] Holley, A. K., et al., *Redox-modulated phenomena and radiation therapy: the central role of superoxide dismutases.* Antioxid Redox Signal, 2014. **20**(10): p. 1567–1589.

[984] Epperly, M. W., et al., *Mitochondrial localization of superoxide dismutase is required for decreasing radiation-induced cellular damage.* Radiat Res, 2003. **160**: p. 568–578.

[985] Velika, B. and I. Kron, *Antioxidant properties of benzoic acid derivatives against Superoxide radical.* Free Radic Antioxid, 2012. **2**(4): p. 62–67.

[986] Murtaza, G., et al., *Caffeic acid phenethyl ester and therapeutic potentials.* BioMed Res Int, 2014. **2014**: p. 145342.

[987] Lin, Y. H., et al., *Antiproliferation and radiosensitization of caffeic acid phenethyl ester on human medulloblastoma cells.* Cancer Chemother Pharmacol, 2006. **57**(4): p. 525–532.

[988] Chavez, K. J., S. V. Garimella, and S. Lipkowitz, *Triple negative breast cancer cell lines: one tool in the search for better treatment of triple negative breast cancer.* Breast Dis, 2010. **32**(1–2): p. 35–48.

[989] Spirou, S. V., et al., *Magnetic hyperthermia and radiation therapy: radiobiological principles and current practice.* Nanomaterials (Basel), 2018. **8**(6): p. 401.

[990] Rajan, A. and N. K. Sahu, *Review on magnetic nanoparticle-mediated hyperthermia for cancer therapy*. J Nanopart Res, 2020. **22**(11): p. 319.

[991] Farzin, A., et al., *Magnetic nanoparticles in cancer therapy and diagnosis*. Adv Healthc Mater, 2020. **9**(9): p. e1901058.

[992] Ma, M., et al., *Size dependence of specific power absorption of* Fe_3O_4 *particles in AC magnetic field*. J Magn Magn Mater, 2004. **268**(1–2): p. 33–39.

Index

https://doi.org/10.1515/9783110605495-008